LAND APPLICATION OF WASTES

Volume II

LAND APPLICATION OF WASTES

Volume II

Raymond C. Loehr
Professor of Agricultural Engineering

William J. Jewell
Associate Professor of Agricultural Engineering

Joseph D. Novak
Professor of Science Education

William W. Clarkson
Research Associate

Gerald S. Friedman
Research Associate

College of Agriculture and Life Sciences
A Statutory College of the State University of New York
Cornell University

Van Nostrand Reinhold Environmental Engineering Series

VNR VAN NOSTRAND REINHOLD COMPANY
NEW YORK CINCINNATI ATLANTA DALLAS SAN FRANCISCO
LONDON TORONTO MELBOURNE

Van Nostrand Reinhold Company Regional Offices:
New York Cincinnati Atlanta Dallas San Francisco

Van Nostrand Reinhold Company International Offices:
London Toronto Melbourne

Library of Congress Catalog Card Number: 78-27646
ISBN: 0-442-21707-2

Manufactured in the United States of America

Published by Van Nostrand Reinhold Company
135 West 50th Street, New York. N.Y. 10020

Published simultaneously in Canada by Van Nostrand Reinhold Ltd.

15 14 13 12 11 10 9 8 7 6 5 4 3 2 1

Library of Congress Cataloging in Publication Data

Main entry under title:

Land application of wastes.

(Van Nostrand Reinhold environmental engineering series)
Includes bibliographies and index.
1. Sewage disposal. I. Loehr, Raymond C.
TD760.L22 628'.36 78-27646
ISBN 0-442-21707-2 (v. 2)

Van Nostrand Reinhold Environmental Engineering Series

THE VAN NOSTRAND REINHOLD ENVIRONMENTAL ENGINEERING SERIES is dedicated to the presentation of current and vital information relative to the engineering aspects of controlling man's physical environment. Systems and subsystems available to exercise control of both the indoor and outdoor environment continue to become more sophisticated and to involve a number of engineering disciplines. The aim of the series is to provide books which, though often concerned with the life cycle—design, installation, and operation and maintenance—of a specific system or subsystem, are complementary when viewed in their relationship to the total environment.

The Van Nostrand Reinhold Environmental Engineering Series includes books concerned with the engineering of mechanical systems designed (1) to control the environment within structures, including those in which manufacturing processes are carried out, and (2) to control the exterior environment through control of waste products expelled by inhabitants of structures and from manufacturing processes. The series include books on heating, air conditioning and ventilation, control of air and water pollution, control of the acoustic environment, sanitary engineering and waste disposal, illumination, and piping systems for transporting media of all kinds.

Van Nostrand Reinhold Environmental Engineering Series

ADVANCED WASTEWATER TREATMENT, by Russell L. Culp and Gordon L. Culp

ARCHITECTURAL INTERIOR SYSTEMS—Lighting, Air Conditioning, Acoustics, John E. Flynn and Arthur W. Segil

SOLID WASTE MANAGEMENT, by D. Joseph Hagerty, Joseph L. Pavoni and John E. Heer, Jr.

THERMAL INSULATION, by John F. Malloy

AIR POLLUTION AND INDUSTRY, edited by Richard D. Ross

INDUSTRIAL WASTE DISPOSAL, edited by Richard D. Ross

MICROBIAL CONTAMINATION CONTROL FACILITIES, by Robert S. Rurkle and G. Briggs Phillips

SOUND, NOISE, AND VIBRATION CONTROL (Second Edition), by Lyle F. Yerges

NEW CONCEPTS IN WATER PURIFICATION, by Gordon L. Culp and Russell L. Culp

HANDBOOK OF SOLID WASTE DISPOSAL: MATERIALS AND ENERGY RECOVERY, by Joseph L. Pavoni, John E. Heer, Jr., and D. Joseph Hagerty

ENVIRONMENTAL ASSESSMENTS AND STATEMENTS, by John E. Heer, Jr. and D. Joseph Hagerty

ENVIRONMENTAL IMPACT ANALYSIS: A New Dimension in Decision Making, by R. K. Jain, L. V. Urban and G. S. Stacey

CONTROL SYSTEMS FOR HEATING, VENTILATING, AND AIR CONDITIONING (Second Edition), by Roger W. Haines

WATER QUALITY MANAGEMENT PLANNING, edited by Joseph L. Pavoni

HANDBOOK OF ADVANCED WASTEWATER TREATMENT (Second Edition), by Russell L. Culp, George Mack Wesner and Gordon L. Culp

HANDBOOK OF NOISE ASSESSMENT, edited by Daryl N. May

NOISE CONTROL: HANDBOOK OF PRINCIPLES AND PRACTICES, edited by David M. Lipscomb and Arthur C. Taylor

AIR POLLUTION CONTROL TECHNOLOGY, by Robert M. Bethea

POWER PLANT SITING, by John V. Winter and David A. Conner

DISINFECTION OF WASTEWATER AND WATER FOR REUSE, by Geo. Clifford White

LAND USE PLANNING: Techniques of Implementation, by T. William Patterson

BIOLOGICAL PATHS TO ENERGY SELF-RELIANCE, by Russell E. Anderson

HANDBOOK OF INDUSTRIAL WASTE DISPOSAL, by Richard A. Conway and Richard D. Ross

HANDBOOK OF ORGANIC WASTE CONVERSION, by Michael W. Bewick

LAND APPLICATIONS OF WASTE (Volume 1), by Raymond C. Loehr, William J. Jewell, Joseph D. Novak, William W. Clarkson and Gerald S. Friedman

LAND APPLICATIONS OF WASTE (Volume 2), by Raymond C. Loehr, William J. Jewell, Joseph D. Novak, William W. Clarkson and Gerald S. Friedman

WIND MACHINES (Second Edition), by Frank R. Eldridge

PREFACE

Land treatment of municipal, agricultural, and industrial wastes and residues can be a cost-effective and efficient means of pollution control and resource recovery. Federal laws and regulations clearly indicate that Congress believes that this alternative approach must be considered, and should be favored wherever public funds are involved. When the first law (P.L. 92-500) emphasizing this technology became effective in 1972, only limited knowledge and experience were available to guide its application. In response to this lack of information, a multidisciplinary team was formed in 1974 in the College of Agriculture and Life Sciences (a statutory unit of the State University of New York) at Cornell University to develop an educational program beginning in 1974. The general goal of the program was to assemble state-of-the-art information to enable engineers, scientists, and pollution-control decision makers rapidly to determine the feasibility of using land application of waste. Financial support to develop this program was provided from 1974 through 1978 by the U.S. Environmental Protection Agency (Grant No. T-900500), and also during 1977 and 1978 by the U.S. Army Corps of Engineers. *Land Application of Wastes* represents a comprehensive summary of scientific and engineering fundamentals and design information of this technology.

The text of this two-volume set represents the bulk of the material developed by the Cornell team. However, it is only one component of a learning program that was produced to provide rapid, intensive, and comprehensive information to decision makers in wastewater pollution control. Additional components of the program include self-paced slide and tape modules ("audio-tutorial lessons") that reinforce and supplement material contained in the written modules. In addition to the written material, support materials that were developed included over 1000 slides, 16 cassette tapes, and an Instructor's Program. These are available at cost (without royalties to the authors). Information may be obtained from Instructional Materials Service, Stone Hall, Cornell University, Ithaca, New York 14853.

In addition, the authors would like to alert the readers to several personal notes. Unlike many large projects involving several disciplines, these results were produced by a team, which included agricultural engineers, environmental engineers, educators, agronomists, and others, who were extremely cooperative and friendly. The result is that *Land Application of Wastes* represents the best output that could be developed with the available skills and knowledge.

The listed order of the authors is arbitrary and is not intended to reflect the level of their contribution. R. C. Loehr, Director of the Environmental Studies Program and Professor of Agricultural and Environmental Engineering, provided the leadership in developing the support from the federal agencies, and assisted in coordinating the development of the later stages of the program. W. J. Jewell, Associate Professor of Agricultural Engineering, developed the engineering design approach as the basis of the program and was the coordinator for the development of the first draft. J. D. Novak, Professor of Science Education, provided the emphasis to search continuously for the major explanatory concepts, and to incorporate these basic concepts into a modular, self-teaching educational program. Primarily through his urging and efforts, a nonprofit agreement to publish this information was developed among Cornell University, U.S. Environmental Protection Agency, U.S. Army Corps of Engineers, other contributors, and Van Nostrand Reinhold Company. W. W. Clarkson and G. S. Friedman were the two major full-time research personnel who prepared the final written drafts of all materials and were responsible for bringing the final product together. W. W. Clarkson, Research Associate in the Department of Agricultural Engineering, is an environmental engineer and drafted much of the final text. G. S. Friedman, Research Associate in the Science Education Department, also drafted final modules, and was responsible for development of the audio-tutorial component and the Instructor's Program.

The authors would like to emphasize that this educational program comes to the practitioner as a pretested program. During the latter stages of development, the material was used in five, week-long workshops, which over 150 individuals attended. It has also been used successfully by others in a number of different formats. The requested feedback and surveys of these efforts indicate that the program is capable of transferring a significant amount of knowledge under a wide variety of conditions and applications. It was the intention of the educational format that this should be so.

The authors would like to acknowledge the high-quality, key agronomic information provided on this project. In particular, J. E. Stone developed the important concepts presented in the module on "Soil as a Treatment Medium," along with several other modules. He was assisted by members of Cornell University's Agronomy Department. In particular, Professor E. L. Stone was especially helpful in interpreting information needs and bridges required between pollution control and agricultural production interests. Other agronomy faculty and staff who provided assistance included W. H. Allaway, G. Armedee, D. R. Bouldin, A. H. Johnson, S. D. Klausner, D. J. Lathwell, D. A. Lauer, G. W. Olson, and P. J. Zwerman. S. M. Dabney was primary author of the "Crop Selection" and "Non-Crop and Forest Systems" modules.

Preparation of the final draft and production of the materials for testing throughout the country depended on having a highly efficient and cooperative Cornell print shop available, as well as many other highly skilled individuals. A. C. Plescia, the final program production coordinator, was assisted by F. Shottenfeld, who worked closely with outstanding individuals such as R. W. Gingras in the Graphic Arts Department. M. M. Bogin edited copy and drafted some material.

J. K. Krizek, A. L. Liebermann, R. H. Nagell, and S. Haeni contributed artwork and graphic design. L. A. Martin, P. M. Comfort, A. C. Plescia, B. C. Littlefair, and others provided secretarial assistance. D. P. Willmot prepared and processed audio-tutorial materials. The audio-tutorial program was narrated by K. Marash.

Technical assistance of various kinds was provided by R. Batkin, W. Messner, D. Payne, D. Way, G. MacCaskill, and R. J. Krizek. Thanks are also due to many other members of the Cornell community who shared time, ideas, and materials.

In addition to the significant effort provided by the Cornell University Agronomy Department personnel, many other individuals were responsible for various aspects of the program. Persons responsible for preliminary drafts of many modules included G. A. Garrigan, C. E. Morris, and D. F. Smith. Part-time assistance was provided by W. Pollard and M. Switzenbaum.

Outside consultants who provided assistance were E. Myers, private consultant; H. J. Ongerth, California State Health Department; R. E. Thomas, and B. L. Seabrook, U.S. Environmental Protection Agency; S. C. Reed, U.S. Army Corps of Engineers; T. D. Hinesly, University of Illinois; D. M. Whiting, National Oceanographic and Atmospheric Administration, National Weather Service. Final modules were also reviewed by N. Urban, H. L. McKim, C. Merry, H. Farquhar, and others of the U.S. Army Corps of Engineers.

The generous assistance of many equipment manufacturing companies, engineering consultants, state pollution-control agencies, and municipalities and industries using land-application systems are too numerous to mention. Nevertheless, they represent a major input.

Finally, as we prepare the final drafts of this material, we know that newer information will make some components obsolete. The focus on the fundamental concepts discussed in this work should provide a useful framework to view future developments in this rapidly advancing field.

THE AUTHORS

INTRODUCTION

STATEMENT OF PURPOSE

The purpose of this module is to acquaint the reader with the program of instruction entitled *Land Application of Wastes: An Educational Program.* This introductory module includes information on the content, structure, and alternative educational uses of the program.

CONTENTS

LAND APPLICATION OF WASTES PROGRAM–A BRIEF OVERVIEW

In the nineteenth century, it became relatively common for municipalities to operate "sewage farms." Some of those facilities were successful as measured by such standards as absence of excessive odors or standing water. Eventually, as urban expansion encroached on available land and technological interests turned to the design of sewage-treatment plants, most sewage farms were abandoned. Since 1972, as a result of the passage of PL 92-500, land application as a waste-management alternative has been reemphasized. Legislative recognition of the importance of land treatment as an alternative technology for waste treatment was further emphasized by the passage in 1977 of the amendments to this law. Critical evaluation of those factors that impose specific limitations on the land application of wastes is needed. Specifically, it is necessary to evaluate:

a. the characteristics of the wastewater and sludge
b. the impact of nitrogen, phosphorus, and potentially toxic elements on the soil system, vegetative cover, runoff water, and groundwater
c. the interactions between the soil system and wastewater constituents

d. the social and legal constraints
e. the effect of the proposed land-application system on present and future regional land-use patterns

Thought must be given to the necessity of designs employing combinations of conventional systems and land application. Such an evaluation must include the broad implications of the limiting factors with respect to both land application in general and to the specific site under consideration. Modification of specific waste characteristics and site conditions to achieve economical treatment may be possible. Examples of such modifications include underdraining certain soils so that larger hydraulic loads could be applied, wastewater pretreatment to reduce heavy metal content, or frequent cropping to achieve a higher rate of nutrient removal.

THE EDUCATIONAL PROGRAM STRUCTURE

The success of a land-application system rests on scientific, legal, and social fundamentals and practices. It is necessary to integrate these, to discuss their implications, and to provide a level of practical experience.

The objective of this educational program is to develop a better understanding of the capacity of the soil to assimilate wastes. The program is designed for engineers, scientists, planners, waste-management specialists, and other practitioners in the area of environmental protection. Land application of municipal, industrial, and agricultural wastes is addressed. Knowledge from several disciplines, including sanitary, environmental, and agricultural engineering, and agronomy, soil science, economics, and law are incorporated into the program. In order to present this diverse array of information in a meaningful way, this program is presented in *modular* format. This format is designed to be useful and understandable to individuals who have had no formal training in application of waste technology as well as persons who have been involved with the design of successful land waste-treatment systems.

The Modules

The program's basic unit of formal instruction is the module. A module is a self-contained learning packet focused on a specific concept or on a set of highly integrated concepts. Within the module, the reader will find information on what they can expect to learn from the module, how the module fits into the entire program, and a discussion of specific aspects of land application of wastes. Modules require from 0.5 to 5 hours of study time, depending in part on the reader's background and on the desired degree of mastery of the study materials.

All modules are presented in a printed format. Fifteen modules have also been adapted as audio-visual presentations known as *audio-tutorials* or *A-T units.* Scripts are provided along with the tape and slide A-T presentations, and several A-T units include supplementary booklets for further study.* The audio-tutorial units are supplementary and need not be included in a study program. They have proved to be particularly valuable in workshops and college courses using the *Land Application of Wastes* materials, where they have been well received by the participants.

Each module includes a set of *learning objectives*, which indicate what the reader can expect to learn from the module. The objectives are usually presented as a series of competencies

*Complete information on and cost of the audio-tutorial program are available from the Instructional Materials Service, Stone Hall, Cornell University, Ithaca, New York 14853.

that the participant will be able to demonstrate after completing the module. For example, an objective for this module might read as follows:

> After completing the module, the reader will be able to:
>
> Describe the organization and content of the program, *Land Application of Wastes*; indicate how the printed modules that follow can be used in workshops, short courses, independent study, or in college courses in environmental or sanitary engineering.

If the reader can meet the objectives without completing the module, then the module can be skipped, or simply skimmed over using italicized summaries.

A table of contents is presented at the beginning of each module. This allows the reader to see precisely what information is covered and where it is found. Readers should carefully peruse the major titles and their immediate subdivisions, and then reflect on how these topics might "fit" with respect to their existing knowledge of the topic. This is the first step in making effective use of the modular format. If the readers cannot complete all of the objectives of a module, the table of contents should allow them to identify which sections contain the material they need to learn in order to complete the remaining objectives, without having to read material already familiar.

Key ideas presented in the text are summarized in italics. This also facilitates efficient learning, review, or both. Following a summary and table of contents of most of the modules, a glossary of terms and concepts used is presented.

Since each module is a self-contained learning packet, readers must be sure they have learned its contents to a sufficient degree. They can do this by returning to the module objectives and assessing their competence in each. If some of the objectives cannot be satisfied, readers can review any or all parts of the audio-tutorial or printed modules or using italicized summaries to aid in locating pertinent sections.

While each module is considered to be an individual unit, it must be remembered that each module is designed as an integral part of the entire educational program. Thus, there is extensive cross-referencing between modules. As the participant proceeds through the program, the individual modules will form an integrated package of information.

The first 10 modules are presented in Volume I of the program, and the remaining 11 modules in Volume II. Let us now look at how the individual modules are integrated into the entire program.

The Design Procedure

In order to bridge the gap between an academic learning experience such as this program, and the real-life situations faced by personnel in the field, a framework that approximates the design procedure used by engineers has been included. The framework is described in "Design Procedures for Land Application of Wastes."

The design procedure identifies two specific levels. At the first level, general information is presented on all topics relevant to land application of wastes. However, this information provides an overview of land-treatment technology, and so many details are not included. Table 1 lists the Level I module titles.

After satisfactorily completing all Level I modules, the reader should be able to arrive at an

Table 1. Introductory Modules Tend to Center Attention on Broad Topics and to Stress Interrelationships Between Fundamental Concepts.

Module	Title
I-1	Soil as a Treatment Medium[a]
I-2	Design Procedures for Land Application of Wastes[a]
I-3	Waste Characteristics[a]
I-4	Treatment Systems, Effluent Qualities, and Costs[a]
I-5	The Role of Vegetative Cover[a]
I-6	Site Evaluation: General Criteria–Information Sources[a]
I-7	Costing Land-Application Systems

[a]Also available as an audio-tutorial module.

initial feasibility decision for a specific design problem. At this point, the reader should be able to state what the limiting factor to land application may be for a given waste input and application site, what social aspects need further investigation, possible alternative sites for a land-application system, and generally whether land application is a feasible waste-management alternative for that particular situation.

The reader will then be ready for Level II. The Level II modules are more detailed, and they contain the necessary information to arrive at more detailed design parameters such as specific hydraulic loads, application rate, land area needed, and so on. After completing Level II modules, a much more detailed problem evaluation is possible. Table 2 lists the titles of Level II modules.

Table 2. Level II Modules Provide More Detailed Information for Completion of Design Problems.

Module[a]	Title[a]
I-8	Societal Constraints
I-9	Legal Aspects
I-10	Case Studies Reviewed
II-1	Nitrogen Considerations[b]
II-2	Phosphorus Considerations[b]
II-3	Organic Matter[b]
II-4	Potentially Toxic Elements
II-5	Pathogens
II-6	Climate and Wastewater Storage[b]
II-7	Crop Selection and Management Alternatives[b]
II-8	Non-crop and Forest Systems[b]
II-9	Waste Application Systems[b]
II-10	Drainage for Land-Application Sites[b]
II-11	Monitoring at Land-Application Sites[b]

[a]Volume I of *Land Application of Wastes* contains Modules I-1 to I-10, and Volume II contains Modules II-1 through II-11.
[b]Also available as an audio-tutorial module.

MODULE AND PROGRAM SEQUENCES

The program consists of over 20 modules.* There is no formal sequence of modules because, as noted, each is self-contained. To start the program, choose one of the seven Level I modules. The modules may be completed in any order desired, although some readers may find it preferable to complete the modules in sequence. The Level I modules can be completed in 3 to 20 hours of study time.

It is suggested that the readers assume that they have been asked to advise a municipality or an industrial firm about a proposed land-application system (such as offered in the sample design problems). With this focus, the reader can ask intelligent questions and gather and relate pertinent information to the information in the modules.

When this material is presented in a workshop or other course the design problems are included only to provide a focus for the fundamental information and to place it in a real rather than abstract context. The course participant should not consider solution of the design problem to be the most important part of the program. The design problem is merely to facilitate the acquisition and use of the basic information.

THE EDUCATIONAL PROGRAM—ALTERNATIVES

Whereas the modules contain a large portion of the current knowledge of land application of wastes and can be read somewhat like a textbook or a symposium volume, they are also designed to be used in a variety of educational programs. The psychological and social factors that were considered in the design of the modules are presented in an Instructor's Program.**

Workshop Programs

Workshop programs draw together a group of participants with varying interests and backgrounds for a short, intensive period of study. Most workshop programs constitute a series of lectures by "experts," and offer some opportunity for questions and discussions. The *Land Application of Wastes* program has been designed to be used effectively in a 2- to 5-day workshop that includes *individual* study of printed and audio-tutorial modules, staff tutoring, small group problem-solving sessions focused on representative design problems, some guest lectures, and special group presentations and correlated individual and group activities. A sample schedule for a 5-day workshop is shown in Table 3.

Sufficient staff must be provided to meet each participant's need for guidance. Guest speakers—experts with various backgrounds in land application—serve as additional resource personnel. Participants are encouraged to interact with the speakers, individually or in groups. The participants represent a wide range of pertinent knowledge and experience. Although each participant moves through the program as an individual, the program format is designed to encourage interaction and support.

Participants will be given realistic design problems based on the scenario of a client approaching a consultant who has some limited information at hand and requesting advice on the feasibility of land application to solve a given problem. Each participant will be asked to select one

*Additional modules will be prepared in the future and will become available either through the Instructional Materials Service at Cornell University or through Van Nostrand Reinhold Company.

**The Instructor's Program consists of three modules describing the organization and conduct of "workshop" educational programs, some alternative programs, and the theoretical rationale for the program as developed. The Instructor's Program is available from the Instructional Materials Service at Cornell University.

Table 3. Schedule for a Typical Workshop

Time	Monday	Tuesday	Wednesday	Thursday	Friday
8:30 A.M.	Registration, welcome, introduction to program	Level I design procedure—Limiting Design factors	Introduction to Level II	Group discussion if needed—alternative systems	Prepare Level II design
9:00	Written and A-T modules, design problems				Presentation of Level II designs
		X	X	X	
9:30		X	X	X	
10:00	Coffee	Coffee	Coffee	Coffee	Coffee
10:30	Level I X	X	X	X	Discussion and evaluation
11:00	X	X	X	X	↓
11:30	X	Complete Level I topics	Discussion—Pathogens and toxic elements (or nitrogen)	X	↓
Noon	Lunch	Lunch	Lunch	Lunch	Adjournment
1:00 P.M.	Speaker	Speaker	X	Speaker	
2:00	X	Prepare Level I design	X	X	
3:00	X	Presentations of Level I designs	Field trip (optional)	X	
4:00	Design problem definition and review of format and early problems	Discussion of cost-effectiveness and pollution-control efficiency	↓	Level II modules, complete public hearing	
5:00					
7:30	Evening program				

X denotes time set aside for individualized study with tapes, slides, written material, and other prepared material. Individual course instruction and tutoring are also available at these times.

"interesting" situation for study, and the participants will be divided into study teams. Each team will meet with a staff member to consider the problem. However, the staff member will only direct the discussion if it strays too far off target, and will act as a resource person to indicate where specific information can be found. The staff member is not to supply factual answers except when necessary to avoid delays in the program. Working together, the team should develop the background information necessary to assess the general feasibility of land application for the specific problem (Level I). A more detailed analysis and feasibility will occur in the subsequent part of the program (Level II).

Each study team presents reports to the total workshop group to indicate progress on their design problem and to solicit comments and criticism. Field trips and other activities may be included in a workshop program.

Independent Study

For most professionals, independent study in the office or home is the most common form of education. *Land Application of Wastes* was designed to be an effective educational tool for use by professionals who wish to study on their own. It can be used with the printed modules alone or in combination with audio-tutorial units that can be purchased through Cornell University or loaned from regional offices of the U.S. Environmental Protection Agency. In the future, we anticipate that some colleges and universities will offer faculty assistance through home study or extension education programs to assist the independent learner. The study time needed to meet the objectives given in all modules will vary depending on an individual's background and experience, with a reasonable range of 30 to 100 hours.

Short Courses and Other College Offerings

Some colleges and universities are offering "short courses" in land-application-of-wastes technology. These courses fit into summer session or other college program offerings and usually carry undergraduate or graduate credit. Both the printed and audio-tutorial modules are used to supplement semester courses in waste-treatment technology. The materials have been designed to be used effectively in conjunction with formal college courses.

To Begin the Program

We advise the reader to begin by thumbing through all of the modules, giving a few minutes to each and stopping occasionally to read italicized summaries or to study a figure or table. Then we advise that you begin with Module I-1, "Soil as a Treatment Medium," and proceed from there. An understanding of fundamental concepts necessary for the solution of waste treatment problems is the final criterion of effective learning, and hence a major goal of this program. The authors would be pleased to learn of your success or difficulties with this program.

Agriculture

CONTENTS

LAND APPLICATION OF WASTES

Volume II

Module II-1

NITROGEN CONSIDERATIONS

SUMMARY

This module expands on the introductory discussion of nitrogen in Level I Modules, I-1, "Soil as a Treatment Medium"; I-3, "Waste Characteristics"; and I-5, "Role of Vegetative Cover". The various chemical forms of nitrogen found in land treatment systems are defined. Inputs from waste application as well as natural sources are quantified for typical situations. A discussion of nitrogen transformations in the soil includes mineralization and immobilization processes (organic-inorganic transformations). The importance of carbon to nitrogen ratio of applied waste is emphasized in determining what reactions will occur in the soil. Possible losses and removals of nitrogen from a land treatment system are discussed. All of the above factors are combined into a mass balance of nitrogen. Examples illustrate the application of this mass balance in determining wastewater and sludge loadings, on a nitrogen basis, for typical situations.

Data on nitrogen loadings and removals from various existing systems are summarized in the final section of the module. Other modules have pointed to nitrogen as the most prevalent limiting factor for waste addition to a given site. This module reiterates that point, while presenting detailed discussion of the components of the nitrogen interactions, or mass balance, of a system and demonstrating the application of this concept to typical systems. An illustration of short-term nitrogen loading calculation, using the decay series concept, is presented in the Appendix.

CONTENTS

GLOSSARY

ammonification–The biochemical process whereby ammoniacal nitrogen is released from nitrogen-containing organic compounds.

denitrification–The biochemical reduction of nitrate and nitrite to gaseous nitrogen either as molecular nitrogen or as an oxide of nitrogen.

evapotranspiration–The combined loss of water from a given area during a specified period of time, by evaporation from the soil and plant surfaces and by transpiration from plants.

immobilization–The conversion of an element from the inorganic to the organic form in microbial or plant tissues, thus rendering the element not readily available to other organisms or other plants.

leaching–The removal of materials in solution from the soil.

mineralization–The conversion of an element from an organic form to an inorganic state as a result of microbial decomposition.

nitrification–The biochemical oxidation of ammonium to nitrate.

nitrogen fixation–The conversion of elemental nitrogen (N_2) to organic combinations or to forms readily utilizable in biological processes; especially by certain bacteria in the nodules of leguminous plants, making the nitrogen available to the plants.

volatilization–The process of vaporizing or becoming gaseous.

OBJECTIVES

Upon completion of this module, the reader should be able to:

1. List the six forms of nitrogen of importance in land treatment.
   a. Indicate which are in gaseous form.
   b. Indicate which is (are) subject to leaching to the groundwater.
   c. Indicate which can be tied up by the soil cation exchange complex.
2. Name three sources of nitrogen input to a land treatment site.
3. Discuss the effect of C:N ratio of organic matter on immobilization and mineralization. Give the approximate cut-off point, in terms of % N or C:N, at which immobilization or mineralization should predominate.
4. List five possible mechanisms for nitrogen loss and removal from a land treatment site. Discuss the conditions necessary for each of these factors.
5. State the generalized nitrogen mass balance equation in descriptive terms, and illustrate its mathematical application to the calculation of nitrogen loading.

INTRODUCTION

The most important factor limiting land application of wastes is often the amount of inorganic nitrogen contained or released from wastewater and sludges. Large inputs of nitrogen, while stimulating plant growth, can lead to increased losses to ground and surface waters. Such increases may result in excessive growth of algae and aquatic weeds. Moreover, a build-up of nitrate in drinking water supplies is thought to pose a health hazard to humans, particularly infants, as well as livestock. Maintenance of the groundwater nitrogen level below the drinking water standard of 10 mg/l NO_3-N requires careful management. Management decisions on nitrogen loading rates must be based on a thorough knowledge of the various forms of nitrogen contained in wastes, their properties, the transformations which occur in soils, and the rates associated with these transformations.

**In order to maintain permissible levels of nitrogen in groundwater, land application site operators must know the various forms of nitrogen in wastes, their properties, the transformations which occur in soils, and the rates associated with these transformations.*

The major forms of nitrogen of importance in waste management are briefly outlined below:

Organic nitrogen is bound in carbon-containing compounds such as proteins. Within the soil, organic-N, a constituent of humus, is not available for direct plant uptake, but is slowly transformed to available forms by means of microbial decomposition. The nitrogen-containing organic compounds are often abbreviated R-NH_2.

Ammoniacal nitrogen includes two forms, the ammonium ion (NH_4^+) and gaseous ammonia (NH_3). Ammonium, by virtue of a net positive charge, can be held in the soil on clay and organic matter cation exchange sites. Ammonium is utilized by both plants and microorganisms as a nitrogen source. Ammonia gas exists in equilibrium with the ammonium ion. At high pH values (>8.5) ammonia predominates and may escape into the atmosphere in significant quantities.

*This and other italicized summaries are intended to highlight key ideas, provide a basis for later review, or to aid in skimming sections that are relatively familiar. They can be ignored in a complete reading of the text.

Nitrite (NO_2^-) is a highly mobile anion formed in soils as an intermediate during the microbial conversion of ammonium to nitrate (see below). Nitrite is generally transitory in soils but may accumulate under certain conditions; it is toxic to higher plants in very small quantities.

Nitrate (NO_3^-) is a highly mobile anion, readily utilized by both plants and microorganisms. Nitrate is of primary interest in waste application systems since it may be readily leached from the soil and, when present in excess in drinking water, is thought to be a health hazard to humans. The U.S. Public Health Service has recommended a drinking water standard of not more than 10 mg/l NO_3-N. The term NO_3-N is read nitrate-nitrogen and refers only to the nitrogen portion of the nitrate anion. Do not confuse NO_3^- with NO_3-N as they are separate entities.

$$10 \text{ mg/l } NO_3\text{-N} = 44.3 \text{ mg/l } NO_3^-$$

$$2.26 \text{ mg/l } NO_3\text{-N} = 10 \text{ mg/l } NO_3^-$$

The form of nitrogen of most concern in land application of wastes is nitrate since it may be leached readily from the soil.

Nitrous oxide (N_2O) is a gaseous form only slightly soluble in water. It is a normal constituent of the atmosphere in small quantities (<1 ppm), and is probably not utilized by plants. Nitrous oxide is formed during the denitrification of nitrate.

Molecular nitrogen (N_2) is a gaseous form comprising nearly 80% of the normal atmosphere. Like nitrous oxide, molecular nitrogen is an end product of denitrification.

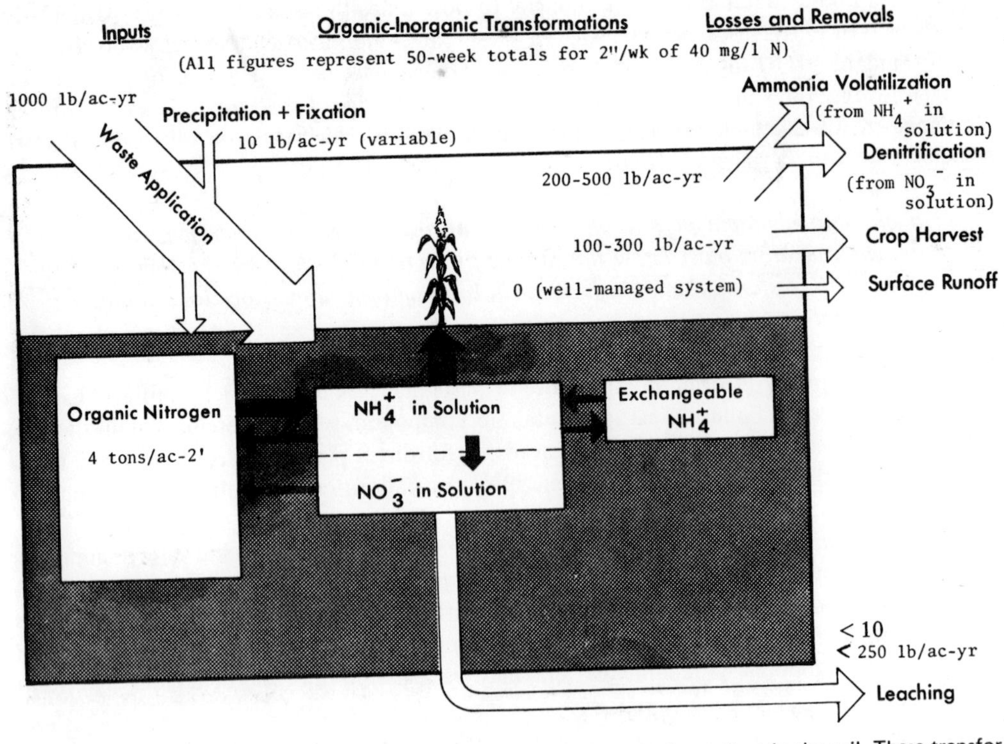

Figure 1. Nitrogenous materials are subject to a series of complex transformations in the soil. These transformations, mediated by microorganisms, environmental factors, and management practices, determine the ultimate fate of nitrogen inputs in terrestrial systems.

NITROGEN BEHAVIOR IN TERRESTRIAL SYSTEMS

The various forms of nitrogen are interconnected through a series of complex transformations, which collectively constitute the nitrogen cycle. For purposes of this discussion, the nitrogen cycle will be divided into three portions: inputs, organic-inorganic transformations, and losses or removals (Figure 1).

Nitrogen Enters System Through Wastes, Precipitation, Atmosphere

Inputs. Nitrogen enters terrestrial systems in wastewater and/or sludge applications, natural precipitation (rain and snow), and through fixation of molecular nitrogen from the atmosphere by specialized microorganisms.

Wastewater and Sludge. Municipal wastewaters and sludges contain variable amounts of nitrogen in the forms of organic-N, ammonium-N, and nitrate-N. Some representative values and common loading rates are presented in Table 1. The data in this table should immediately raise some questions. As was noted in Module I-1, "Soil as a Treatment Medium" and Module I-5, "Vegetative Cover" the annual nitrogen removal by crops seldom exceeds 300 lb/acre. Low rate applications of wastewater (2 in./wk for 30 weeks) add approximately this amount while higher rates considerably exceed this value. The addition of a small amount of sludge (1 acre-inch) may also exceed 300 lb/acre of nitrogen. Thus, with moderate to high application rates of either effluent or sludge, strict nitrogen control will require management methods other than or in addition to crop uptake.

> *Moderate to high application rates of sludge and wastewater require management beyond a reliance on crop uptake.*

Precipitation. Annual additions of nitrogen in precipitation generally occur in the nitrate and ammoniacal forms. Amounts fluctuate markedly depending on season, climate, and surrounding

Table 1. Total Nitrogen Content in Various Wastes and Common Annual Loading to Land.

Waste	Total N	Annual N Loading to Land	
	mg N/l	Assume 20 mg N/l	
		Low Rate[a]	High Rate[b]
Sewage effluent		--------lb/acre-yr--------	
Untreated	20–85		
Primary	20–50	270	2840
Secondary	10–40		
		Assume 5% N (dry wt.)[c]	
Sewage sludge	1–8% (dry wt.)	450	

[a]Assumes application rate of 2 in./wk for 30-week period (representative of irrigation and overland flow systems)
[b]Assumes application rate of 6 in./day for 14 days followed by 14-day rest period over a 30-week season (representative of rapid infiltration system).
[c]Assumes one annual application of liquid sewage sludge (96% moisture; 4% solids) in amount of one acre-inch

Sources: Larson, *et al.*, 1973; Menzies, 1973; Pound and Crites, 1973; Reed, *et al.*, 1972 industrial activity. Within the U.S., precipitation inputs vary from 1-15 lb N/acre-yr (Sepp, 1971).

Nitrogen input from rain and snow varies from 1 to 15 pounds per acre per year.

Nitrogen Fixation. Specialized bacteria which develop in the roots of legumes and certain other plants are capable of "fixing" atmospheric nitrogen; that is, they can convert molecular nitrogen into a form usable by plants. This mutually beneficial relationship between microorganisms and plants is termed *symbiosis.* Where appropriate plants are present, symbiotic nitrogen fixation may range from approximately 50–300 lb N/acre-yr. A certain amount of nonsymbiotic fixation also occurs through the activity of free living microorganisms, and can contribute from 20–100 lb N/acre-yr (Brady, 1974). There is evidence that nitrogen fixation is much reduced by the presence of readily available nitrogen from other sources such as waste additions. Nitrogen fixation may therefore constitute only a minor input on land receiving waste applications.

Nitrogen input from nitrogen-fixing bacteria is minor on land receiving waste applications.

Waste Additions Affect Organic-Inorganic Transformations of Nitrogen in Soil

The amount of native nitrogen present in soils ranges from approximately 2,000-15,000 lb/acre, much of which is concentrated in the surface layers. Most of this nitrogen is bound in organic forms and is neither available to plants nor for leaching. Decomposition of organic matter by soil microorganisms slowly transforms organic-N to inorganic forms, a process termed *mineralization.* The reverse of mineralization, that is, the transformation from inorganic-N to organic forms, is called *immobilization.*

Most native nitrogen in soils is bound in organic forms; it is unavailable to plants and will not leach to groundwater.

Mineralization. The general term mineralization covers a whole series of reactions, the most important of which are ammonification and nitrification. Ammonification is the conversion of organic nitrogen to ammonia as shown below.

$$\underset{\text{(organic nitrogen)}}{R\text{-}NH_2 + H_2O} \xrightarrow[\text{enzymes}]{\text{microbial}} R\text{-}OH + \underset{\text{(ammonia)}}{NH_3} + \text{energy}$$

This reaction can be carried out by a wide variety of microorganisms.

Ammonification, which can be carried out by a variety of microorganisms, is the conversion of organic nitrogen to ammonia.

Within the pH range of most soils (i.e., 4.5-7.5), ammonia exists as the ammonium ion (NH_4^+). This positively charged ion may react with the cation exchange complex in the soil. Because of competition with other cations such as calcium (Ca^{+2}), magnesium (Mg^{+2}), and potassium (K^+), however, only about 5% of the cation exchange capacity is available for ammonium adsorption (Lance, 1972). For soil with a CEC of 15 meq/100 g, this results in an ammonium adsorption capacity of 1 g/ft^2 of soil, or about 100 lb N/acre. Clearly, this capacity could be saturated in a matter of months, even at low ammonium application rates. Ammonium, however, undergoes a further transformation: nitrification. Nitrification essentially regenerates the capability of the soil to adsorb ammonium ions. So far as we now know, nitrification is carried out in two steps by specialized microorganisms.

$$\textit{Step 1:}\quad 2NH_4^+ + 3O_2 \xrightarrow{\text{Nitrosomonas}} 2NO_2^- + 2H_2O + 4H^+ + \text{energy}$$

$$\textit{Step 2:}\quad 2NO_2^- + O_2 \xrightarrow{\text{Nitrobacter}} 2NO_3^- + \text{energy}$$

Both steps require oxygen and are thus favored in well aerated soils. Nitrification is also affected by temperature, soil moisture, and pH. At temperatures below approximately 30°F, nitrification essentially stops; the optimum temperature range is between 80–90°F (Brady, 1974). The process is inhibited in very wet (poorly aerated) or very dry soils, but readily occurs at moisture levels within and slightly below the acceptable range for higher plants. Nitrification is greatest at neutral to slightly alkaline pH values and is slow in strongly acid soils. The effect of pH is probably indirect, resulting from an association with the availability of exchangeable bases such as potassium, calcium and magnesium.

The ability of the soil to adsorb ammonia is limited; only about 5% of CEC is available for such adsorption.

Nitrification regenerates the capability of soil to adsorb ammonia; it is favored by good aeration of soil, 80-90°F temperatures and a neutral to slightly alkaline pH.

Under most conditions favoring the two reactions, Step 2 closely follows Step 1, precluding any significant accumulation of nitrite. Large amounts of ammonia, however, as might occur when ammonified wastes are added to highly alkaline soils, are apparently toxic to Step 2 organisms (Nitrobacter), resulting in a temporary build-up of nitrite which is very toxic to higher plants. Assuming near ideal temperature, moisture, and pH conditions, nitrification occurs very rapidly. Nitrate-nitrogen generation rates of from 6-22 lb/acre-day have been reported in surface soil after the addition of 100 lb/acre of ammonium nitrogen (Broadbent and Tyler, 1957).

Addition of ammonified wastes to highly alkaline soils can result in a build-up of nitrite; nitrite accumulation is very toxic to higher plants.

The major significance of nitrification in waste application systems lies in the fact that a relatively immobile form (NH_4^+) is transformed into a highly mobile form (NO_3^-) which may then readily leach from the soil. Furthermore, nitrate is required for denitrification, a process to be discussed later.

Through nitrification immobile ammoniacal nitrogen is converted to the highly mobile form of nitrate which may readily leach from soil.

Immobilization. Immobilization is essentially the reverse of mineralization, that is, inorganic nitrogen is converted into organic nitrogen as a component of microbial tissue. Immobilization occurs when the nitrogen content of decomposing organic material is insufficient to supply microbial needs. Inorganic nitrogen from the soil is then incorporated into the microbial cells.

The ratio of carbon to nitrogen (C:N) in the organic matter indicates whether mineralization

Table 2. Relative C:N Status for Various Wastes.

Low C:N Ratio (N content greater than 1.5% by dry wt)
Sewage sludges
Fresh animal feces
Some legume straws
Most young plant residues
High C:N Ratio (N content less than 1.5% by dry wt)
Vegetable cannery residues
Timber processing waste
Mature cereal straw
Some manures with included bedding straw

Source: Viets, 1974.

or immobilization of nitrogen is likely to predominate. When materials with a high C:N ratio (low nitrogen concentration) are added to soils, immobilization occurs initially. With time, however, carbon is lost as carbon dioxide (CO_2), while nitrogen is conserved. The C:N ratio thus narrows and mineralization eventually predominates. If the added organic matter has a low C:N ratio (high nitrogen concentration), mineralization occurs from the outset.

The carbon-nitrogen ratio of the organic matter in soil indicates whether organic nitrogen will be transformed to inorganic forms or the reverse will predominate.

As a general rule, mineralization occurs when the N content of the organic matter exceeds approximately 1.5% by dry weight; this corresponds to a C:N ratio of between 20 and 25. Where the N content is less than 1.5%, immobilization occurs until decomposition with accompanying carbon loss sufficiently narrows this ratio. Some representative wastes in both high and low C:N categories are presented in Table 2.

When materials with a high C:N ratio are applied to soils immobilization occurs; mineralization predominates when materials with a low C:N ratio are added.

Several Mechanisms Remove Nitrogen from System

Nitrogen is lost or removed from terrestrial systems through several means including ammonia volatilization, denitrification, crop harvest, surface runoff, and leaching. Lance (1972) provides a useful review of these processes.

Ammonia Volatilization. In moist, alkaline soils, ammonium salts undergo a chemical reaction resulting in the evolution of ammonia gas (NH_3). When liquid sludge is surface applied, 60% or more of the NH_4-N may thus be lost, depending on the application rate and soil properties. In general, such losses decrease with increasing clay content, and increase with application rate (Ryan and Keeney, 1975). Volatilization losses can be reduced or prevented by injecting or otherwise incorporating sludge into the soil so that ammonium can react with the soil cation exchange complex.

Sixty percent of ammonium nitrogen may be lost in the form of ammonia gas in liquid sludge application; incorporation of sludge into soil reduces this loss.

Denitrification. Denitrification is the process(es) whereby nitrate-nitrogen is converted into gaseous forms, N_2O or N_2. The reactions involved may be purely chemical (chemodenitrification) or brought about by microorganisms (biodenitrification). Chemodenitrification is generally restricted to highly acid soils and will likely be of only minor significance in land application systems. Therefore, the following discussion focuses on biodenitrification.

> *Biodenitrification is the conversion by microorganisms of nitrate into gaseous forms of nitrogen, beneficial because nitrate may be readily leached from soil to groundwater.*

Under anaerobic conditions, where molecular oxygen is lacking, bacteria use the chemically combined oxygen in nitrate. Nitrous oxide (N_2O) and molecular nitrogen (N_2) thus form and escape into the atmosphere; nitrous oxide generally predominates under field conditions.

$$\underset{\text{nitrate}}{NO_3^-} \longrightarrow \underset{\text{nitrite}}{NO_2^-} \longrightarrow \underset{\substack{\text{nitrous}\\\text{oxide}}}{N_2O} \longrightarrow \underset{\substack{\text{molecular}\\\text{nitrogen}}}{N_2}$$

The bacteria involved in these reactions require organic carbon as an energy source. Thus, to encourage biodenitrification, land application systems must be managed so as to bring nitrate and organic carbon together under anaerobic conditions. This is no simple task.

> *Land application systems must bring nitrate and organic carbon together in order to encourage biodenitrification.*

Sewage effluents often contain most of their total nitrogen in organic or ammoniacal forms. Transformation of these forms to nitrate requires aerobic conditions which are not conducive to denitrification. Thus, alternate aerobic and anaerobic periods or zones must be achieved. Moreover, in many secondary effluents, the relative amounts of total N and biodegradable organic carbon are of the same general magnitude, 20-30 mg/l. The loss of carbon as CO_2, however, often occurs at a higher rate than nitrification. This results in organic carbon removal prior to the formation of appreciable nitrate. Thus, by the time nitrogen is converted to a form that can be denitrified, much of the required organic carbon has been consumed. Approximately 0.9-1.3 grams of biodegradable organic carbon are needed to denitrify one gram of nitrate-nitrogen (Viets and Hageman, 1971). This means that if 70% of the total organic carbon is lost as CO_2 in aerobic zones, then less than 30% of the total N can be denitrified even under optimal denitrifying conditions. For these reasons, the extent of denitrification is extremely difficult to predict and generalized values are often of minimal value without on-site demonstrations using a specific waste.

> *Denitrification is difficult to predict because by the time nitrogen is converted to a form that can be denitrified, much of the required organic carbon has been consumed.*

Crop Uptake and Harvest. Nitrogen uptake by harvestable crops constitutes a significant mechanism for N removal. Some representative values are presented in Table 3. Without harvest, the annual nitrogen uptake will simply return to the soil in an organic form. During initial years of operation, some of this nitrogen may accumulate in the soil. As soon as a new equilibrium level is reached, annual additions could be offset by annual mineralization so that no further accu-

Table 3. Nitrogen Removal by Selected Crops when Irrigated with Municipal Sewage Effluent.

Crop	Yield per acre	N Uptake lb/acre-yr
Corn Grain	114 bu	83
Corn Silage	6.0 tons	161
Wheat Grain	54 bu	83
Reed Canarygrass (3 cuttings)	4.3–7.0 tons	272–356

Sources: Sopper and Kardos, 1973; Kardos, *et al.*, 1974.

mulation occurs. However, data indicating N accumulation in the range of 120 lb/acre-yr over several decades of operation are available from Werribee Farm, Melbourne, Australia (Johnson, *et al.*, 1974).

Significant amounts of nitrogen are removed by harvest of cover crop.

Nitrogen Loss from Surface Runoff. Surface runoff occurs whenever water application rates exceed the soil infiltration capacity. Most states require that surface runoff be contained from land application systems. The magnitude of nitrogen loss in surface runoff is highly variable from one site to another depending on natural precipitation, topography, season, and application rates and methods. Such loss can be made minimal in a well designed and managed system.

Nitrogen loss from surface runoff can be minimized by good management of land application site.

Leaching of Nitrogen. After the water holding capacity of a soil is reached, additional water percolates through the soil. The resulting leaching losses of nitrogen are generally small in natural systems, due to limited amounts of available N. Inputs of wastewater or sludge need not result in excessive leaching loss of N where sound management is practiced. Obviously, where nitrogen inputs exceed the capacity for crop uptake and/or denitrification, leaching of nitrate must be expected. Careful assessment of nitrogen inputs versus estimated losses and removals can limit nitrate leaching to acceptable levels. "Acceptable levels" should mean that the NO_3^--N level in groundwater under or near the land application site do not exceed 10 mg/l. The exact interpretation of this standard (i.e., whether the water to be sampled is the groundwater or only the leachate from the site, or whether the sampling site relates to nearby areas where groundwater is used for drinking supply, or sampling must be performed directly under the application site) is left to the state regulatory authority involved. They must be consulted in dealing with this question.

If nitrogen inputs exceed the capacity of crop harvesting and denitrification to take up nitrogen, leaching of nitrogen can be expected.

GOOD MANAGEMENT REQUIRES CAREFUL DETERMINATION OF NITROGEN LOADING

Nitrogen Mass Balance

In any land application system, nitrogen additions must be balanced against acceptable losses and removals. Estimates of inputs and losses are thus required, and must be assessed on both a short- and long-term basis.

Table 4. Nitrogen Mass Balance Equation for Terrestrial Systems.

Inputs		*Removals*
Total N in waste		N removal in crops
+		+
N in precipitation	=	N loss by leaching and surface runoff
+		+
N fixation		Denitrification
+		+
Mineralization of native organic N		Ammonia volatilization
		+
		N accumulation in soil

From Sopper and Kardos, Recycling treated municipal wastewaters and sludge through forest and cropland. Pennsylvania State University Press, University Park, Pa., 1973, p. 282.

The nitrogen mass balance equation helps to determine how nitrogen additions can be balanced against losses and removals.

A nitrogen mass balance equation for a terrestrial system is presented in Table 4. Potential N inputs include waste additions, precipitation, N fixation, and mineralization of native soil nitrogen. Potential outputs consist of crop removal, leaching and runoff losses, denitrification, ammonia volatilization and net storage or accumulation of soil N. Again, the latter includes both organic and inorganic forms. Not all terms may be needed in every case nor are all terms now quantifiable for all situations.

In the initial few years of operation, waste application may cause a net accumulation of nitrogen in soil organic matter or on soil colloids. Thus, not all nitrogen inputs would be accounted for in losses or removals. Net storage from one year, however, would become a potential input the following year. Failure to consider such short term accumulations can lead to an over-estimate of permissible loading rates.

Normally, additions of ammonium-N result in the rapid attainment of a new equilibrium. The cation exchange complex may be saturated during the first year or even after the first application, after which no net accumulation occurs. Accumulation of organic-N may continue for several years in the case of highly organic wastes, but eventually the microbial population will adjust so that annual mineralization equals annual additions. At this point the system has reached a new steady-state condition. The storage term in the righthand portion of the mass balance equation (Table 4) thus equals release from storage in the left-hand portion. The two terms then cancel one another on a long-term basis.

Calculation of Loading

The following discussion assumes that any short-term increases in nitrogen storage in the soil have already occurred and that there are no significant additions by fixation (i.e., the cover crop is nonleguminous).* The mass balance equation thus reduces to:

Total N in waste + N in precipitation

= N removal in crops + Leaching loss + Denitrification + Ammonia volatilization

This equation can be used to calculate nitrogen loading rates for either wastewater or liquid sludge applications.

*Additional information regarding short-term considerations is presented in the Appendix.

Wastewater Loading. The mass balance equation may be written as:

$$N + \frac{cP}{4.43} = C + \frac{aQ}{4.43} + dN + vN \tag{1}$$

where

N = total nitrogen in applied wastes (lb/acre-yr)
c = concentration of nitrogen in precipitation (mg/l)
P = precipitation (acre-in./acre-yr)
C = removal of nitrogen in crop (lb/acre-yr)
a = allowable nitrogen concentration in percolating or runoff water (mg/l)
Q = volume of water leaving site via percolation or runoff (acre-in./acre-yr)
d = fraction of N which is denitrified (% × 10^{-2})
v = fraction of N which is volatilized as ammonia (% × 10^{-2})
4.43 = conversion factor

The volume of water leaving the site (Q) is a function of precipitation, evapotranspiration, and wastewater additions such that:

$$Q = P + W - ET \tag{2}$$

where

P = precipitation
W = wastewater additions
ET = potential evapotranspiration (assumes that $P + W$ will allow potential ET to be realized in all cases)
(acre-in./acre-yr)

Furthermore, for wastewater containing a total nitrogen concentration y (mg/l):

$$N = \frac{yW}{4.43} \tag{3}$$

$$\frac{yW}{4.43} + \frac{cP}{4.43} = C + \frac{a(P + W - ET)}{4.43} + \frac{d(yW)}{4.43} + \frac{v(yW)}{4.43} \tag{4}$$

Solving for wastewater application (W):

$$W = \frac{4.43\,C + a\,(P - ET) - cP}{y - a - y(d + v)} \tag{5}$$

The following assumptions and conditions are suggested for general usage of Equation 5:

1. Ammonia volatilization (v) is set equal to zero unless known to be otherwise.
2. Denitrification (d) is also set equal to zero. This is a failsafe approximation and can be relaxed only on the basis of on-site demonstrations or future research findings.
3. The allowable nitrogen concentration in runoff and percolating water (a) is set equal to the Environmental Protection Agency recommended NO_3-N drinking water standard, 10 mg N/l.
4. Crop removal (c) is set equal to a constant for a given crop. There will be some variability among years which may at times result in the runoff or percolate water containing more nitrogen than the recommended standard. The same will occur in years when the precipitation is below average.

5. The annual excess of nitrogen is assumed to be dissolved in the total annual excess of water, that is, without seasonal highs and lows.
6. The net effect of conditions 4 and 5 is that some water leaving the site may at times contain nitrogen in excess of the 10 mg/l standard. This condition should be clearly understood and some agreement reached with regulatory agencies as to what is or is not acceptable. Careful monitoring will be required to substantiate the overall performance of the system. The conservative nature of condition 2 may partially offset the vagaries of conditions 4 and 5. These conditions provide a very conservative approach to the N loading calculation. They should be evaluated according to the principles outlined in this module and elsewhere, and relaxed if conditions warrant at a particular site, for example when the design provides for conditions conducive to ammonia volatilization or denitrification.

The following two examples are included to illustrate the use of these equations in calculating the acceptable loading rates for wastewater.

PROBLEM #1 (Wastewater Loading)

A. *Given:* An area in the northeastern U.S. with an average annual precipitation of 35 inches and a potential evapotranspiration of 20 inches. Determine how much sewage effluent, containing 20 mg/l total N, can be applied so that the soil leachate will average $\leqslant 10$ mg/l NO_3-N. Assume that reed canarygrass, harvested from the site, will remove 250 lb/acre-yr and that precipitation contains an average of 0.5 mg N/l.

Solution: By Equation 5 and suggested conditions for usage

$$W = \frac{4.43\,(250) + 10\,(35 - 20) - 0.5\,(35)}{20 - 10}$$

$$= 124 \text{ acre-in./acre-yr}$$

Note: By Equation 3, total N applied would equal 560 lb/acre-yr.

B. Suppose the application season equals 30 wk/yr. Determine the weekly liquid loading needed to dissipate 124 acre-in./acre-yr.

Solution:

$$\frac{124 \text{ acre-in./acre-yr}}{30 \text{ wk/yr}} = 4.1 \text{ acre-in./acre-wk}$$

Note: This value must be compared with the hydraulic capabilities of potential sites to see if it is acceptable.

C. Assuming that 4.1 acre-in./acre-wk is compatible with the hydraulic properties of potential sites, how many acres would be required for a 1 mgd wastewater flow?

Solution:

$$1 \text{ mgd} = 365{,}000{,}000 \text{ gal/yr}$$

$$1 \text{ acre-in} = 27{,}154 \text{ gal}$$

$$\text{acreage requirement} = \frac{365{,}000{,}000 \text{ gal/yr}}{(124 \text{ acre-in./acre-yr})\,(27{,}154 \text{ gal/acre-in.})}$$

$$= 108 \text{ acres}$$

Note: This value represents effective land application area; storage areas, buffer zones, etc., will be additional.

PROBLEM #2 (Wastewater loading)

A. *Given:* Same conditions as in problem #1 except a semi-arid area where annual precipitation equals 20 inches and annual potential evapotranspiration equals 20 inches. Determine effluent loading on both an annual basis and a weekly basis; determine the acreage required.

Solution:

$$W = \frac{4.43\,(250) + 10\,(20 - 20) - 0.5\,(20)}{20 - 10}$$

$$= 110 \text{ acre-in./acre-yr}$$

By Equation 3, this amounts to 496 lb N/acre-yr. Assuming a 30-wk application season, the weekly application rate is:

$$\frac{110 \text{ acre-in./acre-yr}}{30 \text{ wk/yr}} = 3.67 \text{ in./wk}$$

For a 1 mgd flow, the acreage requirement is:

$$\frac{365{,}000{,}000}{(110)\,(27{,}154)} = 122 \text{ acres}$$

Compare the answers in this problem with those obtained for Problem #1. The lower allowable loading rate in Problem #2 is due to the decrease in precipitation as compared to evapotranspiration, resulting in a lower dilution factor for nitrate in percolating wastewater.

Liquid sludge loading:

$$\mathrm{N} + \frac{cP}{4.43} = C + \frac{aQ}{4.43} + d\mathrm{N} + v\mathrm{N} \tag{1}$$

where

N = total nitrogen in applied wastes (lb/acre-yr)
c = concentration of nitrogen in precipitation (mg/l)
P = precipitation (acre-in./acre-yr)
C = removal of nitrogen in crop (lb/acre-yr)
a = allowable nitrogen concentration in percolating or runoff water (mg/l)
Q = volume of water leaving site via percolation or runoff (acre-in./acre-yr)
d = fraction of N which is denitrified ($\% \times 10^{-2}$)
v = fraction of N which is volatilized as ammonia ($\% \times 10^{-2}$)
4.43 = conversion factor

The volume of water leaving the site (Q) is a function of precipitation, evapotranspiration, and the water added in liquid sludge such that:

$$Q = P + zW - ET \tag{6}$$

where

P = precipitation
W = liquid sludge additions
ET = potential evapotranspiration
(all three: acre-in./acre-yr)
z = moisture fraction in liquid sludge ($\% \times 10^{-2}$)

Furthermore,

$$N = \frac{(t)\,(s)\,10^6\,W}{4.43} \tag{7}$$

where

t = total N fraction in waste (% of solids $\times\ 10^{-2}$)
s = total solids (% $\times\ 10^{-2}$)

Therefore

$$\frac{ts\,(10^6)\,W}{4.43} + \frac{cP}{4.43} = C + \frac{a(P + zW - ET)}{4.43} + \frac{dts\,10^6\,W}{4.43} + \frac{vts\,10^6\,W}{4.43} \tag{8}$$

Solving for liquid sludge application (W):

$$W = \frac{4.43\,C + a(P - ET) - cP}{ts10^6 - az - ts\,10^6\,(d + v)} \tag{9}$$

The same conditions suggested for the use of Equation 5, apply to the use of Equation 9.

The following example is included to illustrate the use of these equations in calculating the acceptable loading rates for liquid sludges.

PROBLEM #1 (Liquid sludge loading)

A. *Given:* An area in the northeastern U.S. with an average annual precipitation of 35 inches and a potential evapotranspiration of 20 inches; a liquid sludge with the following characteristics: 96% moisture, 4% solids, 3% total nitrogen (dry solids basis).

Determine how much liquid sludge can be applied so that the soil leachate will average $\leqslant$10 mg/l NO_3-N. Assume that a grass crop on the site will remove 250 lb N/acre-yr, and that precipitation contains an average of 0.5 mg N/l.

Solution: By Equation 9 and suggested conditions for usage:

$$W = \frac{4.43\,(250) + 10\,(35 - 20) - 0.5\,(35)}{(.03)\,(.04)\,10^6 - 10\,(.96)}$$

$$= 1.04 \text{ acre-in./acre-yr}$$

Note: By Equation 7, total N applied would then equal 282 lb/acre-yr.

EXISTING LAND APPLICATION SYSTEMS PROVIDE DATA ON NITROGEN LOADING

Irrigation

Wastewater spray irrigation systems have commonly used application rates ranging from 1-3 in./wk. With secondary effluents, corresponding quantities of nitrogen are of the same general magnitude as crop requirements over the growing season. Under these application rates, only small amounts of nitrogen are lost through leaching.

Only small amounts of nitrogen are lost through leaching in most spray irrigation systems using secondary effluent.

Table 5. Partial Nitrogen Mass Balance for a Wastewater Irrigation Site Under Reed Canarygrass Cover in Pennsylvania. Secondary Effluent Was Applied at a Rate of 2 in./wk on a year-round Basis. Values in the Table are Annual Averages Over a 6-Year Period.

Annual average lb/acre-yr	*Inputs*	=	*Removals*	Annual average lb/acre-yr
354	Total N in waste +		N removal in crops +	345
at least 66	{ N in precipitation + N fixation + Mineralization of native organic N }		N loss by leaching +	75
			{ Denitrification + Ammonia volatilization + N accumulation in soil }	?

From Sopper and Kardos, Recycling treated municipal wastewaters and sludge through forest and cropland. Pennsylvania State University Press, University Park, Pa., 1973, p. 282.

Table 5 presents a partial nitrogen mass balance for 6 years of operation of a spray irrigation field under hay cover. Nearly all of the applied nitrogen can be accounted for in the harvested reed canarygrass. When crop removals and estimated leaching losses are combined, however, they exceed the N content of applied wastewater. The excess presumably resulted from precipitation inputs, fixation of atmospheric nitrogen, and/or mineralization of native organic nitrogen. The latter source was likely the major contributor.

Overland Flow

Rates of wastewater application and nitrogen loading in existing overland flow systems are roughly similar to irrigation systems. The principal mechanisms for nitrogen renovation include crop removal and denitrification, often followed by dilution upon discharge into surface waters.

Table 6. Partial Nitrogen Mass Balance for An Overland Flow System Under Reed Canarygrass Cover. Experimental Flats (10′ × 5′ × 7.5″; 2% slope) Received Secondary Effluent at a Rate of 2.5 in./wk Over a 12-Week Period.

12-wk total lb/acre	*Inputs*	=	*Removals*	12-wk total lb/acre
137	Total N in waste[a] +		N removal in crops +	104
?	{ N in precipitation + N fixation + Mineralization of native organic N }		N in runoff[b] +	23
			{ Denitrification + Ammonia volatilization + N accumulation in soil }	at least 10

[a]Applied wastewater contained the equivalent of 60 lb organic-N/acre, 51 lb ammonium-N/acre, and 26 lb nitrate-N/acre
[b]Runoff water contained the equivalent of 14 lb organic-N/acre, 3.5 lb ammonium-N/acre, and 5.5 lb nitrate-N/acre

Source: Hoeppel, *et al.*, 1974.

Table 7. Nitrogen Relationships for Summer (May-September) and Winter (November-April) Periods in An Overland Flow System in Ada, Oklahoma. The Experimental Plots, 36′ X 118′ with 2-4% Slope, Supported Native Grasses; Raw Comminuted Sewage Was Applied at a Rate of 3.9 in./wk.

	Summer		Winter	
Nitrogen in applied waste	mg/l	lb/acre	mg/l	lb/acre
Org-N	5.8	111	5.8	131
NH_4-N	17.0	326	17.0	385
NO_3-N	0.8	15	0.8	18
Total-N	23.6	452	23.6	534
Nitrogen in runoff total-N	2.2		6.8	
% Renovation on concentration basis	91		71	

Source: Thomas *et al.*, 1974.

Renovation efficiencies, whether on a nitrogen mass or concentration basis, are typically high. Two examples are presented in Tables 6 and 7.

Nitrogen renovation is high in most overland flow systems, the principal mechanisms being crop removal and denitrification.

Table 6 data result from the application of secondary sewage effluent. On a mass basis, nitrogen removal equals nearly 83%. This appears to be due mostly to crop uptake; it is not clear, however, how much native N became available during the application period. Therefore, N loss due to denitrification is probably underestimated.

Table 7 shows results from the year-round application of raw comminuted sewage. The data do not allow construction of a mass balance, although on a concentration basis, N renovation percentages are quite high. The observed winter decrease in overall removal efficiency was likely due to decreased rates of nitrification and denitrification as well as the reduction in plant requirements.

Rapid Infiltration

Rapid infiltration systems generally utilize high rates of wastewater application with the major portion of the effluent moving to groundwater at some depth or distance. To the extent that nitrogen renovation occurs before entering the groundwater, it must be due chiefly to denitrification since nitrogen additions often far exceed vegetation requirements. Additional apparent renovation results from dilution in the groundwater. With the kinds of data now available from currently operating systems, it is very difficult to differentiate between denitrification and dilution effects.

In rapid infiltration systems, nitrogen renovation occurs chiefly through denitrification; some apparently results from dilution in the groundwater.

A system in Lake George, New York, applies approximately 65-100 acre-ft of wastewater annually to deep sand beds. This amounts to between 3000 and 4600 lb N/acre-yr. Apparent nitrogen renovation varies considerably depending on sampling location and season. Winter values, based on samples from shallow wells in the immediate vicinity of the sand filter beds, show a high NO_3^--N concentration of up to 21 mg/l (Aulenbach, *et al.*, 1975). This amounts to a renovation efficiency of about 25%. In a seepage area some 200 feet away from the filter beds, however, NO_3^--N concentrations in all seasons fall below the 10 mg/l drinking water standard. Further reduction, apparently through dilution, occurs when the leachate enters West Brook before flowing into Lake George. According to Aulenbach, *et al.* (1975), "in most instances, the nitrate-nitrogen content of West Brook exceeded recommended limits of 0.3 mg/l for control of excess algal growth. However, because of the cold water and the low phosphorus content of the stream, no problems of excessive algal growth have occurred in West Brook itself."

Wastewater application at Flushing Meadows, near Phoenix, Arizona, ranges from 300-400 acre-feet per year giving an annual nitrogen load of between 24,000 and 32,000 lb/ acre-yr. Nearly 90% of the nitrogen in the original wastewater occurs as ammonium. System management relies on alternate wetting and drying cycles in an attempt to maximize denitrification. Under short flooding and drying periods of two days each, aerobic soil conditions seemed to predominate. The highly ammonified effluent thus underwent nearly complete nitrification, resulting in low ammonium-N but consistently high nitrate-N ($\approx$20 mg/l) levels in the percolating water. Subsequently, wetting and drying periods were extended to two weeks each. This schedule resulted in the appearance of nitrate-N peaks of up to 71 mg/l in the leachate a few days after the start of each successive wetting period (Figure 2). Apparently, added ammonium was retained in the surface soil by cation exchange reactions during the wetting cycle. The ammonium was then nitrified during drying periods and flushed into the groundwater early in the next flooding cycle. As the alternate 2-week cycles were continued, however, ammonium concentration in the percolating water gradually increased, while nitrate behavior remained essentially unchanged. Apparently incomplete nitrification during drying cycles led to a gradual ammonium saturation of the cation exchange complex. Increasing amounts of applied ammonium were thus subject to leaching. Because of these results, Bouwer (1973) has suggested that long wetting-drying cycles (e.g., 2 weeks each) should be periodically alternated with shorter cycles (e.g., 2 days each) to increase aerobic conditions and "rejuvenate" the exchange complex in terms of ammonium retention. Other alternatives include lower loading rates and/or periodic, long rest periods of several weeks to allow complete drying of the upper soil layers.

In rapid infiltration systems long wetting-drying cycles of 2 weeks each alternated with shorter cycles of 2 days each have been suggested to increase aerobic conditions and aid in ammonium retention.

The presence of high nitrate pulses has also been shown in other wastewater application studies (Lance, 1972) and offers some interesting management alternatives. For example, winter application of ammonified effluent could result in efficient removal of nitrogen as long as the adsorption capacity of the soil is not exceeded. In the spring, when warmer temperatures promote biological nitrification, periodic high nitrate concentrations in the soil leachate should be expected. Essentially, the original wastewater nitrogen has been concentrated in a relatively small volume of leachate. The bulk of leachate moving between the pulses is low in nitrogen. Recovery of the nitrified leachate pulses through underdraining or strategic well placements will likely be mandatory. Upon collection, however, this water can be recycled to croplands for its nitrogen fertilizer value. Two and one-half acre-feet of wastewater containing 30 mg/l of NO_3^--N would supply about 200 lb N/acre for a demanding crop such as reed canarygrass; this amount of wastewater is equivalent to 1 acre-in./week over a 30-week period. If it is not pos-

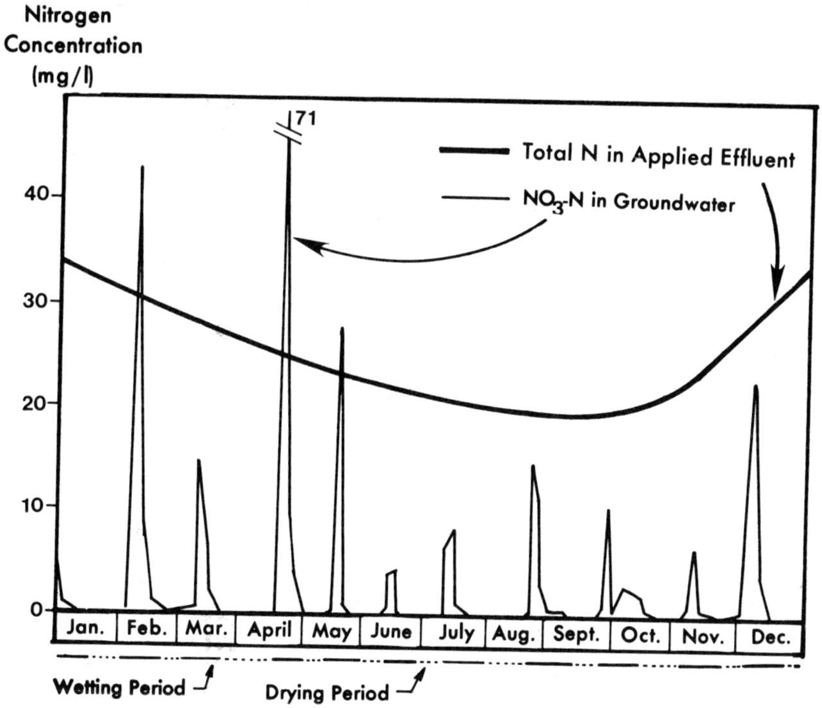

Figure 2. The high-rate application of ammonified effluent, under alternate wetting and drying cycles, results in periodic pulses of nitrate release to the groundwater. Following each pulse, the nitrate concentration in the leachate falls off sharply yielding highly renovated wastewater. From Bouwer in Sopper and Kardos, Recycling treated municipal wastewater and sludges through forest and cropland, Pennsylvania State University Press, University Park, Pa., 1973, p. 169.

sible to utilize water for crop irrigation, an inexpensive method of renovation may be to recycle it back to a point in the system where the presence of organic carbon would allow for denitrification.

Where high nitrate pulses are a problem, the nitrified leachate pulses should be recovered through underdraining or strategic placement of wells.

Barriered Landscape System Maximizes Denitrification

An experimental system, termed a Barriered Landscape Water Renovation System (BLWRS), is being tested at Michigan State University (Erickson, *et al.*, 1972). As shown in Figure 3, it consists of a mound of soil underlain by an impervious polyvinyl water barrier. The thin bed of limestone or slag on top of the mound removes phosphorus. The aerobic zone allows for organic matter decomposition and nitrification of ammonium. The downward movement of nitrified effluent, however, is stopped by the water barrier and forced to move laterally in the anaerobic zone immediately above the barrier. Denitrification in this zone is stimulated by adding a readily degradable carbon source such as crushed grain. Renovated water then moves off the edge of the water barrier and through deeper soil layers to the water table. Such a system can maximize denitrification, even under relatively low waste-water application rates. The nitrogen removal potential was demonstrated with liquid swine waste (Figure 3), but remains to be tested for domestic wastewaters.

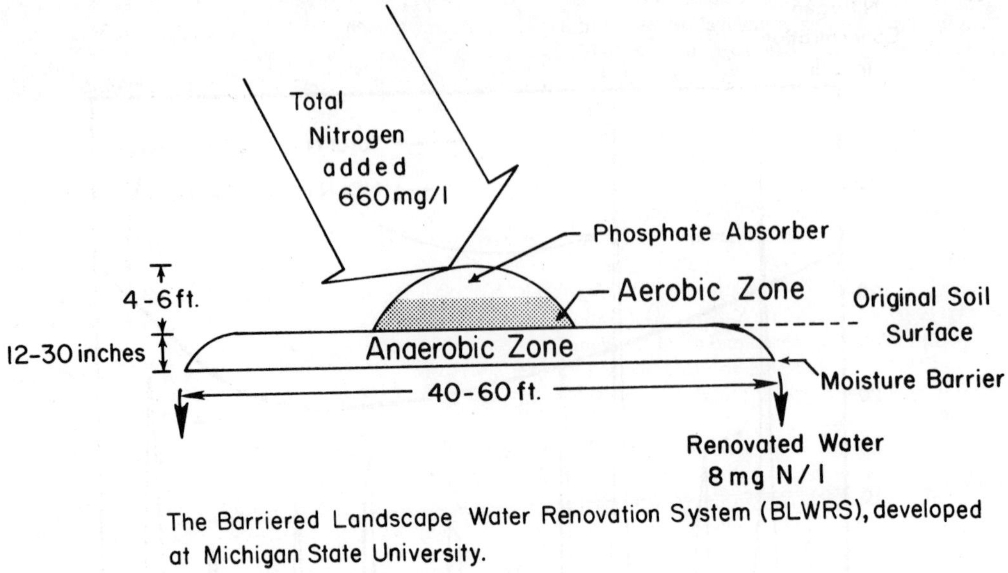

Figure 3. The Barriered Landscape Water Renovation System (BLWRS), developed at Michigan State University, provides a means for maximizing denitrification of nitrogenous wastes. Applications of liquid swine waste, equivalent to about 9,500 lb N/acre, have resulted in highly renovated water (after Erickson *et al.* 1972).

BIBLIOGRAPHY

Aulenbach, D. B., N. L. Clesceri, T. J. Tofflemire, S. Beyer, and L. Hajas. 1975. Water renovation using deep natural sand beds. *J. Amer. Water Works Assn.*, 67:510–515.

Bouwer, H. 1973. Renovating secondary effluent by groundwater recharge with infiltration basins. pp. 164–175. *In* Sopper, W. E. and L. T. Kardos, eds. Recycling treated municipal wastewater and sludges through forest and cropland. Penn State Univ. Press, University Park, PA. 479 p.

Brady, N. C. 1974. The nature and properties of soils. 8th ed. Macmillan Co. Inc., New York. 639 p.

Broadbent, F. E. and K. B. Tyler. 1957. Nitrification of ammoniacal fertilizers in some California soils. *Hilgardia*, 27:247–267.

Brown, R. E. 1975. Significance of trace metals and nitrates in sludge solids. *J. Water Pollut. Contr. Fed.*, 47:2863–2875.

Erickson, A. E., J. M. Tiedje, B. G. Ellis and C. M. Hansen. 1972. Initial observations of several medium sized barriered landscape water renovation systems for animal wastes. pp. 405–410. *In* Proc. 1972 Cornell Agr. Waste Mgmt. Conf. 580 p.

Hoeppel, R. E., P. C. Hunt, and T. B. Delaney, Jr. 1974. Wastewater treatment on soils of low permeability. Misc. Paper Y-74-2. U.S. Army Corps Eng. Cold Regions Res. Eng. Lab. Hanover, N.H. 84 p.

Johnson, R. D., R. L. Jones, T. D. Hinesly, D. J. David. 1974. Selected chemical characteristics of soils, forages, and drainage water from the sewage farm serving Melbourne, Australia. Dept. of the Army, Corps of Engineers. 54 p.

Kardos, L. T., W. E. Sopper, E. A. Myers, R. R. Parizek and J. B. Nesbitt. 1974. Renovation of secondary effluent for reuse as a water resource. Environ. Protect. Technol. Serv., EPA-660/2-74-016. U.S. Govt. Print. Office, Washington, D. C.

Keeney, D. R., K. W. Lee and L. M. Walsh. 1975. Guidelines for the application of wastewater sludge to agricultural land in Wisconsin. Tech. Bull. 88. Dept. Natl. Resources, Madison, Wisc. 36 p.

Lance, J. C. 1972. Nitrogen removal by soil mechanisms. *J. Water Pollut. Contr. Fed.*, 44:1352–1361.

Larson, W. E., C. E. Clapp, and R. H. Dowdy. 1973. Research efforts and needs in using sewage wastes on land. Paper presented at Soil Conserv. Soc. Amer. Meeting, Hot Springs, Ark.

Menzies, J. D. 1973. Composition and properties of sewage sludge. Paper presented at Soil Conserv. Soc. Amer. Meeting, Hot Springs, Ark.

Pound, C. E. and R. W. Crites. 1973. Characteristics of municipal effluents. pp. 49–61. *In* Proc. joint conf. on recycling municipal sludges and effluents on land. EPA, USDA, National Assn. State Univ. Land-Grant Coll., Champaign, Ill. July 9–13. 244 p.

Pratt, P. F., F. E. Broadbent, and J. P. Martin, 1973. Using organic wastes as nitrogen fertilizers. *Calif. Agr.*, **27**:10–13.

Reed, S. C. (Coordinator) 1972. Wastewater management by disposal on the land. Special Rep. 171. U.S. Army Corps of Eng. Cold Regions Res. Eng. Lab., Hanover, N.H. 183 p.

Ryan, J. A. and D. R. Keeney. 1975. Ammonia volatilization from surface applied wastewater sludge. *J. Water Pollut. Contr. Fed.*, **47**:386–393.

Sepp, E. 1971. The use of sewage for irrigation—a literature review. Bureau of Sanitary Eng., Calif. Dept. Pub. Health.

Sopper, W. E. and L. T. Kardos. 1973. Vegetation responses to irrigation with treated municipal wastewater. pp. 271–294. *In* Sopper, W. E. and L. T. Kardos, eds. Recycling treated municipal wastewater and sludge through forest and cropland. Penn State Univ. Press, University Park, Pa. 479 p.

Thomas, R. E., K. Jackson, and L. Penrod. 1974. Feasibility of overland flow for treatment of raw domestic wastewater. Environ. Protect. Technol. Ser. EPA-660/2-74-087. U.S. Govt. Print. Office, Washington, D. C. 31 p.

Viets, F. G., Jr. and R. H. Hageman. 1971. Factors affecting the accumulation of nitrate in soil, water and plants. Agr. Handbk. 413. ARS, USDA. U.S. Govt. Print. Office, Washington, D.C.

Viets, F. G., Jr. 1974. Nitrogen transformation in organic waste applied to land. pp. 51–65. *In* National Program Staff. Factors involved in land application of agricultural and municipal wastes. (DRAFT) ARS, USDA. Beltsville, Md. 200 p.

APPENDIX

Determination of Short-Term Nitrogen Loading

During the initial years of waste application, particularly with sludges, organic-N may accumulate in the soil. Since organic-N must mineralize before it is available for crop uptake or appreciable leaching, the rate of mineralization determines the rate of organic-N application.

Annual rates of mineralization are often expressed as a series of fractional mineralization for a given application and residuals from previous applications. Such a series is referred to as a decay series. For example, a decay series of .10, .05 has been suggested for municipal sewage sludges (Brown, 1975). Thus, for a given application, 10% of the organic-N is mineralized during the first year, and 5% of the residual (that which was not mineralized previously) is mineralized the second and all subsequent years until complete transformation has occurred. The same approach can be applied individually to successive yearly applications.

A decay series represents annual rates of mineralization expressed as a series of fractional mineralization for given applications and residuals from previous applications.

Suppose 100 lb/acre of organic-N were applied for each of three years. Using the above mentioned decay series, the amount of N mineralized the first year would be:

$$N_{(1)} = .10\,(100) = 10 \text{ lb/acre}$$

with 90 lb/acre remaining as a residual. During the second year, another 10 lb/acre would mineralize from the new application plus 5% of the residual from the first year. Therefore,

$$N_{(2)} = .10\,(100) + .05\,(90) = 10 + 4.5 = 14.5 \text{ lb/acre}$$

with a 90 lb residual from the second year application and an 85.5 lb/acre residual from the first year.

During the third year:

$$N_{(3)} = .10\,(100) + .05\,(90) + .05\,(85.5) = 18.8 \text{ lb/acre}$$

The above calculations constitute a "constant input" approach. However, in land application systems, the major concern is to maintain a constant output, consistent with crop requirements and acceptable leaching losses. A further consideration exists in that sewage sludges will generally contain inorganic-N (mostly in the ammoniacal form) in addition to organic-N. On a practical basis, all of the inorganic-N will become available during the year of application.

Consider a liquid sludge with the following characteristics:

$.96 = z =$ moisture fraction ($\% \times 10^{-2}$)
$.04 = s =$ total solids fraction ($\% \times 10^{-2}$)
$.03 = t =$ total N (% of solids $\times 10^{-2}$)
$.017 = i =$ inorganic-N (% of solids $\times 10^{-2}$)
$.013 = o =$ organic-N (% of solids $\times 10^{-2}$)

This sludge is to be applied in an area with an average annual precipitation (P) of 35 inches and potential evapotranspiration (ET) of 20 inches. A corn crop on the site will take up approximately 100 lb/acre-yr (C); nitrogen input via precipitation is 0.5 mg/l (c). We want to determine the allowable application of liquid sludge so as not to exceed 10 mg/l NO_3^--N (a) in the leachate. Further assume that the sludge will be injected so as to eliminate any ammonia volatilization (v) and that denitrification (d) is not known, and thus is set equal to zero as a safety factor. From equation 9 in a previous section,

$$W = \frac{4.43\ C + a\ (P - ET) - cP}{t\ s\ 10^6 - az}$$

However, this equation was derived for long-term conditions where no net accumulation of organic-N is occurring. Under short-term conditions:

W_1 = application during first year (acre-in./acre-yr)
t_1 = N available during first year (% of solids $\times 10^{-2}$)
$= (i) + (o)\ M_1$

where

M_1 = mineralization rate during first year (i.e., first term of decay series).

Therefore

$$W_1 = \frac{4.43\ C + a(P - ET) - cP}{i\ s\ 10^6 + o\ s\ 10^6\ M_1 - az}$$

Solving for the given waste and site characteristics:

$$W_1 = \frac{4.43\ (100) + 10\ (35 - 20) - 0.5\ (35)}{(.017)\ (.04)\ 10^6 + (.013)\ (.04)\ 10^6\ (.10) - 10\ (.96)}$$

$= 0.8$ acre-in./acre during the first year

Note: by adapting equation 7, this amounts to

$$N_1 = \frac{t_1\ s\ 10^6\ W}{4.43} = 132 \text{ lb N/acre}$$

To maintain a constant output of available N, however, less sludge can be applied in successive years due to mineralization of residual organic-N. Figure 4 shows the reduction in annual sludge application needed to maintain a constant output of 132 lb/acre-yr. Projected results for two application methods are illustrated. Where sludge is injected into the soil, preventing ammonia volatilization, the required reduction amounts to about 30–50% over the initial 5–10 years. For surface applications, assuming 100% ammonia volatilization, a 10-fold reduction

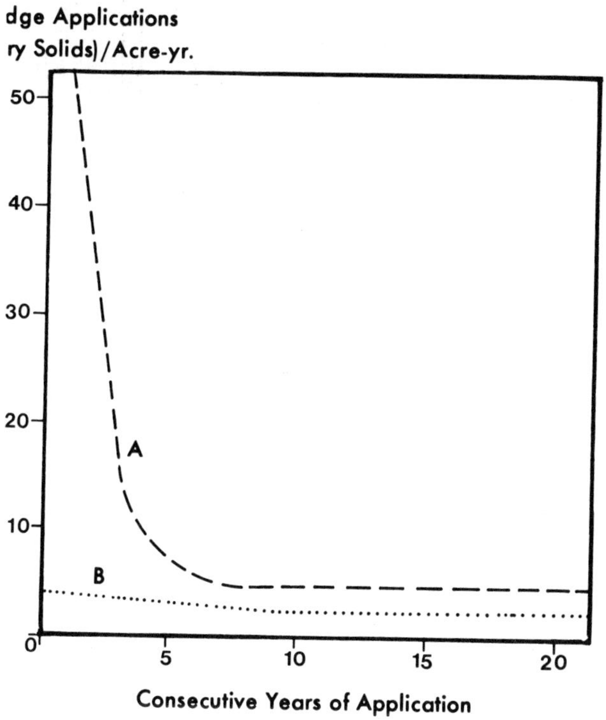

Figure 4. Initial sludge applications may result in a short-term accumulation of organic-N in the soil. In order to compensate for subsequent increased mineralization, successive sludge applications must be reduced until a new equilibrium is established. The amount of reduction depends on the content and form of waste nitrogen, the method of application, and the organic-N decay series (here assumed to be 0.10, 0.05). A = all ammonia nitrogen volatilized during storage and surface application of sludge; B = no ammonia loss, application by soil injection (after Brown, 1975).

is projected because mineralization of accumulated residuals becomes the dominant source of plant-available nitrogen.

The results in Figure 4 emphasize the necessity for analysis of inorganic versus organic nitrogen, as well as the effects of surface versus injected application. Note also that after the initial years of application, the calculated amounts tend to level off, approaching a new steady state over the long-term.

However, even for sludges with characteristics similar to those shown in Figure 4, the calculated loadings cannot necessarily be extrapolated to all sludge application sites. The decay series used for this example (i.e., .10, .05) constitutes a general estimate for a particular climate, in this case Ohio (Brown, 1975). Pratt *et al.* (1973) have suggested a decay series of .35, .10, .05 for liquid sludge applications in California. Still another series (.15, .06, .04, .02) is suggested for Wisconsin (Keeney *et al.*, 1975). Further differences will likely occur in other areas since the parameters which determine mineralization rates are strongly influenced by climatic as well as soil conditions.

Module II-2

PHOSPHORUS CONSIDERATIONS

SUMMARY

The treatment of wastewater phosphorus via land application includes both chemical and biological mechanisms. Chemically, phosphorus reacts with iron, aluminum, and calcium compounds in the soil providing efficient removal over a wide range of pH values. Phosphorus is also absorbed by rooted plants which, upon harvest, constitute a further removal mechanism.

The general principles of phosphorus interactions between soil and wastewater are illustrated, and the mass balance is derived for a typical loading rate and application season. Soil P movement and retention are discussed, with particular reference to the chemical environment necessary to produce movement of phosphorus among the three possible categories: dissolved, labile, and fixed. Crop uptake and surface runoff are examples of other mechanisms of phosphorus removal discussed in this module.

Adsorption isotherms provide a means of assessing short-term phosphorus retention capabilities in different soils. Such isotherms, however, do not account for relatively slow reversion reactions which render soil phosphorus compounds increasingly unavailable for leaching with time. Reversion reactions are favored by alternate application and resting periods.

Application of domestic sewage effluents at a rate of 2 in./wk for 30 weeks results in annual phosphorus additions of approximately 140 lb P/acre. At this rate, many soils are capable of efficient phosphorus removal for periods ranging from 20 to 50 years or more. Rapid infiltration systems, with annual applications of 1500 lb P/acre or more, are generally less efficient in phosphorus removal but vary considerably depending on specific site characteristics. Results from existing land application systems are cited to illustrate the basic principles of phosphorus removal.

CONTENTS

OBJECTIVES

After completing this module, the reader should be able to:

1. Describe common loading rates of phosphorus in land application systems and where the various forms of phosphorus found in wastewater come from.
2. Describe those soil conditions which favor the retention of phosphorus.
3. Discuss the availability of phosphorus in the soil and how it is retained in the soil in various forms.
4. Discuss how the land application site should be managed to enhance phosphorus removal.
5. Explain the usefulness and the limitations of adsorption isotherms in determining phosphorus retention of soil.
6. Discuss the relative effectiveness of spray irrigation, overland flow, and rapid infiltration systems for phosphorus removal.

INTRODUCTION

Phosphorus, like nitrogen, is an essential nutrient for the growth of plants and animals. Both nutrients are of major importance in determining levels of eutrophication in surface waters, with phosphorus often being the limiting nutrient. Increases in the phosphorus concentration of water entering lakes and streams will therefore tend to increase the probability and severity of "nuisance" plant growth.

Phosphorus and nitrogen, however, differ in their behavior within the soil. Under certain conditions, nitrogen is lost to the atmosphere through ammonia volatilization and denitrification, but there is no analogous gaseous loss of phosphorus. Even more important are differences in mobility of the two nutrients. Both exist in anionic forms (NO_3^- and $H_2PO_4^-$ or HPO_4^{-2}) which are not subject to cation exchange reactions, but nitrate ions are readily leached from the soil while phosphate ions are largely retained. The soil mechanisms responsible for phosphorus retention are discussed in this module.

PHOSPHORUS BEHAVIOR IN TERRESTRIAL SYSTEMS

Figure 1 shows a general scheme for phosphorus interactions in a land application system. Waste inputs of phosphorus occur in both organic and inorganic forms. Organic forms predominate in sludges and manures while inorganic forms are more prevalent in wastewater effluents. Various forms occur in the soil including phosphorus bound in organic matter, largely humus, dissolved phosphorus in the soil solution, and phosphorus bound in inorganic forms. Both freshly added and native organic forms are slowly mineralized by soil microorganisms through normal decomposition processes (1) (numbers refer to Figure 1). Some of this mineralized P may be reabsorbed and thus temporarily immobilized by the microbes (2). Immobilization is most significant immediately following waste application when the microbial population is rapidly increasing. Dissolved P comprises only a small fraction of the total soil phosphorus. Nevertheless, it is from this fraction in the soil solution that plants obtain needed phosphorus for growth and development (3), and from which any leaching loss occurs. Dissolved P is in equilibrium with the much larger amount of phosphorus bound in inorganic forms (4), primarily iron-, aluminum-, and calcium-phosphates. The equilibrium relationship between these forms and dissolved P is of considerable importance in the overall performance of a land treatment system.

**Phosphorus occurs in the soil in several forms-it is bound in organic matter, dissolved in the soil solution and bound in inorganic forms. Plants obtain the necessary phosphorus for growth and development from dissolved P which comprises only a small fraction of the total phosphorus in soil. Any leaching loss also occurs from dissolved phosphorus.*

Phosphorus in Wastes Comes from Several Sources

Phosphorus in wastes arises from a variety of sources. Organic forms occur in microbial tissue, plant residues, and metabolic by-products from living organisms. Inorganic forms exist as condensed phosphates from detergent residues, and as ortho-phosphate anions ($H_2PO_4^-$, HPO_4^{-2}). Most wastes, however, are characterized in terms of total P without further distinction. Some representative values for different wastes along with common loadings to land are listed in Table 1.

Organic phosphorus found in waste comes from mircobial tissue plant residues, and metabolic by-products of living organisms; inorganic phosphates arise from detergent residues, fertilizers, etc.

Figure 2 assigns typical values for the input and output terms of the P balance through a crop irrigation system, using an application schedule of 2 in./wk for a 30-week season. The influent concentration of 10 mg/l as P is an average value for raw domestic wastewater and primary or secondary effluent, and is suitable for these demonstration purposes. As in any

*This and other italicized summaries are intended to highlight key ideas, provide a basis for later review or to aid in skimming sections that are relatively familiar. They can be ignored in a complete reading of the text.

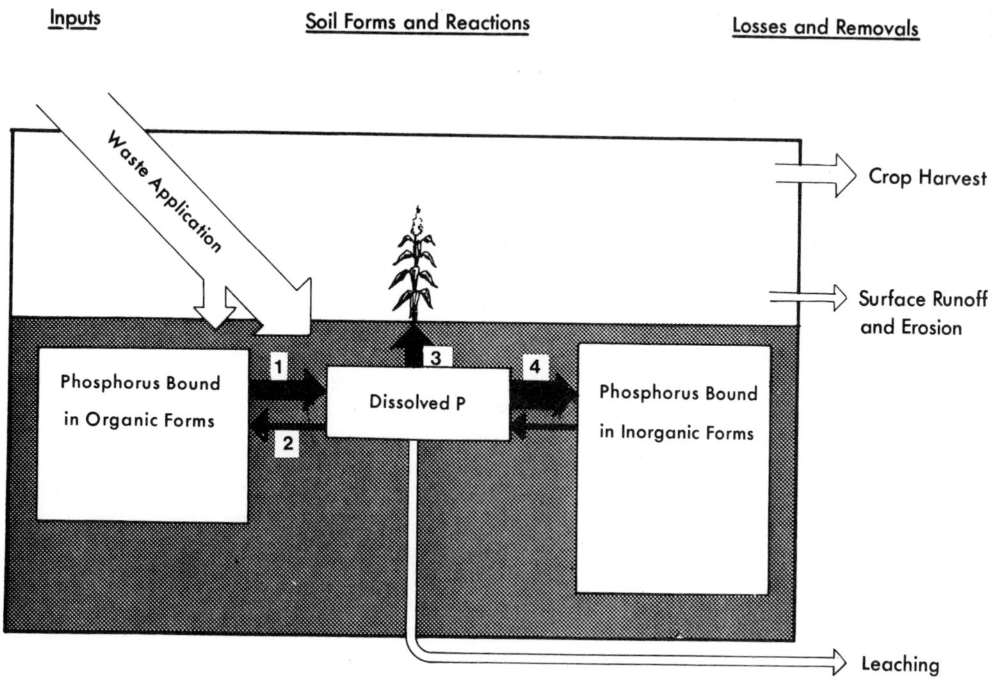

Figure 1. Phosphorus additions to soil are largely held either in organic combination or by reaction with soil minerals. Dissolved phosphorus, in equilibrium with bound forms, supplies plant needs and is subject to leaching.

Table 1. Total Phosphorus Content in Various Wastes and Common Annual Loading to Land.

Waste	Total P	Annual P Loading to Land	
	mg P/l	Assume 10 mg P/l	
		Low rate[a]	High rate[b]
Sewage effluent		---------- lb/acre-yr ----------	
Untreated	6–20		
Primary	8–14	135	1420
Secondary	4–17		
		Assume 3% P (dry wt)[c]	
Sewage sludge	1–6% (dry wt)	270	

[a]Assumes application rate of 2 in./wk for 30-week period (representative of irrigation and overland flow systems).

[b]Assumes application rate of 6 in./day for 14 days followed by 14-day rest period over a 30-week season (representative of rapid infiltration system).

[c]Assumes one annual application of liquid sewage sludge (96% moisture; 4% solids) in amount of one acre-inch.

Sources: Keeney, *et al.*, 1975; Peterson, *et al.*, 1973; Pound and Crites, 1973a; Reed, *et al.*, 1972.

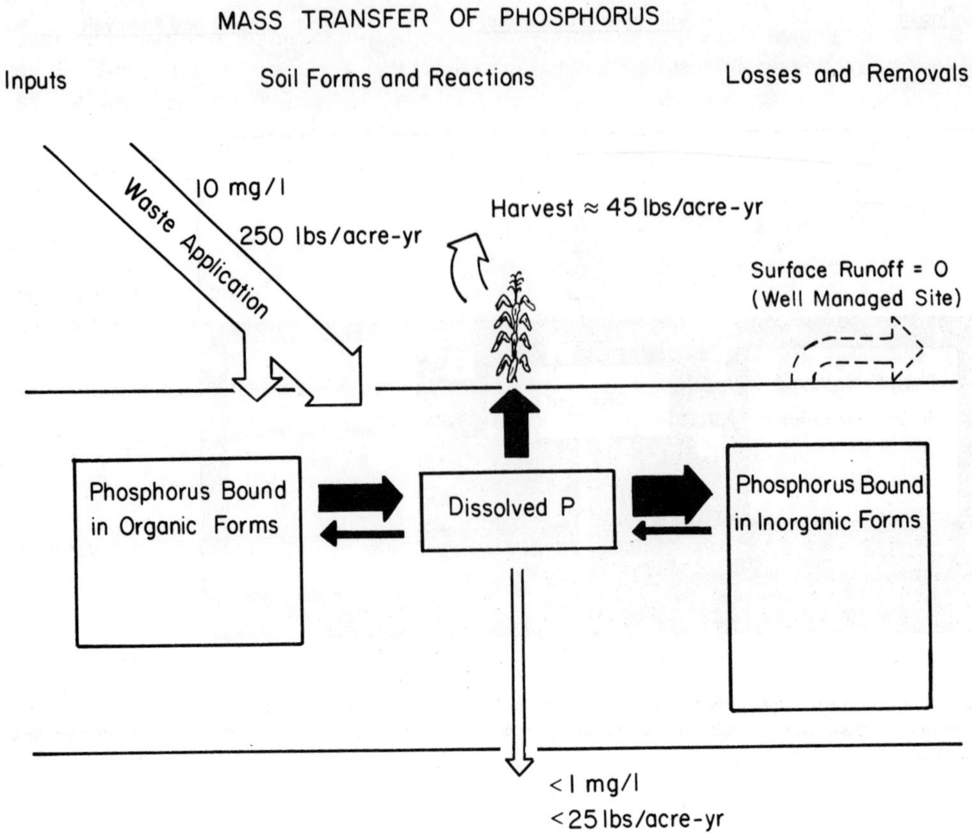

Figure 2. Mass flow diagram shows P concentration and mass interactions which could be expected at typical loading rates.

characterization of waste inputs to land treatment, laboratory studies would be performed on the specific influent to be treated.

Soil Retains Phosphorus by Adsorption, Precipitation

Unlike nitrogen, phosphate anions are largely retained in the soil through specific adsorption and precipitation reactions. Both reactions were introduced in general terms of Module I-1 "Soil as a Treatment Medium." The particular type of reaction and the end products formed, however, are dependent on the mineralogical composition of the soil medium, the pH of the soil solution, and the residence time of phosphorus in the soil.

Phosphate anions are retained in the soil by specific adsorption and precipitation reactions.

Fine Textured Soils Retain Phosphates Better

Fine textured soils in general have a greater capability to retain phosphates than do soils of coarser texture. Texture alone, however, does not describe the actual mineralogical composition of the clay fraction. Of primary interest are the iron (Fe), aluminum (Al), and calcium

(Ca) bearing minerals since they are highly instrumental in phosphorus reactions. The principal iron and aluminum minerals are the hydrous oxides such as gibbsite, goethite, and limonite; principal calcium minerals are the various forms of apatite, and calcium carbonate (calcite).

> *Fine textured soils generally retain phosphorus to a greater extent than coarse soils do.*

Form of Phosphorus and Its Retention Vary with pH

Figure 3 shows the general relationships between soil pH and phosphorus reactions. The form of phosphorus in solution is determined by pH. In the acid range $H_2PO_4^-$ predominates, while in the alkaline range HPO_4^{-2} is more abundant. A mixture of the two forms occurs around neutrality (pH 7). The actual mechanisms for phosphate retention are also a function of pH. In the very acid range (pH < 5), the solubility of iron and aluminum increases. Some soluble phosphate then reacts with the iron and aluminum to form insoluble precipitates (A) (letters refer to Figure 3). Much larger quantities of phosphate are adsorbed onto iron and aluminum hydrous oxides (B), and this occurs over a wide pH range. Additional adsorption occurs on other clay minerals. Precipitation of insoluble calcium phosphates begins at a pH of around 6.

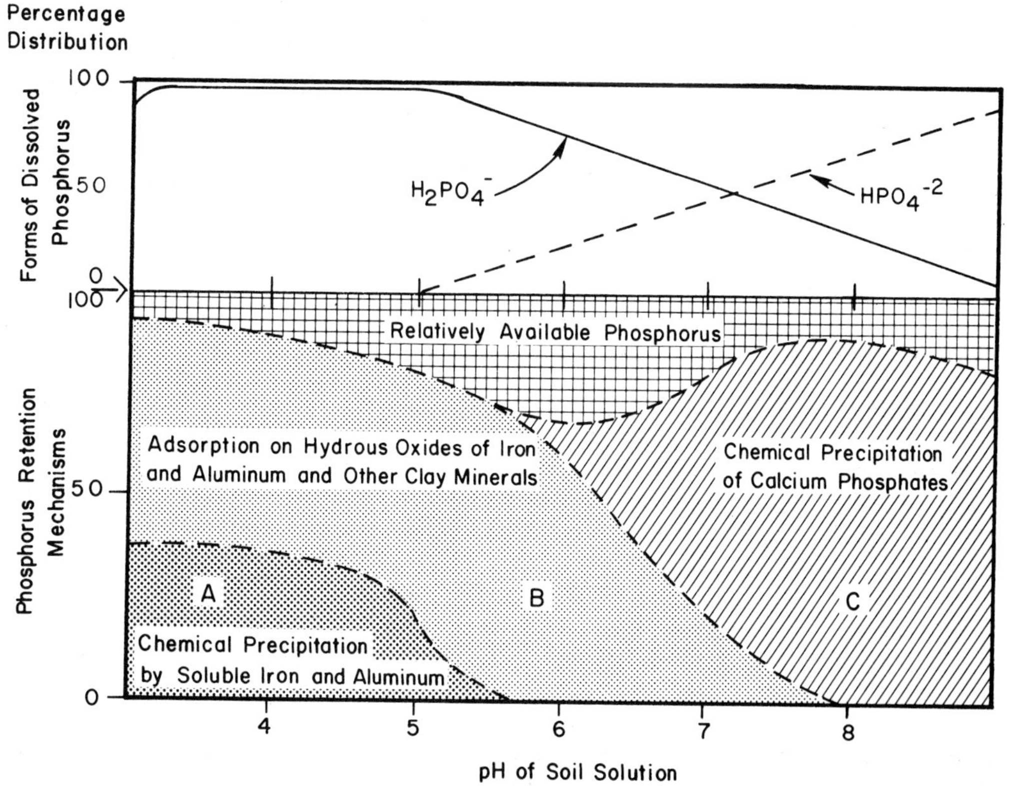

Figure 3. Forms of dissolved phosphorus (3a) and phosphorus retention mechanisms (3b) vary with pH. In the acid range, phosphates are bound with iron and aluminum compounds (A, B) while in the alkaline range, calcium phosphates predominate (C). Maximum availability of phosphorus for plant uptake as well as leaching occurs between pH 6 and 7 (Adapted from Brady, 1974.)

As pH increases further, reactions with calcium become the dominant mechanism for P removal (C). Thus, some mechanism for efficient phosphorus removal functions over a wide range of soil pH values.

The pH of the soil determines the form of phosphorus in solution and the way it will be retained. Efficient phosphorus removal mechanisms function over a wide range of pH values.

At the top of Figure 3b is an area termed "relatively available phosphorus." This refers primarily to the availability to plants, but also indicates the availability for leaching from the soil. Maximum availability exists over the pH range of 6–7. From the standpoint of phosphorus retention, it would seem advantageous to maintain a pH of less than 6 or perhaps around 8. This would be true if phosphorus were the only waste constituent of concern; however, other factors must also be considered. For example, heavy metals are less firmly held in acid soils and are thus more likely to move into the drainage water. See Module II-4, "Potentially Toxic Elements," for discussion of this point. This can be minimized by maintaining a pH value at or above neutrality. In addition, microbial activity generally increases in neutral to alkaline soils, and thus decomposition of organic wastes is more rapid in this pH range. Moreover, plant requirements for various nutrients are more apt to be met in the mid-range of pH values. Thus, while it is true that acid soils have a large capacity to retain P, there are other reasons to favor a higher pH environment. Many wastewaters themselves have a pH of around 7 and repeated applications of such water tend to raise the pH of an initially acid soil.

Although acid soils are most favorable for retaining phosphorus, other factors must be considered when deciding on the most favorable pH for the site, including the retention of potentially toxic elements, microbial activity, and availability of nutrients to plants.

Availability of Phosphorus in Soil Depends on Residence Time

Although Figure 3 indicates the compounds formed in soil, it oversimplifies the actual dynamics of reaction. The availability of phosphorus compounds in soil is highly dependent on time. Equilibria between adsorbed and precipitated forms and those in solution may be established within a few hours after addition of wastewater. However, readjustments occur over periods of weeks, months or longer, resulting in the formation of even more insoluble forms. A schematic of this concept is presented in Figure 4.

Availability of phosphorus compounds in soil varies over time. Readjustments occurring over periods of weeks, months, or longer result in the formation of more insoluble forms.

Within the soil, dissolved phosphorus is in equilibrium with loosely adsorbed materials and/or the more soluble precipitates. These loosely bound forms are termed labile P and will readily dissolve should the solution P concentration decrease due to either plant and microbial uptake or dilution with rainwater. Likewise, if the solution P concentration should increase, as with wastewater additions, P would initially move from the soil solution into the labile reserve.

Moving downward in the diagram, successive boxes represent more insoluble or more tightly

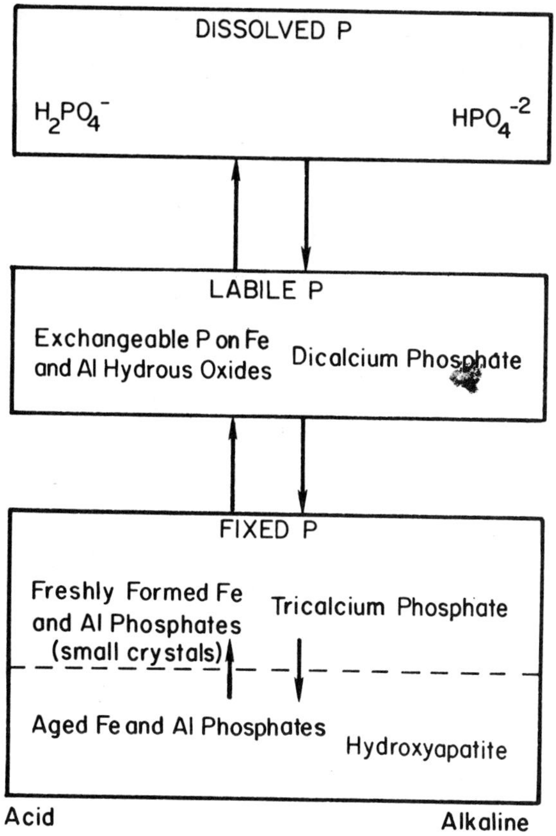

Figure 4. Phosphorus compounds in soil can be classed in three general groups according to their availability both for plant uptake and for leaching. Highly available dissolved P is in equilibrium with less available labile P which in turn is in equilibrium with essentially unavailable fixed forms. At any given time, most soil phosphorus, perhaps 80–90%, occurs in a fixed form.

adsorbed forms, often said to be "fixed." Conversion to these less available forms takes place much more slowly than changes between the dissolved and labile forms.

Formation of compounds in the lower categories is not well understood. In the case of the iron and aluminum hydrous oxides, fixation is thought to involve a diffusion of phosphorus deeper into the mineral itself thus rendering the P less available. Calcium phosphates in the least soluble range apparently arise from dissociation-reprecipitation reactions.

The differences in availability of calcium phosphates are indicated by their solubility product constants (K_{sp}). As described in Module I-1 "Soil as a Treatment Medium," a solubility product constant is obtained by multiplying together the concentrations of reactants (moles/l) in a saturated solution. The smaller the K_{sp} value for a given compound, the more insoluble it is. K_{sp} values for the insoluble calcium phosphates presented in Figure 4 are as follows:

Compound	Chemical Formula	Solubility Product (K_{sp})
Dicalcium phosphate	$CaHPO_4$	2.8×10^{-7}
Tricalcium phosphate	$Ca_3(PO_4)_2$	1.0×10^{-27}
Hydroxyapatite	$3Ca_3(PO_4)_2 \cdot Ca(OH)_2$	2.0×10^{-114}

Comparison of these values indicates the extreme insolubility and thus the high degree of fixation that occurs in the aged calcium compounds.

The aged calcium compounds are the least soluble form of phosphorus and have the highest degree of fixation; at any given time most soil phosphorus occurs in fixed form, perhaps 80 to 90%.

Phosphorus must remain in the soil for periods ranging from weeks to years for these least available compounds to form. With continuous applications of wastewater, residence time does not permit such formation in many cases. For this reason, resting periods between applications will generally increase the lifetime of the system in terms of phosphorus retention capabilities.

Continuous applications of wastewater prevent the formation of the most insoluble compounds; resting periods between applications increase the lifetime of the site in terms of phosphorus retention.

Heavy applications of wastewater not only decrease the residence time in the soil, but also increase the chance of soil saturation with water, producing anaerobic conditions. Under such conditions, iron changes from the ferric ion to the ferrous ion. This change increases phosphate solubility, reduces P retention, and may even release previously fixed P to the soil solution. Thus, it is important to keep the soil well aerated if it is to be an effective phosphorus sink.

Soil must be kept well-areated to function as an effective phosphorus sink.

Phosphorus May Be Removed from the Site by Various Means

The principal pathways for loss or removal of phosphorus are surface runoff and erosion, crop harvest and leaching. Loss via surface runoff would be highly variable from one site to another depending on natural precipitation, topography, season and application rates and methods. As recommended in Module 6 (Vol. 2), "Climate and Wastewater Storage," a land treatment site should be designed to contain the flow from a 25-year, 24-hour storm with no surface runoff outside the site boundaries. Minimizing surface runoff simultaneously minimizes erosion. Since erosion is the only way that fixed phosphorus is lost, its control is very important.

Because fixed phosphorus is only lost through soil erosion, surface runoff control is effective in minimizing this loss.

Crop uptake of phosphorus is rather low in contrast to nitrogen. A few representative values are presented in Table 2. Nevertheless, crop uptake can account for a significant portion of applied phosphorus under low-rate loading conditions (135 lb P/acre-yr as given in Table 1). Under high-rate conditions, crop uptake, even though remaining the same in lb/acre-yr, would reflect a much lower percentage of applied P. Vegetative cover, however, is important for other reasons including erosion control and maintenance of infiltration. Moreover, transpiration by plant cover reduces leaching and thus increases the residence time of wastewater phosphorus in the soil. The overall significance of a cover crop is thus magnified.

Keeping the hydraulic loading rate low enables crops to take up a considerable portion of the phosphorus applied plus yielding other benefits.

Table 2. Phosphorus Removal by Selected Crops When Irrigated With Municipal Sewage Effluent.

Crop	Yield (per acre)	P Removal (lb/acre-yr)
Corn grain	114 bu	26
Corn silage	6.0 tons	43
Wheat grain	54 bu	20
Reed canarygrass (3 cuttings)	4.3–7.0 tons	33–56

Sources: Sopper and Kardos, 1973, Kardos, *et al.*, 1974.

Leaching losses of phosphorus are very small in natural systems. Inputs of wastewater or sludge need not result in excess P leaching since the fixing power of many soils is quite high. Nevertheless, careful management is necessary to prolong the life of the system. Coarse sandy or gravelly soils have a high hydraulic loading potential, but often only limited ability to retain phosphorus. Moreover, since water moves readily, such soils may not wet uniformly and thus the P fixation will be further reduced. A short residence time will also reduce fixation since formation of the more insoluble compounds is precluded.

Leaching loss of phosphorus can be kept to a minimum through careful management of the land treatment site. Special care must be taken with coarse soils.

ASSESSMENT OF PHOSPHORUS RETENTION IN SOILS

When wastewater is applied to a soil, initial reactions occur in the surface layer. The phosphorus concentration in the water moving out of this layer will be reduced in accordance with the amount retained. Sequential changes in P concentration at various soil depths in response to repeated wastewater applications are illustrated in Figure 5.

In the initial condition prior to wastewater additions, the native fraction of bound phosphorus in many soils is sufficient to produce P concentrations in the soil solution of 5–250 parts per billion (ppb) (Taylor and Kunishi, 1974). As wastewater is applied, the phosphorus concentration in the surface layer will increase and water of steadily decreasing concentration will move into lower layers. Continued application results in the gradual saturation of soil layers from the surface downward. Different soil layers, however, will have different capacities to retain phosphorus. Therefore, in the case depicted at the bottom of Figure 5, the two saturated layers could contain considerably different amounts of phosphorus, instead of equal concentrations as shown.

There is no simple method to determine the amount of phosphorus that a soil can retain before saturation. Some indication, however, is obtained by so-called adsorption isotherms, which describe the relationship between that quantity of phosphorus held in the soil and that remaining in the soil solution (Ellis, 1973; Powers, *et al.*, 1975; Taylor and Kunishi, 1974). To construct an adsorption isotherm, soil samples from a given layer are combined with a series of solutions containing known amounts of phosphorus. After initial equilibrium is reached (12–18 hours), the solution is analyzed; the difference between initial and final phosphorus concentrations indicates the amount retained. When data from several such trials are plotted, the resulting curves quantitatively describe the relationship between adsorbed and dissolved phosphorus.

Adsorption isotherms help indicate the amount of phosphorus which a soil can retain.

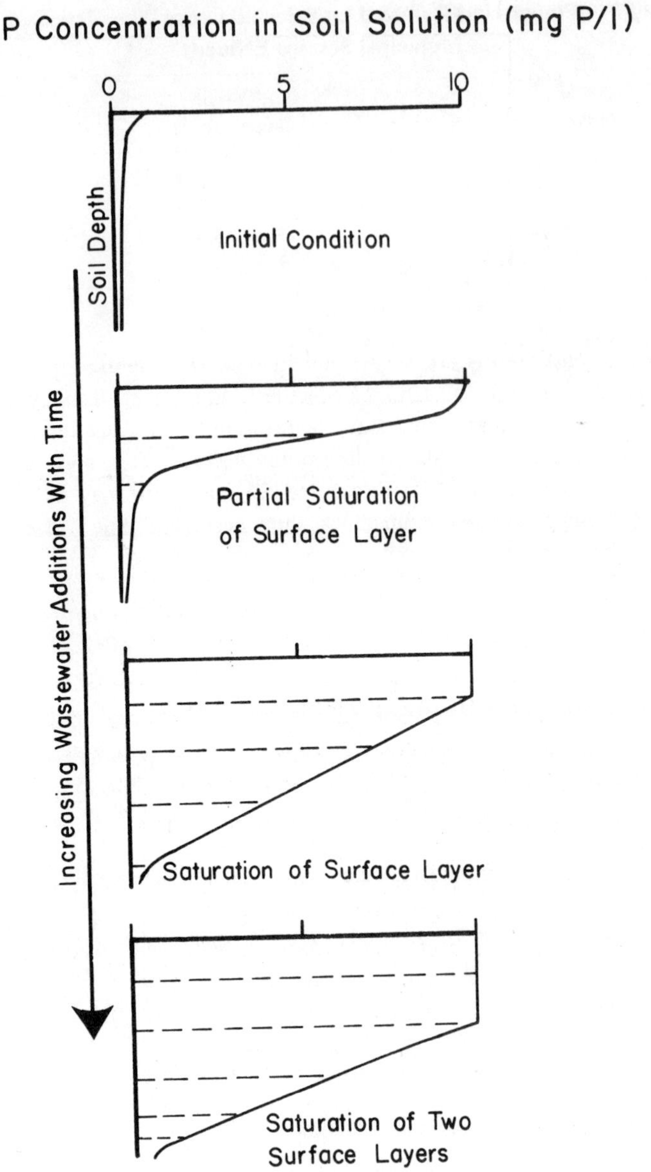

Figure 5. Dissolved phosphorus concentration in most unamended soils is less than 1 mg/l. Initial additions of wastewater (10 mg P/l) cause a readjustment in the equilibrium between dissolved and bound phosphorus in the surface soil layer. As more and more phosphorus is applied and retained, successive soil layers become saturated; at this point, the phosphorus concentration in the soil solution equals that of the wastewater itself. The time required for saturation of any given layer (s) varies from a few weeks to several decades depending on soil characteristics and wastewater application rates. (After Taylor and Kunishi, 1974.)

Sample adsorption isotherms for two different soil layers are shown in Figure 6. The horizontal axis represents the P concentration in solution in equilibrium; the vertical axis indicates the corresponding amount of P retained by the soil. Suppose a wastewater containing 10 mg P/l is applied to a sample from the soil layer described by isotherm A. As the first wastewater is added, this soil layer will transmit solution containing less than 0.3 mg P/l until approxi-

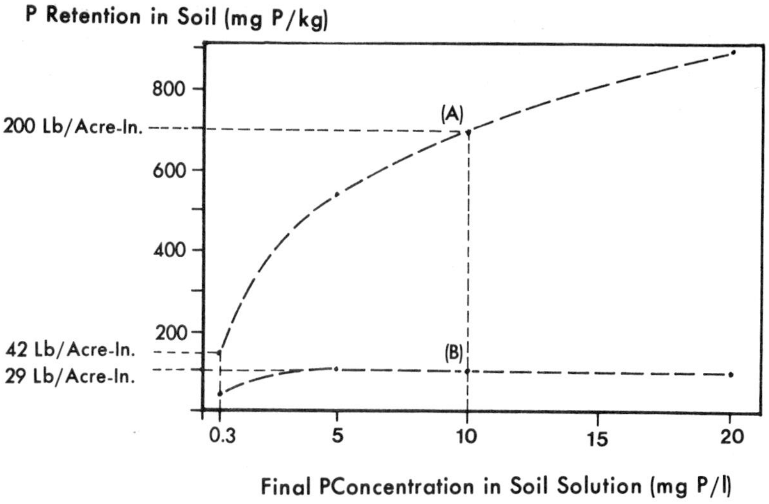

Figure 6. Adsorption isotherms for two different soil layers indicated a wide variation in retention capabilities. Such analyses are useful to show relative differences within and among soils; however, isotherms provide conservative estimates since they do not account for the slower reversion-type reactions. (After Taylor and Kunishi, 1974.)

mately 145 mg P/kg of soil have been retained. Phosphorus retention is then equivalent to about 42 lb/acre-inch of soil, assuming a soil bulk density of 1.3 g/cm^3. With continued application, the P concentration in the water passing out of this layer will slowly rise to the 10 mg/l level of the wastewater itself. At this point, the soil is saturated and as defined by the adsorption isotherm, will contain approximately 700 mg P/kg or 200 lb P/acre-inch. In contrast, the soil described by isotherm B has much lower retention capacity and at saturation will contain only about 29 lb P/acre-inch (100 mg P/kg of soil).

Table 3 presents phosphorus retention capacities for several Michigan soils as predicted by adsorption isotherms. Note the variation within as well as among the different soils. Although phosphorus retention is often reflected by texture, the variation shown in these soils is not necessarily attributable to textural differences. The dune sand clearly shows the least overall ability to retain phosphorus, and the sand and gravel layer in the Kalamazoo loam shows the lowest retention of any single layer. The highest retention, however, occurs in the Rubicon sand, even greater than the silty clay layers on the Ontonagon series. This is due to appreciable amounts of iron and aluminum within the Rubicon sand profile. All sands are obviously not comparable in this respect.

It should be emphasized that adsorption isotherms give only a partial indication of phosphorus retention since they do not account for the slow, yet very important, reversion-type reactions that occur in soils over time. Nevertheless, the data in Table 3 indicate minimum site lifetimes ranging from 13 to 30 years for low-rate applications of domestic sewage effluent. For example, assuming a wastewater with 10 mg P/l applied at a rate of 2 in./wk for 30 weeks (135 lb/acre-yr), 4 feet of Rubicon sand would take over 25 years to reach phosphorus saturation. Reversion reactions would further prolong the useful lifetime of the site. In contrast, adsorption isotherms indicate only a limited site lifetime where application rates are employed as in rapid infiltration systems. An annual phosphorus application of 1500 lb/acre would seemingly saturate the same depth of Rubicon sand in less than three years. Alternate wetting and drying cycles, however, may allow for considerable reversion reactions so that the actual phosphorus

Table 3. Phosphorus Retention Capacity in Selected Soils as Predicted by Adsorption Isotherms.

Soil Type	Textural Class by Layer	Depth (in.)	Phosphorus Retention Capacity (lb/acre-ft)	Retention Capacity to 48 Inches (lb/acre)
Uniform dune sand	sand	0–60	77	308
Spinks loamy sand	loamy sand	0–20	360	
	loamy sand	20–50	518	1809
	sand	50–60	277	
Kalamazoo loam	loam	0–12	340	
	sandy clay loam	12–36	908	2173
	sand and gravel	36–60	17	
Miami loam	loam	0–12	791	
	clay loam	12–30	546	2786
	loam	30–60	784	
Rubicon sand	sand	0–12	455	
	sand	12–24	1524	3437
	sand	24–60	729	
Ontonagon silty clay loam	silty clay loam	0–12	1337	
	silty clay	12–36	1160	4240
	silty clay	36–60	583	

Source: Schneider and Erickson, 1972.

retention would be greater. Under such circumstances, actual on-site demonstrations are necessary to fully determine the effectiveness of soil in phosphorus removal.

Because adsorption isotherms do not show the slow, reversion type reactions that occur in soils over time, they give only a partial indication of phosphorus retention. Alternate wetting and drying cycles allow for reversion reactions, increasing the phosphorus retention of the soil.

EXPERIENCE WITH PHOSPHORUS IN EXISTING SYSTEMS

Irrigation Systems Can Almost Completely Remove Phosphorus

Existing crop irrigation systems, where wastewater applications range from 1 to 3 in./wk, generally show nearly complete removal of phosphorus. Table 4 shows data based on 11 years of effluent irrigation in the Pennsylvania State University Wastewater Renovation and Conservation Project. Over these years, no site has shown leaching losses greater than 3% of the total phosphorus applied (Kardos and Hook, 1976).

Crop irrigation systems of land treatment that apply 1 to 3 inches per week show almost complete removal of phosphorus.

Roughly comparable results are reported at a sewage farm in the Netherlands, where untreated sewage effluent has been applied to a sandy soil for over 50 years. Application rates have ranged from 6 to 14 in./month. Fifty-year totals (Table 5) indicate nearly 93% combined phosphorus removal by crops and soil to a depth of 2.6 feet (Beek and deHaan, 1974). Major

Table 4. Approximate Phosphorus Balance for An 11-year Period (1963–1973) of Effluent Irrigation on the Pennsylvania State Wastewater Renovation and Conservation Project

Vegetative Cover	Total P Applied	Removal by Harvest	Estimated Loss to Drainage	Soil Retention
	lb/acre			
Corn[a]	706	309	6	391
Reed canarygrass	1975	438	12	1525
Abandoned field (white spruce and herbaceous species)	931	0	31	900
Hardwood forest	1611	0	18	1530

[a]Corn, reed canarygrass and abandoned field areas were all on Hublersburg clay loam soil; hardwood forest was on Morrison sandy loam soil.

Source: Kardos and Hook, 1976.

Table 5. Approximate Phosphorus Balance for a 50-Year Period on a 247-Acre Sewage Farm in the Netherlands.

Total P Applied	Harvest Removals (Hay, Milk, Meat)	Estimated Loss by Leaching	Soil Retention in 2.6 ft
lb/acre			
8400	779 (10.5%)	390 (5.3%)	6115 (82.6%)

Source: Beek and deHaan, 1974.

retention of phosphorus in the surface soil layers is indicated by present distribution as shown in Figure 7. Note the similarity of this data to the general concept depicted in Figure 5.

Overland Flow is Less Effective in Phosphorus Removal

Currently operating overland flow sites are often less effective in phosphorus removal than irrigation systems. Percent renovation values of 50 and 35% have been reported for systems at Ada, Oklahoma, and Melbourne, Australia, respectively (EPA, 1975). The surface flow path on overland flow sites does not provide much opportunity for contact with soil particles except at the immediate ground surface. A few inches of permeable topsoil overlaying nearly impermeable subsoil, however, can increase the contact by allowing for subsurface flow within this upper layer. The more contact with soil particles, the greater the opportunity for renovation. Resting periods can also increase P renovation in the overland flow approach. At the Campbell Soup facility at Paris, Texas, adoption of an intermittent application schedule increased P removal from 50% to nearly 90% (Gilde, *et al.*, 1971).

Although overland flow systems are less effective in phosphorus removal than irrigation systems, removal can be increased by overlaying impermeable subsoil with a few inches of permeable topsoil to increase wastewater contact with soil particles.

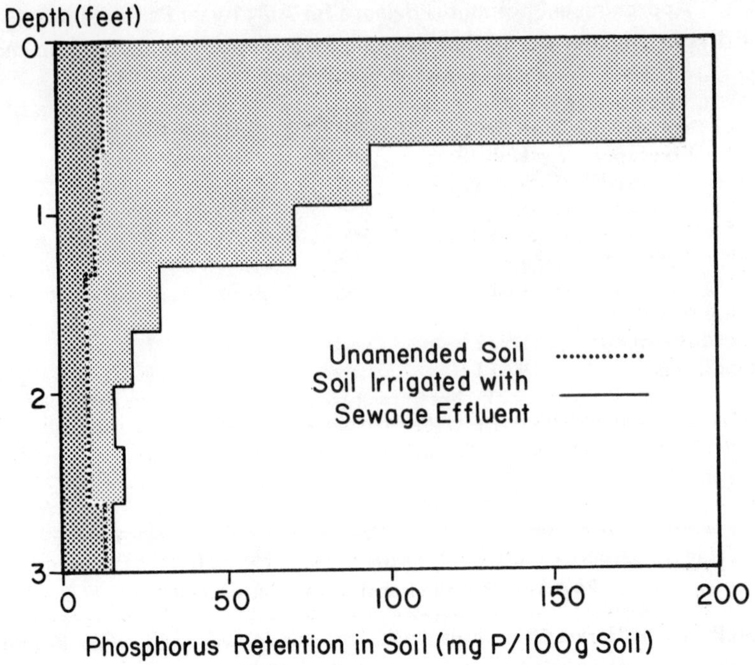

Figure 7. Distribution of phosphorus with depth in a sandy soil profile after 50 years of irrigation with untreated sewage effluent. The difference between amended and unamended profiles indicates that over 80% of the applied phosphorus is held in the upper 2.6 feet of soil (after Beek and deHaan, 1974).

Rapid Infiltration is the Least Effective in Phosphorus Removal

Rapid infiltration systems provide the least opportunity for phosphorus renovation of any land application method. Nevertheless, these systems vary considerably because of different phosphorus loading rates and the nature of the receiving soil.

The Flushing Meadows project near Phoenix, Arizona, exemplifies the upper extreme in phosphorus loading. Wastewater applications in excess of 300 acre-ft/yr with initial P concentrations ranging from 7-12 mg/l produce an annual phosphorus loading of between 5,700 and 9,800 lb/acre. Moreover, the coarse soil, a loamy sand overlying deep sand and gravel deposits, contains only small amounts of phosphate-fixing minerals such as iron or aluminum oxides. A static water table exists at about 10 feet. Not surprisingly, phosphorus removal is relatively low. Samples from a 30-foot well (i.e., 20 feet below the water table), have shown phosphorus concentrations of 5 mg/l (Bouwer, 1973). Estimates of the respective roles of soil retention and groundwater dilution in reducing phosphorus concentrations are not available.

In contrast, deep sand beds at Lake George, New York, seem to provide considerable renovation of wastewater phosphorus under applications ranging from approximately 65 to 100 acre-ft/yr. (Pound and Crites, 1973b; Aulenbach, *et al.*, 1975). Initial phosphorus concentrations in the wastewater, however, are low (1-4 mg P/l) so that annual P loading ranges from 175 to 1,080 lb/acre, far less than at Flushing Meadows. Here again, the effect of groundwater dilution complicates any specific analysis of phosphorus removal by the soil.

Rapid infiltration systems are the least effective form of land treatment in terms of phosphorus removal, but the success of systems varies depending on the rate of phosphorus loading and the nature of the receiving soil.

BIBLIOGRAPHY

Aulenbach, D. B., N. L. Clesceri, R. J. Tofflemire, S. Beyer, and L. Hajas. 1975. Water renovation using deep natural sand beds. *J. Amer. Water Works Assn.*, **67**:510–515.

Beek, J. and F. A. M. deHaan. 1974. Phosphate removal by soil in relation to waste disposal. pp. 77–86 *In* Proc. Interna'l. Conf. on Land for Waste Management, Ottawa, Canada, Oct. 1973. 388 p.

Bouwer, H. 1973. Renovating secondary effluent by groundwater recharge with infiltration basins. pp. 164–175. In Sopper, W. E. and L. T. Kardos, eds. Recycling treated municipal wastewater and sludge through forest and cropland. Penn. State Univ. Press. University Park, Pa. 479 p.

Brady, N. D. 1974. The nature and properties of soils. 8th ed. Macmillan Co. Inc., New York. 639 p.

Ellis, B. G. 1973. The soil as a chemical filter. pp. 46–70. *In* Sopper, W. E. and L. T. Kardos, eds. Recycling treated municipal wastewater and sludge through forest and cropland. Penn State Univ. Press. University Park, Pa., 479 p.

EPA. 1975. Evaluation of land application systems. Tech. Bull. EPA-430/9-75-001. U.S. Govt. Print. Office, Washington, D.C.

Gilde, L. C., A. S. Kester, J. P. Law, C. H. Neeley, and D. M. Parmlee. 1971. A spray irrigation system for treatment of cannery wastes. *J. Water Pollut. Contr. Fed.*, **43**:2001–2025.

Kardos, L. T. and J. E. Hook. 1976. Phosphorus balance in sewage effluent treated soils. *J. Environ. Qual.*, **5**:87–90.

Kardos, L. T., W. E. Sopper, E. A. Myers, R. R. Parizek, and J. B. Nesbitt. 1974. Renovation of secondary effluent for reuse as a water resource. Environ. Protect. Technol. Ser. EPA-660/2-74-016. U.S. Govt. Print. Office, Washington, D.C.

Keeney, D. R., K. W. Lee, and L. M. Walsh. 1975. Guidelines for the application of wastewater sludge to agricultural land in Wisconsin. Tech. Bull. 88, Dept. Natl. Resources. Madison, Wisc. 36 p.

Peterson, J. R., C. Lue-Hing, and D. R. Zenz. 1973. Chemical and biological quality of municipal sludge. pp. 26–37. *In* Sopper, W. E. and L. T. Kardos, eds. Recycling treated municipal wastewater and sludge through forest and cropland. Penn. State Univ. Press. University Park, Pa. 479 p.

Pound, C. E. and R. W. Crites. 1973a. Characteristics of municipal effluents. pp. 49–61. *In* Proc. joint conf. on recycling municipal sludges and effluents on land. EPA, USDA, Nat. Assn. State Universities and Land-Grant Colleges, Champaign, Ill. July 9–13, 1973. 244 p.

Pound, C. E. and R. W. Crites. 1973b. Wastewater treatment and reuse by land application. Vol. II. Environ. Protect. Technol. Ser. EPA-660/2-73-066b. U.S. Govt. Office, Washington, D.C.

Powers, R. F., K. Isik, and P. J. Zinke. 1975. Adding phosphorus to forest soils: storage capacity and possible risks. *Bull. Environ. Contam. Toxicol.*, **14**(3):257–264.

Reed, S. C., coordinator, 1972. Wastewater management by disposal on the land. Special Rep. 171. U.S. Army Corps of Eng. Cold Regions Res. Eng. Lab. Hanover, N.H. 183 p.

Schneider, I. F., and A. E. Erickson. 1972. Soil limitations for disposal of municipal wastewaters. Res. Rep. 195. Mich. State Univ., Agr. Expt. Sta. East Lansing, Mich. 54 p.

Sopper, W. E. and L. T. Kardos. 1973. Vegetation responses to irrigation with treated municipal wastewater. pp. 271–294. *In* Sopper, W. E. and L. T. Kardos, eds. Recycling treated municipal wastewater and sludge through forest and cropland. Penn State Univ. Press. University Park, Pa. 479 p.

Taylor, A. W. and H. M. Kunishi. 1974. Soil adsorption of phosphates from wastewater. pp. 66–96. *In* Factors involved in land application of agricultural and municipal wastes. (DRAFT). National Program Staff. ARS, USDA. Beltsville, Md. 200 p.

Module II-3

ORGANIC MATTER

SUMMARY

This module identifies the impact of sewage organic matter on soils. For convenience, that organic matter is separated into the readily decomposable compounds (measured as BOD_5 or COD) and the more resistant material (volatile suspended solids, refractory organics, and sludges). The fates of those organics are reviewed along with loading rates and recommended soil conditions.

Laboratory studies indicate that some soil systems are capable of oxidizing as much as 8,000 lb COD/acre-day if surface clogging is avoided, macropore space is sufficient, optimum moisture levels are maintained, and adequate soil aeration is assured.

More stable organic matter generally is not present in quantities that are significantly greater than the normal losses from productive soils. Usually the actual amount of organic matter that is added through land application is small compared to the amount of native organic matter present. Also, the newly added organic matter decomposes more rapidly than native organic matter. These factors, combined with the restrictive load limits imposed by the nitrogen content of wastes, keep the net accumulation of organic matter quite low in highly productive soils.

CONTENTS

GLOSSARY

acclimated mircoorganisms–Any population of microorganisms which is adapted to certain environmental conditions. Used to refer to bacteria which are exposed to a particular substrate before being used in the BOD_5 analysis on a sample of that substrate.

biodegradation–The breakdown, by microorganisms, of complex organic compounds in a substance to their constituent elements or compounds.

BOD–Biochemical oxygen demand. The amount of dissolved oxygen required to meet the metabolic needs of microorganisms utilizing organic substrate in a water sample of known volume. Used as an indirect measure of the biodegradable organic matter content of the sample.

BOD_5–Five day biochemical oxygen demand test. Oxygen consumption is measured over a 5-day period. The standard laboratory BOD analysis for most applications.

bulk density–The mass of powdered or granulated solid material (such as soil) per unit of volume.

COD–Chemical oxygen demand. The oxygen equivalent of that portion of organic matter in a sample susceptible to oxidation by a strong chemical oxidant.

denitrification–The breakdown of nitrate or nitrite to gaseous products such as nitrogen, nitrous oxide, and nitric oxide. Brought about in anaerobic conditions by denitrifying bacteria.

field capacity–The moisture content which a soil will retain against the force of gravity, i.e., when all macropores have drained but micropores remain saturated.

macropore space–The total volume of soil pores which are of large enough size that water is not held in them by capillary attraction.

micropore space–The total volume of soil pores which are small enough to hold water by capillary force over the force of gravity.

nitrification–Oxidation of ammonium salts to nitrites and oxidation of nitrites to nitrates by certain bacteria.

NOD–Nitrogenous oxygen demand. Utilization of free oxygen by certain autotrophic bacteria to oxidize ammonia to nitrite and nitrate, usually exerted from 6 to 10 days after the start of a BOD analysis. Also known as second-stage BOD.

PCB–Polychlorinated biphenyl. A synthetic organic compound used in several manufacturing processes. Persistent in the environment–resists degradation.

substrate–Source of nutrients used by a microbial population for biological synthesis.

OBJECTIVES

Upon completion of this module, the reader should be able to:

1. List several management practices which can have beneficial and harmful effects on soil aeration capacity.
2. State the major problem, and one factor which minimizes its harmful effects, related to most synthetic organic compounds released to the soil in waste treatment sites.
3. Explain the relationship between carbon to nitrogen ratio and decomposition of organic matter.
4. Discuss briefly the typical impact of applying waste organics to the soil, in terms of the proportion of applied mass to the mass of native soil organics.
5. List some types of areas which may be reclaimed or stabilized through application of waste organic matter.

INTRODUCTION

Organic material is always present in raw domestic sewages. In general, this organic matter is responsible for odors and for oxygen depletion in receiving waters or soils. Therefore, organic matter must be accounted for and managed properly if a given treatment system is to be successful.

The designation "organic matter" includes a wide variety of chemicals. To simplify the discussion here, organic matter is separated into two categories: that which is generally water soluble and readily biodegradable and that which is insoluble and more resistant to biodegradation. The first category is loosely identified as the BOD_5 component while the second is the volatile suspended solids, refractory organics and sludges.

In this module, a short review of the distribution and chemical classification of sewage organic matter is followed by more specific comments regarding the impact of BOD_5 organics in soil systems. Consideration is given to soil conditions, microbial populations, nutrient ratios, loading rates and application practices.

A final section considers the more resistant organic matter of sewages and sludges. These will be reviewed in terms of their implications to soil systems and their fates after land application. Attention is given to the effects of sewage organic matter on soil aggregation, humus buildup, toxic element retention and soil clogging.

Adoption of "acceptable" organic loading rates based on design parameters used in oxidation lagoons and other treatment processes is not good practice. The assimilative capacity of acclimated soil should be determined for the particular waste under study. Recent treatability studies with high organic wastes have shown assimilative capacities far in excess of common empirical guidelines for waste loadings.

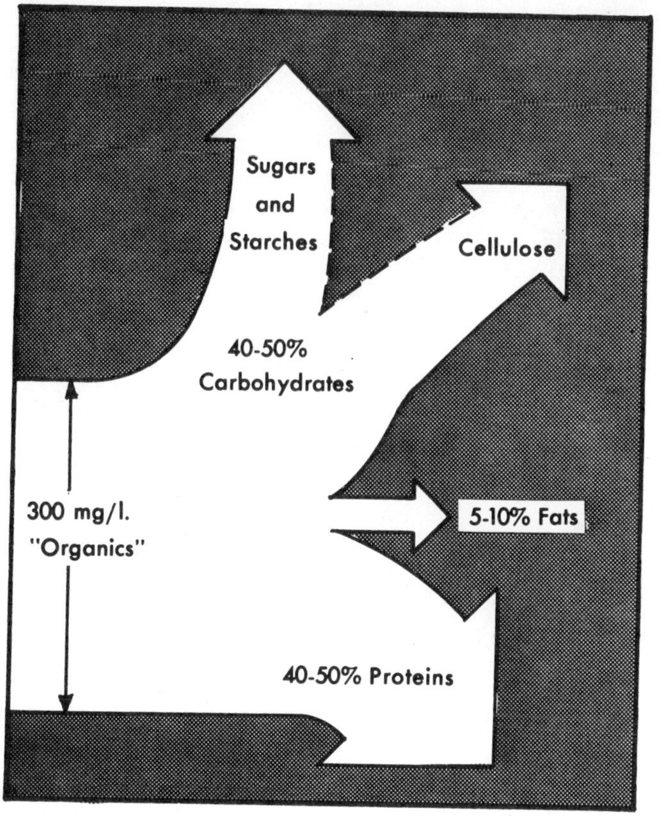

Fig. 1. The organic matter of untreated wastewater is primarily carbohydrates, proteins, and fats (adapted from McKinney, 1962).

ORGANIC MATTER OF UNTREATED WASTEWATERS

Various readers can be expected to differ in their interpretations of what constitutes the "organic matter" of untreated wastewaters. A rather abbreviated listing of specific components appears in Figure 1.

Categories of Organic Matter

For the purposes of this module, some rather simplistic generalizations will be used to avoid misunderstandings and to center the discussion on more pertinent issues. One of these will be to divide organic matter into soluble and insoluble components as illustrated in Figure 2.

The second generalization will be to divide organic matter in terms of the relative demand for oxygen which the different forms impose on the environment. In this respect, three categories of wastewater components are easily identified.

The first category includes the organics that exert a high biochemical oxygen demand. These are often soluble and are measured by the standard BOD_5 methods. Occasionally wastewaters of municipalities or food processing industries carry substances that may or may not be detectable in the BOD_5 analysis yet which exert a high chemical oxygen demand (COD). The second category includes the proteins. These organics are reduced to ammonia (NH_3) which, in turn,

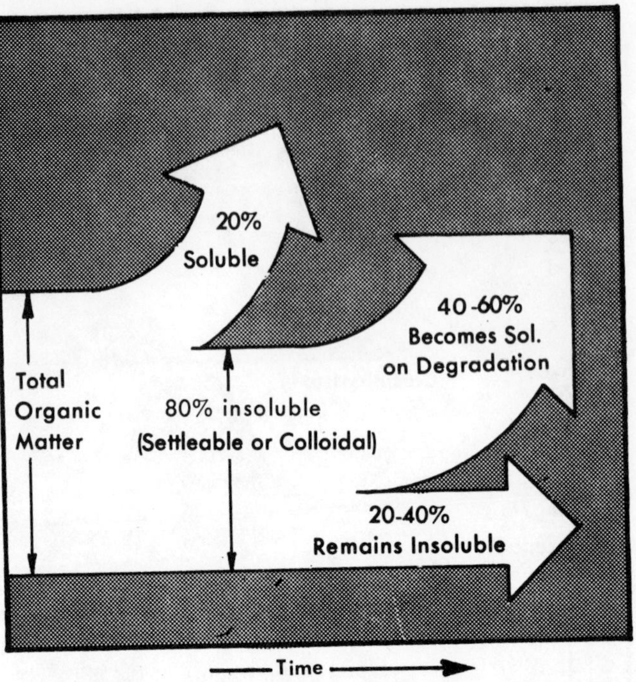

Figure 2. The amount of organic matter in untreated wastewater that is or becomes soluble exceeds the amount that remains insoluble (adapted from McKinney, 1962).

exerts an oxygen demand measured as the nitrogenous demand portion of the BOD, occurring usually after 5 days and sometimes referred to as NOD. The third category includes the less active, usually insoluble organics with no rapid demand for oxygen. When necessary, the impact of these organics may be estimated indirectly by measurements such as Theoretical Oxygen Demand (ThOD), Total Oxygen Demand (TOD), Theoretical Organic Carbon (ThOC), and Total Organic Carbon (TOC).

**Organic compounds can be categorized by the analytical techniques used to measure them.*

Oxygen Uptake Patterns

For the purposes of this module, the familiar BOD_5 data are useful for estimating the impact of readily degradable organic matter in land systems. The reason for this is that decomposing matter demands the same amount of oxygen from soils as it does from aquatic systems. Therefore, as an initial point, consider the BOD_5 oxygen uptake pattern reproduced in Figure 3. Note that the most rapid uptake of oxygen occurs in about the first two days. Then there is a marked decrease in uptake rate, but, of course, a continuing increase in total demand.

If the oxygen-demanding organic matter is placed in an acclimated environment (one already populated with microorganisms ideally suited for the substrate) the initial rate of oxygen up-

*This and other italicized summaries are intended to highlight key ideas, provide a basis for later review, or to aid in skimming sections that are relatively familiar. They can be ignored in a complete reading of the text.

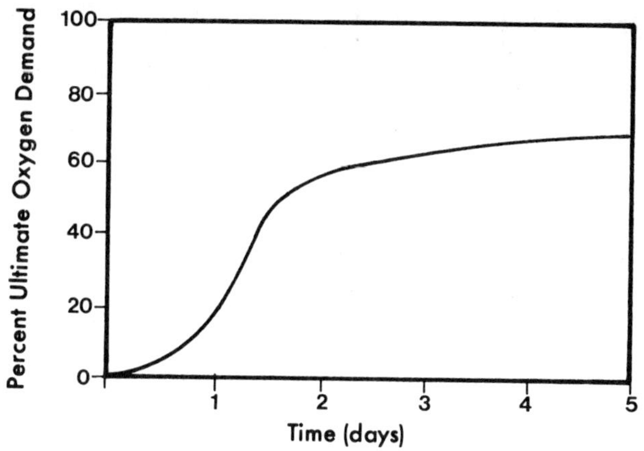

Figure 3. Between 65 and 70% of the total oxidizable organic matter in typical municipal wastewater is represented by the BOD_5 value (adapted from McKinney, 1962).

take may be increased dramatically. The ultimate total oxygen demand, however, will not be substantially different (Figure 4). Note that acclimated systems place substantial immediate demands on the oxygen reservoir of an environment. Unacclimated systems respond more slowly and have a reduced rate of oxygen uptake. This is the reason that the laboratory analysis for BOD_5 is always done with bacteria which are acclimated to the wastewater under investigation. In this way a much closer approximation to ultimate BOD can be determined, as shown in Figure 4.

Microorganisms acclimated to the organics will show more rapid uptake of oxygen than those which are unacclimated. This is a basic premise used in the laboratory analysis of BOD.

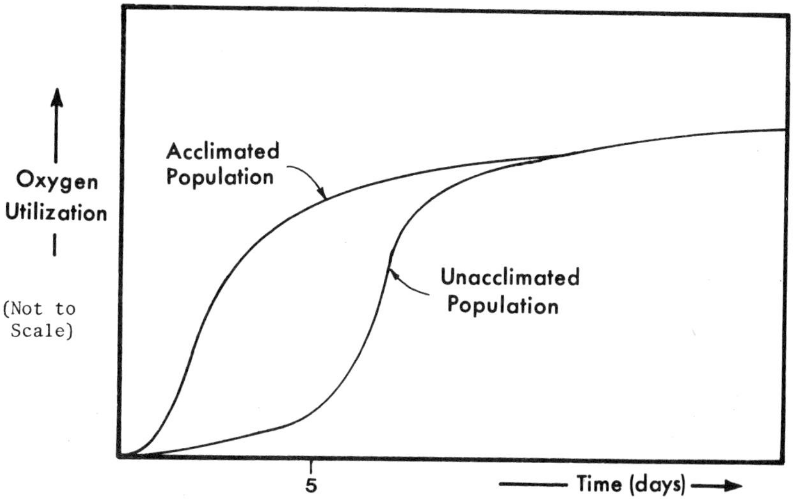

Figure 4. Microbial populations unacclimated to the substrate exhibit significant lag times before oxygen uptake occurs (adapted from McKinney, 1962).

Significance to Land Application

The BOD_5-exerting organics consume large volumes of oxygen very rapidly, so it is not unreasonable to wonder if the soil environment can satisfy that demand. Laws that limit BOD_5 application rates to about 5,000 mg/1 (600 lb/acre-day) apparently reflect the thought that the capacity of soil to meet oxygen demands approximates that of common lagoons (Jewell and Loehr, 1975). In the following section, however, it will be shown that the premise may be unnecessarily conservative for at least some wastewaters. There is an appreciable body of evidence indicating that soils can supply sufficient oxygen to compete directly with secondary treatment systems with a variety of wastewaters.

The third section of this module concentrates on the fate of the more slowly decomposable organic matter of wastewater and sludges. It will be shown that their accumulation on productive soils is not excessive.

RESPONSE OF SOIL SYSTEMS TO ORGANICS

Amending soils with high BOD_5 wastes poses a stress on the available oxygen within the soil system. As long as that oxygen demand can be met, the BOD_5 load will not be excessive. But meeting that load requires the interaction of several complex events. In the following paragraphs we will first focus on the response of soils to BOD_5 loading, then on those factors that influence that response.

Organic Loadings and Soil Responses

Wallace (1976) reported that land disposal systems existing at the time of his publication handled industrial wastes ranging from 10 to 2,020 lb BOD_5/acre-day. He did not stipulate the adequacies of treatment achieved. Wallace did cite lysimeter studies that indicated 100% BOD removal from sulfite liquor wastes at low concentrations (<150 lb BOD_5/acre-day) under aerobic conditions. Gross organic overloads of that same industrial waste appeared at surface loadings approximating 1,000 lb BOD_5/acre-day (Figure 5).

Jewell and Loehr (1975) reported results of a study in which food processing wastewater

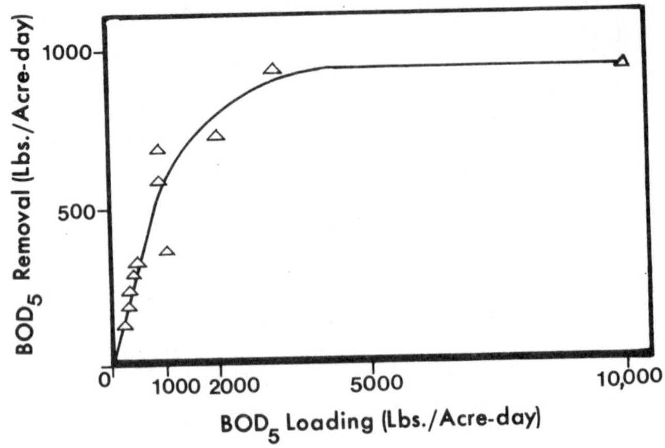

Figure 5. BOD_5 removal from sulfite liquor loadings onto 10-foot soil columns had a marked decrease in efficiency at loadings greater than about 1000 lb/acre-day (adapted from Wallace, 1976).

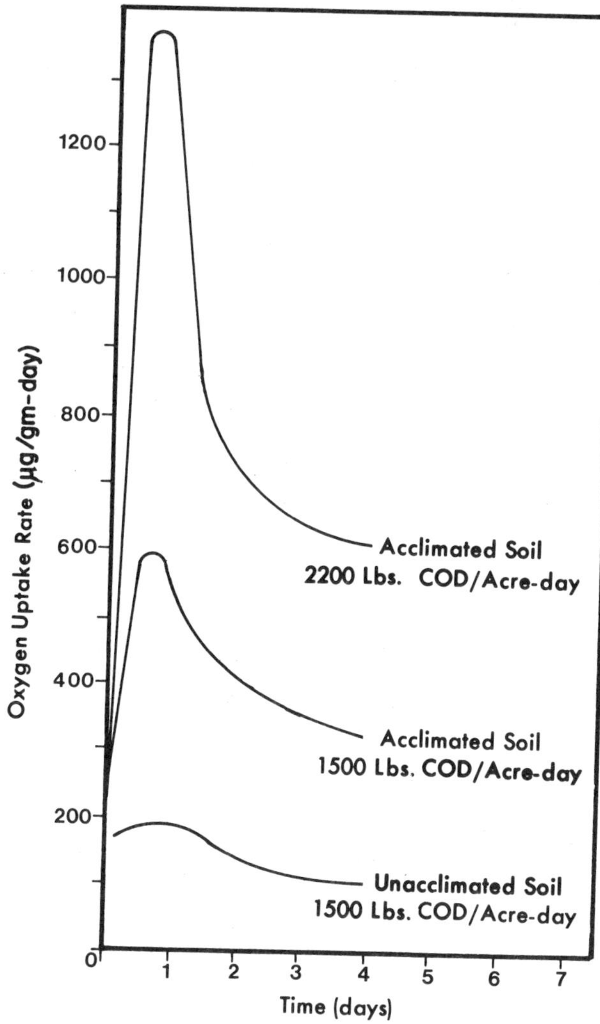

Figure 6. Oxygen uptake rates increase rapidly when soil is acclimated to large COD loadings of food processing wastewater, yet no point of organic overloading is reached at less than 2200 lb COD/acre-day (adapted from Jewell and Loehr, 1975).

was added to soil in a Warburg apparatus. The rate of oxygen uptake was determined for both acclimated and unacclimated soils, each under various COD loadings. Partial results are given in Figure 6.

Laboratory studies show that acclimation of soil bacteria to a particular waste greatly increases the soil's capacity to treat organic wastes effectively.

Here again the effects of acclimation are to increase the rate of oxygen uptake and even allow a rapid adjustment to excess loading. Note also that in each case the respiration has passed its maximum rate in less than two days.

Jewell and Loehr concluded that the soils studied retained the potential of accepting food processing wastes from vegetable canning at concentrations of 10,000 mg/1 COD without suf-

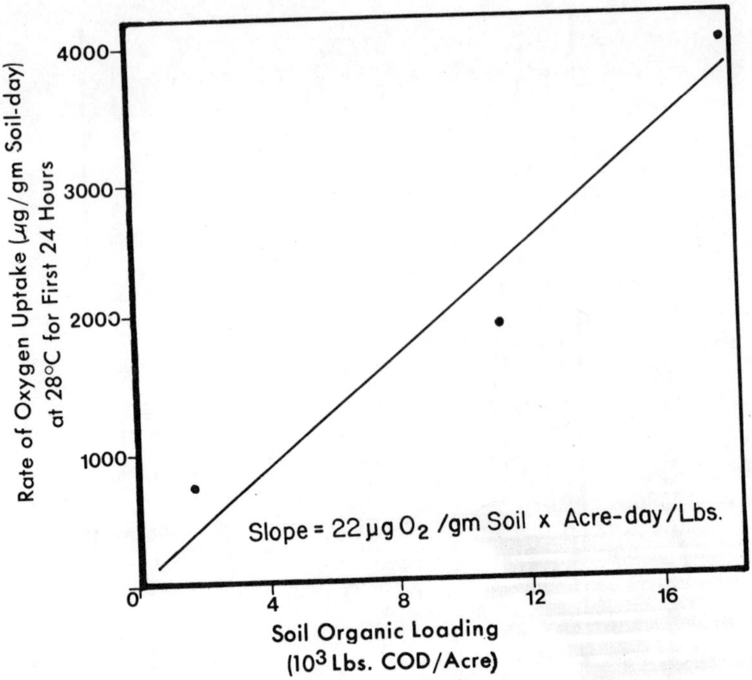

Figure 7. Soil organic loadings of food processing wastes led to no organic overloads even with loads exceeding 16,000 lb COD/acre-day in laboratory studies (adapted from Jewell and Loehr, 1975).

fering organic overloading. The reported oxygen uptake rate at various soil organic loadings is given in Figure 7. Note the massive loadings reported in these data. Recall that some laws prohibit BOD_5 loads in excess of 600 lb/acre-day from land application. Also, compare with Wallace's data for industrial waste (Figure 5) and consider the implications suggested regarding the appropriateness of specific wastes for land application.

Ultimate Fate of Organics in Soils

Although usually the maximum exertion of BOD by organic wastes occurs in the first 48 hours or so after application (Jewell and Loehr, 1975), that does not indicate that total oxidation has occurred. Total oxidation means that all organics are reduced to the ultimate respiratory products: carbon dioxide and water. It is clear from Figure 6 that even after 5 days the respiration rates in the test samples exceeded initial rates, hence substantial amounts of substrate survive at least that long. In fact, some of the original oxygen-demanding organic matter still may be present. Most of the organic matter that is oxidized, however, is converted to gums, slimes, protoplasm, and other metabolic by-products by the soil microorganisms. These by-products, particularly the gums and slimes, have important if temporary roles in the formation of soil aggregates (Allison, 1968) before they, too, are degraded (Martin, *et al.*, 1965).

Metabolic by-products of microbial decomposition can help form soil aggregates before they also undergo decomposition.

The degradation of waste organics and metabolic by-products increases the percentage of soluble organics (review Figure 2). This raises concern for the possibility of contaminating groundwater with waste organics. However, field studies from a vegetable processing plant reported by Jewell and Loehr (1975) show that this should not be of concern in a well-acclimated system. Even at an initial one-day loading of 8,000 lb COD/acre (1,300 lb/acre-day average rate with 5 days rest between applications) a 99.2% removal of COD was achieved. Less than 35 mg/l total COD reached the groundwater from that loading. Wallace (1976) reviewed data (presented in Figure 5) showing that if soils are forced to anaerobic conditions, only about 10% of the total BOD_5 load may be oxidized (although 100% of the simple sugars are oxidized) and presumably, the remainder remains mobile enough to pose a hazard to groundwater.

Oxygen Reserve of Soils

The bio-oxidation of waste organics relies on the reservoir of oxygen within the soil. That oxygen supply is extremely variable, of course, but something of its magnitude can be suggested by an illustration.

A well structured silt-loam surface soil at optimum moisture for plant growth may contain around 50% pore space shared equally by air and water. If it is assumed that 30% of that pore space is macropore space and that the soil is homogeneous throughout the first foot, then one acre-foot contains about 3,300 cubic feet of soil gases. If the oxygen content of those gases averages, say, 19%, there are some 44 pounds of oxygen present. This clearly is inadequate for even the 600 lb BOD_5/acre-day limited by some laws.

Therefore, even though the work of Jewell and Loehr (1975) shows that soils have a capacity for biodegrading large BOD_5 loadings, *in situ* soils face limitations imposed by available oxygen. Soils cannot practically be mechanically aerated. Therefore, soil systems must rely heavily on the diffusion of oxygen from the atmosphere. Despite its extreme importance, however, that diffusion cycle itself is quite simple.

The respiration of soil organisms reduces the partial pressure of oxygen in macropores and increases the partial pressure of carbon dioxide. Since the atmosphere is connected to the macropores, the higher partial pressure of oxygen in the atmosphere acts as the driving force to transport oxygen through the soil barrier into the macropore space. Similarly, the partial pressure of carbon dioxide within the macropores exceeds the partial pressure of carbon dioxide in the atmosphere, and carbon dioxide is driven across the soil barrier to the atmosphere (Brady, 1974).

Soils depend on diffusion of oxygen from the atmosphere to replenish available oxygen for soil organisms.

The oxygen diffusion rate (ODR) varies with soils and soil depth but a representative soil and its ODR are given in Figure 8. Note that the ODR varies with atmospheric oxygen concentration. Clearly, this is not a major concern over the ranges in oxygen variation likely to be encountered in field work. Generally an ODR of 1.7×10^{-2} pounds of oxygen per acre-minute is needed to sustain vigorous root growth, so the depth at which that limit is reached indicates the approximate maximum depth of root growth and, by implication, the maximum depth for major aerobic respiratory processes for microorganisms.

Oxygen diffusion rate varies with soils and soil depth. ODR affects root depth.

It must be recognized, however, that suspended organic matter does not tend to be distributed uniformly in soils. Rather, it is filtered out in the initial few centimeters. This means that the maximum oxygen demand will have to be met relatively near the soil surface where ODR's are

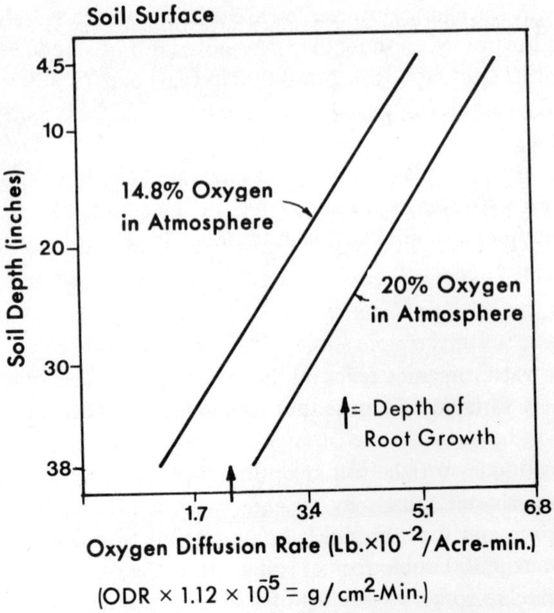

Figure 8. The oxygen diffusion rate (ODR) decreases with soil depth, and root growth terminates where the ODR approximates 1.7×10^{-2} lb oxygen/acre-min (adapted from Brady, 1974).

the highest. Therefore, while the ODR may decrease more or less linearly with depth (Figure 8) the actual oxygen demand imposed by wastes may be very high near the surface but rapidly diminish with depth.

Management for Organic Loadings

The above discussions emphasize the need for maintaining aerobic conditions in soils. This, in turn, requires management which encourages the generation and maintenance of macropore space and a maximum rate of diffusion. Specific practices are reviewed in the following paragraphs.

Resting Periods. It is common practice to alternate periods of surface application with periods of rest. The rest period is necessary to allow the soil to relieve its hydraulic load, of course, but is also a period for soil aeration. A common schedule is 1-day application and 4 to 7 days rest (Jewell and Loehr, 1975). Nevertheless, resting schedules are site specific. It may be necessary to determine the optimum oxygen loading and oxygen consumption, and to rely on oxygen diffusion rates to estimate the length of needed resting periods. The consultation of qualified soil scientists would be needed for such work.

Resting periods in an application cycle are necessary for soil aeration to maintain aerobic conditions.

A number of mathematical models have been postulated to help determine loading rates and schedules, but verification data from field studies are scarce (Wallace, 1976).

Pore Space. One of the most important soil features affecting aeration is micropore space. In this respect, fine textured soils (in which micropore spaces predominate) are least desirable. A well-granulated silt loam surface soil at optimum moisture for plant growth is the most desirable (Brady, 1974).

Aeration is directly related to the proportion of macropores in the soil.

Loosening and granulating fine-textured soils promotes aeration by raising the proportion of the macropores. On the other hand, continuous cropping, particularly of soils originally high in soil organic matter, often results in a reduction of large macropore spaces by as much as one-half (Brady, 1974). This is attributed to the loss of soil organic matter—a topic considered in greater detail later in this module. Compaction of soil by machinery or even large grazing animals can further reduce soil macropore space. The sodium and potassium ion concentrations in some untreated wastewaters may be sufficient to cause the destruction of the soil aggregates. Since soil aggregation is so critical to soil structure, destruction leads to massive loss of macropore space. Tillage may have good short term effects on surface macrospace, but over time it promotes the destruction of soil granules and the oxidation of soil organic matter, both of which reduce macropore space. Moreover, the appearance of good tilth after tillage is often achieved at high cost in overall performance of the soil profile. Unless the subsurface layers have dried beyond the point of compaction, tillage operations and associated wheel traffic can easily destroy macropore space beneath the tilled layer, impairing downward movement of air and water. Such damage is not easily overcome.

Continuous cropping, grazing, Na and K application, and tillage, can, over time, contribute to the breakdown of macropores and reduction of aeration capacity.

Hydraulic Loading. Excessive hydraulic loadings lead to filling of soil pore space, a displacement of soil gases and a great reduction in gas diffusion. The plant growth can be severely retarded if this condition persists for any length of time and removal of BOD_5 and COD waste organics becomes impractically slow. Clearly, rapidly drained soils are best suited for the maintenance of adequate aeration.

Excessive hydraulic loading and soil surface clogging impede aeration.

Surface Clogging. The formation of a compacted, tightly sealed surface layer will prevent gas diffusion and the subsequent aeration of soil. Surface clogging is a problem with some wastes (such as from milk products) and some raw sewages (Kirkham, 1974). Resting periods often promote the drying, crumbling and dissipation of such surface caking (Mioduszewski and Hinesly, 1972).

Sludges often are plowed under to control odor and to prevent surface caking. Sludges also may be applied through subsurface injection procedures. These practices maintain soil infiltration rates and offer odor control, but biological degradation of organics is impeded somewhat. This obstruction is caused by the lower temperature and oxygen concentrations of sub-surface soils.

Nonbiodegradable Organics. The organic matter associated with food processing and municipal wastes usually poses no fundamental biochemical problem to soil systems. Manufactured organics, on the other hand, may be considerably less compatible with the biochemical processes of soils. Petroleum and tar-based synthetics including crankcase oils, pesticides, PCB, detergents, solvents, and so on may persist for a long time. Nevertheless, these organics appear to adhere to surfaces of soil particles. They remain relatively immobile on all but very sandy soils. Any migration occurs through erosion.

Synthetic organic compounds are often not readily decomposed like natural organics. They may persist for long periods in soil but are generally immobile.

For most such products (PCB being a notable exception), biodegradation does occur with time, leading to secondary or tertiary products which are most likely to enter into natural cycles of soil organic matter (Mansell, *et al.*, 1970; Kirkham, 1977; Fink, *et al.*, 1970; Brady, 1974). Instances have been reported, however, where a "tar-like" layer about 1.5 feet thick developed about 1 foot beneath the surface of high rate infiltration beds. Such deposits had to be dug out before soil infiltration capacity could be restored (Satterwhite, 1974). Apparently such oily layers are, as yet, relatively rare and their causes remain unknown. Nevertheless, the presence of appreciable quantities of persistent organics might better be avoided.

Recapitulation

Although data are limited, it appears that very large BOD_5 loadings can be degraded in soils so long as soil aeration is maintained. It is quite likely that nitrogen, toxic elements, or hydraulic loadings become limiting in land application before organics. Adequate aeration is associated with well drained soils of sufficient macropore space with no surface clogging and adequate resting periods. Even high loads of food processing wastes (>8,000 lb COD/acre-day) have been renovated without excessive losses (<35 mg/1) of organics to groundwater.

INCORPORATION OF STABLE ORGANIC MATTER IN SOIL

It is well known that organic matter tends to accumulate in and disappear from soils. For example, mineral surface soils carry native organic matter in concentrations ranging from trace quantities to 15 or 20%. Organic soils (peats, mucks, etc.) may range upward to 90% organic matter (Brady, 1974). It is natural, therefore, to consider the organic matter in wastes and its effect on both the quality of the soil and the net accumulation of organic matter in soils.

Mineral soils contain between a trace and 15 or 20% organic matter. Organic soils can have up to 90%.

Before such questions can be addressed, a perspective is needed on the relation of native organic matter to the soil system. Waste organic matter then can be considered in terms of its ability to be integrated into the natural system.

Soils and Native Organic Matter

The major source of native organic matter is the decomposition of vegetation. The components of vegetation vary widely in their rates of decomposition as is indicated in Figure 9. Compare these with components of untreated domestic wastewaters given in Figure 1. Notice that many of the same classifications of compounds appear in each list. Also, note the longevity of some of the forest litter compounds.

Most native organic matter in soils comes from decayed vegetation, which can vary in decomposition rate.

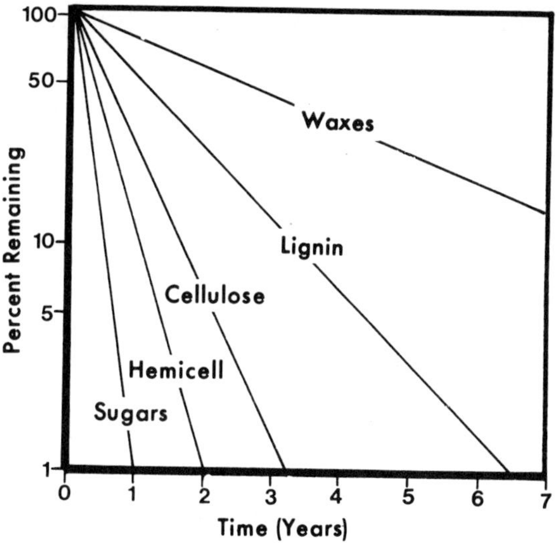

Figure 9. The rate of decomposition of organics of forest litter varies with the nature of the organic matter (adapted from Wallace, 1976).

Decomposition of Organic Matter. The decomposition of organic matter, whether from vegetation or wastes, is related to the ratio between carbon and nitrogen in the organic substrate. Normally, optimum decomposition rates occur when the C:N ratio is around 10:1. Other parameters also must be favorable, of course, such as temperature (about 30°C), moisture (60 to 80% field capacity), and aerobic conditions.

If a substrate such as corn stalks is incorporated into the soil, its high C:N ratio (50:1) may lead to slow decomposition (Parr, 1974) and nitrogen deficiency for rooted plants. This is particularly true if no residual nitrogen fertilizer is in the soil. The effect is even more pronounced when straw is used. What happens is that the small amount of available nitrogen cycles continually within the microbial population while organic carbon is respired as carbon dioxide. No nitrogen is available for rooted plants (which are poorer competitors for nitrogen than microorganisms) and carbon is gradually lost, narrowing the C:N ratio down until it approximates the optimum range.

> *Under aerobic conditions, organic matter decomposition is faster at a C:N ratio of about 10:1.*

On the other hand, substrates such as manures or municipal wastewaters generally have smaller C:N ratios. The decomposition of organic matter occurs while nitrogen is mineralized. This process leads to the release of ammonia (NH_3) and nitrate ion (NO_3^{-1}). If this supply of nitrogen is in excess of the needs of rooted plants, nitrogen toxicity of vegetation and contamination of groundwater may result. At any rate, nitrogen losses lead to a widening of the C:N ratio and, again, an approach to optimum conditions.

Accumulation of Native Organic Matter. In some 550 surface mineral soils summarized by Brady (1974), the average concentrations of organic matter ranged from 1.55 to 3.83%. Those concentrations represent from 15.5 to 38.3 tons of organic matter per acre-furrow slice (6 inches deep). Mineral subsoils generally carry substantially less organic matter. These values represent a reasonably steady-state balance of organic matter in which annual additions approximate annual losses.

Normally, annual organic matter additions equal depletion in mineral soil.

The magnitude of annual losses are estimated at roughly 450 pounds of organic matter per acre-year from untreated, low-producing soil and some 6,800 lb/acre-yr from more productive soil (Brady, 1974; Russell and Russell, 1950). It is only under anaerobic conditions, peat formation, or initial colonization of a denuded landscape that carbon accumulation (from natural vegetative growth) exceeds decomposition.

Production of Humus. The most persistent carbonaceous forms appear, in time, as chemically complex matter composed of oils, fats, waxes and especially lignin on one hand and metabolically produced polysaccharides and polyuronides on the other. Altogether, these substances are called humic matter and the resulting chemical complex is known as humus (Brady, 1974). Humus is brown to dark brown in color, a collection of amorphous and colloidal substances with mean residence times in soils reported from >300 to >1500 years (Campbell, *et al.*, 1967; Brady, 1974). Humus has C.E.C. values considerably in excess of mineral colloids (150–300 meq/100 g for humus compared to only 8–150 meq/100 g for minerals) (Brady, 1974). Humus is important in the removal of toxic elements from the soil solution, particularly copper and nickel (Chaney, 1974). This is discussed in more detail in Module II-4 "Potentially Toxic Elements."

Influence of Organic Matter on Aggregation. Various kinds of organic matter (chiefly polysaccharides and breakdown products of humic substances) and numerous organic substances act as binding agents for soil aggregation. Aggregate stabilization, however, depends on the presence of non-aggregating matter between aggregates. In this respect, large spherical humus molecules (among other soil substances) act as important aggregate stabilizers by keeping aggregates separate and preventing their fusion into clods (Allison, 1967).

Humus is high in C.E.C. and is very important in binding metal ions. Humus also stabilizes soil structure by preventing clod formation.

Soil and Waste Organics

The above discussion presents an overview of the sources, magnitudes and fates of native organic matter in soil systems. Since much of the organic matter in untreated municipal wastewaters is chemically similar to native organics, the following discussion on waste organics is closely related to what already has been covered. We will now focus on questions of organic matter accumulations, effects on soils, and hazards.

Magnitude of Organic Matter Application. While there is no maximum load of oxygen-demanding organics established by empirical studies, sludges and manures are usually applied at from 10 to 20 tons/acre-yr (Loehr, 1974). Parr (1974) stresses the need to seek optimum application rates of sludges rather than "maximum" rates, yet he suggests that 10 to 20 tons dry sludge/acre-yr is reasonable while 50 tons dry sludge/acre-yr would probably create severe hazards from nitrogen pollution (not organic overload).

The most important factor that prevents any sizable accumulations of waste organics is that the actual amount of organic matter applied to the soil is relatively low. This may seem surprising if one thinks of adding, say, 10 tons of moist sludge per acre. Consider, however, that a mineral soil carrying 4% organic matter will contain 160 tons of native organic matter per acre-2 feet, assuming a uniform distribution of organic matter throughout the soil. On the other hand, this 10 tons of moist sludge contains only about 500 pounds of dry organic matter. Thus, the applied organic matter represents only about 1.6% of the organic matter already present in the soil and some of that addition will decompose. Plowing under residue from cereal crops is no different. The estimated return ranges to about 4,460 pounds of organic matter per acre-year (Shields and Paul, 1973). Yet these values may be compared to Brady's estimate that a productive soil of the type described here may respire an equivalent of some 8,000 pounds of organic matter per *year.* Also, recall the previously cited estimate of Jewell and Loehr (1975) that at least some soils have the capacity of oxidizing single slugs of certain food processing wastes containing organic matter loadings of equivalent to some 8,000 lb COD/acre-day.

Recommended sludge application rates represent a small percentage addition annually to the native soil organic content.

Sludge organic matter addition may be comparable to plowing under cereal crop residue, and less than that lost in a productive soil annually.

Of course, real soils cannot be expected to have a uniform distribution of organic matter through a 2-foot column, nor will respiration processes be uniform throughout that column, but it is clear that, at the very best, those applications of organic matter barely balance losses from active soils. But as much as 80% or so of the added organic matter may be oxidized rapidly as described earlier. Thus, 80% of the organic matter would have no significant residence time and therefore would make accumulations of organic matter even less likely. This is illustrated in Figure 10. Consider how much Figure 10 would be altered if the application was, say, equivalent to 20, 50, or even 200 tons of manure or sludge (at 25% organic matter).

Organic Loss through Carbon Respiration. Respiration of organic matter is not limited to native material. Newly added organic matter soon will begin to be lost. Jenkinson (1971) indicated that about two-thirds of carbon added in one year as cereal straw (a substrate of high C:N ratio) was respired by the following year. After 5 years about four-fifths of the original carbon was gone. Shields and Paul (1973) studied C^{14} labeled wheat straw and grasses on two different soils and showed that soils, crop, and cropping practices influenced the decomposition rates but, in general, the half-life of straws was 24 months and 48 months whereas resistant grass had a decomposition half-life of about 96 growing months. Jenkinson (1965) also provided other evidence that newly applied carbon was less resistant to degradation than native soil organic matter.

Nitrogenous Wastes and Their Decay Rates. Sludges and manures usually are substrates of low C:N ratios. Their rates of decay are given as the familiar decay series discussed in the Appendix of Module I-2 "Nitrogen Considerations." Some typical decay series values are given in Table 1.

It is clear from Table 1 that two factors—nitrogen content and mineralization rate—contribute to the release of inorganic nitrogen from organic waste material. A comparison of fresh bovine waste and sewage sludge indicates that if the same amount of each material containing equal percent nitrogen were incorporated into the soil, the fresh bovine waste would produce more

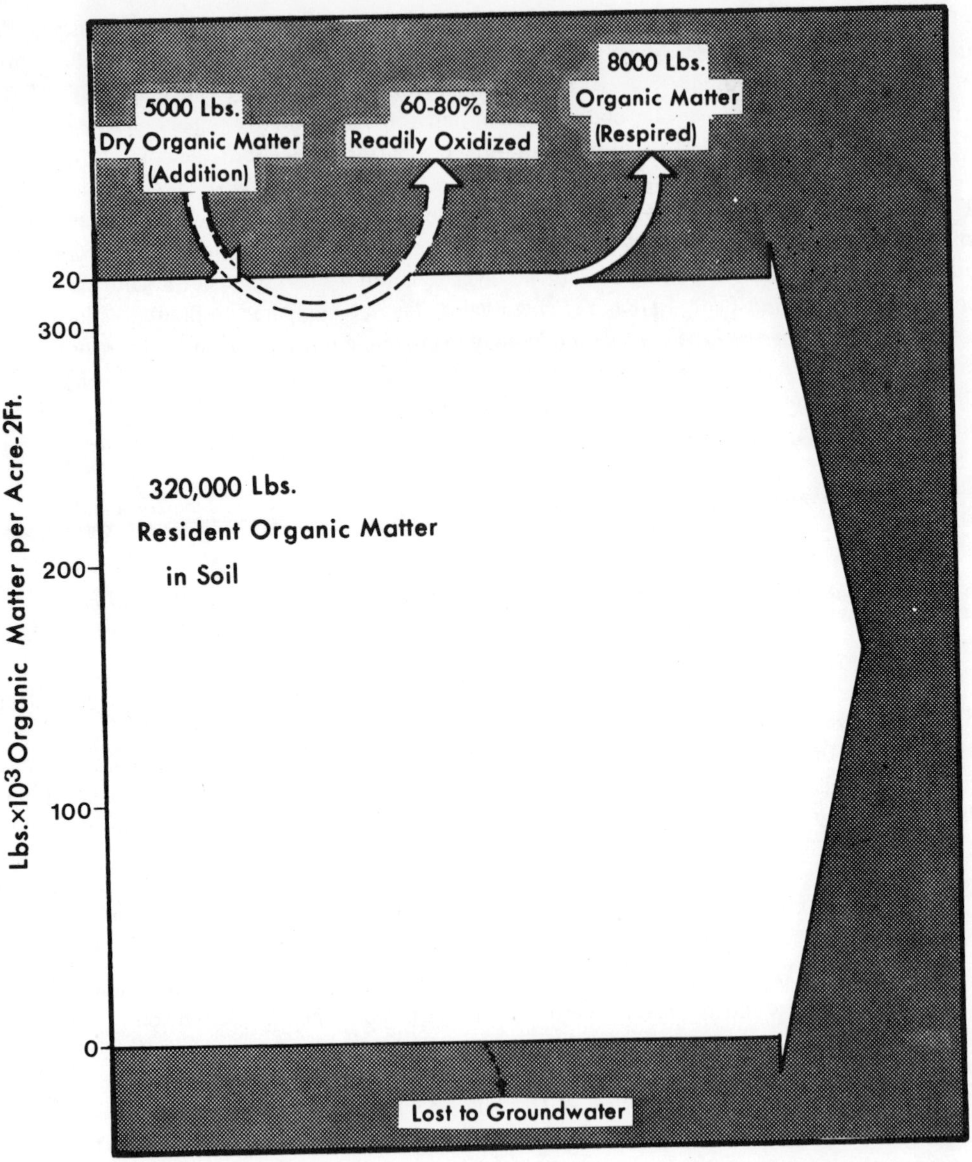

Figure 10. The magnitude of the organic matter added to soil is small compared to quantities of native organic matter, and losses due to normal respiration in the waste make net accumulation unlikely.

than twice the mineralized nitrogen in the first year. Thus it is important to be aware of the behavior of various waste materials and apply proper management techniques to minimize potential problems of nitrate leaching to groundwater. In this case, by surface spreading the fresh bovine waste before incorporation in the soil, ammonia volitilization would be greatly enhanced and driven off. The decay series shown in Table 1 are representative of local conditions in California studies by Pratt, *et al.* (1973). Other decay series for similar organic wastes have been proposed by other researchers in various parts of the country, indicating that the site planner should be familiar with local variations in mineralization rates, which are affected by climatic factors. Restricting application rates due to nitrogen limitations will in almost all instances keep organics loading at acceptable levels.

Table 1. The Rate of Mineralization of Nitrogen from Substrates Varies with Each Substrate as Indicated by Their Respective Decay Series.[a]

Substrate	Range of % N in Dry Material[b]	% N in Sample of Given Decay Series[c]	Decay Series
Sewage sludge	2.0–6.0	2.5	0.35, 0.10, 0.06, 0.05, 0.04, 0.03
Dry feedlot manure	1.2–2.0	1.5	0.35, 0.15, 0.10, 0.05
Fresh bovine waste[d]	2.4	3.5	0.75, 0.15, 0.10, 0.05

[a]Decay series is the ratio of the fraction lost/fraction immobilized in successive years.
[b]Parr, 1974.
[c]Pratt, *et al.*, 1973.
[d]~50% of N is in urea or uric acid.

Changes in the physical condition of soils are favorable under proper sludge/manure application schedules. Notable among those changes is the enhancement of soil aggregation even in some tight clay soils (Kirkham, 1974). This in turn increases macropore space, improves soil aeration and creates favorable conditions for root growth. These same improvements are reported for sandy soils with the additional benefit of providing good binding material to prevent the sand from blowing away (Kirkham, 1974). Mine spoils, and sand dunes and other barren matter having no significant indigenous organic matter will accumulate organic matter under wastewater/sludge irrigation. The final equilibrium which is approached will be site specific, and there is no evidence that it will reach the fertility of highly productive farming soils. Nevertheless, reasonably hardy grasses, trees, shrubs and cover crops have been grown on these irrigated soils, sand dunes have been stabilized, and erosion attenuated.

Sludge and manure are capable of altering physical characteristics of soils. Mine spoils, sand dunes, etc. can accumulate organics from applied wastes and support vegetation.

In general, careful husbandry of soils, including the proper use of fertilizers, a judicious selection of crops, and the use of correct soil amendments can improve soil organic content somewhat, but even productive fields are likely to remain considerably lower in organic matter than nearly virgin soil (Brady, 1974).

BIBLIOGRAPHY

Alexander, M. 1961. Introduction to soil microbiology. John Wiley & Sons, Inc., New York. 472 p.

Allison, F. E. 1967. Soil aggregation–some facts and fallacies as seen by a microbiologist. *Soil Sci.*, **106**(2): 136–143.

Brady, N. C. 1974. The nature and properties of soils. 8th ed. Macmillan Publishing Co., Inc., New York. 639 pages.

Broadbent, F. E. 1973. Organics. *In* Proc. joint conf. recycling municipal sludges and effluents on land. EPA/USDA/Nat'l Assoc. of State Universities and Land Grant Colleges, Champaign, ILL. July 9–13, pp. 97–102.

Campbell, C. A., E. A. Paul, D. A. Rennie, and K. J. McCallum. 1967. Applicability of the carbon-dating method of analysis to soil humus studies. *Soil Sci.*, **104**:217–224.

Chaney, R. L. 1974. Recommendations for management of potentially toxic elements in agriculture and municipal wastes. *In* National Program Staff, Factors involved in land application of agricultural and municipal wastes. (DRAFT). Agricultural Research Service. Soil, Water, and Air Sciences, USDA, Beltsville, Md. pp. 97–120.

Fink, D. H., G. W. Thomas, and W. J. Meyer. 1970. Adsorption of anionic detergents by soils. *J. Water Pollut. Contr. Fed.*, **42**(2) part 1:265–271.

Jenkinson, D. S. 1965. Studies on the decomposition of plant material in soil. I. Losses of carbon from ^{14}C labelled ryegrass incubated in soils in the field. *J. Soil Sci.*, **16**: 104–115.

Jenkinson, D. S. 1971. Studies on the decomposition of C^{14}-labelled organic matter in soil. *Soil Sci.*, **111**: 64–70.

Jewell, W. J. and R. C. Loehr. 1975. Land treatment of food processing wastes. Presented at American Society of Agricultural Engineers Winter Meeting, Chicago, Paper No. 75–2513, 38 p.

Kirkham, M. B. 1974. Disposal of sludge on land: effect on soils, plants, and groundwater. *Compost Sci.*, Mar.-Apr. pp. 6–10.

Lejcher, T. R. and S. H. Kunkle. 1973. Restoration of acid spoil banks with treated sewage sludge. *In* W. E. Sopper and L. T. Kardos, eds. Recycling treated municipal wastewater and sludge through forest and cropland. Penn State Univ. Press. University Park, Pa. pp. 184–199.

Loehr, R. C. 1974. Agricultural waste management: Problems, processes and approaches. Academic Press, New York. 576 pages.

Mansell, R. S., D. Kirkham, and D. R. Nielsen. 1970. Nitrate and detergent recovery in aerated soil columns. *Soil Sci. Soc. Amer. Proc.*, **34**:883–889.

Martin, J. P., J. O. Erwin, and R. A. Shepherd. 1965. Decomposition and binding action of polysaccharides from *Azobacter indicus* (*Beijerinokia*) and other bacteria in soil. *Soil Sci. Soc. Am. Proc.*, **29**: 397–400.

McKinney, R. E. 1962. Microbiology for sanitary engineers. McGraw-Hill Book Co., Inc., New York. 293 pages.

Mioduszewski, W. and T. D. Hinesly. 1972. Digested sludge dewatering on soils. *In* T. D. Hinesly, O. C. Braids, J. A. E. Molina, R. I. Dick, R. L. Jones, R. C. Meyer, and L. F. Welch. Agricultural benefits and environmental changes resulting from the use of digested sludge on field crops. Draft Report Prepared for the Environmental Protection Agency. Grant No. DO1-UI-00080.

Parr, J. F. 1974. Organic matter decomposition and oxygen relationships. *In* National Program Staff. Factors involved in land application of agricultural and municipal wastes. (Draft). Soil, Water and Air Sciences. USDA, Beltsville, Md. pp. 12–139.

Pound, C. E., R. A. Crites, and D. A. Griffes. 1975. Evaluation of land application systems. EPA-430/9-75-001. Mar. U.S. EPA, Washington, D.C. 182 p.

Pratt, P. F., F. E. Broadbent, and J. P. Martin. 1973. Using organic wastes as nitrogen fertilizers. *Calif. Agr.*, **27**:10–13.

Russell, E. J. and E. W. Russell. 1950. Soil conditions and plant growth. Longmans, Green Ltd., London. 194 p.

Russell, E. W. 1973. Soil conditions and plant growth. 10th ed. Longmans, Green Ltd., London. 849 p.

Shields, J. A. and E. A. Paul. 1973. Decomposition of C^{14}-labelled plant material under field conditions. Can. J. Soil Sci., **53**:297–306.

Statterwhite, M. B., G. L. Stewart, B. J. Condike, and E. Vlach. 1974. Rapid infiltration of primary sewage effluent at Fort Devens, Mass. USA CRREL project 4A062112A891, Task 05 (Draft report).

Wallace, A. T. 1976. Land disposal of liquid industrial wastes. *In* R. L. Sanks and T. Asano, eds. Land treatment and disposal of municipal and industrial wastewater. Ann Arbor Science Publishers, Inc., Ann Arbor, Mich. pp. 147–162.

Module II-4

POTENTIALLY TOXIC ELEMENTS

SUMMARY

Potentially toxic elements enter sewages from residential, urban, and industrial sources. The major portion of toxic element loads carried by sewages is transferred to sludges in conventional primary and secondary wastewater treatment. Secondary effluents generally contain less potentially toxic element concentrations than maximum permissible levels set by the EPA for drinking or irrigation waters. Dissolved salts and nitrogen in effluents can produce toxic conditions.

When sludges are applied to soils, these elements interact with both the organic and inorganic matter of soil. Chemical reactions also occur within the soil pore space. The net result is that these elements may become available to vegetation and to the animals consuming that vegetation. It is at this point that the toxic potential of these elements may be realized. The effect can be manifested as reduced crop yields, diminished vegetative cover, or sickly animals.

Certain crops exhibit varying response to potentially toxic elements. Brief mention of crop responses is made in this module. Hazards to plants and animals posed by selected elements are discussed in some detail. There are five elements indicated which may exert potentially serious effects: B, Cd, Cu, Mo, and Ni. The toxic effects of boron and nickel are most important in terms of their phytotoxicity—or toxicity to plants—rather than detrimental effects on animals and humans.

Methods of computing allowable sludge loading rates to land are discussed and compared. Sludge loading on the basis of potentially toxic elements is compared to the nitrogen basis and the specific limitations which cadmium may exert is also discussed.

CONTENTS

GLOSSARY

cation exchange capacity (CEC)–A measure of the ability of the soil to retain positively charged ions (cations). Generally serves as a rough index of all reactions occurring between charged species and colloidal surfaces. CEC is a function of both the relative intensity of attraction between ions and soil surfaces and the relative concentration of exchangeable ions present in the soil solution.

chelate–A molecular structure in which a heterocyclic ring can be formed by the unshared electrons of neighboring atoms.

electrostatic bond–A chemical bond in which two atoms are kept together by forces caused by transferring one or more electrons from one atom to the other.

heavy metal–A metal whose specific gravity is approximately 5.0 or greater.

ligand–The molecule, ion, or group bound to the central atom in a chelate or a coordination compound. Example: the ammonia molecules in $[Co(NH_3)_6]^{+3}$.

methylation–Formation of an organic compound in which the hydrogen of the hydroxyl group (OH^-) of methyl alcohol is replaced by a metal.

phytotoxic–Poisonous to plants.

precipitation–The production of a solid or solid phase chemically separated from a solution.

reversion–Immobilization of ions by substitution in the crystal lattice of soil solids.

solubility product constant (K_{sp})–A constant expressed as the product of the molar concentrations of ions in equilibrium with their electrolyte, or combined form.

toxic–Harmful, destructive or deadly; poisonous.

zinc equivalent (ZE)–A coefficient expressing the concentration of zinc, copper, and nickel, assuming copper to be two times as toxic and nickel eight times as toxic as zinc. Thus: $ZE = [Zn^{+2}] + 2[Cu^{+2}] + 8[Ni^{+2}]$, usually expressed in μg/gm.

OBJECTIVES

Upon completion of this module, the reader should be able to:

1. Name five elements considered to pose a potentially serious hazard in land application systems. Indicate which are phytotoxic.

2. Describe the general types of crops which are most susceptible and least susceptible to damage from potentially toxic elements.
3. Generally relate the various contributors of potentially toxic elements to municipal sewage.
4. Compare the various approaches to calculating permissible sludge application rates. Indicate the strengths and weaknesses of each approach.

INTRODUCTION: POTENTIALLY TOXIC ELEMENTS AND THEIR SOURCES

Identification of Potentially Toxic Elements

The elements listed as potentially toxic vary somewhat with authors, although the inconsistencies are comparatively minor. The eighteen elements identified in Table 1 are those included in the lists of maximum contaminant levels in drinking water (Train, 1975) and irrigation water (EPA, 1975b) set by the Environmental Protection Agency. Most of these elements, along with a few others, are discussed in this module in the section "Hazards to Plants and Animals Posed by Specific Elements." These elements also share two general characteristics:

1. They appear in significant concentrations in at least some untreated wastewaters of municipalities.
2. They accumulate in soils amended with municipal wastes and, in some cases, produce toxic symptoms either in vegetation grown on that soil or in animals that eat that vegetation.

List of Elements. As Table 1 is examined, note how necessary it is to designate the expected use of the water before toxic limits can be specified. This represents the diverse sensitivities of vegetation and animals.

**Before toxic limits can be specified, the expected use of the water must be determined.*

"Heavy Metals" and "Trace Elements." Metals having densities in excess of 5.0 are arbitrarily designated as "heavy metals." The list of potentially toxic elements dealt with in this module includes elements that are neither "heavy" nor "metals." Yet, sloppy nomenclature often includes all the toxic elements together under the title "heavy metals."

It is also useful to note that potentially toxic elements occur naturally throughout our environment, albeit (in some instances) in extraordinarily small concentrations. All organisms, therefore, have evolved certain tolerances for these elements and, for some, a definite requirement in life processes. Because of their wide occurrence in small concentrations in plants and animals, the listed potentially toxic elements also are known as "trace elements."

**Because potentially toxic elements occur in nature, all organisms have developed tolerances for them to some extent.*

The elements listed in Table 1, along with others, are called "heavy metals" by some, "trace elements" by others, and "toxic elements" by still others. The term "potentially toxic elements" is considered more descriptive, and hence is used throughout this module.

While all the toxic elements are potentially dangerous, some receive special attention because of circumstances surrounding their occurrence, their greater concentrations, or their impact.

*This and other italicized summaries are intended to highlight key ideas, provide a basis for later review or to aid in skimming sections that are relatively familiar. They can be ignored in a complete reading of the text.

Table 1. Eighteen Potentially Toxic Elements Identified in Two Sources.

Element		Max. Level in Drinking Water (mg/1)[a]	Max. Level For Continuous Use Irrigation Water, All Soils (mg/1)[b]
As	Arsenic	0.05	0.01
Al	Aluminum		5.00
B	Boron		0.75
Ba	Barium	1.00	
Be	Beryllium		0.50
Cd	Cadmium	0.01	0.01
Cr	Chromium	0.05	0.10
Co	Cobalt		0.05
Cu	Copper		0.20
Hg	Mercury	0.002	
Mn	Manganese		0.020
Mo	Molybdenum		0.21
Ni	Nickel		0.20
Pb	Lead	0.05	5.0
Se	Selenium		0.02
Ag	Silver	0.05	
Zn	Zinc		2.0
F	Fluorine	varies with mean temp.	

[a]Train, 1975.
[b]EPA, 1975b.

For example, if cadmium exceeds 1.0% of the zinc content of municipal sludges, its uptake by crops may lead to cadmium poisoning in animals eating those crops (Chaney, 1973). Similarly, high zinc or copper levels are sometimes toxic to plants ("phytotoxic"), and composted municipal refuse may be a significant source of highly phytotoxic boron. On the other hand, an element such as silver may be of only academic interest in most land application systems merely because it does not appear in significantly high concentrations.

It is generally agreed that the potentially toxic elements of untreated wastewaters most likely to pose a threat of phytotoxicity or animal toxication are *boron*, *cadmium*, *zinc*, *copper*, *molybdenum*, and *nickel* (Chaney , 1973; Leeper, 1972; Page, 1974; CAST, 1976).

Boron, cadmium, zinc, copper, molybdenum, and nickel are the elements in untreated wastewater most likely to be toxic for plants and/or animals.

Sources of Toxic Elements

As mentioned earlier, toxic elements occur naturally throughout the environment. Toxic elements also may be added directly in the form of, say, metal-based pesticides (viz. arsenate of lead used in the past) if the site is used for growing appropriate crops. However, most toxic element increase in soils at land application sites comes from the general community through its wastewater and, particularly, its sludges.

Increase in toxic elements at land applications sites comes from the wastewater and sludges of the general community.

Residential Contributions. Toxic elements are present in significant concentrations even in domestic sewage (Davis and Jacknow, 1975). Presumably this represents contributions from the brass, copper, and lead of plumbing, from cosmetics, hobby materials, and from internal and external medicines. This latter category is exemplified by mercurial diuretics and dandruff preparations employing selenium. The concentration of elements commonly found in sludge from communities without excessive industrial waste inputs (or with adequate source abatement) should be no higher than 2,000 mg/l for zinc and 1000 mg/l for copper, lead, and chromium. (Menzies and Chaney, 1974).

Community residents discharge significant amounts of potentially toxic elements probably originating from cosmetics, hobby materials, medicines and particularly from metal plumbing pipes.

Special Industries. Some industries discharge potentially toxic elements in high concentrations. Foundries, electroplating and photoengraving shops, and similar concerns often discharge iron, copper, nickel, zinc, chromium and/or cadmium. Mercury is often associated with wastes from the electrolytic production of chlorine and caustic soda, from the manufacture of electrical equipment and antifouling paint (Hammond, 1971), and from the combustion of coal (Billings and Matson, 1972; Joensuu, 1971). The manufacture of viscose rayon, the vulcanization of rubber, or even some processes of paper production can discharge significant quantities of zinc.

Potentially toxic elements discharged by some industries are more highly concentrated than those discharged by other sources.

Urban Environment. The major source of potentially toxic elements in municipal sewages of the larger cities results from the more general environment. For example, lead from burning gasoline and other fuels, petrochemical industries, and incinerators is rainwashed into sewers in large quantities. Other contributions are from the smoke and dusts associated with smelters (copper and arsenic), from losses due to tire wear, from corrosion (especially in areas subject to acid rains), from the leaching of plant food and fertilizers from gardens, lawns, and parks, and so on. These sources contribute more heavily to treatment plants served by combined storm/sanitary sewer systems.

Major source of potentially toxic elements in municipal sewage of larger cities is from the general environment.

Relative Contribution. To assess the relative contributions of residential, special industrial, and urban environmental sources, refer to Figure 1. Note that the general pattern shown in Figure 1 does not hold for nickel. Substantial amounts of nickel do not come from residential or urban environments.

While the data of Figure 1 apply to New York City (where electroplating and photoengraving are the main industrial discharges of toxic elements to the municipal treatment system), the general pattern has wide applicability. The general urban environment contributes 10% to >50% more potentially toxic elements than residential areas, and residential areas often contribute more than industries. The discharges from industries to the wastewater collection system, however, will be much more concentrated than those from the other sources. Pretreatment of industrial wastes is often the most effective means of correcting an imbalance in sludge com-

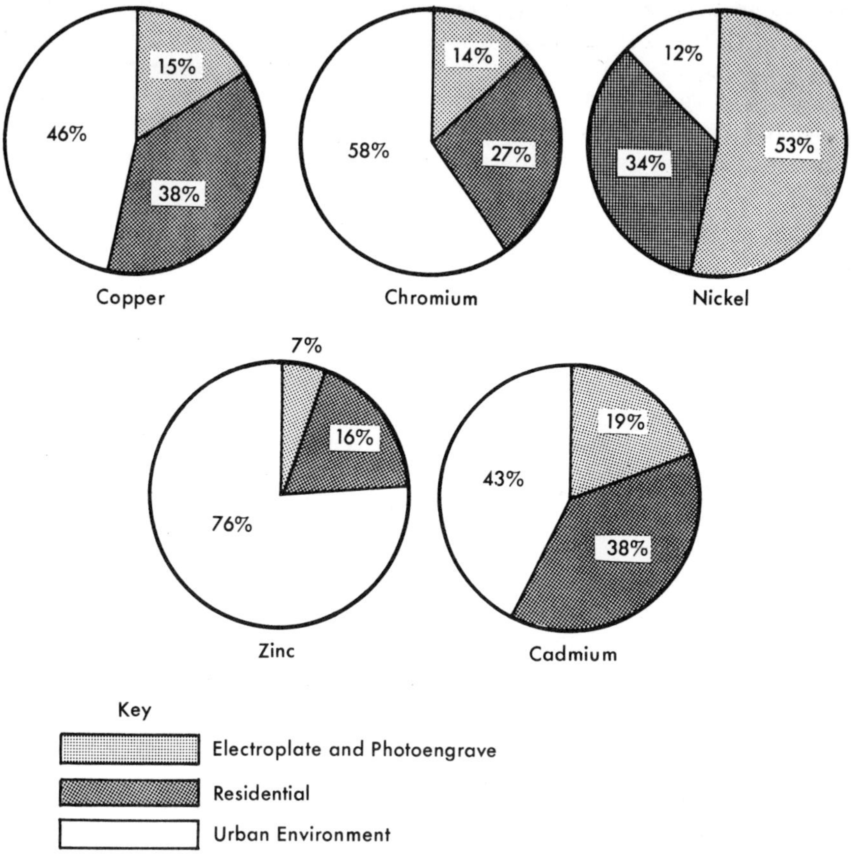

Figure 1. In municipal sewage from New York City, the major source of Cu, Cr, Zn, and Cd is from the general urban environment, with less contributed by residential outfall or even special industrial concerns (adapted from Davis and Jacknow, 1975).

position to be applied to the land. However, industrial pretreatment programs will lead to significant toxic element reductions only if non-point discharges are comparatively low.

> *Pretreatment of industrial wastes can be an efficient way of reducing potentially toxic element concentrations in municipal sludge, but only if the industrial contribution is significant.*

SEWAGES, SLUDGES, SOILS: CONCENTRATING TOXIC ELEMENTS

The potentially toxic element load carried by raw sewage varies with the size of the city, the season, and even the time of day. Nevertheless, the major proportion of such an element load is removed in the sludge-forming processes associated with primary and secondary treatment. Several physical and chemical factors account for this, but what is particularly interesting is that those same general processes account for toxic element sorption by soils. By reviewing the simpler event, the uptake of potentially toxic elements by sludge, it is easier to make the transition to such element uptake by soils.

Potentially Toxic Element Uptake by Sludges

The uptake of potentially toxic elements by sludges is influenced by many factors, five of which will be discussed here: (1) pH, (2) ligand bonding, (3) retention time, (4) toxic element concentration, and (5) ion competition for adsorption sites. Each factor influences the attachment made by toxic elements to the solid-phase floc of sludge.

Potentially toxic elements tend to make attachments to the solid-phase floc of sludge.

The Influence of pH. The flocculant matter of sludges offers many surfaces which are negatively charged and capable of forming electrostatic bonds with cations. However, toxic elements are not the only cations in wastewaters. At low pH's the dominant cation may be the hydrogen ion (H^+). That would lead to a competition for bonding sites between the prevalent hydrogen ions and the less common ions of trace metals. In that competition, toxic element ions are at a statistical disadvantage and are not readily taken up by the sludge. Higher pH values (lower hydrogen ion concentrations) in the sewage, of course, allow for less competition and favor toxic element uptake. However, at pH's greater than 6.5, conditions also favor the direct precipitation of toxic elements as metal hydroxides. These precipitates become fixed in the biological floc and end up in the sludge. At pH's greater than 10 (common in lime precipitation for phosphorus removal) substantial precipitation of metal hydroxides occurs, particularly with zinc, cadmium, cobalt, ferrous, and ferric ions (Wood and Tchobanoglous, 1975). Representative data showing the effect of pH on the removal of nickel (Ni^{+2}) in activated sludge are given in Figure 2. Note that toxic element uptake also is dependent on the amount of active surface available as represented by the concentration of mixed liquor suspended solids.

Higher pH values in sewage during conventional treatment favor precipitation of metal ions into sludge.

Ligand bonding—chelation. Both dissolved and solid phases of the sewage organic matter may have chemically active sites capable of forming one or more coordinate covalent bonds with potentially toxic elements.

A single organic molecule capable of forming two or more coordinate covalent bonds is called a *chelate*. Chelating agents may react with toxic elements forming complexes that may be either insoluble or soluble. Thus, chelating agents may serve to retain toxic elements in the wastewater, blocking their accumulation in sludges, and preventing further chemical reactivity. After application to soil, the organic chelate itself may be degraded and the toxic element released.

Chelates in sewage may act to retain toxic elements in the wastewater and prevent their accumulation in sludge.

Retention Time. The uptake of potentially toxic elements by biological floc is somewhat dependent on the nature of the element, but in general, reaches a near steady state after from one to two hours of retention. This is illustrated in Figure 3.

Toxic Element Concentration. The uptake of toxic elements by biological floc increases with ion concentration and is adversely affected only by very high ion concentrations (>300 mg/l). Data are scarce, but the effect (if any) of very high concentrations seems to be a decrease in quantity of uptake. This is also illustrated in Figure 3.

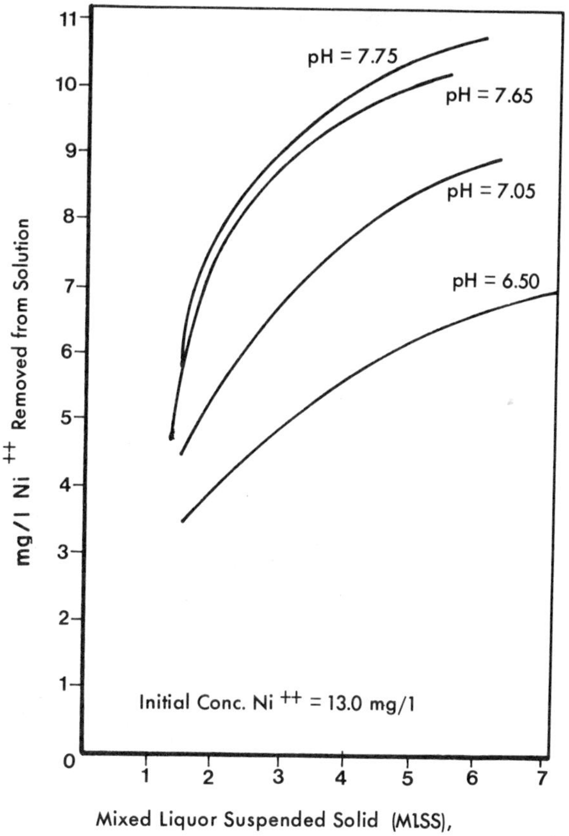

Figure 2. The uptake of Ni^{+2} by sludge increases with pH and available reactive surface (adapted from Cheng, *et al.*, 1975).

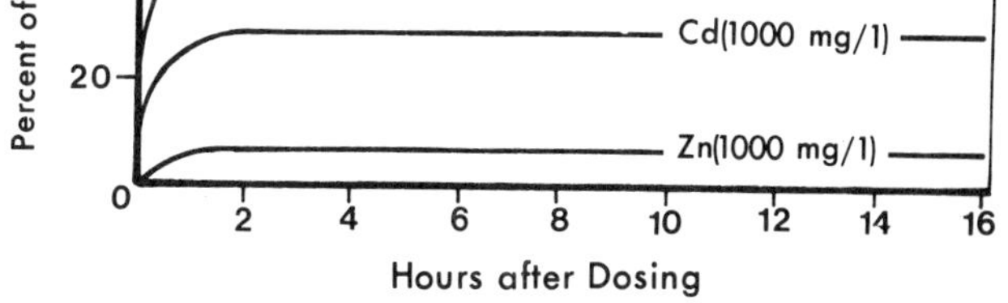

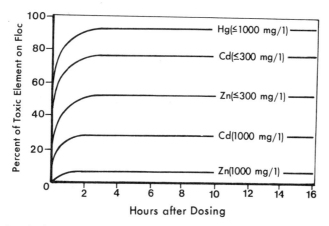

Figure 3. Uptake of toxic elements by biological floc is near steady state after two hours at any concentration, but is suppressed for some ions after higher dosing concentrations (adapted from Neufeld and Hermann, 1975).

Ion Competition for Adsorption Sites. Toxic elements compete for reaction sites on the biological floc, with some ions showing stronger affinities than others. This can be specific to the system under investigation, and underscores the need for chemical analysis of any wastewater or sludge which must be treated.

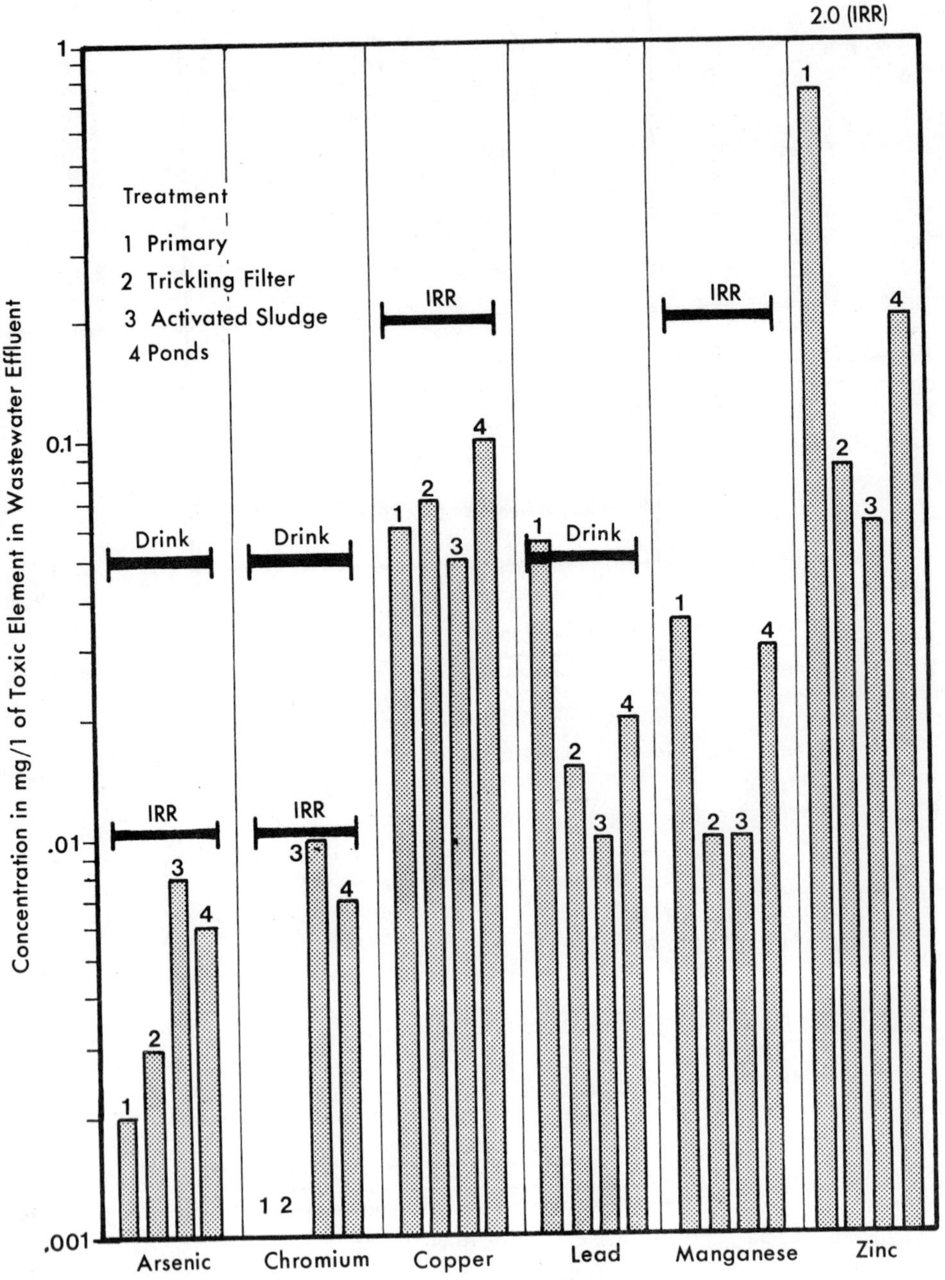

Figure 4. Toxic element concentrations in primary and secondary effluents are below EPA standards for drinking and irrigation water (adapted from Pound and Crites, 1973; Train, 1975).

Table 2. The Potentially Toxic Element Concentrations of Dry Sewage Sludge Greatly Exceeds that Found in Secondary Effluent. (Adapted from Baker and Chesin, 1975)

	Sewage Sludge (mg/kg dry wt)			
Element	Range[a]	Range[b]	Range[c]	Secondary Effluent Range[d] (mg/1)
Arsenic	1–18			0.003–0.008
Chromium	20–41,000	143–1,498	200–9,100	0.001–0.01
Copper	52–11,700	872–1,718	360–10,300	0.05–0.1
Lead	15–26,000	239–3,407	220–3,500	0.01–0.02
Manganese	60–3,900			0.01–0.03
Zinc	72–49,000	1,053–6,540	1,300–21,200	0.06–0.2

[a]Page, 1974.
[b]Baker and Hornick, 1974.
[c]Regan and Peters, 1972.
[d]Pound and Crites, 1973.

Toxic Element Retention in Secondary Effluent

As the factors discussed in the preceding paragraphs indicate, most of the toxic element loads of untreated wastewaters accumulate in sludges. A question remains: What is the toxic element load carried by primary and secondary effluents? The answer, assembled in Figure 4, is quite revealing. With only one minor exception (primary effluent, lead ion), the reported toxic element concentrations in both primary and secondary effluent are below maximum contaminant standards set by EPA for both drinking and irrigation waters. Clearly, the major concern for potentially toxic element concentrations lies in sludges, not in secondary effluents. Table 2 illustrates this point. On occasion the toxic element concentration of sludges is almost as rich as ores. However, commercial reclamation may be precluded by the chemical form of the elements.

Concentration of potentially toxic elements is not generally a concern in primary and secondary effluent; the concern is with sludges.

Accumulation of Potentially Toxic Elements in the Soil

After sludge application, potentially toxic elements become available for soil reactions through a variety of processes. For example, the environmental change may cause certain chemical complexes of the sludge to dissolve. Also, the organic matter of biological flocs and chelates may decompose, releasing these elements from the sludge. Even competitive reactions, involving soil constituents, may occur, forcing the desorption of these elements from the sludge. Whatever the mechanism of release from sludges, the elements then react with soils.

Potentially toxic elements are released into the soil by a number of mechanisms following sludge application.

It is important to note that the solid matter of soil is analogous to the surface presented by sewage sludge. As is the case with sludges, effective uptake of these elements in soils depends on the pH of the soil solution, the availability of adsorption sites and ligand bonding (soil clay and humus), the relative competition that occurs between these elements, and the rate of flow of solutions through the soil. Other significant reactions such as precipitation and methylation also occur in soils.

For convenience, this discussion will separate reactions of toxic elements into two categories: (1) reactions with soil solid matter and (2) reactions within the soil solution.

Uptake of potentially toxic elements by soil depends on pH of soil, availability of adsorption sites, competition among such elements, and rate of flow of solutions through soil.

Reactions with Soil Solid Matter. Three specific interactions between soil solid matter and toxic elements will be reviewed here: (1) selective adsorption, (2) covalent and ligand bonding, and (3) reversion processes.

Selective adsorption is a process of adsorption of cations that occurs on the surfaces of soil colloids (clay and humus). An overview of this process is given in Module I-1, "Soil as a Treatment Medium." The total amount of such element removed from the soil solution by this process is not great, but is most efficient when the soil pH is maintained at 6.5 or greater.

This first adsorption of cations is neither permanent nor sufficiently protective of vegetation. Roots take up some fraction of adsorbed cations. Further, desorption, or displacement from the colloids, may occur when the ion concentrations of soil solutions change. However, such desorption is not likely to lead to toxic element "breakthrough" to groundwater unless the soil is acid, of low C.E.C. and consists of sandy subsoils (Chaney, 1974).

A small amount of toxic elements is taken up by soil solids through selective adsorption of cations. Such adsorption is not permanent nor does it prevent some uptake by plant roots.

Covalent and ligand bonding incorporates ions in the chemically complex organic matter of the soil. This seems to be the major process removing toxic elements. Some organometallic complexes, such as Cu-organic matter, are insoluble and immobile, but others seem quite the opposite. For example, it is thought that some potentially toxic elements are made more available to vegetation after forming chelate complexes with soil organic matter.

Potentially toxic elements are removed primarily by covalent and ligand bonding; the organic complexes formed may be soluble or insoluble.

Reversion is a slow process, often extended over a period of years, by which these elements are converted to highly stable chemical forms accessible to neither the soil solution nor vegetation uptake. In general, the amount of toxic element that a soil can immobilize depends on the cation exchange capacity of the soil, but because clay minerals differ in structure, specific predictions are hard to make. Probably reversion does not account for more than 10 to 20% of the total adsorption capacity attributed to the mineral portion of the soil. The adsorption capacity due to soil organic matter is irrelevant with respect to this process; humus lacks the crystal lattice needed for immobilization by substitution. Whatever reversion procedure is experienced, there is ample evidence that a radical alteration in environmental conditions (for example a significant drop of soil solution pH) may release at least some of the immobilized cations.

Through the slow process of reversion, potentially toxic elements are rendered inaccessible to both the soil solution and vegetative uptake; they are converted to highly stable chemical forms.

Interactions in Soil Solutions. Two specific interactions which occur outside the solid particles and hence in soil pore space are: (1) precipitation and (2) methylation.

Precipitation was discussed earlier as a mechanism by which many elements are removed from sewage and incorporated into sludges. The principal cause of the precipitation there was the high pH which allowed the formation of insoluble hydroxides. Similarly, in highly alkaline soils, potentially toxic elements may be precipitated as hydroxides. However, other chemical forms, such as sulfides and phosphates (to name a couple), also are precipitated within soil pores.

In highly alkaline soils, many elements may be precipitated as insoluble hydroxides and as other chemical forms within the soil pores.

It is important to recognize the limitations of precipitation as a mechanism for immobilizing toxic elements. First of all, standard solubility constants (K_{sp}) only describe equilibrium conditions, giving no information on the rate of precipitate formation. A salt may have a very low K_{sp} yet the time for reaching equilibrium may be much longer than the time required to pass through the land application site. Finally, it must be remembered that soil organic matter will interact with many elements in solution, chelating them and thereby preventing the formation of inorganic salts.

Contrary to "common sense," the availability of toxic elements to vegetation is not correlated with solubility in the soil solution. Data of Lindsay (1972), Boawn, *et al.* (1957), Giordano and Mortvedt (1969), and others indicate that even highly insoluble compounds can provide as much toxic element to plants as highly soluble compounds. For example, if equal amounts of zinc as the soluble sulfate and insoluble carbonate are made available to vegetation, 8% more zinc is absorbed from the zinc carbonate even though it is about five orders of magnitude less soluble than zinc sulfate. Zinc oxide is six orders of magnitude less soluble than zinc sulfate yet delivers 67% more zinc to vegetation. Similar apparent discrepancies are reported for copper and nickel.

There is no correlation between availability of toxic elements to vegetation and their solubility in the soil solution; highly insoluble compounds can provide more of a given element to plants than highly soluble forms.

Methylation is one of many possible reactions associated with the metabolism of soil bacteria. The familiar example usually cited is methylation of mercury. At one time it was thought that discharges of free mercury to rivers and lakes were harmless. Since the mercury was insoluble, it was expected to settle in the mud and remain immobilized. However, under anaerobic conditions at around pH 7.0, specific microorganisms metabolize free mercury and produce monomethyl (CH_3Hg^+) and dimethyl [$(CH_3)_2Hg$] mercury, phenyl mercurials [$C_6H_5Hg^+$ and $(C_6H_5)_2Hg$], and methoxyethyl mercury [$CH_3O(CH_2)Hg^+$), all of which are very toxic and very soluble (Goldwater, 1971). Since other toxic elements also are known to undergo methylation, and since methylated elements often are both water soluble and very toxic (especially to higher animals), the real hazards of this event relative to land application will have to be more completely resolved. At present, not even the extent of mercury methylation in soil, much less its potential hazard, has been fully documented. It is felt however, that this process poses a much greater threat to aquatic than terrestrial systems.

The potential hazards of toxic element methylation must be determined in land application; many elements which undergo methylation are water soluble and can be very toxic in an aquatic environment.

POTENTIALLY TOXIC ELEMENTS IN SOIL, VEGETATION, AND ANIMALS

Earlier, while introducing potentially toxic elements, two very significant points were touched upon lightly. One was that such elements occur naturally in soils. The second was that organisms take in these elements but have evolved a certain tolerance for them. The following discussion returns to those themes.

Potentially Toxic Element Load of Natural Soils

In Module II-11, "Monitoring at Land Application Sites," it is argued that monitoring of a site should begin months before the site goes into operation. Clearly, since all soils naturally carry certain amounts of various elements, pre-operation monitoring of soils provides baseline data against which future accumulations can be gauged. While each site is, of course, different with respect to actual content of potentially toxic elements, some representative values are given in Figure 5.

Soils at a land application site should be monitored for potentially toxic elements months before operation begins to determine naturally occurring amounts.

Uptake of Potentially Toxic Elements by Vegetation

Solubility vs. availability. Earlier in this module it was argued that there simply is no definite way to specify what percentage of the total potential toxic element concentration is available for vegetation uptake. However, it has become popular to use solubility in 0.1*N* HCl as an approximate estimate of availability of toxic elements to vegetation. Also, it is generally agreed that the addition of toxic elements alone would be more likely to cause plant damage than equivalent amounts in sludges, at least on a short-term basis (Brown, 1975).

The percentage of the total potentially toxic element that is available for vegetation uptake cannot be specified.

Variables affecting uptake. The uptake of potentially toxic elements by vegetation is not consistent from species to species or even in varietal forms within a species. Nor are these elements evenly distributed throughout the plant. Depending on the plant species and the element involved, high concentrations may accumulate preferentially in leaves, stalks, roots, or less commonly in grain. Finally, the increased concentrations in plant tissue usually are not proportional to the amount of the element added to the soil.

Increased concentration of a potentially toxic element in plant tissue usually is not proportional to the amount of the element added to the soil.

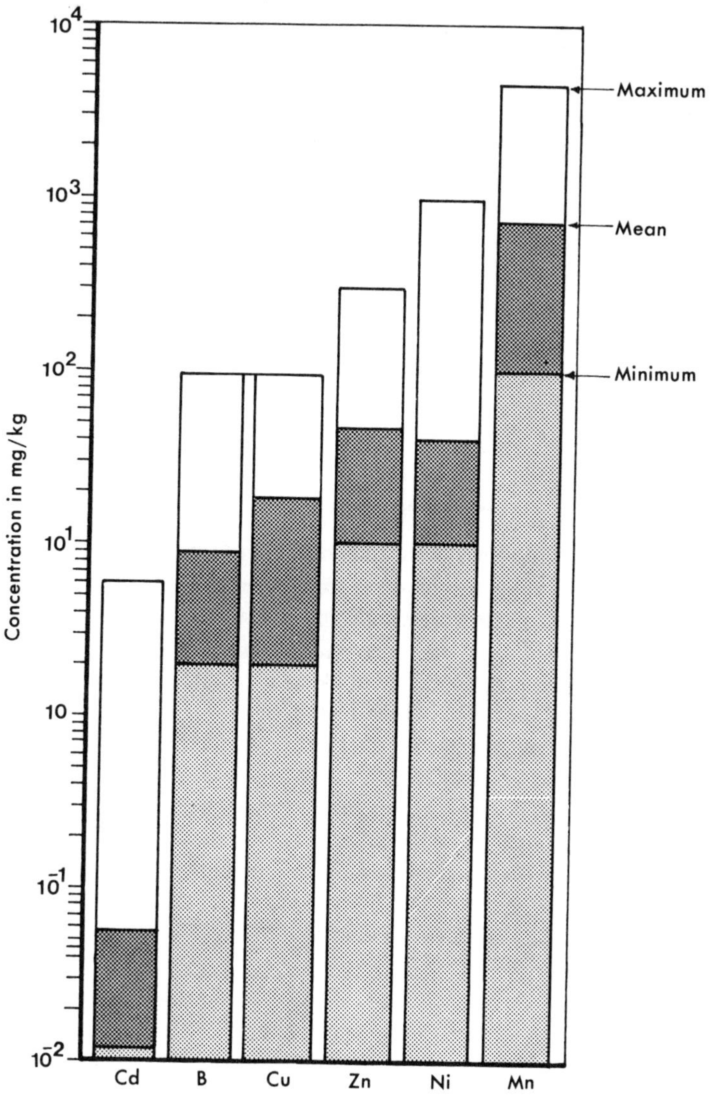

Figure 5. The range of naturally occurring toxic elements in soils (extractable with 0.1*N* HCl) extends over six orders of magnitude (adapted from Alloway and Davies, 1971).

Crops Associated with Land Application Practices

The crops most likely to be grown must be considered in evaluation of alternative sites. For residues containing fairly high levels of potentially toxic metals, use of the site for production of annual seed crops such as corn harvested for grain with the crop residues left in the field tends to minimize potential hazards. Use of the site for production of leafy vegetables provides much less protection from potential hazards.

Plant species and even varieties differ widely in tolerance to these metals. Vegetable crops that are very sensitive to toxic metals include the beet family (chard, spinach, redbeet, and sugarbeet), turnip, kale, mustard, and tomatoes. Beans, cabbage, collards, and other vegetables are less sensitive. Field crops such as corn, small grains, and soybeans are moderately tolerant. Most grasses, (fescue, lovegrass, Bermudagrass, orchard grass, perennial ryegrass, etc.) are tolerant to high amounts of metals.

Uptake of toxic elements varies according to plant species and even plant variety.

At excess levels, the grain, fruit, or edible root of many crops contains two- to tenfold lower levels of Zn than the leaves. Thus, corn grain from a site could contain less toxic elements than mixed pasture or silage. Whether the relative exclusion of toxic metals by grasses would allow more nutrient removal, yet safer levels of Zn, Cd, and Pb, than growth of corn grain is unclear. A further advantage of grain crops vs. leafy crops is the two- to tenfold greater exclusion of Cd relative to Zn during grain formation.

Toxic elements may accumulate preferentially in leaves, stalks, roots, and less commonly in grain.

Some have suggested that we should breed varieties of crops that can tolerate and exclude undesired toxic elements. Although more tolerant varieties probably could be developed, only an effort to produce metal-tolerant perennial grasses that would exclude toxic elements seems reasonable for using field crops in waste application sites. This would allow maintenance of an economic crop on sites dedicated to receive sludge or effluent after other economic crops would no longer be grown due to the soil toxic element levels. Properly designed advance studies and monitoring can be used to divert some sludges that are high in potentially toxic elements to sites that are used for ornamental plants, fiber crops, or other uses that do not involve a potential hazard to food chains. The problem of land application of sewage sludge will be most severe for large industrialized cities with a limited range of potential application sites. In extreme cases, some method such as incineration may be more feasible than land application. It should be recognized, however, that the hazard to food chains from potentially toxic elements in ashes may be even more severe than from these same elements in the unburned organic material.

One management option for sludges with very high contaminant levels is to apply these wastes to non-food crops such as fibers or sod.

A more detailed discussion and tabulated data on crop tolerances for potentially toxic elements and salts, as well as other factors, are included in Module II-7, "Crop Selection and Management Alternatives."

Table 3 summarizes some known beneficial effects and potentially toxic effects of many of the elements likely to be encountered in land treatment of wastewater and sludges. The next section focuses on several of these elements for which toxic effects and/or potentials have been documented. Some management tools to minimize such effects are mentioned.

SPECIFIC ELEMENTS POSE HAZARDS TO PLANTS AND ANIMALS

In this section potentially toxic elements are grouped according to the relative risk of toxicity to plants or animals. Conditions which favor or hinder their availability to plants and animals are also discussed. Most of the information given here comes from two reports, "Food Chain Aspects of the Use of Organic Residues" (Allaway, 1977) and "Application of Sewage Sludge to Cropland" (Council for Agricultural Science and Technology–CAST–1976).

Elements posing relatively little hazard are:

- Aluminum
- Antimony
- Arsenic
- Chromium
- Fluorine
- Lead
- Mercury
- Nitrogen
- Selenium
- Zinc

Elements posing a potentially serious hazard are:

- Boron
- Cadmium

Table 3. Effects of Selected Elements on Plant and Animal Nutrition.

		Behave as Essential or Beneficial Elements		Behave as Toxic Elements	
Element	Comment	Plant	Animal	Plant	Animal
B	narrow margin	yes	no	yes	
Ca		yes	yes		
Cl		yes	yes	yes	
Cr	activates insulin	no	yes		
Co	B_{12}, n-fixation		yes		
Cu	toxic to sheep if Mo <0.1 ppm	yes	yes	yes	10–20 ppm[a]
F	tea only crop with uptake	no	yes	yes	yes
I	once common animal deficiency	no	yes		
Fe	most common animal deficiency	yes	yes	pH < 5	
Mg	deficiency causes grass tetany	yes	yes		
Mn	toxic in acid soils	yes	yes	pH < 5	
Mo	induces Cu deficiency	yes	yes		5–20 ppm[a]
P	most critical mineral deficiency	yes	yes		
K	contributes to Mg deficiency	yes	yes		
Se	may be deficient or toxic	no	yes		4 ppm[a]
Si	may interfere with digestion	yes	yes		yes
Na	vegetable diets often lack Na	Some	yes	yes	
S	cystine, lysine, methionine	yes	yes		
Zn	lacking in some U.S. diets	yes	yes	yes	
Ni		no	maybe	yes	possible
Sr			maybe		
Sn			maybe		
V			maybe		
Al		no	no	pH < 5.5	
As	phytotoxic before animal toxic	no	no	yes	yes
Cd	counteraffected by Zn and Se	no	no	yes	yes
Pb	hazard primarily surface deposition	no	no		yes
Hg	aquatic accumulation	no	no		yes

[a]Concentration (ppm) at which vegetative dry matter may be toxic to consuming animals.

Source: Adapted from Allaway (1968), Walsh, *et al.* (1976), and Allaway (1977).

Copper
Molybdenum
Nickel

This classification of the potentially toxic elements into groups of "relatively little hazard" and "potentially serious hazard" assumes that good management practices are in effect at the application site. It applies to elements entering the food chain through plant roots rather than by ingestion from plant or soil surfaces by grazing animals.

Relatively Little Hazard

Aluminum, Iron and Manganese. The levels of these three elements in sludge usually will not be of any environmental concern. Levels of Al, Fe, or Mn may be high in sludges from tertiary treatment processes, but they will not be a limiting factor in sludge application rates if the pH is maintained above 5.5 and the soil is well aerated. Soil pH is the most important control. Below pH 5.0, aluminum toxicity is common in plants, and excessive manganese may be taken up if it is present in the soil. At low pH levels, high availability of iron can induce plant manganese deficiency. Well-aerated soils are important to avoid reducing conditions. Both manganese and iron are more soluble in a reducing environment.

Aluminum, iron, and manganese will not be toxic as long as low pH is avoided and good aeration of soil is maintained.

Antimony. Antimony is not known to be an essential element for plants, but it has been reported to be moderately toxic. Antimony is sorbed very strongly by kaolinite and sesquioxides under acid conditions, but appears to move freely in neutral to alkaline soils. Antimony generally occurs in sewage sludge in very low concentrations, and although it could be a hazard to plants and animals at high concentrations, there is no evidence currently available of a hazard.

Arsenic. The use of organic arsenicals in poultry feeds is the principal source of As in organic wastes. Concentrations of 15 to 30 ppm As have been reported in poultry house litter (Morrison, 1969). Where poultry litter of this type had been applied to soils over a period of 20 years, no increase in the As concentration in alfalfa and clover was detected.

There are fewer data on As concentration in municipal sewage sludges than for many of the other trace elements, however, Page (1974) gives a range of 1-18 ppm As in some sludges from Michigan. Predictions of the hazard from As in organic materials must be derived from experiments on As residues from pesticides. The fate of As from pesticide residue has been discussed by Woolson, et al, (1971) in terms of accumulation of As in soils. Crop failure, or sharply diminished yields, usually result before the production of crops that have hazardous levels of As in the edible portion.

Soil tests for predicting the hazard of phytotoxicity of As residues have been developed (Woolson, *et al*, 1971). Use of these tests to monitor As accumulations and prevent development of phytotoxic levels offers substantial protection against food chain damage from As in organic materials.

Tests to monitor arsenic accumulations in soil provide protection against its damaging the food chain.

Chromium. Chromium is not expected to be a limiting factor in determining the quantity of sludge that may be applied to agronomic soils because (a) plants can tolerate relatively high levels of chromium applied in sludge and (b) plants do not accumulate chromium even when it is present in the soil at high levels (Underwood, 1971).

Little soluble chromium is found in soils, because when added in the soluble trivalent form it rapidly is transformed to insoluble forms, probably hydroxides or oxides. Hexavalent chromium is toxic to plants, but it is reduced to the trivalent form in sludge digestion. In view of the limited possibility for hexavalent Cr to be present in organic materials, it appears that Cr in these sludges presents no hazard to food chains.

Fluorine. Fluorine is probably essential to animals, but can be toxic to plants and animals at high concentrations. The hazard of fluorides in organic residues is apparently minor and confined to sludges containing soluble fluorides when applied to strongly acid soils. Liming to higher pH can substantially correct any F difficulties.

Lead. The most surprising fact about Pb in soil is its apparent concentration in greatest amounts at the surface with the organic matter. This is the product of two causes; (1) some plants absorb it in substantial amounts; (2) it is very strongly fixed by soil. The chemical mechanism of fixation by soul is that Pb^{2+} is hydrolyzed and polymerized, and is more tightly bound as the pH rises. In accordance with this tight holding by the soil, the Pb extracted with dilute acetic acid does not show the high values shown by the total mass balance of Pb through the soil profile.

Where the organic residue is incorporated into the soil, Pb contained in the residue appears to offer very little hazard as far as toxicity to humans and animals is concerned, and only a modest toxicity to plants (Baumhardt and Welch, 1972). Lead added with sludge is not phytotoxic because the sludge contains a large amount of phosphate which ties up Pb and prevents injury to plants, which would result from their uptake of Pb. On soils that have been treated with as much as 3200 kg/ha of Pb in the form of soluble salts, the concentration of Pb in corn leaves was increased but the Pb concentration in the grain was unchanged. Therefore, production of crops where only the edible seed is harvested may be an effective way of preventing Pb toxicity to humans and animals. The uptake of Pb by plants may also be reduced by liming acid soils. In general, potential hazards to Pb toxicity are reduced whenever the Pb is incorporated in soil.

Hazards of lead toxicity are reduced when it is incorporated into a neutral soil.

Mercury. Experimental evidence available for mercury shows that the metallic ion is strongly held by soil and is not absorbed by plants. There is much less mercury in circulation than available adsorption sites in the soil. Therefore, the adsorption sites for this element should never approach saturation before another toxic element poses a hazard.

Concern over hazard to food chains from Hg in sewage sludge has stemmed from evidence of methylation of Hg in aquatic systems, and from recent episodes of methylmercury poisoning from inadvertent human consumption of treated seeds. More recent research has indicated that the Hg concentration in above-ground parts of plants is generally very low, even though Hg seed treatments or other additions of Hg to soils had been made. Chaney (1973) indicates that hazards from Hg in terrestrial food chains are much less than those in aquatic food chains.

Organic and inorganic mercury compounds may decompose to elemental mercury, then volatilize or form compounds with sulfide and chlorides. Because of its high affinity for solid soil surfaces, mercury persists in the upper soil layer and is not a threat to groundwater.

Mercury concentration in above-ground parts of plants is generally very low. Land application of mercury poses no threat to groundwater.

Nitrogen. Problems associated with N in organic wastes include the hazard of high nitrate concentrations in crops, drainage water, and groundwater, including the possibility of nitrosamine formation. Cattle can be affected by grass tetany, as described in Module II-7, "Crop Selection and Management Alternatives."

There is no evidence that N in organic wastes presents either more or less hazard than equivalent amounts of nitrifiable or plant-available N from inorganic sources. The objectives of management of N in organic waste application are to provide needed amounts of N to the crop, to avoid production of crops high in NO_3^-, and to minimize NO_3 in water. The hazard from high NO_3^- crops can be minimized by growing corn or other cereals for grain. Cereal seeds have not been found to contain high levels of NO_3^- even when the leaves and stems contain hazardous levels. Where appropriate crops and crop uses are selected, potential for NO_3^- accumulation in waters constitutes the major criterion for safe application of N in organic residues to farm lands.

Management objectives for nitrogen are to provide needed amounts for crops, avoid producing crops high in NO_3^-, and minimize NO_3^- accretion in water.

Nitrogen is the limiting factor in most land application situations. It is discussed in depth in Module II-1, "Nitrogen Considerations," and mentioned in connection with system design at several other points in the modules.

Selenium. Selenium is essential for animals and probably for people, but not for plants. Concentrations of about 0.04 ppm Se or more in diets are required in order to prevent Se deficiency in young animals. Much higher concentrations of dietary Se are toxic to animals, and a dietary concentration of less than 4 to 5 ppm is often suggested as the maximum amount that will avoid Se toxicity. Both Se deficiency and Se toxicity are problems of practical importance in livestock production in the United States and in other countries. There are a number of recent reviews describing nutritional effects and environmental cycling of Se (Underwood, 1971; National Academy of Sciences, 1971; Allaway, 1973).

Both selenium toxicity and deficiency pose problems to livestock producers.

The potential value and potential hazard of Se in organic residues are influenced by the stability and biological activity of different forms of Se. The Se in manures and sewage sludge is very likely to be present as elemental Se, selenites strongly bound to hydrous iron oxides, heavy metal selenides, or trimethylselenonium salts. All of these compounds are of very low availability to plants or animals, and when incorporated into soils they are rarely taken up in potentially toxic concentrations by plants. Special care must be taken to avoid concentrations of Se which are phytotoxic for such conditions would be a serious threat to the food chain. There appears to be a minimal hazard to food chains from Se in organic residues, and minimal opportunity to improve the nutritional quality of very low Se crops through organic residue utilization practices. Selenium is, however, most readily available in alkaline soils.

The selenium compounds found in sludges are of low availability to plants and animals; selenium is most readily available in alkaline soils.

Zinc. Zinc is the most abundant, and most valuable in terms of nutritional quality to crops, of the trace elements in organic residues. Zinc deficiency in crops has been noted with increasing frequency in recent years, and it has been suggested that some of this increase is due to feed crop production without use of animal manures (Viets, 1966). Application of animal manure to the cut areas in fields leveled for irrigation in order to correct Zn and iron deficiency is a common practice.

Zinc deficiency in livestock, especially hogs, is a problem of long standing and there is evidence of Zn deficiency in people in many countries, including the United States (National Academy of Sciences, 1974). Zinc deficiency in people and animals is frequently associated with low availability or digestibility of dietary Zn rather than with low total Zn intake. Zinc from plant sources, and especially Zn in seed plants seems to be of lower digestibility than the Zn in foods of animal origin. However, the concentration of digestible Zn in seed crops can be increased through increase in the supply of available Zn to the plant (Welch, *et al.*, 1974).

Zinc is a nutritionally valuable element which is deficient in many animals and humans.

Zinc is more toxic to plants than to animals or humans. One of the recent examples of phytotoxic effects of Zn contained in liquid sewage sludge has been described by King and Morris (1972). In this instance, the yield of rye declined whenever rye foliage contained concentrations of Zn of about 500 ppm or more. Levels of dietary Zn of this order and higher have been fed to animals with very little evidence of ill effects (Underwood, 1971). Therefore, it appears safe, as pointed out by Chaney (1973), to assume that phytotoxicity of Zn provides a safeguard against extensive production of food or feed crops that may contain Zn concentrations that would be hazardous to consumers of these crops.

Phytotoxicity provides a reliable barrier to toxic Zn levels in food and feed crops.

The factor of phytotoxicity is particularly significant to the control of another metal, Cd, in the food chain. By maintaining a Cd/Zn ratio of 1 : 100 in sludges or irrigation waters, heavy metal accumulations of Cd and Zn in soil will lead to phytotoxic conditions due to Zn before Cd concentration can become detrimental to the food chain. Cd is also toxic to plants, but Cd can reach unacceptable levels in the food chain before its phytotoxic levels are reached.

Heavy applications of organic residues that are high in P and relatively low in Zn may lead to Zn deficiency in crops from phosphate-induced Zn deficiency (Viets, 1966). According to studies by Giordano and Mays (1976) and Touchton, *et al.* (1976), the availability of Zn is reduced more by liming than is that of Cd, Cu, or Ni.

Maintaining a Cd:Zn ratio of 1:100 is a widely used management practice designed to prevent the production of food crops with toxic Cd concentrations.

Other elements posing potential hazards. In addition to the elements discussed above, there are many other elements that can be present in municipal sewage sludge or effluents. Some of these elements undoubtedly pose potential hazards to food chains, but the extent of and possible procedures for minimizing these hazards have not been worked out. The potentially hazardous elements include strontium, lithium, beryllium, antimony, vanadium, bromine, iodine, tin, and probably others. Some of these elements, especially iodine, may have beneficial effects on the nutritional quality of crops unless application rates of residue and iodine uptake by plants reach excessive levels.

It must be recognized that most of the elements could be present in sewage from industrialized cities. Therefore, an effective monitoring system that measures concentration of a very large number of elements in sewage sludge is essential to the use of these sludges in food crop production.

Many other elements that can be found in sewage from industrial cities may pose hazards to the food chain; monitoring of a large number of elements in sludge is recommended.

Potentially Serious Hazard

Boron. Problems with B are likely to be confined to soluble B compounds in sewage effluents. Composted municipal rubbish is another source of B. Boron is not required by animals and is of low toxicity to them. The first impact of application of high B effluents to agricultural soils is reduced crop growth due to B toxicity. Boron injury to plants caused by irrigation with high B sewage effluents was noted in the U.S. as early as 1928. Techniques for evaluating the B status of soils and plants are described by Bradford (1966). Alkaline soils used for production of B-tolerant crops represent the best conditions for minimizing the hazard of phytotoxicity from high B effluents.

The first effect of high levels of boron in effluents is to reduce crop growth due to boron toxicity. Alkaline soils and B-tolerant crops minimize the hazard of boron phytotoxicity.

Boron in excess can injure plants. Boron in effluents could become phytotoxic in soils where the concern is not about the B content of normal irrigation water.

Use of high B effluents as sources of B to meet plant requirements on low B soils may be hazardous due to the narrow margin between B requirements and B toxicity for plant species. Only effluents containing very constant levels of soluble B could be used in this way.

Cadmium. Concern over Cd in foods stems primarily from the experiments and writings of Schroeder and his associates (1964), plus the itai-itai disease occurrence in the Jintsu Valley of Japan (Kobayashi, 1971). Schroeder's work carried the implication that dietary Cd might accumulate in the kidneys and lead to hypertension. In itai-itai disease, a very painful disintegration of bone was evident.

Recent reviews of the Cd problem (Fleischer, *et al.*, 1974; National Academy of Sciences, 1974) show that adverse effects of dietary Cd upon human health are not clearly established. Although workers in Cd industries are subject to respiratory diseases, excessive incidence of hypertension among these workers has not always been found. Itai-itai disease appeared to have affected only that part of the population most likely to be deficient in calcium or Vitamin D.

Even though detrimental effects upon people or animals of Cd contained in organic materials have not been established, there is still a high level of concern over this problem, and more research effort has probably been directed toward Cd than any other trace element in municipal sewage sludge. Even so, differing opinions on the potential hazard of Cd in sewage sludges continue to appear. The following points are relevant to analysis of this problem:

Cd is of great concern as a potential health hazard in humans and animals, but opinion differs on the potential hazard of cadmium in sludges.

1. Although many municipal sludges are low in Cd, some from industrialized cities contain more than 500 ppm Cd (dry basis) (Page, 1974). Sludges containing more than 500 ppm Cd are probably quite rare in that levels in this range may be associated with sterilization of digesters

(Regan and Peters, 1970). Chaney (1973) reports an observed range of 5-2000 ppm Cd in sewage sludge.

2. Current background levels of Cd are probably about 0.05 to 0.40 ppm in human diets and the average daily human intake of Cd is probably 40 to 150 μg/day. Cadmium intake of people affected by itai-itai disease may have been 600 to 1000 μg/day. Vegetables containing over 40 ppm Cd have been collected in places where pollution by airborne Cd has occurred. (National Academy of Sciences, 1974).

3. Levels of Cd found in plants grown on field soils treated with sewage sludge, including some sludges containing over 200 ppm Cd, have been substantially lower than those in plants exposed to airborne Cd, or in plants grown in soils or culture solutions containing added organic Cd (Chaney, 1973; John, *et al.*, 1972; Jones, *et al.*, 1973; Page, *et al.*, 1972). Similarly, Cd contained in foods of plant origin is very likely less digestible, and therefore less hazardous, than equal amounts of Cd in soluble inorganic forms contained in water or inhaled as dusts.

4. The ratio of Zn to Cd in plant material grown in unpolluted areas, and in human diets is usually in the range of 100 to 1. Under some circumstances Zn competes with Cd in uptake and translocation processes in plants. Dietary Zn may also decrease the toxicity of Cd in animals. Similar protection against Cd toxicity may be afforded by dietary selenite (National Academy of Sciences, 1974).

Zinc, and possibly selenite, can compete with Cd in plant uptake. Cd toxicity may also be lessened in animals by this competition.

5. Plant species vary widely in their tendency to accumulate Cd added as soluble inorganic Cd compounds to soils or culture solutions. The concentration of Cd in edible seeds or fruits is generally lower than in leafy vegetables. The concentration of Cd in the kernels of corn has been especially low, even when the corn is grown in soils treated with heavy rates of high Cd sludge (Chaney, 1973).

Cd tends to concentrate much less in fruit and seeds than in leafy vegetable tissue.

The most important feature of any program for minimizing the hazard of Cd from sewage sludge is probably the need for continuous monitoring of Cd concentrations in sludges and in plants growing on sludge-treated areas. Although safe upper limits of Cd in food and feed crops have not been established, uncontrolled increases in the concentrations of Cd in these crops are undesirable.

It is most important to monitor cadmium concentrations in sludges and in plants growing on sludge-treated areas.

Monitoring of Cd/Zn ratios in wastes and adjustment of these ratios to keep the level of Cd at less than 1.0% of the Zn present, selection of finer textured neutral or alkaline soils as disposal sites, regulation of annual and total application rates, and cropping the disposal site to corn used for grain are measures that help to minimize the hazard to food chains from Cd contained in sludges. In the absence of well defined values for the permissible limit of Cd concentrations in human and animal diets, it is uncertain whether or not phytotoxicity of Cd will act as a safeguard against detrimental effects of Cd in food and feed crops.

Keeping the Cd : Zn ratio at less than 1.0%, use of fine textured alkaline soils, regulation of application rates, and growing corn for grain are management suggestions where Cd may pose a problem.

Copper. Cu is adsorbed on negatively charged surfaces in soils like the other heavy metals. In two studies that have been made of its adsorption on clay minerals and quartz, the adsorption from dilute solution has been shown to fit the Freudlich formula, $y = ac^{1/n}$, where y is the Cu adsorbed per unit adsorbent and c is the concentration of Cu in solution, a and n being constants. The adsorption is greatly increased on raising the pH, so that the Cu is then more firmly fixed and thereby less toxic to plants. While added Cu is so strongly retained on acid soils that it barely moves toward the subsurface, it is still reactive enough to be toxic to plants.

Most of the present concern over Cu in organic residues stems from proposals to use relatively high (>200 ppm) levels of Cu in rations for hogs to promote growth and for antibiotic effects. This use of Cu in pig rations is reviewed by Underwood (1971). The disposal of manure slurries from Cu-supplemented pigs has been studied by Batey, *et al.* (1972). They concluded that there was little danger from Cu toxicity in animals from use of manure from Cu-supplemented pigs unless the slurry was topdressed on a growing sward of grass. Incorporation of manure from Cu-supplemented pigs into the soil did not consistently increase the Cu concentration in forage grasses.

The Cu concentration in municipal sewage sludges is reported to range between 250 and 17,000 ppm (Chaney, 1973). Median or average values as reported by Page (1974) are frequently about 1,000 ppm. Thus the Cu concentration in municipal sludges approximate maximum values expected in manure from Cu-supplemented pigs. Copper concentrations in crops produced on land treated with municipal sludges have not been elevated to levels indicative of a hazard from Cu toxicity to consumers of the plants (Peterson, *et al.* 1971).

Crops grown on sludge application sites have not shown copper levels that would pose a hazard to consumers of plants.

Monogastric animals including man are not highly susceptible to Cu toxicity and the primary ill effect of diets containing 200 ppm Cu or more is ordinarily a correctable interference with absorption of zinc or iron (Underwood, 1971). Phytotoxicity of Cu will ordinarily inhibit plant growth before Cu concentrations in this range would be found in plants.

Ruminants, especially sheep, are more susceptible to Cu toxicity and may be adversely affected by dietary Cu concentrations found in plants growing on Cu-polluted areas. This Cu toxicity can be controlled by careful use of dietary supplements of molybdenum.

Monogastric animals including man are not highly susceptible to copper toxicity; ruminants are.

Applications of Cu in organic wastes may help to prevent Cu deficiency in plants, and possibly in animals, but this is a crude and inefficient method of correcting these deficiencies.

Molybdenum. The potential hazard from Mo in organic residues is very closely related to the nature of the soils available for treatment with organic residues, to the crops produced, and to the use of these crops. Molybdenum is required by plants, but the requirement is low in terms of normal nutrient application rates. Where levels of available Mo in soils are very high, plants may accumulate concentrations of Mo that are detrimental to ruminants that eat these plants. The effect of excessive Mo on animals is primarily a disturbance of Cu metabolism, and most instances of Mo toxicity can be corrected by supplementation of the animal's diet with Cu, or by use of injectable forms of Cu. Ruminants are much more susceptible to Mo toxicity than are monogastric animals. Molybdenum-Cu relationships in animal nutrition are discussed by Underwood (1971).

Molybdenum toxicity has been a serious practical problem to livestock producers in the intermountain area of the Western United States. All of the areas producing forages containing toxic concentrations of Mo are occupied by wet, alkaline or neutral soils containing relatively high levels of organic matter. Nearby well drained soils, even though formed from high Mo parent materials do not produce high Mo forages.

Molybdenum toxicity affects ruminants and is related to Cu toxicity. Wet, highly organic, neutral to alkaline soils produce the highest forage Mo content.

On the basis of the naturally occurring instances of Mo toxicity, it seems safe to assume that Mo in organic residues presents no hazard to the food chain provided that these residues are used on well drained soils with pH values of 6 or less. Where it is necessary to use high Mo residues on poorly drained alkaline soils, the use of these areas as pastures for ruminants should be discouraged, or else careful attention should be directed to Cu supplementation of the animals.

Molybdenum in sewage sludges poses no hazard to the food chain when applied to well-drained soils with a pH of 6 or less.

Nickel. Ni behaves generally like Zn, though it forms stronger chelate links with organic groups and so has some analogies to Cu. But its main characteristic in land treatment is to be held on the negative surfaces, the more strongly as pH rises. As with Zn, when an added Ni salt is equilibrated with a soil for a month, the Ni can be recovered only in small amounts with the conventional ammonium acetate, but this means little. Soon after adding a Ni salt, 50% of the Ni may be recovered by ammonium acetate extraction. If the soil was brought from pH 5.2 to 7.0 by adding lime, only 20% would be recovered, which is in keeping with the lowered toxicity after heavy liming.

The role of Ni in animal nutrition has been reviewed by Underwood (1971). Nickel may soon be established as an essential element for animals, but required dietary levels are likely to be very low. Dietary Ni levels of 500 ppm or more have been fed to experimental animals without acutely toxic effects, but chronic effects of lower levels of Ni ingestion are possible. Vanselow (1966) has summarized data on Ni in plants and indicates that Ni concentrations in plants of 50 to 100 ppm dry basis are usually indicative of Ni toxicity to the plant. Most of the information on Ni toxicity to plants has been developed from studies of plants growing on soil derived from serpentine, and other factors in addition to high Ni concentrations may have been responsible for plant symptoms observed. Phytotoxicity appears to provide a substantial barrier against Ni toxicity to humans and animals. Monitoring of Ni levels in sewage sludge and in crops grown on sludge-treated areas is highly desirable however, since Page (1974) reports Ni concentrations in sludges of 3,000 ppm or more and cites evidence that plants grown on sludge-treated soils are substantially higher in Ni than controls.

Although crop failure or sharply diminished yields precedes the production of crops containing hazardous levels of nickel in the edible portion, it is desirable to monitor sludge-treated crops for nickel concentrations.

SPECIFIC RECOMMENDATIONS FOR SLUDGE APPLICATION

Soil Conditions

Not every soil is suited for sludges high in potentially toxic element concentrations. Three important soil variables are texture, pH, and organic matter.

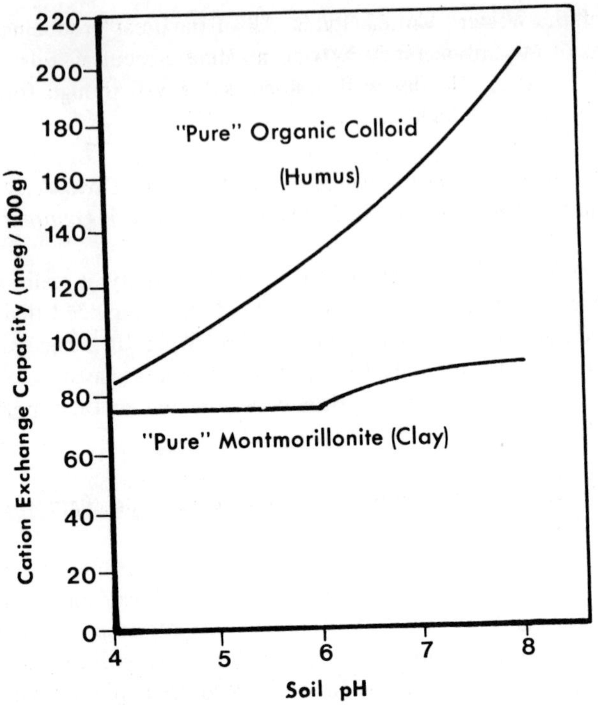

Figure 6. The cation exchange capacity of organic colloids increases with pH more rapidly than that of inorganic colloids (adapted from Brady, 1974).

Usually, the preferred soil will be medium to fine textured and either neutral or alkaline. In all cases, soils used for *food crops* are expected to be maintained at pH 6.5 or above at all times during, and for two years after, application (Chaney, 1974). This recommendation will protect against toxic element mobilization through redissolving and leaching as discussed earlier. If extensive liming is needed to maintain this level of soil pH (especially pH 7 or above), there is the possibility of a Ca/Mg imbalance which must be avoided. Leeper (1972) considers soil pH to be the most important single aspect of the reaction between soils and toxic elements.

The organic matter of soil is, weight for weight, more significant to toxic element retention than inorganic matter, yet our specific knowledge of the role and fate of organic matter in soils is most inadequate (Leeper, 1972). Unamended fields of mineral soils which are considered acceptable for toxic element loadings by the state of Wisconsin application guidelines range from 5 tons organic matter per acre (0.5%) to over 70 tons per acre (7%) (Keeney, *et al.*, 1975). Evidence cited by Leeper indicates that the power of organic matter to retain toxic elements is about the same as for inorganic matter (approximately 2% cation exchange capacity, C.E.C.), but organic matter tends to have a much higher C.E.C. value. This relation is readily observed when comparing "pure" humus and "pure" clay as illustrated in Figure 6 (Brady, 1974). Unfortunately, organic matter decays at rates somewhat dependent on climate, and the protective effect of organic matter then is lost (Chaney, 1973; Mann and Barnes, 1957). A method for calculating organic matter accumulation and decay in soil is given by Leeper (1972). The effect of organic matter accumulation from sludge amendments is expected to be slight. The factors accounting for this are discussed in Module II-3, "Organic Matter."

Medium to fine textured soil with pH > 6.5 and 3% to 7% organic matter are best for application of sludges high in potentially toxic elements.

Table 4. Sludge Loadings Vary Widely, Yet All Have Been Reported as Causing "No Significant Effect" on Vegetation. (Adapted from Kirkham, 1974)

Authority	Recommended Load Rate
National Swedish Board of Health and Welfare (Emmelin, 1973)	0.43 short ton/acre-yr
Hinesly (1972)	27 short tons/acre-yr (5-yr average)
Metropolitan Sanitary District of Greater Chicago (Lyman, *et al.*, 1972)	"similar to Hinesly"
Le Riche (1968) (England)	29 short tons/acre-yr (19-yr average)

Sludge Loading Rates

There are various opinions as to what sludge loading constitutes an acceptable rate. Several of these are summarized in Table 4.

Zinc Equivalent (Z.E.)

Chumbley (1971) gathered a considerable body of experience with potentially toxic elements through his research in England. To arrive at a usable, broad generality, Chumbley neglected any toxic effects due to elements other than zinc (Zn), nickel (Ni), and copper (Cu) (Chaney, 1973; 1974). He then defined "Zinc Equivalent" (Z.E.) as a coefficient expressing a concentration of toxicants, weighted in terms of zinc (assuming copper and nickel to be 2 times and 8 times more toxic than zinc, respectively). Thus:

$$\text{Z.E. } \mu g/g = Zn^{+2}\ \mu g/g + 2(Cu^{+2}\ \mu g/g) + 8(Ni^{+2}\ \mu g/g) \quad (1)$$

The major concept which has been used for toxic element calculations over the past several years is Zinc Equivalent.

For example, if a sludge was found to have the maximum levels of toxic elements (dry weight basis) recommended by Chaney (1974) it would contain:

Zn = 2000 μg/g
Cu = 1000 μg/g
Ni = 200 μg/g
Cd = 1% of Zn
Pb = 1000 μg/g
Hg = 10 μg/g
Cr = 1000 μg/g

The zinc equivalent of that contaminated sludge would be:

$$\text{Z.E. } \mu g/g = 2000\ \mu g/g + 2(1000\ \mu g/g) + 8(200\ \mu g/g)$$

$$\text{Z.E.} = 5600\ \mu g/g$$

Cation Exchange Capacity (C.E.C.)

The C.E.C. of soils is a parameter readily evaluated by soil chemists and used as an index of toxic element retention (Chaney, 1974; Page, 1974; Leeper, 1972; Chumbley, 1971; Allaway,

1977). It must not be thought however, that significant toxic element retention necessarily occurs on the exchange sites measured by the C.E.C. analyses. Rather, C.E.C. values only give the net effect of all negatively charged surfaces, whether crystalline or amorphous (Leeper, 1972). Toxic element concentrations are always too low to compete extensively with normal soil Na^{+1} and Ca^{+2} ions. Nevertheless, soil C.E.C. is the best ordinarily measured character of soils which relates to the ability of a soil to hold excess toxic elements (Chaney, 1974). It has been the practice of various researchers to assume a maximum loading of potentially toxic elements equal to some percentage of the soil C.E.C. As a general rule, Leeper (1972) recommends that the total toxic element load which could be added safely to unamended soils should not exceed 5.0% of the soil C.E.C. at pH greater than 6.5. This is similar to Chaney's (1973) conclusion and the recommendation of the British Ministry of Agriculture, Foods and Fisheries reviewed by Chumbley (1971). Chaney (1974) makes a very strong effort to emphasize that Leeper's generality is decidedly unsafe for general farming needs at pH values less than 6.5.

Soil C.E.C. is the best ordinarily measured characteristic of soils which relates to the ability of soils to hold excess toxic elements.

Before continuing, it would be informative to glance at C.E.C. values associated with various soils (Table 5).

Table 5. C.E.C. Values of Surface Soils of the United States Have a Large Range. (From Brady, 1974, p. 102)

Soil Type	C.E.C. (meq/100 g)
Sand	
Sassafras (N.J.)	2.0
Plainfield (Wisc.)	3.5
Sandy Loam	
Greenville (Ala.)	2.3
Cecil (S.C.)	5.5
Colma (Calif.)	17.1
Loam	
Sassafras (N.J.)	7.5
Dover (N.J.)	14.0
Silt Loam	
Delta (Miss.)	9.4
Dawes (Nebr.)	18.4
Grundy (Ill.)	26.3
Clay and Clay Loam	
Cecil clay loam (Ala.)	4.0
Coyuco sandy clay (Calif.)	20.2
Sweeney clay (Calif.)	57.2

The range of C.E.C. values is significant. Clearly sandy soils have very low values whereas some organic soils may have C.E.C. values extending from 250–400 meq/100 g (Leeper, 1972).

Table 6. Maximum Sludge Metal Addition to Farmland. (Over the Site Lifetime)

Metal	Maximum Metal Addition (kg/ha) to Soil with a C.E.C. (meq/100 g) of:		
	Less than 5	5–15	Greater than 15
Pb	500	1000	2000
Zn	250	500	1000
Cu	125	250	500
Ni	50	100	200
Cd	5	10	20

Source: Dowdy et al. (1976); Jorling (1977).

USDA Guidelines

The U.S. Department of Agriculture has developed guidelines for cumulative metals addition in sludge to cropland, based on soil C.E.C. (Table 6). These guidelines also form part of the recommendations for toxic element additions issued by the EPA (Jorling, 1977). Table 6 relates to privately owned farmland. EPA suggests that higher application rates may be acceptable on publicly owned farmland or land dedicated to waste disposal, where adequate monitoring safeguards are employed.

These recommendations are reported in Dowdy, *et al.* (1976), who suggest three other safeguards to be used concurrently: (1) initial soil pH of 6.5 (determined on samples of 1:1 soil/water ratio) or greater, and maintained at 6.2 or greater when application begins; (2) leafy vegetables not to be grown; and (3) Cd additions not to exceed 2 kg/ha-year (this is very conservative).

Table 6 represents an attempt to limit application, rationally, on the basis of the most important soil property for metal ions, C.E.C. As is noted in CAST (1976), the zinc equivalent theory was based on limited data, although the best available at the time, and new information shows that toxicities of Zn, Ni, and Cu are not necessarily additive. The Z.E. concept is felt to underestimate greatly the amounts of metals which can be applied to soils of near-neutral to higher pH. Z.E. also does not apply to a broad range of plants.

The Z.E. theory is presented here as a method of determining applications which has been used extensively over the past several years and may continue to find use in some quarters. The USDA guidelines are presented as illustrative of the trend at present and as a method which may be used extensively in the near future. Both concepts should be familiar to the designer of sludge application sites.

Present State of Sludge Guidelines

The differences between the Z.E. and USDA recommendations are further illustrated by the following discussion involving Z.E. and two other more recent and widely used methods of quantifying sludge application rates based on metal additions. The entire field of sludge limiting parameters is just developing and is in a state of flux. Other guidelines, such as those being developed at Penn State University, are being proposed. The designer will have to remain closely attuned to the current literature for guidance.

The remaining part of this module presents sample computations of sludge application rates using different guidelines, and examining different possible limiting parameters. Potentially toxic element limits are compared to nitrogen and cadmium limits to determine an overall limiting factor for land application of a given sludge.

The equations which will be used to illustrate these points are those of Chaney, Chumbley (Z.E.), and the State of Wisconsin.

New guidelines for determining sludge application rates are being developed at a rapid rate. The designer is urged to keep abreast of this information.

Other theories have been put forward which have attempted to incorporate soil C.E.C. This has resulted in other sludge application formulas which have been used as follows.

Leeper suggested assigning a rough value of 15 meq/100 g to the C.E.C. of "surface soils."

Chumbley chose to center attention on Leeper's 15 meq/100 g, and calculated the amount of zinc needed to equal 5% of 15 meq/100 g as follows:

Given: A soil of C.E.C. = 15 meq/100 g

Find: Concentration of Zn^{+2} = 5% soil C.E.C.

Method: Atomic weight Zn = 66

$$\text{eq. wt. } Zn^{+2} = \text{at. wt.} \div 2 = 33 \text{ g } Zn^{+2}$$

$$\text{meq. wt. } Zn^{+2} = 10^{-3} \times \text{eq. wt.} = 0.033 \text{ g } Zn^{+2}$$

$$15 \text{ meq. wt. } Zn^{+2} = 15 \times 0.033 \text{ g} = 0.495 \text{ g } Zn^{+2}$$

$$5\% \text{ of } 15 \text{ meq. wt. } Zn^{+2} = 0.05 \times 0.495 \text{ g} = 0.025 \text{ g } Zn^{+2}$$

Therefore:

$$5\% \text{ soil C.E.C. (in } Zn^{+2}) = 0.025 \text{ g } Zn^{+2}/100 \text{ g soil} = \underline{250\ \mu g/g \text{ or } 250 \text{ ppm } Zn^{+2}}$$

The conclusion reached and widely reported is that unamended soils can be loaded up to 250 ppm with zinc with no particular danger.

A value which is more useful comes from determining the number of pounds per acre of zinc represented by, say, 250 μg Zn^{+2}/g soil. The calculation is as follows:

Let: 1 acre - 6″ = 2×10^6 lb and 454 g = 1 lb

$$\text{Then: } \frac{250\ \mu g\ Zn^{+2}}{\text{g soil}} \times \frac{\text{g}}{10^6\ \mu g} \times \frac{2 \times 10^6 \text{ lb}}{\text{acre-6}''} = \frac{500 \text{ lb } Zn^{+2}}{\text{acre-6}'' \text{ soil}}$$

$$\text{And: } \frac{500 \text{ lb } Zn^{+2}}{\text{acre-6}'' \text{ soil}} \times \frac{454 \text{ kg}}{10^3 \text{ lb}} \times \frac{2.47 \text{ acres}}{\text{ha}} = \frac{561 \text{ kg } Zn^{+2}}{\text{ha-6}'' \text{ soil}}$$

Application Rate of Dry Sludge to Unamended Soils

Assuming that a given soil has a C.E.C. of about 15 meq/100 g (so that it would be advisable to avoid loading more than 500 pounds Z.E./acre on unamended soil at pH >6.5), how can that be translated to an application rate of an actual sludge?

Three approaches are presented which are not mutually consistent. Each will be reviewed:

Application rates (total lifetime load or annual loads) are calculated differently by three separate equations in common use.

Chaney's (1974) Approach. Chaney proposed that the annual application rate be calculated from a predetermined total application. His basic equation for determining the total site application for unamended soils of pH >6.5 is:

$$\text{Total sludge (dry tons/acre)} = \frac{16{,}300 \times \text{C.E.C.}}{\text{Zinc Equivalent of sludge} - 300} \qquad (2)$$

(the - 300 adjusts for the addition to the soil of some exchange capacity in the sludge matter). Thus, for a soil of C.E.C. = 15 and sludge Z.E. = 1400,

Total sludge (dry tons/acre) = 222 tons/acre.

For a 30-year span, that is about 7.4 tons per acre per year or about 30 tons/acre for a single application once every four years.

Chaney: 30 tons/acre once every 4 years under given conditions. (C.E.C. = 15; A.E. = 1400)

Chumbley's (1971) Approach. Chumbley chose to utilize a chart, reproduced as Table 7, to approach this problem. He illustrated its use by assuming that a sludge batch had a zinc equivalent of 1,400 mg/kg in the dry matter. From Table 7, that sludge could be applied safely to unamended soil at pH >6.5 over a 30-year lifespan at a rate of one 24.2 tons/acre dry matter application once every four years.

Chumbley: 24.2 tons/acre once every four years under given conditions. (C.E.C. = 15; Z.E. = 1400)

Wisconsin's Approach. Wisconsin has adopted guidelines for sludge application based on an interim guide recommended by the EPA. It is similar to Chaney's approach, but is more liberal. The total sludge loading is calculated from the formula:

$$\text{Total sludge (dry tons/acre)} = \frac{32{,}500 \times \text{C.E.C.}}{(Zn^{+2}\ \mu g/g) + 2(Cu^{+2}\ \mu g/g) + 4(Ni^{+2}\ \mu g/g)} \qquad (3)$$

Table 7. The Tons of Dry Sludge per Acre Safely Applied in a Single Application to Unamended Soil of pH > 6.5 Depends on the Number of Years Between Applications and the Zinc Equivalence of the Sludge. (Adapted from Chumbley, 1971)

Proposed Number of Years Between Single Applications	Proposed Rate of Application (in tons/acre dry matter)[a] for Sludge of Given Z.E. Assuming a 30-Year Site Lifespan. Z.E. 5.6	11.2	22.4	33.6	44.8	56.0
1	1510	750	370	250	190	150
2	3020	1510	750	500	380	300
3	Tons 4530	2260	1110	750	570	450
4	6040	3020	1510	1010	750	600
5	7550	3770	1880	1260	940	750

[a]It is assumed that Chumbley's values were in the customary English weight unit, the long ton, so the values of this table have been converted to the U.S. short ton (long ton × 1.12 = short ton).

If our sludge contains 500 μg/g Zn^{+2}, 250 μg/g Cu^{+2} and 50 μg/g Ni^{+2} (ZE = 1400), then total sludge by the Wisconsin formula becomes:

$$\frac{32{,}500 \times 15}{500 + 2(250) + 4(50)} = 406 \text{ tons/acre}$$

406 tons/acre translates to 13.5 tons per acre per year, or 54 tons/acre once every 4 years.

Wisconsin: 54 tons/acre every four years under given conditions. (CEC = 15; Z.E. = 1400).

Equation (3) contains several conversion factors and is based on the following premises: (a) CEC is related to soil factors controlling availability of metals in soil; (b) Cu is twice and Ni is 4 times as toxic to plants as Zn. Metal additions are limited to 10% of soil CEC, rather than 5% as in Chaney. The equation is presented as empirical and subject to revision (Keeney, *et al.* 1975).

Equation (3) may be stated in terms of lb./acre as follows:

(a) Total metal equivalent loading (lb/acre) = 65 × CEC

(b) $\text{Sludge metal equivalent/ton} = \dfrac{(Zn^{+2}\ \mu g/g) + 2(Cu^{+2}\ \mu g/g) + 4}{500}$

and $\dfrac{(a)}{(b)}$ = Total sludge loading (dry tons/acre)

The Wisconsin guidelines also provide a table relating total metal equivalent loadings to soil texture and organic matter content, for use in preliminary planning or on small sites where complete soil characterization cannot be economically justified. However, analytical values of soil CEC should be used whenever possible. Cadmium is specifically limited to a maximum of 2 lb/acre/year and to a total site lifetime maximum of 20 lb/acre. Annual and site lifetime sludge loading rates must be computed on the basis of Cd and compared to the metal equivalent rates to fulfill the Wisconsin guidelines. Cadmium guidelines are discussed further in the last section of this module.

A comparison of these three guidelines (Chumbley, Chaney, and Wisconsin) in sludge application rates to previously amended soils (which have already received sludge application) shows that the approaches differ much more widely in permissible loadings to previously amended than to unamended soils.

Nitrogen Versus Potentially Toxic Element Loading

The application of sludge to land often is recommended. Some see it as a simple matter of recycling natural products. Others may cite benefits in increased soil aggregation (Epstein, 1973; Lunt, 1953) and soil condition (Olds, 1960) while many see sludge as a cheap fertilizer. Chaney (1974) notes that if the toxic element concentration of sludge does not exceed his recommended values for "domestic" sewage, and if soil pH is maintained at pH >6.5, trace element fertilizer values will be realized for food crops. However, more commonly, much is made of the nitrogen fertilizer value of sludges (Kirkham, 1974).

Using sludges for fertilizers raises questions about the limiting factors inherent in N, potentially toxic elements, and Cd loads of sludges.

The complex role of nitrogen of wastewaters and its interaction with soils is reviewed in Module II-1, "Nitrogen Consideration." Nitrogen loadings are discussed and a rough estimate is made that agricultural soils may require 300 to 400 pounds of nitrogen per acre-year for fertilization. The question suggested by this is whether toxic element loads or nitrogen loads act as limiting factors in application schedules. Similarly, it would be expected that there should be some maximum lifespan for sites and certain minimum values for soil C.E.C. that could be associated with a given sludge.

The equations used here for these determinations are equations (4) and (5), based on Chaney's recommendations and the Wisconsin guidelines, respectively. Other guidelines could also be used.

$$\frac{n\ \text{lb}}{\text{acre-yr}} = \frac{16{,}300 \times \text{C.E.C.}}{\text{Z.E.} - 300}\ \frac{\text{Tons}}{\text{acre}} \times \frac{1}{\text{L yr}}\ \frac{2000\ \text{lb}}{\text{Ton}} \times \frac{\phi}{10^6} \tag{4}$$

$$\frac{n\ \text{lb}}{\text{acre-yr}} = \frac{32{,}500 \times \text{C.E.C.}}{\text{Z.E.}_{\cdot(w)}}\ \frac{\text{Tons}}{\text{acre}} \times \frac{1}{\text{L yr}} \times \frac{2000\ \text{lb}}{\text{Ton}} \times \frac{\phi}{10^6} \tag{5}$$

where, in each equation:

n = lbs N/acre-yr desired for optimum nitrogen fertilization
Z.E. = Zn^{+2} µg/g + 2(Cu^{+2} µg/g) + 8(Ni^{+2} µg/g)
$\text{Z.E.}_{\cdot(w)}$ = Zn^{+2} µg/g + 2(Cu^{+2} µg/g) + 4(Ni^{+2} µg/g) (Wisconsin)
L = site lifespan in years
ϕ = maximum concentration of N (in mg/kg) in sludge which will give the desired N/acre-yr loading
C.E.C. = cation exchange (in meq/100 g)

To illustrate how these equations may be used to answer the questions stated earlier, use is made of the data from two anaerobically digested sludges of the Metropolitan District of Chicago as reported by Hinesly (1972). Their assay values are reproduced in part in Table 8.

Table 8. Concentrations of Selected Chemicals in Chicago Sludges are Given as Reported by Two Treatment Plants. (Adapted from Hinesly, 1972)

	Stickney		Calumet	
	Given (ppm)	Calc.[a] (mg/kg)	Given (ppm)	Calc.[a] (mg/kg)
Cd	14	321	3.0	146
Zn	223	5,115	83.0	4,049
Cu	67	1,537	16.0	780
Ni	15	344	3.0	146
N	2,860[b]	65,596	2,150[c]	104,878
Z.E.		10,941		6,777
$\text{Z.E.}_{\cdot(w)}$[d]		9,565		6,193
Total Solids	43,600		20,500	

[a]Calc. mg/kg = ppm element × 10^6/ppm solids.
[b]N = "Total-N" + "NH_4 – N". These are 1560 ppm and (assumed) 1300 ppm respectively.
[c]N = Total-N" + "NH_4 – N". These are 1500 ppm and 650 ppm, respectively.
[d]$\text{Z.E.}_{\cdot(w)}$ = Zinc equivalent modified according to the Wisconsin guidelines.

Method for determining limiting factors: Solve equations 4 and 5 for ϕ and rearrange to get:

$$\phi_1 = \text{Z.E.}\ (m) - k \tag{6}$$

$$\phi_2 = \text{Z.E.}_{(w)}\ (\gamma) \tag{7}$$

where

ϕ_1 = maximum concentration of N (in mg/kg) in sludge needed to achieve desired fertilizer value, based on Chaney's recommendation

ϕ_2 = same as ϕ_1 except based on Wisconsin's recommendations

Z.E. and Z.E.$_{(w)}$ assume values previously defined

$$m = \frac{\text{nL}}{32.6 \times \text{C.E.C.}}$$

$$\gamma = \frac{\text{nL}}{65 \times \text{C.E.C.}}$$ where n, L, and C.E.C. have values as defined earlier

$k = 300\ (m)$

On substitution of values, if Total N (σ) is considered to be Kjeldahl N of sludge, then if:

1. ϕ_1 or $\phi_2 >$ total N, toxic elements are limiting,
2. ϕ_1 or $\phi_2 =$ total N, toxic elements and nitrogen are simultaneously limiting.
3. ϕ_1 or $\phi_2 <$ total N, nitrogen is limiting.

Example:

Assume C.E.C. = 15 meq/100 g
L = 30 years
Z.E. = 10,941 μg/g (Stickney sludge, Table 8)
Z.E.$_{(w)}$ = 9565 μg/g
n = 400 lbs N/acre-yr
σ = Total N = 65,596 mg/kg

$$\phi_1 = 10941\ \frac{\mu g}{g} \times \frac{400\ \text{lbs}}{\text{acre-yr}} \times 30\ \text{yr} \times \frac{1}{32.6 \times 15} = 268{,}491\ \text{mg/kg}$$

$\phi_1 >$ Total N, therefore, for given conditions, toxic element is limiting based on Chaney's recommendations.

$$\phi_2 = 9565\ \frac{\mu g}{g} \times \frac{400\ \text{lbs}}{\text{acre-yr}} \times 30\ \text{yr} \times \frac{1}{65 \times 15} = 117{,}723\ \text{mg/kg}$$

$\phi_2 >$ Total N, therefore, for given conditions, toxic element is limiting based on the Wisconsin guidelines.

Effective Lifespan

Method for determining the effective lifespan of a site: Substitute σ for ϕ equations 4 and 5 and solve for L (lifespan).

$$L_1 = \frac{32.6 \times \text{C.E.C.} \times \sigma}{n(\text{Z.E.} - 300)}$$

$$L_2 = \frac{65 \times \text{C.E.C.} \times \sigma}{n(\text{Z.E.}_{w})} \tag{8}$$

where:

L_1 = lifetime based on Chaney's recommendation
L_2 = lifetime based on Wisconsin guidelines
σ = total N of sludge, dry wt (Kjeldahl - N + NH_4 - N)
C.E.C., n, Z.E. and Z.E.$_{(w)}$ values remain as defined earlier.

Example:

σ = 65,596 mg/kg (Stickney sludge, Table 8)
C.E.C. = 15 meq/100 g
n = 400 lb N/acre-year
Z.E. = 10,941 μg/g
Z.E.$_{(w)}$ = 9,565 μg/g

$$L_1 = \frac{32.6 \times 15 \times 65{,}596}{400\,(10{,}941 - 300)} = 7.54 \text{ yr}$$

$$L_2 = \frac{65 \times 15 \times 65{,}596}{400\,(9{,}565)} = 16.72 \text{ yr}$$

C.E.C. Value Computed

Method for determining the C.E.C. value needed in a soil: Solve equations 4 and 5 for C.E.C. and substitute σ for ϕ as described earlier:

$$\text{C.E.C.}_1 = \frac{nL(\text{Z.E.} - 300)}{32.6 \times \sigma}$$

$$\text{C.E.C.}_2 = \frac{nL(\text{Z.E.}_{(w)})}{65.5 \times \sigma}$$

where

C.E.C.$_1$ and C.E.C.$_2$ are values based on the recommendations of Chaney and Wisconsin, respectively.
n, L, Z.E., Z.E.$_{(w)}$ and σ remain as defined earlier.

Example:

σ = 65,596 mg/kg
n = 400 lbs N/acre-yr
Z.E. = 10,941
Z.E.$_{(w)}$ = 9,565
L = 20 years

$$\text{C.E.C.}_1 = \frac{400 \times 20 \times (10{,}941 - 300)}{32.6 \times 65{,}596} = 39.8 \text{ meq/100 g}$$

$$\text{C.E.C.}_2 = \frac{400 \times 20 \times 9{,}565}{65 \times 65{,}546} = 17.9 \text{ meq/100 g}$$

Confusion over the use of "ppm" may cause serious problems. This is particularly true when looking at data expressing nitrogen concentrations. Care must be taken to be assured that:

1. "Nitrogen values are expressed in dry weight values (mg/kg) before substituting into the above equations.

2. "Total N" is the sum of organic N and NH_4^+-N (also Kjeldahl N alone is called "total N").

Cadmium Loading Rate

Chaney (1974) recommends that no sludges be added to soils in which the cadmium concentration exceeds 1.0% of the zinc. The Wisconsin guidelines differ in this respect. There, cadmium additions must not exceed 2 lb/acre/yr and the total site lifetime maximum of 20 lb/acre (Keeney, *et al.*, 1975). Therefore, the Chaney guidelines may exclude the use of a certain sludge for land application if its composition does not fit the Cd:Zn restriction. Wisconsin guidelines provide for the use of this sludge, but restrict its application rate in terms of Cd, as illustrated in the following example.

> *Cadmium is a serious limiting parameter because of the possibility of harmful health effects. Guidelines differ on the method of calculating acceptable Cd loadings.*

To illustrate these guidelines, reconsider Hinesly's (1972) data (Table 8). Based on Chaney's recommendations, the Stickney sludge contains cadmium at a level of 6.3% zinc while the Calumet sludge contains cadmium at a level of 3.6% zinc. Clearly, neither sludge is suitable for land application according to Chaney.

The Wisconsin guidelines would not exclude the use of these sludges on soils. However, the application rate would have to be calculated in terms of cadmium limitations. The method recommended is as follows:

Annual load limit due to Cd = 2 × 500/Cd mg/kg = 3.1 tons (Stickney) and 6.8 tons (Calumet) sludge per acre per year. The total *lifespan* cadmium load, however, must not exceed 20 lb/acre according to the Wisconsin guidelines. That amounts to 31 tons (Stickney) and 68 tons (Calumet) sludge per acre. To determine whether Cd is limiting, it is necessary to evaluate both *annual* and *lifespan* parameters.

This amounts to a simple comparison between these permissible cadmium loadings and loadings set by zinc equivalent values or nitrogen. Previous calculations in this module have shown that Chaney's recommendations would allow a lifetime loading (30 years, C.E.C. = 15 meq/yr, pH > 6.5, unamended soil) of 23 and 38 tons/acre for Stickney and Calumet sludges respectively. These become 0.77 and 1.26 tons/acre-yr. Clearly toxic element loading is limiting with respect to cadmium on an annual basis.

In these cases, it is apparent that toxic element also is limiting with respect to lifetime loading and with respect to nitrogen ($\phi > \sigma$ in each case).

BIBLIOGRAPHY

Allaway, W. H. 1968. Agronomic controls over environmental cycling of trace elements. *Adv. Agron.*, **20**:235–274.

Allaway, W. H. 1973. Selenium in the food chain. *Cornell Vet.*, **63**:151–170.

Allaway, W. H. 1977. Food chain aspects of the use of organic residues. pp. 283–298. *In* Soils for management of organic wastes and wastewaters. SSSA, ASA, CSSA, Madison, Wisc. 650 p.

Alloway, B. J., B. E. Davies. 1971. Trace element content of soils affected by base metal mining in Wales. *Geoderma.*, 5(3):197–208.

Baker, D. E. and S. B. Hornick. 1974. Unpublished data as reported in D. E. Baker and L. C. Chesin. 1975. Chemical Monitoring of Soils for environmental quality and animal and human health. *Adv. Agron.*, Vol. 27. Academic Press, New York. pp. 305–375.

Baker, D. E. and L. C. Chesin. 1975. Chemical monitoring of soils for environmental quality and animal and human health. *Adv. Agron.*, Academic Press, New York. 380 pages.

Batey, T., C. Berryman, and C. Line. 1972. The disposal of copper-enriched pig-manure slurry on grassland. *J. Br. Grassl. Soc.*, **27**:139–143.

Baumhardt, G. R. and L. F. Welch. 1972. Lead uptake and corn growth with soil applied lead. *J. Environ. Qual.*, **1**:92–94.

Billings, C. E. and W. R. Matson. 1972. Mercury emissions from coal combustion. *Science*, **176**:1232–1233.

Boawn, L. C., F. Viets, Jr. and C. L. Crawford. 1957. Plant utilization of zinc from various types of zinc compounds and fertilizer materials. *Soil Sci.*, **83**:219.

Boawn, L. C. and Rasmussen. 1971. Crop response to excessive zinc fertilization of alkaline soil. *Agron. J.*, **63**:874–876.

Bradford, G. R. 1966. Boron. p. 33–61. *In* H. D. Chapman (ed.) Diagnostic criteria for plants and soils. Div. of Agric. Sci., Univ. of California, Riverside.

Brady, N. C. 1974. The nature and properties of soils (8th Ed.). Macmillan Publishing Co., Inc., New York, 639 pages.

Brown, R. E. 1975. Significance of trace metals and nitrates in sludge soils. *J. Water Pollut. Contr. Fed.*, **47**(12):2863–2875.

CAST-Council for Agricultural Science and Technology. 1976. Application of sewage sludge to cropland. Office of Water Program Operations. EPA 430/9-76-D13, Washington, D.C. 63 p.

Chaney, R. L. 1973. Crop and food chain effects of toxic elements in sludges and effluents. *In* Proceedings of the joint conference on recycling municipal sludges and effluents on land. EPA/USDA/Nat'l. Assoc. of State Universities and Land Grant Colleges, Champaign, Ill., July 9–13, pp. 129–141.

Chaney, R. L. 1974. Recommendations for management of potentially toxic elements in agricultural and municipal wastes. *In* Factors involved in land application of agricultural and municipal wastes. (Draft). National Program Staff; Soil, Water, and Air Sciences; Agricultural Research Service, USDA, Beltsville, Md. pp. 97–120.

Cheng, M. H., J. W. Patterson, and R. A. Minear. 1975. Heavy metals uptake by activated sludge. *J. Water Pollut. Contr. Fed.*, **47**(2):362–376.

Chumbley, C. G. 1971. Permissible levels of toxic metals in sewage used on agricultural land. A.D.A.S. Advisory paper No. 10, 12 pp.

Davis, III, J. A. and J. Jacknow. 1975. Heavy metals in wastewater in three urban areas. *J. Water Pollut. Contr. Fed.*, **47**(9):2292–2297.

Dowdy, R. H., R. E. Larson, and E. Epstein. 1976. Sewage sludge and effluent use in agriculture. *In* Land Application of waste materials. Soil Conservation Society of America. Ankeny, Iowa.

Ellis, B. G., and B. D. Knezek. 1973. Adsorption reactions of micronutrients in soils. *In* Mortvedt *et al.* (ed.) Micronutrients in agriculture. Soil Science Society of America, Madison, Wisc.

EPA. 1975. Evaluation of land application systems. Tech. Bull. of U.S. EPA, Office of Water Program Operations. EPA-430/9-75-001.

EPA. 1975b. Manual for evaluating public drinking water supplies. Office of Water and Hazardous Materials. EPA-430/9-75-011. pp. 15, 58.

Epstein, E. 1973. The physical processes in the soil as related to sewage sludge application. *In* Proceedings of the joint conference on recycling municipal sludges and effluents on land. EPA/USDA/Nat'l. Assoc. of State Universities and Land Grant Colleges, Champaign, Ill. July 9–13, pp. 67–74.

Fiskell, J. G. A., P. H. Everett, and S. J. Locascio. 1964. Minor element releases from organo-N fertilizer materials in laboratory and field studies. *J. Ag. Food Chem.*, **12**:363–367.

Fleischer, M., A. F. Sarofim, D. W. Fassett, P. Hammond, H. T. Shacklette, I.C.T. Nisbet, and S. Epstein. 1974. Environmental impact of cadmium: a review by the Panel on Hazardous Substances. Environ. Health Perspect. Exp. Issue No. 7. p. 253–323.

Flick, D. F., H. F. Kraybill and U. M. Dimitroff. 1971. Toxic effects of cadmium: a review. *Environ. Res.*, **4**:71–85.

Fuehring, H. D. and G. S. Scofi. 1964. Nutrition of corn on a calcareous soil: II. Effects of zinc on yields of grain stover in relation to other micronutrients. *Soil Sci. Soc. Amer. Proc.*, **28**:79.

Giordano, P. M. and J. J. Mortvedt. 1969. Response of several corn hybrids to level of water soluble zinc fertilizers. *Soil Sci. Soc. Amer. Proc.*, **33**:145.

Giordano, P. M., and D. A. Mays. 1976. Yield and heavy metal content of several vegetable species grown in soil amended with sewage sludge. *In* Biological implications of metals in the environment. 15th Annual Hanford Life Sci. Symp., Richland, Wash.

Goldwater, L. J. 1971. Mercury in the environment. *Scientific American*, **224**(5):15–21.

Hambridge, K. M., C. Hambridge, M. Jacobs, and J. D. Baume. 1972. Low levels of zinc in hair, anorixia, poor growth, and hypogensia in children. *Pedia. Res.*, **6**:868–874.

Hammond, A. L. 1971. Mercury in the environment: Natural and human factors. *Science*, **171**:788–789.

Helling, C. S., G. Chester and R. B. Corey. 1964. Contribution of organic matter and clay to soil cation exchange capacity as affected by the pH of the saturating solution. *Soil Sci. Amer. Proc.*, **28**:517–520.

Hinesly, T. D. *et al.* 1972. Effects on corn by application of heated anaerobically digested sludge. *Compost Sci.*, **13**:26.

Hinesly, T. D. 1972. Introduction, Chapter I. *In* T. D. Hinesly, O. C. Braids, J. A. E. Molina, R. I. Dick, R. L. Jones, R. C. Meyer, and L. F. Welch. Agricultural benefits and environmental changes resulting from the use of digested sludge on field crops. Draft report prepared for the EPA, Grant No. 001–UI–00080.

Joensuu, O. I. 1971. Fossil fuels as a source of mercury pollution. *Science*, **172**:1027–1028.

John, M. K., C. J. Van Laerhoven, and H. H. Chuah. 1972. Factors affecting plant uptake and phyto-toxicity of cadmium added to soils. *Environ. Sci. Technol.*, **6**:1005–1009.

Jones, R. L., T. D. Hinesly, and E. L. Ziegler. 1973. Cadmium content of soybeans grown on sewage-sludge amended soil. *J. Environ. Qual.*, **2**:351–353.

Jorling, T. C. 1977. Federal register. Vol. 42, No. 211. pp. 57420–57427. *Re*. Technical bulletin–Municipal sludge management: environmental factors. MCD-28 (EPA 430/9–77–004).

Keeney, D. R., K. W. Lee, and L. M. Walsh. 1975. Guidelines for the application of wastewater sludge to agricultural land in Wisconsin. Tech. Bull. No. 88. Dept. of Natural Resources. 36 pages.

Kehoe, R. A. 1966. Under what circumstances is ingestion of lead dangerous? *In* Symposium on Environmental Lead Contamination. Publ. No. 1440:51–58. Public Health Service, U.S. Dept. of H.E.W.

King, L. D., and H. D. Morris. 1972. Land disposal of liquid sewage sludge. II. The effect of soil pH, manganese, zinc and growth and chemical composition of rye (*Secale cereale* L.). *J. Environ. Qual.*, **1**:425–429.

Kirkham, M. B. 1974. Disposal of sludges on land: effects on soils, plants, and groundwater. *Compost Science*, **15**(2):6–10.

Kirkham, M. B. 1975. Trace elements in corn grown on long-term sludge disposal site. *Environ. Sci. and Technol.*, **9**(8):765–768.

Kobayashi, J. 1971. Relation between the "itai-itai" disease and the pollution of river water by cadmium from a mine. Proc. 5th Int. Water Pollut. Res. Conf., July–August 1970. San Francisco, Calif. p. 1–7.

Lagerwerff, J. V. 1974. Current research in heavy metals in soil and water. NSF-RANN, *Trace Contaminant Abst.*, **2**:20.

Leeper, G. W. 1972. Reactions of heavy metals with soils with special regard to their application in sewage wastes. Dept. of the Army, Corps of Engineers under Contract No. DACW73–73–C–0026. 70 p.

Lindsay, W. L. 1972. Inorganic phase equilibrium of micronutrients in soils. In J. J. Mortvedt, P. M. Giordano, and W. L. Lindsay, eds. Micronutrients in agriculture. *Soil Sci. Soc. America.* 41.

Lunt, H. A. 1953. The case for sludge as a soil improver. *Water and Sewage Works*, **100**:295–301.

Mann, H. H. and T. W. Barnes. 1957. The permanence of organic matter added to soil. *J. Agr. Sci.*, **48**:160–163.

Menzies, J. D. and R. L. Chaney. 1974. Waste characteristics. *In* Factors involved in land application of agricultural and municipal wastes. (Draft). National Program Staff; Soil, Water and Air Sciences; Agricultural Research Service; USDA, Beltsville, Md. pp. 18–36.

Miller, R. J. and D. E. Koeppe. 1970. Accumulation and physiological effects of lead in corn. Proc. 4th Ann. Conf. on Trace Substances in Environ. Health, Columbia, Mo. p. 186–193.

Mitchell, R. L., and J. W. Reith. 1966. The lead content of pasture herbage. *J. Sci. Food Agr.*, **17**:437–445.

Murrmann, R. P., and F. R. Koutz. 1972. Role of soil chemical processes in reclamation of wastewater applied to land. *In* S. Reed. Wastewater management by disposal on the land. Corps of Engineers, U.S. Army, Hanover, N. H.

National Academy of Sciences. 1971. Selenium in nutrition. Natl. Acad. of Sci., Washington, D.C.

National Academy of Sciences. 1974. Geochemistry and the environment. I. The relation of selected trace elements to health and disease. Natl. Acad. of Sci., Washington, D.C.

Neufeld, R. D. and E. R. Hermann. 1975. Heavy metal removal by acclimated activated sludge. *J. Water Pollut. Contr. Fed.*, 47(2):310–329.

Olds, J. 1960. How cities distribute sludge as a soil conditioner. *Compost* Science, *Autumn*: **60**–63.

Page, A. L. 1974. Fate and effects of trace elements in sewage sludge when applied to agricultural lands. A literature review study. U.S. EPA–670/2–74–005.

Page, A. L., F. T. Bingham, and C. Nelson. 1972. Cadmium absorption and growth of various plant species as influenced by solution cadmium concentration. *J. Environ. Qual.*, **1**:288–291.

Patterson, J. B. D. 1971. Metal toxicities arising from industry. Trace elements in soils and crops. *Ministry of Agr. Fish Food Tech. Bull.*, **21**:193–207.

Peterson, J. R., T. M. McCalla, and G. E. Smith. 1971. Human and animal wastes as fertilizers. pp. 557-596. *In* R. A. Olson, T. J. Army, J. J. Hanway, and V. J. Kilmer (ed.) Fertilizer technology and use. Soil Sci. Soc. of Am., Madison, Wis.

Pound, C. E. and R. W. Crites. 1973. Wastewater treatment and reuse, Volume I–Summary. Office of Research and Development, United States Environmental Protection Agency. EPA-660/2-73-006a.

Regan, T. M. and M. M. Peters. 1970. Heavy metals in digesters: Failure and cure. *J. Water Pollut. Contr. Fed.*, **42**:1832-1839.

Regan, T. M. and M. M. Peters. 1972. *In* Proc. Annu. Environ. Water Res. Eng. Conf., 10th, Vanderbilt University, Nashville, Tenn. p. 157.

Ryan, J. A., D. R. Keeney, and L. M. Walsh. 1973. Nitrogen transformation and availability of an anaerobically digested sewage sludge in soil. *J. Environ. Quality*, **2**:489-492.

Schnitzer, M. and S. I. M. Skinner. 1966. Organometallic interactions in soils: V. Stability constants of Cu^{2+}-, Fe^{2+}-, and Zn^{2+}-fluoric acid complexes. *Soil Sci.*, **102**:361.

Schnitzer, M. and S. I. M. Skinner. 1967. Organometallic interactions in soils: VII. Stability constants of Pb^{2+}-, Ni^{2+}-, Co^{2+}-, and Mn^{2+}-fluoric acid complexes. *Soil Sci.*, **103**:247.

Schroeder, H. A. 1964. Cadmium hypertension in rats. *Am. J. Physiol.*, **207**:62-66.

Touchton, J. T., L. D. King, H. Bell, and H. D. Morris. 1976. Residual effect of liquid sewage sludge on coastal bermudagrass and soil chemical properties. *J. Environ. Qual.*, **5**:161-164.

Train, R. E. 1975. Alternate waste management techniques for best practicable waste treatment. *Federal Register*, **41**(29):6190-6191.

Underwood, E. J. 1971. Trace elements in human and animal nutrition. 3rd ed. Academic Press, New York.

Vanselow, A. P. 1966. Nickel. pp. 302-309. *In* H. D. Chapman (ed.) Diagnostic criteria for plants and soils. Div. of Agric. Sci., Univ. of Calif., Riverside.

Viets, F. G., Jr. 1966. Zinc deficiency in the soil-plant system. pp. 90-128. *In* A. S. Prasad (ed.) Zinc metabolism. C. C. Thomas Publisher, Springfield, Ill.

Walsh, L. M., M. E. Sumner, and R. B. Corey. 1976. Consideration of soils for accepting plant nutrients and potentially toxic nonessential elements. *In* Land application of waste materials. Soil Conservation Society of America. Ankeny, Iowa.

Welch, R. M., W. A. House, and W. H. Allaway. 1974. Availability of zinc from pea seeds to rats. *J. Nutr.*, **104**:733-740.

Wojtalik, T. A. 1971. The accumulation of mercury and its compounds. *J. Water Pollut. Contr. Fed.*, **43**:1280-1292.

Wood, D. K. and G. Tchobanoglous. 1975. Trace elements in biological waste treatment. *J. Water Pollut. Contr. Fed.*, **47**(7):1933-1945.

Woolson, E. A., J. H. Axley, and P. C. Kearney. 1971. The chemistry and phytotoxicity of arsenic in soils: I. Contaminated field soils. *Soil Sci. Soc. Am. Proc.*, **35**:938-943.

Module II-5
PATHOGENS

SUMMARY

This module is intended to help engineers evaluate the relative health risks from pathogens at land treatment sites versus conventional waste treatment systems. It is recognized that more research is needed on the relative risks of the two systems. It should be stressed that conventional systems, such as trickling filters, may pose similar problems of possible disease spread. The engineer must consider the infective dose of various pathogens and the ultimate use of the wastewater being applied to the land.

This module reviews the following topics:

- the relationship between survival time of pathogens and the chance of disease transmission to humans
- factors that favor survival of pathogens in the soil
- survival of pathogens on vegetation
- factors which affect the movement of pathogens through the soil and their uptake and retention on soil particles
- the possibility of pathogens moving to the groundwater and how long they live in groundwater
- the risks of crop contamination and how to minimize them
- effects of various conventional treatment processes on pathogen removal
- factors that affect efficiency of chlorination on pathogen removal
- ways to minimize disease risks to animals grazing on wastewater-irrigated pasture
- insect control at the land treatment site
- specific pathogens contained in wastewater, including most common bacteria, viruses, protozoa, and worms
- infective doses of various pathogens
- ways to reduce pathogens in aerosols
- factors that influence the survival and dispersion of pathogens in aerosols
- summary of field tests on spread of pathogens via aerosols
- epidemiological study of disease spread from a land treatment site

CONTENTS

GLOSSARY

aerosol–a gaseous suspension of ultramicroscopic particles of a liquid; particles are in the size range of 0.01 to 50 micrometers (μm).

bacteria–any of numerous unicellular microorganisms traditionally classified with fungi as Schizomycetes. Bacteria occur in a wide variety of forms, existing either as free-living organisms or as parasites, and having a wide range of biochemical, often pathogenic, properties.

coliform bacteria–colon bacilli, or forms which resemble or are related to them. The coliform group of bacteria possess the faculty of fermenting lactose and milk sugar to produce gas, and therefore, can be simply diagnosed as being present by a simple visual observation.

cyst–a sac with a distinct wall. Many microorganisms form cysts under adverse environmental conditions. These cysts may contain either the egg, the immature organism, or the mature organism. Some pathogens are transmitted primarily by being released as cysts which are then ingested by the new host.

enteric–of or within the intestine. Enteric bacteria are microorganisms found in the intestinal tract of humans and animals.

epidemic–a sudden increase in the incidence rate of a disease to a value above normal, affecting large numbers of people and spread over a wide area.

epidemiology–the study of the mass aspects of disease.

fecal coliforms–the type of coliform bacteria present in virtually all fecal material produced by mammals. One of the common kinds of fecal coliforms is *Escherichia coli* (*E. coli*). (Non-fecal coliforms habitate soil, grain, and decaying vegetation.) Since the fecal coliforms may not be pathogens, they only indicate the potential presence of human disease organisms.

gastroenteritis–inflammation of the mucous membrane of the stomach and intestine, one form of which is called food poisoning. This disease may be caused by organisms which are contained in human wastes and are transmitted by ingestion of contaminated food or water.

helminth and helminth ova–a worm, especially a parasitic intestinal nematode or trematode worm. There are two phyla of helminths, the flatworms and the roundworms. Among the flatworms are the fluke and the tapeworm. Among the roundworms are the intestinal roundworm and the hookworm. The helminth ova or eggs are of particular interest to sanitary engineers because it is often the ova which must be eliminated from sewage. For instance, helminth ova are quite resistant to chlorination; they must be retained in anaerobic digesters for one month to ensure destruction; they can remain viable in sludge following primary treatment. The ova can be passed out with the feces of infected organisms and then ingested with food or water. One helminth ovum is capable of hatching and growing when ingested.

indicator organism–an organism that when measured in water indicates pollution or the presence of pathogens but which does not necessarily cause disease in itself. For instance, the presence of fecal coliforms in water indicates that human disease organisms may be present, but coliforms themselves are not usually pathogens.

MPN (most probable number)–a statistical approximation of the number of organisms per unit volume of sample. Used to measure density of coliform bacteria, computed from the number of positive observations of gas formation in a nutrient broth, where multiple tests are performed at different sample dilutions.

pathogen–a disease-producing agent, usually refers to living organisms.

protozoa–single-celled, usually microscopic organisms of the phylum or subkingdom Protozoa, which include the most primitive forms of animal life. Few protozoa cause disease in man. The most common protozoan of interest to waste management personnel is *Entamoeba histolytica*, which causes amoebic dysentery. *E. histolytica* cannot exist in its active form outside of its host, but is capable of forming cysts which are excreted and may resist waste treatment processes.

vector—an agent, such as an insect, capable of mechanically or biologically transferring a pathogen from one organism to another. A mosquito which carries organisms that cause malaria is such a vector.

virus—a large group of infectious agents ranging in size from 10 to 250 namometers in diameter, composed of a protein sheath surrounding a nucleic acid core and capable of infecting all animals, plants, and bacteria; having the ability only to replicate inside a living cell. Characterized by total dependence on living cells for reproduction and by lack of independent metabolism.

OBJECTIVES

Upon completion of this module, the reader should be able to:

1. List four major types of organisms commonly found in wastewater which may be pathogenic.
2. Name potential means of pathogen spread shared by land application and conventional biological treatment plants.
3. Discuss the factors affecting survival of the various pathogens in soil.
4. Discuss what is known about pathogen removals in conventional treatment processes. Factors to be considered are:
   a. reliability of removal
   b. percent removal of each pathogen by each treatment process
5. Discuss factors affecting movement of microorganisms through soil.
6. Name two types of underlying rock which are undesirable due to ease of pathogen movement to groundwater. Name the physical property they have in common which facilitates the movement.
7. Define "infective dose" and compare the infective dose for typical organisms in each of the four groups of pathogens.
8. Summarize what is known about pathogen survival in aerosols. Include the percentage of the wastewater aerosolized by spray, survival times, distance of travel, and effect of chlorination vs. atmospheric desiccation on bacterial survival. List several methods of controlling aerosol transport of microorganisms.

INTRODUCTION

Several factors affect the perceived and actual public health problems posed by the land application of wastes. Land treatment must be compared equitably to conventional wastewater treatment systems in facilities planning. This makes it necessary to evaluate the relative health risks associated with land treatment and conventional waste treatment and disposal systems. This is a difficult task due to the wide range of parameters that may influence these risks. Much more research is needed in this area.

**Health risks from land treatment must be compared to those from conventional treatment.*

*This and other italicized summaries are intended to highlight key ideas, provide a basis for later review or to aid in skimming sections that are relatively familiar. They can be ignored in a complete reading of the text.

An important consideration is how the community views the system. Conventional systems, such as trickling filters, may pose similar, if less obvious, problems of possible disease spread. Sufficient data are not available to show whether or not land treatment is a greater health hazard than conventional treatment and discharge systems. The paucity of information available on disease caused by wastewater treatment processes may reflect:

- the absence of a problem,
- lack of intensive surveillance, or
- insensitivity of present epidemiological tools to detect recurrent small-scale incidence of disease.

The general health of the community is important in that if a disease is not present in the population, it will not be present in the population's wastes. Since the level of enteric disease in the United States is relatively low, the wastewater would be expected to contain low levels of pathogens compared with that in many regions of Asia, Africa, and South America. However, the continual occurrence of waterborne disease outbreaks caused by sewage-contaminated water in the United States indicates that sufficient numbers of pathogens are present to be a public health concern. Precautions must be taken even in the healthiest of communities as some persons are capable of transmitting disease, while not themselves exhibiting the disease. Typhoid Mary is perhaps the best known example of this phenomenon.

General health of community is important in determining risk of disease from sewage.

Because there is always a danger of pathogens in sewage, varying degrees of treatment are used to lessen the probability that viable pathogens will be released to the environment in concentrations large enough to cause disease. The major groups of pathogens vary in their resistance to sewage treatment processes.

Waterborne diseases persist in the United States, giving evidence of potential danger of pathogens in sewage.

It must be accepted that some pathogens remain viable after waste treatment. If the wastes are applied to the land, then pathogens may be present on the soil and vegetation. However, the probability of these pathogens causing health problems is quite low. No epidemics related to land application of treated, disinfected sewage have been reported (Dunlop, 1968).

It should be emphasized, too, that no incidents of disease have been documented from any planned and properly operated land treatment system in the United States. Comparative epidemiological studies on human populations associated with conventional as well as land treatment systems are needed to provide sufficient and reliable data on which uniform regulations can be based.

No disease transmission has been documented from any planned, properly operated land treatment system in the U.S.

Thus, evaluation of potential health hazards must rest upon our knowledge of the occurrence of pathogens in wastewater and their fate during land treatment.

If the wastewater is to be used to irrigate a crop, several precautions should be taken. Most states in the United States have their own regulations which must be observed, or guidelines which are not legally binding, but strongly recommended. Below are several which are broadly applicable.

- Raw wastewater should not be used for irrigation with spray application.
- Raw wastewater may only be used to irrigate nonedible or orchard crops (without spray application).
- Total coliform density should not exceed 5000 MPN organisms/100 ml for most irrigation water uses.
- Crops to be eaten raw should not be irrigated with wastewater unless the MPN is below a stated value.
- Livestock feed crops should be dried and stored to reduce possibility of pathogen transmittal.

These are examples of attempts by various state regulatory bodies to interpret and put into effect information such as that presented in this module.

The ultimate use of the wastewater applied to land is also of concern in terms of pathogen content. If the water is to be reclaimed for drinking water, additional precautions such as chlorination may be necessary. At present, this is not looked on favorably by a majority of workers in the field. In emergencies, reuse of wastewater for drinking can be a viable alternative (Bernarde, 1973). In most cases, however, the water will eventually recharge the groundwater. Movement of pathogens to groundwater is generally quite restricted. Many pathogens are filtered from the percolating water at or near the soil surface.

The ultimate use of wastewater will determine how much its pathogenic content must be reduced. Movement of pathogens to the groundwater is generally restricted in land application systems.

Different groups of pathogens need different environmental conditions to remain active. A bacterium may remain alive in a dry soil only 2 days, while a helminth egg, protected by its cyst, can remain viable for months or even years under drought conditions.

There are two concerns which the engineer must face in terms of pathogens applied to land. First, there is the real threat—the actual possibility of disease transmission to the human community involved in applying sewage that is treated to varying degrees to the land. A much more difficult problem is in quantifying the degree of danger that may result, and in satisfying the public that the danger has been minimized.

No treatment process can eliminate 100% of the pathogens. The difficulty lies in quantifying the risk to the public and convincing them that risk has been minimized.

The literature reveals few epidemic outbreaks related to wastewater irrigation. Where these have occurred, they were most often caused by application of raw, untreated sewage to foods eaten raw (Bryan, 1974). Secondly, although many studies have revealed possible or potential hazards, none have claimed a means to eliminate 100% of the pathogens in wastes. Therefore, we must assume that we are working with a possible source of disease.

Much material in this section has been adapted from the EPA *Process Design Manual for Land Treatment of Municipal Wastewater*, Appendix D, "Pathogens" (1977).

PATHOGEN SURVIVAL

The engineer must understand the persistence and movement of pathogens in the soil, in overland runoff, in the groundwater, on crops, and in aerosols in order to control their spread to humans and animals following land application of wastewater. The survival of pathogens in soil contributes to the chance of pathogen transfer to a susceptible host, for those microorganisms

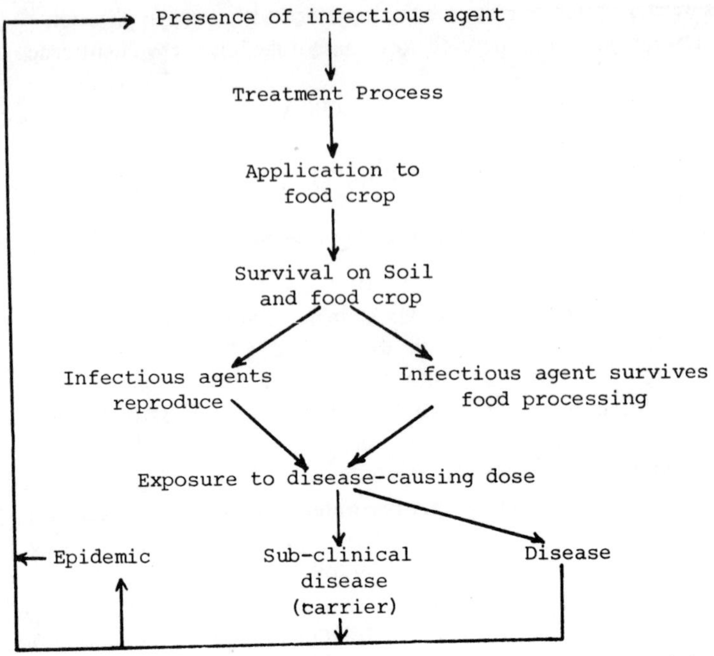

Figure 1. Pathogens must overcome several barriers in order to result in disease transmission from infectious agent to healthy populations via land application of wastes.

not intercepted before reaching the soil. It has generally been found that land treatment using intermittent application and drying periods results in die-off of enteric bacteria retained in soil.

Chances of disease transmission are greater the longer pathogens survive in the soil because they are more likely to be transferred to a susceptible host.

There are several ways pathogens may be spread when applied to land. They may be enclosed in a droplet of water and spread as aerosols. Bacteria and viruses, due to their small size, are particularly susceptible to being spread as aerosols. Helminth ova and protozoan cysts may become imbedded in the soil where they can remain viable if environmental conditions are conducive. Or, pathogens may be spread to vegetation and cause disease upon ingestion. Pathogens do not necessarily stay in one place, but tend to travel with water movement through the system.

Despite the pathways of survival and movement just outlined, pathogens must overcome many barriers to transmit disease following land treatment. This cycle is shown in Figure 1.

Pathogen Survival in Soil Depends on Many Factors

Factors that favor the survival of native soil organisms tend to be destructive to pathogens in the soil. The native organisms, adapted to the soil environment, can eliminate the pathogens by competing with them for food and living space.

Many pathogens are destroyed by the natural environmental conditions which favor native soil organisms.

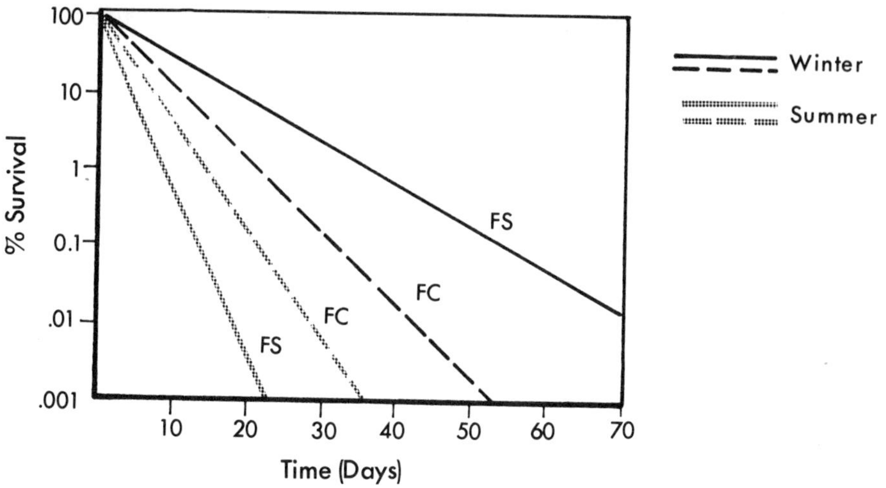

Figure 2. Fecal coliform (FC) and fecal streptococci (FS) show typical dieback curves for bacteria in soils (from Von Donsel, in Miller, 1973).

The chance of sewage-borne pathogens moving through the soil to the groundwater, or off the site in surface runoff, depends partially on how long the pathogens survive in the soil. Initial reactions between pathogens in wastewater and the soil matrix are physical entrapment and chemical adsorption at the soil surface. Bacterial pathogens appear to undergo rapid die-off in the soil matrix. The data presented in Figure 2 are typical of this die-off (Miller, 1973). However, these curves are not applicable to all species of bacteria in all situations. For example, Rudolfs, *et al.* (1950) found fecal streptococci from digested sewage sludge to survive up to six months in clay soil, though not in silt loam or sandy soil.

Some pathogens are physically entrapped and chemically adsorbed at the soil surface, undergoing rapid die-off in the soil matrix.

In general, bacteria may survive in the soil for a period varying from a few hours to several months, depending on the type of organism, type of soil, moisture-retaining capacity of soil, moisture and organic content of wastewater being applied, and predation and antagonism from the resident microbial flora of the soil. The key to eliminating pathogens from soil appears to lie in ensuring adequate detention time in the soil for the pathogens to be destroyed by the above factors. Soils which are very coarse textured are often not very effective in eliminating pathogens, as the pathogens pass too quickly through the soil matrix.

Pathogen destruction is usually enhanced if pathogens are detained long enough in the soil matrix.

Enteric bacteria in reduced numbers usually persist in the soil for 2 to 3 months, although survival times as long as 5 years have been reported. (Rudolfs, *et al.*, 1950) Under certain favorable conditions, applied organisms may actually multiply and increase in numbers. But most pathogenic bacteria cannot reproduce in the soil and will slowly die off. Table 1 lists typical survival times for certain pathogens.

Most pathogenic bacteria cannot reproduce in soil and slowly die off.

Table 1. Survival of Selected Pathogens on Soils (Parsons, *et al.*, 1975).

Organism	Range of Survival Time
Salmonella	15–more than 280 days
Salmonella typhi	1–120 days
Tubercle bacilli	more than 180 days
Entamoeba histolytica cysts	6–8 days
Enteroviruses	8 days
Ascaris ova	Up to 7 years
Hookworm larvae	42 days

The influence of soil type on bacterial survival is important insofar as its moisture content, moisture-retaining capacity, pH, and organic matter content are concerned. The survival of *E. coli*, *S. typhi*, and *M. avium* is greatly enhanced in moist rather than in dry soil. Survival time is less in sandy soil than in soils with greater water-holding capacity, such as moist loam and muck. Survival time is shorter in strongly acid peat soil (pH 2.9 to 3.7) than in limestone soil (pH 5.8 to 7.7). Increasing the pH of peat soil has resulted in extended survival of enterococci. The effect of higher soil pH on bacterial persistence is often ascribed to an increased availability of nutrients or to an effect on inhibitory agents.

Soil moisture, pH, and organic matter content all have an effect on bacterial survival.

Bacteria survive longer as the organic content of the soil increases. *Salmonella* declined rapidly when they were applied to pastures with water containing no fecal matter, as compared to wastewater contaminated with feces (Tannock and Smith, 1971). Under natural conditions, the buildup of organisms may be greater in soils with high moisture and high organic content.

Both pathogens and indicator organisms survive longer in winter than in summer. *S. typhi* survived as long as 2 years at constant freezing temperatures in one study. In fact, the self-cleansing property of soil is slowed down in the Russian Arctic where winters are prolonged (Mirzoev, 1968). Microorganisms disappear more rapidly at the soil surface than below the surface, apparently because of desiccation, effects of sunlight, and other factors at work at the soil surface.

Studies have shown that bacteria survive longer in winter than in summer.

Table 2. Factors that Affect the Survival of Enteric Bacteria in Soil. (Metcalf and Eddy, 1977)

Factor	Remarks
pH	Shorter survival in acid soils (pH 3 to 5) than in neutral and alkaline soils
Antagonism from soil microflora	Increases survival time in sterile soil
Moisture content	Longer survival in moist soils and during periods of high rainfall
Temperature	Longer survival at low (winter) temperatures
Sunlight	Shorter survival at the soil surface
Organic matter	Longer survival (regrowth of some bacteria when sufficient amounts of organic matter are present)

The competition and antagonism the alien enteric bacteria face from the resident soil microflora is another important factor. Organisms applied to sterilized soil survive longer than they would in unsterilized soil. Table 2 lists factors that influence the survival of bacteria in soil.

Pathogens Survive for Shorter Times on Vegetation

Vegetation irrigated with wastewater is likely to have a varying number of pathogens associated with it. This is true even if the wastewater has received pretreatment, since, as has been seen, primary and secondary treatment processes do not eliminate all the pathogens.

In some cases, large numbers of pathogens, particularly protozoan cysts and helminth eggs, may remain viable after treatment. No state which has developed guidelines for land application of wastes allows vegetables or fruits which are to be eaten raw to be spray-irrigated with wastewater. California does allow soil surface irrigation of fruit orchards with primary effluent, however, because fruit does not contact the wastewater. These and other regulations promulgated by the states to control land application of wastes are reviewed in "Legal Aspects," Module I-9.

Wastewater pathogens do not generally survive as long on vegetation as they do in soil because they are more exposed to adverse environmental conditions.

In general, pathogens survive for shorter periods on vegetation than in soil. This is due to the lack of protection given by vegetation from ultraviolet radiation, dessication, and temperature extremes. Coliform bacteria may survive up to 35 days on vegetation, *Salmonella typhi* up to 68, and tubercle bacilli up to 49 days on grass, if they can adhere to the underside of plant parts where they are protected from direct sunlight.

Information Sketchy on Virus Survival

There is little information available on the survival of enteric viruses in the soil, on crops, or in the groundwater. The little available information is limited to only a few types of viruses. There is no information on the survival of viruses in aerosols generated by sprinkler irrigation of wastewater or aeration tanks and trickling filters. Available information on virus survival is summarized below:

- Virus survival in soil depends on the nature of the soil, temperature, pH, moisture, and possibly antagonism from soil microflora.
- Viruses readily adsorb to soil particles. Such viruses bound to solids are as infectious as free viruses.
- Viruses survive for times as short as 7 days and as long as 6 months in soil. Climatic conditions, particularly temperature, have a major effect on survival time.
- Enteric viruses can survive from 2 to more than 188 days in fresh water, temperature being the most important factor with survival greater at lower temperatures.
- Virus survival on crops is shorter than in soil, because viruses are more exposed to deleterious environmental effects.
- Contamination of crops most commonly occurs when wastewater comes into contact with the surface of the crop.
- In rare cases, the translocation of animal viruses from roots of the plants to the aerial parts can occur.
- Sunlight is believed to be a major factor in killing viruses.
- Viruses cannot reproduce at all in the soil, and they slowly die off.

Table 3. Several Factors Promote Virus Removal in Soil (Sproul, 1975).

Type of subsoil	Most desirable–deep
	Adequate–sandy soils; require greater depth for removal
	No removal–fractured layers
Soil depth	At least 5 to 10 ft to ledge or fragipan
Depth to groundwater	At least 5 ft
Application rate	Maximum rate of 0.05 to 0.10 ft/day (.6 to 1.2 in.)
Water quality	As clean as possible
Adsorption characteristics of viruses	No control possible

Information on the survival of enteric viruses in soil, on crops, in groundwater is limited to a few types of virus. Table 3 summarizes current thinking on virus removal from wastewater in various soil conditions.

Movement and Retention of Pathogens in the Soil

Retention time in the soil must be related to survival time to determine the probability of some concentration of viable bacteria and viruses passing through to the groundwater. This can be thought of as a treatment efficiency of microbial removal by the land treatment system.

The longer pathogens are retained in the soil, the more die off and the chance of their contaminating the groundwater is reduced.

Before assessing the movement and retention characteristics of microorganisms in the soil, all the inactivation mechanisms in effect prior to infiltration of wastewater to the soil must be taken into account. This includes die-off in holding ponds and on the surface of the ground and crop due to sunlight and desiccation. Desiccation in aerosol droplets from spray systems is minor because of the very small fraction of the applied wastewater which is aerosolized (< 1.0%).

Once the relative level of microorganisms surviving to infiltrate is determined, the mechanisms acting on them must be evaluated to estimate the treatment efficiency. A discussion of the various forces acting to retain or facilitate the movement of microorganisms through soil follows.

Most bacteria are subject to filtration by the soil at the soil-water interface. Heavy-textured clays and consolidated sands are the most efficient soils for filtration. As a result of mechanical and biological straining, and the accumulation of wastewater solids and bacterial slimes, an organic mat is formed in the top 0.2 in. (0.5 cm) of soil. This mat is capable of removing even finer particles by bridging or sedimentation before they reach and clog the original soil surface. Butler *et al.* (1954) observed the greatest removal of bacteria on the mat that formed on the soil surface, followed by a subsequent buildup of bacteria at lower levels. Filtration by the soil is the primary means of retaining bacteria in the soil, or in some cases in an additional biological mat, but other forces also act on bacteria.

Filtration by the soil at the soil-water interface is the primary way bacteria are removed from wastewater and retained in the soil.

Wastewater bacteria are effectively removed by percolation through a few feet of fine soil by the process of straining at the soil surface and at intergrain contacts, sedimentation, and adsorption by soil particles. Adsorption of bacteria to sand depends on pH and zeta potential of the soil and is reversible. Factors that reduce the repulsive forces between the two surfaces, such as the presence of cations, would be expected to allow closer interaction between them and allow adsorption to proceed.

Other mechanisms that retain bacteria in the top few feet of fine soil are intergrain contacts, sedimentation and adsorption by soil particles.

Adsorption plays an important role in the removal of microorganisms, as well as other wastewater constituents, in soils containing clay because of the very small size of clays, their generally platy shapes, and the occurrence of a large surface area per given volume. The substitution of lower valence metal atoms in their crystal lattices make them ideal adsorption sites for bacteria in soils (Krone, 1968). Filtration effects predominate in sandy soils due to their granular structure.

Soils containing clay remove most microorganisms through adsorption. Soils containing sand remove them through filtration at soil-water interface.

It is apparent from the foregoing discussion that the upper layers of the soil are most efficient for removing microorganisms. Once these organisms are retained, the primary consideration is the length of their survival in the soil matrix, where they are inactivated following exposure to sunlight, oxidation, desiccation, and antagonism from the soil microbial population.

Upper layers of soil are most efficient for removing microorganisms from wastewater.

Other complex and interlocking factors determine the distance of travel. Generalizations are difficult, but movement is related directly to the hydraulic infiltration rate and inversely to the particle size of the soil and to the concentration and cationic composition of the solute. Retention and subsequent survival also depend on the rate of groundwater flow, oxygen tension, temperature, and availability of food. Krone *et al.* (1958) developed mathematical equations to predict movement based on size, shape, density, viscosity of the liquid, soil type, surface characteristics, the process of filtration, and physical and biological characteristics of the pathogens, including their size, shape, surface properties, and die-off rates. One point that must not be overlooked is that any microorganism will travel quickly through any fissured underlying zone such as limestones or basalts. This condition should be avoided if the potential for groundwater contamination is deemed significant by a high application concentration, lack of significant retention time in the overlying soil matrix, or evidence that rapidly moving groundwater passing under the site will be tapped for drinking water nearby.

Movement of microorganisms through soil relates directly to hydraulic infiltration rate and inversely to the size of soil particles and the concentration and composition of cations in the solute.

Microorganisms will travel quickly through fissured zones such as limestone and basalt to the groundwater.

The first major field studies on bacteria removal during wastewater percolation through soil were performed at Whittier and Azusa, California. At Whittier, coliform concentrations were reduced from 110,000/100 to 40,000/100 ml after percolation through 3 ft (0.9 m) of soil in

12 days, and none appeared at greater depths. When treated wastewater effluent containing 120,000 organisms/100 ml was allowed to percolate in Azusa soil, the percolates produced at 2.5 and 7 ft (0.75 and 2.1 m) contained 6000 organisms/100 ml (McGauhey and Krone, 1967). At Lodi, California, coliform levels were observed to decrease below drinking water standards within 7 ft (2.1 m) of the surface when wastewater effluent was applied to sandy loam soil, but in one case, coliforms were detected at a depth of 13 ft (3.9 m) (Gerba, *et al.*, 1975).

Several full-scale, detailed studies have demonstrated the high bacterial removal that can be achieved by land treatment.

In a thorough study at the Santee Project near San Diego, California, impressive removal of indicator bacteria occurred within 200 ft (60 m) of horizontal travel; little additional removal occurred in the next 1300 ft (396 m). The medium consisted of coarse gravel and sand confined in a river bed. At the Flushing Meadows Project near Phoenix, Arizona, wastewater was applied to infiltration basins that consisted of 3 ft of fine loamy sand underlain by a succession of coarse sand and gravel layers to a depth of 250 ft (75 m). With a wastewater infiltration rate of 330 ft/yr (99 m/yr), the total coliforms decreased to a level of 0 to 200 organisms/100 ml at 30 ft (9 m) from the point of application when basins were inundated for 2 weeks followed by a dry period of 3 weeks. When 2- to 3-day inundation periods were used, however, the total coliform levels were reduced to 5/100 ml, a reduction of 99.9% (Gerba, *et al.*, 1975).

Virus Retention is Primarily a Function of Adsorption Phenomena

Unlike bacteria, where filtration at the soil-water interface appears to be the main factor in limiting movement through the soil, adsorption is probably the predominant factor in virus removal by soil. Thus, factors influencing adsorption phenomena will determine not only the efficiency of short-term virus retention but also the long-term behavior of viruses in the soil. Viruses tend to exhibit colloidal properties, and their attraction to soil particles is strongly influenced by pH of the soil, presence and concentration of cations, and the ionizable groups on the virus itself. pH and cationic effects are readily reversible, so once adsorption is achieved, a change in conditions can wash virus through the soil layer.

Viruses are primarily removed from wastewater by adsorption by the soil; thus factors enhancing adsorption increase virus removal.

The pH is of considerable importance relative to adsorption. At the pH at which the isoelectric point of the virus occurs, the net electric charge is zero. The virus has a positive charge below the isoelectric point, and a negative charge above the isoelectric point. Viruses are strongly negatively charged at high pH levels and strongly positively charged at low pH levels. The isolectric pH for enteric viruses is usually below pH 5; thus, in the pH range of most soils enteroviruses as well as soil particles retain a net negative charge. In general, virus adsorption to surfaces is enhanced at a pH below 7 and reduced at a pH above 7 (Wallis, *et al.*, 1972). It is important to note that viruses once adsorbed to solids at a low pH are readily desorbed by a rise in pH.

Virus adsorption is enhanced at pH below 7 and reduced at pH above 7. Since pH is changeable, once adsorption is achieved, a change can reverse it and viruses can be washed through the soil layer.

The action of cations on viruses has been well documented, although the exact chemical mechanism at work has not been determined. The greater the positive charge of cations in solution with viruses, at a given concentration, the greater the attraction of the virus for clay soil particles. Studies on the effect of organic matter content of the soil on virus retention have produced conflicting results.

The greater the positive charge of cations in solution with viruses, the greater the attraction of viruses for clay particles.

Laboratory studies have indicated that rainfall can have a dramatic effect on the migration of viruses through soil. Alternating cycles of rainfall and effluent application result in ionic gradients that enhance the movement of virus. Rainfall acts to drop the ionic concentration of salts in the soil after wastewater application. Such changes in ionic strength have been found to be closely linked with the elution of viruses near the soil surface (Duboise, *et al.*, 1976). This is seen as a burst of released viruses in soil columns when the specific conductance of the water in the soil column begins to decrease after the application of rainwater (simulated in the laboratory by the use of distilled water). This same elution effect can also be seen if a rise occurs in the pH of the water applied to the surface of the soil; that is, a rise in pH from 7.2 to 8 or 9 results in the elution of viruses adsorbed to soil. Viruses are also capable of elution even after remaining in columns saturated with wastewater for long periods of time.

Alternating cycles of heavy rainfall and effluent application decrease virus retention by the soil.

Studies have indicated that certain management practices may prove useful in limiting virus migration through soil (Lance, *et al.*, 1976). Using 98 in. (250 cm) columns of sandy loam soil, they found that many of the viruses eluting near the soil surface after addition of 4 in. (10 cm) of distilled water were later adsorbed near the bottom of the column and that migration of the viruses could be minimized if the columns were flooded with wastewater shortly after the simulated rainfall. In addition, allowing the columns to drain (i.e., soil not saturated with effluent) for at least 5 days before application of the distilled water resulted in no apparent virus movement through the soil. This led the authors to suggest that if a heavy rainfall occurred at a land treatment site within 5 days after application of wastewater, the area could be reflooded with wastewater to restrict subsurface virus migration through the soil. This could only be attempted if it fit in with other wastewater renovation parameters, and at this time, the practice of flooding a site to contain virus movement cannot be endorsed without qualifications.

Studies of virus movement suggest that reflooding a site with wastewater after a heavy rainfall is advantageous if wastewater had been applied within 5 days before the rainfall. Such a practice cannot be endorsed without considering other effects of such flooding.

Only recently have advances been made in the concentration and sampling of viruses from large volumes of water, enabling more precise field experiments on virus movement through soils to proceed. Therefore, very little lab work has been paralleled in full scale. There is some field data corroborating the foregoing theory of rainfall effects on virus movement.

Wellings *et al.* (1974) reported on the travel of viruses through soil at a wastewater reclamation pilot project near St. Petersburg, Florida. At a 10 acre (4 ha) site, chlorinated secondary effluent was applied by a sprinkler system at the rate of 2 to 11 in./wk (5 to 28 cm/wk). The soil consisted of Immokalee sand with little or no silt or clay. On one side of the test plot, an underdrain of tiles was placed at a depth of 5 ft (1.5 m) on top of an organic impermeable layer. This subsurface drain directed the percolated water through a weir where gauze pads were placed for the collection of viruses. Both polioviruses and echoviruses were isolated from the weir water, demonstrating that viruses must survive aeration and sunlight during spraying as well as percolation through 5 ft (1.5 m) of sandy soil. Viruses were also isolated in wells 10 and 20 ft (3 and 6 m) below the soil surface when 50 to 150 gallons (189 to 567 l) of percolate were sampled. No viruses were detected in these wells for the first 5 months of study. Only after two heavy rains was poliovirus type 1 isolated. The viruses were first detected in the 10 ft (3 meter) well and some time later appeared in the 20 ft (6 m) well, indicating that viruses were migrating through the soil. The authors reasoned that the high rainfall resulted in a large increase in the water/soil ratio which led to increased solubility of portions of the organic layer and thus desorption of attached viruses. The observation of viruses as a "burst" after the rainfall was cited as evidence that the rainfall was responsible for the presence of viruses in the wells.

One study showed that heavy rainfall caused desorption of viruses because of increased solubility of organic layer. Viruses migrated through soil to contaminate a 20-ft well.

This same group of investigators (1975) also reported on the detection of viruses in groundwater after the discharge of secondary effluent into a cypress dome in Florida. The soil under the dome consisted of black muck and layers of sand and clay. Viruses were isolated from wells 10 ft (3 m) deep, again after periods of heavy rainfall. The viruses had traveled 23 ft (7 m) laterally subsurface to reach the observation wells. Another important observation made during this study that is not generally recognized is the failure to detect fecal coliform bacteria in the well samples found to contain viruses.

One study detected viruses but no fecal coliform bacteria in well water; thus the presence or absence of fecal coliform does not always indicate the presence or absence of pathogens.

A virus study was conducted at a rapid infiltration site at Fort Devens, Massachusetts, where primary effluent was applied. High concentrations of viruses were added to the effluent. Virus travel through the coarse sand containing little clay and a low surface area was observed (Schaub *et al.* 1975).

In contrast to these findings, field studies at the Flushing Meadows rapid infiltration project near Phoenix, Arizona, indicate limited virus movement through the soil (Gilbert, *et al.*, 1976). At this site, basins in sandy loam underlain at a 3-ft depth of coarse sand and gravel are intermittently flooded with secondary effluent at an average hydraulic loading rate of 300 ft/yr (90 m/yr). Although viruses were detected in the wastewater used to flood the basins, no viruses were detected in wells 20-ft (6 m) deep, located midway between the basins. These results indicated that at least a 99.99% removal of viruses had occurred during travel of secondary treated wastewater through 30 ft (9 m) of sandy soil–20 ft (6 m) vertically and 10 ft (3 m) laterally. The low rainfall, fine sandy loam, and the practice of intermittent flooding of the soil during land treatment at this site may have resulted in better conditions for virus removal than at other land treatment sites studied to date.

Table 4. Factors that Influence the Movement of Viruses in Soil. (Metcalf and Eddy, 1977)

Factor	Remarks
Rainfall	Viruses retained near the soil surface may be eluted after a heavy rainfall because of the establishment of ionic gradients within the soil column.
pH	Low pH favors virus adsorption; high pH results in elution of adsorbed virus.
Soil composition	Viruses are readily adsorbed to clays under appropriate conditions and the higher the clay content of the soil, the greater the expected removal of virus. Sandy loam soils and other soils containing organic matter also are favorable for virus removal. Soils with a low surface area do not achieve good virus removal.
Flowrate	As the flowrate increases, virus removal declines, but flowrates as high as 32 ft/d (9.6 m/d) can result in 99.9% virus removal after travel through 8.2 ft (2.5 m) of sandy loam soil.
Soluble organics	Soluble organic matter competes with viruses for adsorption sites on the soil particles, resulting in decreased virus adsorption or even elution of an already adsorbed virus. Definitive information is still lacking for soil systems.
Cations	The presence of cations usually enhances the retention of viruses by soil. An increase in valence enhances retention.

A land treatment site with low rainfall, fine sandy loam, and the practice of intermittent flooding with wastewater achieved a high percentage of virus removal.

Table 4 is a general summary of properties of land application sites relating to virus retention or movement in soil.

Knowledge Limited on Groundwater Survival

If pathogenic organisms pass through coarse soil and fractured rocks like limestone, they may reach the groundwater. Although few studies have been made on bacterial survival in groundwater, it appears that bacteria may persist in underground water for months.

Although more studies are needed, data show that bacteria may survive in groundwater for months.

E. coli have been found to survive up to 1,000 days in subsoil water, whereas a 50% reduction in numbers occurred within 12 hours in well water. It should be pointed out that there is wide variation in the reported length of bacterial viability in underground waters, and more careful studies should be made before definite conclusions are drawn.

CROP CONTAMINATION VARIES

In general, the extent of crop contamination can be determined by the type of crop and the kind of irrigation practice that is used. Wastewater irrigation of fiber crops presents the least health risk and that of food crops, especially those to be eaten raw, the greatest risk. Subsurface irrigation or certain flooding techniques avoid direct contact between wastewater and the vegetation, whereas sprinkling and flooding wet the low-growing vegetation as well as the soil.

Table 5. Survival of Selected Pathogens on Vegetation (Parsons, *et al.*, 1975)

Organisms	Media	Survival Times (days)
Salmonella	Vegetables, fruits	3–49
	Grass or clover	12–more than 42 (and over winter)
Tubercle bacilli	Grass	10–49
Entamoeba histolytica cysts	Vegetable	less than 1–3
Enteroviruses	Vegetables	8
Ascaris ova	Vegetables, fruits	27–35

The greatest health concern is with low-growing crops, such as vegetables, which have a greater chance of contamination and are often eaten raw. Table 5 summarizes reported survival times for relatively common pathogenic organisms.

Subsurface irrigation and certain flooding techniques on food crops reduce the risk of crop contamination by wastewater.

Disease Risk to Grazing Animals can be Minimized

Pathogens may be transmitted to animals grazing on pasture land recently irrigated with wastewater. Some state laws prohibit grazing on such areas while other state laws require that dairy animals not be allowed to graze on such areas for a certain number of days following irrigation.

Although a number of cases of disease in animals have been attributed to their unintentional exposure to wastewater, less is known about the risks to animals grazing on pastures irrigated with wastewater. Grazing land has been irrigated with untreated wastewater on a large scale in Europe and Australia and with treated wastewater in the United States. There seems to be little threat to the health of farm animals under normal conditions. It also appears that cattle which are fed effluent-irrigated forage or silage show no ill effects nor does the milk or meat from them appear to be infected.

The risk of transmitting disease to animals grazing on wastewater-irrigated pasture depends on how long pathogens persist, their concentration, the animals' health and the interval between irrigation and grazing.

However, diseases have been transmitted to domestic animals from sewage-contaminated water and pasture. Carefully controlled experiments and field data need to be conducted to develop effective guidelines in this area. The chances of infection for animals grazing on wastewater-irrigated pasture depends on: the persistence of pathogens, the concentration of pathogens, the health of the animals, the interval between irrigation and grazing. The disease threat to animals from eating wastewater-irrigated crops and fodder can be minimized by:

- Allowing animals to graze only after a certain interval of time has passed following irrigation. A number of pathogens in wastewater are inactivated during desiccation and when exposed to sunlight.
- Drying forage crops.
- Storing the dried forage before feeding it to animals.

Smart (in Parsons, *et al.* 1975) noted that there is little to indicate that feeding hay irrigated with sewage effluent to cattle poses a health hazard if the hay is baled when dry, and cured for 14 days. He also noted that there is no health hazard to humans from eating the meat of these animals.

Precautions Decrease Occupational Risks

Workers handling wastewater are exposed more to pathogens than is the general public. The full impact of this occupational risk has not been determined in epidemiological studies. More studies are needed to evaluate the hazard both to the worker and his contacts, however the following safeguards should be practiced. Employees of land treatment sites should:

- Receive regular typhoid and tetanus inoculations and poliovirus and adenovirus vaccinations.
- Avoid direct contact with mists as much as possible; this is especially a problem at sprinkler irrigation sites.
- Change working clothes regularly to avoid not only personal infection, but the risk of spreading infection to the worker's family.
- Take care that their food and drinking water does not become contaminated by wastewater.
- Practice reasonable hygiene, such as washing hands after work and before meals.

Careful hygiene can minimize risk of disease to workers at land treatment sites.

Administering prophylactic doses of immune serum globulin to a worker who accidentally ingests wastewater should be considered. Immune serum globulin is prepared from large pools of normal adult plasma and gives passive protection in 80 to 90% of persons who have been exposed to infectious hepatitis within a period of one to two weeks.

Workers at conventional secondary treatment plants are also exposed to some continuous levels of enteric bacteria and viruses. Aerosol transport mechanisms disseminate microorganisms in trickling filter and activated sludge plants, as do such hands-on activities as bar screen cleaning and clearing of obstructions from mechanical gear. It is not felt that this exposure has presented any particular risk to these workers. A study some years ago in New York City showed that wastewater treatment plant workers had a lower rate of absenteeism than any other segment of the city's staff. This led some to propose a theory that the workers' proximity to continual low levels of pathogens had a beneficial effect by building their immunity to infectious diseases.

Exposure to pathogenic aerosols at conventional secondary treatment plants is not believed to put plant workers at any particular risk.

Insect Control Needed at Site

Control of mosquitoes at land treatment sites is needed to protect both the animal and human population from the nuisance and disease threat associated with large numbers of mosquitoes. Although there is no conclusive evidence that insect-transmitted disease increases in wastewater irrigation areas, more studies are needed to evaluate this problem.

Studies by Parizek, *et al.*, (1967) did show, however, that the diversity and abundance of mosquitoes increases in areas where spray irrigation of wastewater is practiced. An increase in avian blood parasites which are transmitted by mosquitoes also was noted.

Control measures would include the avoidance of standing water conditions as much as possible, especially very shallow pockets of warm water remaining on the ground for more than a few days, since these pockets are favored breeding sites. Appropriate mosquito control efforts for land treatment systems would be the same as those practiced in any other area where standing water may occur, and local health authorities should be contacted for assistance as needed.

Because irrigation can increase the mosquito population at the site, control measures such as not letting water stand and other preventive approaches should be used.

PATHOGENS COMMON TO WASTEWATERS, AND THE DISEASES WHICH THEY CAN CAUSE

This part of Module II-5 contains a discussion of the diseases which can be spread by improperly treated and handled sewage.

Although the general public may fear "disease" from a land treatment system, most persons are unaware of the specific diseases, the pathogens which cause them, and the degree of threat each poses to the community. The following sections provide a brief overview of the types of disease organisms which may be found in sewage, and a discussion of the most prevalent diseases associated with improperly handled sewage. A large variety of disease-causing microorganisms and parasites are present in domestic wastewater. These include pathogenic bacteria, viruses, protozoa, and parasitic worms. The number of individual species of pathogens is high. For example, over 100 different types of viruses are known to be excreted in human feces. The relative concentrations of these pathogens are highly variable, being dependent on a number of complex factors, but pathogens are almost always present in human wastes in sufficient numbers to be a public health concern.

A large variety of pathogens are present in domestic wastewater, posing a concern to public health.

The planner and engineer must:

- Identify and put into perspective any potential routes of disease transmission involved in land treatment of wastewater so that appropriate safeguards can be assured.
- Identify the relative risk of infection for humans and animals from these and other sources.

The engineer must identify routes of disease transmission and the risk of infection from these sources.

It should be noted that most of the studies cited in this section involve applying untreated wastewater to the land in an unplanned manner, or deliberate artificial seeding of bacteria or viruses in high concentrations to the soil to determine their survival or movement under various conditions. Each study had different objectives, types of soil, climatic factors, types of organisms, and methods of detection. Consequently, the results must be interpreted accordingly. Safeguards should be established on a case-by-case basis so that the relative risk of disease transmission in each situation can be evaluated individually.

Safeguards against disease transmission must be created on a case-by-case basis.

Bacteria Common to Wastewater

The most common bacterial pathogens found in wastewater include strains of *Salmonella, Shigella,* enteropathogenic *Escherichia coli* (*E. coli*), *Vibrio*, and *Mycobacterium.* The genus *Salmonella* includes over 1,200 different strains, many of which are pathogenic for both man and animals. Members of this group are commonly isolated from wastewater and polluted receiving waters.

The Salmonella bacteria are responsible for several diseases, including typhoid, paratyphoid, and salmonellosis. In all cases, disease is transmitted by fecal contamination of food or water. *Salmonella typhosa* has been responsible for incidents of typhoid fever associated with wastewater contaminated drinking water and with the eating of raw vegetables grown on soil fertilized with untreated wastewater. Infected persons excrete the organisms in the feces and urine, thus posing the threat of rapid dissemination of the disease.

Typhoid, paratyphoid, and salmonellosis can be transmitted by fecal contamination of food or water.

Paratyphoid fever is caused by Salmonella bacteria other than *Salmonella typhosa.* The most common cause in the United States is *S. schottmullerei.* Although similar to typhoid fever, paratyphoid is milder and not as long-lasting. The infectious dose in man for *S. typhi* and *S. schottmullerei* is more than 100,000 organisms.

It takes 100,000 bacteria to infect a person with paratyphoid fever, a relatively high infectious dose.

Gastroenteritis, one form of food poisoning, may be caused by *Salmonella typhimurium.* The organism is excreted with feces, and is transmitted by ingestion of contaminated food or water. Many animals can be infected with Salmonella and may have these organisms in their meat, eggs, or feces. Rodents are among these, and should be controlled at land treatment sites.

The incidence of typhoid fever in the United States decreased markedly with widespread municipal water treatment in the early 1900s (N.Y. State Health Comm., 1932). However, the incidence of other Salmonella infections has increased. Although no cases of Salmonella infections directly related to land application sites have been reported, Salmonella diseases might be spread by land application of sewage if not properly handled.

Salmonella infections other than typhoid fever can be spread by land application of sewage if it is not properly handled.

Shigella organisms are the most commonly identified cause of acute bacterial diarrheal disease in the United States. These bacteria normally live in the intestinal tract of man. During the active disease, they are excreted in large numbers but remain viable in the feces only a short time. Waterborne spread of the organisms can cause outbreaks of shigellosis, commonly known as bacillary dysentery, which occur frequently in undeveloped countries and occasionally in developed countries. Unlike Salmonella, Shigella organisms are rarely found in animals other than man.

Controlling Shigella organisms includes proper sewage disposal and adequately treated drinking water supplies. In addition, flies can transmit the disease, and therefore, must be controlled around the land treatment sites. Bacillary dysentery is a growing problem in Central and South America.

Proper sewage disposal and adequately treated drinking water control spread of Shigella, the most common cause of acute bacterial diarrhea in the U.S.

Cholera is caused by the organism *Vibrio cholerae.* In Israel in 1970 cases of cholera were attributed to the practice of irrigating vegetable crops with untreated wastewater. This practice is contrary to regulations of the Ministry of Health (Cohen, *et al.* 1971). There were no reported cases of cholera in the United States between 1911 and 1973, until a single case occurred in Texas with no known source. Although individual cases may arise in international travelers, they are unlikely to become a source of disease through wastewater land application projects.

Wastewater is unlikely to be a source of cholera pathogens.

Viruses are Smallest Pathogens

Viruses, the smallest wastewater pathogens, consist of a single nucleic acid surrounded by a protective protein coat. Viruses that are shed in fecal matter, referred to as enteric viruses, can infect tissues in the throat and gastrointestinal tract, but they are also capable of replicating in other organs of the body. They include the true enteroviruses (polio-, echo-, and coxsackie-viruses), reoviruses, adenoviruses, rotaviruses, and infectious hepatitis virus.

Viruses, the smallest wastewater pathogen, cause diseases ranging from meningitis to infectious hepatitis to eye infections. Many viral infections are latent, making it difficult to trace their source.

These viruses can cause a wide variety of diseases, such as paralysis, meningitis, respiratory illness, myocarditis, congenital heart anomalies, diarrhea, eye infections, rash, liver disease, and gastroenteritis. Almost all of these viruses also produce inapparent or latent infections. This makes it difficult to recognize them as being waterborne. Documented outbreaks of waterborne viral disease have largely been limited to infectious hepatitis, mainly because of the explosive nature of the outbreaks and the characteristic nature of the disease.

Poliovirus (the cause of polio) is a stable virus and can remain infectious for relatively long times in water. Infection is primarily by ingestion of contaminated water or food. Waterborne epidemics have been reported. Vaccines could theoretically eliminate polio if the entire world were vaccinated. This is unfeasible, so attention must be paid to the possible introduction of poliovirus from an unvaccinated population.

Poliovirus may be spread by unvaccinated persons via contaminated water; land application sites must still be concerned about its spread.

Coxsackieviruses are responsible for a wide range of diseases. The mechanism of transmission is uncertain, although viruses are often found in feces, which would indicate that oral ingestion may be a causative factor.

Echoviruses induce symptoms ranging from minor respiratory disease to afflictions of the central nervous system. Commonly this type of virus is transmitted orally, although some may infect the respiratory system by inhalation. Again, since these viruses may be excreted with the feces, proper sewage treatment is necessary.

Proper sewage treatment is needed to control echoviruses.

Hepatitis virus is a cause of liver disease only in man. Infectious hepatitis can be spread through the fecal-oral route. These viruses are quite stable though various chemical processes have been used to destroy them.

Infectious hepatitis occurs in two forms, Types A and B, of which only Type A is transmitted by the fecal-oral route. Although convalescence may be prolonged, recovery is complete in over 85% of the cases. The fatality rate in the United States is between 1.5 and 2% out of an incidence rate in 1972 of 30.5/100,000 (Jawetz, *et al.*, 1974). However it is a major public health problem world-wide. Sudden epidemics resulting from fecal contamination of drinking water or food occur periodically. Consumption of shellfish from sewage-contaminated waters also accounts for outbreaks of hepatitis. The organism responsible for the disease is quite resistant to heat, acid, and chemical treatment. Because sewage contamination of foods has been linked to outbreaks of the disease, it is of the utmost importance to control the hepatitis virus through proper handling of sewage.

Outbreaks of infectious hepatitis have been linked with ingestion of sewage-contaminated foods. It is of utmost importance to control this virus.

Knowledge of the actual number and concentration of human pathogenic viruses in wastewater is inadequate because sampling and analytical procedures are neither standardized nor adequate. Furthermore, the methodology for detecting and monitoring many of these agents has not yet been developed. This probably accounts for the fact that almost 60% of all documented cases of disease attributable to drinking water in the United States have been reported to be caused by agents as yet not isolated in the laboratory. It appears now that many of the virus sampling difficulties encountered in the past are being overcome, and some researchers are beginning to call for routine viral analysis in areas like treatment plant control. Great difficulties remain in the analytical work necessary to identify and quantify viruses, however, and this procedure is certainly out of range of almost all laboratories as a routine measure. It is important to note that viruses as a group are generally more resistant to environmental stresses than pathogenic bacteria.

Scientific procedures to detect viruses are inadequate, and 60% of all disease cases attributed to drinking water have been caused by agents not identified in laboratories.

Viruses are generally more resistant to environmental stresses than pathogenic bacteria.

Protozoa, Worms Can Cause Disease

Protozoa pathogenic to man and capable of transmission in wastewater are *Entamoeba histolytica*, the agent of amoebic dysentery; *Naegleria gruberi*, which may cause fatal meningoencephalitis; and *Giardia lamblia*, which produces a variety of intestinal symptoms. Waterborne outbreaks of *Giardia lamblia* have increased in the United States in recent years.

Entamoeba histolytica cannot exist in its active form outside of its host. However, it is capable of forming cysts which are excreted. These cysts protect the protozoan from adverse environmental conditions outside the host. The cysts may also resist waste treatment processes and are capable of causing disease when ingested with contaminated food or water.

The agent of amoebic dysentery can form cysts which resist treatment and can cause disease when ingested.

The eggs of several intestinal parasitic worms have been found in wastewater and have been shown to be a potential health problem to wastewater treatment plant operators and laborers employed on farms in India and East Germany where wastewater is used for irrigation (Clark, *et al.*, 1976). Modern water treatment methods have proved a very effective barrier against the waterborne spread of disease caused by protozoa and parasitic worms in developed countries.

Modern water treatment in developed countries prevents the waterborne spread of disease from protozoa and parasitic worms.

VARIOUS TREATMENTS REDUCE PATHOGEN LEVELS

The planner or engineer needs to know the number of pathogens in untreated and treated wastewater to evaluate the relative risk of disease transmission associated with land application of wastewater. He also must know the number of pathogens that are necessary to cause an infection in man or other animals.

It is important to know the number of pathogens in untreated wastewater and the infective dose of various pathogens.

Table 6 estimates the relative concentrations of pathogens in untreated wastewater and the relative efficiency of removal by primary and secondary treatment. Unfortunately, data on the removal efficiency of all wastewater treatment methods for many pathogens are either nonexistent or largely based on laboratory studies by researchers who overestimate the efficiency that can be obtained in actual practice. (Foster, Engelbrecht, 1973).

Data on removal efficiency of wastewater treatment methods for many pathogens are inadequate.

Conventional Treatment Efficiencies

Figures 3 through 6 show the efficiencies of various forms of conventional treatment for the removal of bacteria, viruses, protozoa and helminths. These figures are summaries of data from

Table 6. Estimated Concentrations of Selected Wastewater Pathogens[a].

	Number of Organisms/gal (3.78 l)			
Pathogen	Untreated Wastewater	Primary Effluent	Secondary Effluent	Disinfection[b]
Salmonella	2.0×10^4	1.0×10^4	5.0×10^2	5×10^{-1}
E. histolytica	1.5×10^1	1.3×10^1	1.2×10^1	1.2×10^{-2}
Helminth ova	2.5×10^2	2.5×10^1	5.0×10^0	5×10^{-3}
Mycobacterium	2.0×10^2	1.0×10^2	1.5×10^1	1.5×10^{-2}
Human enterovirus (poliovirus, etc.)	4.0×10^4 [c]	2.0×10^4	2.0×10^3	2×10^2

[a]Adapted from Foster and Engelbrecht in recycling treated municipal wastewater and sludge through forest and cropland, Soppers and Kardos, eds. the Pennsylvania State University Press, University Park, Pa., 1973.
[b]Conditions sufficient to yield a 99.9% kill.
[c]As high as 4×10^6 per gal (3.78 l) have been reported.

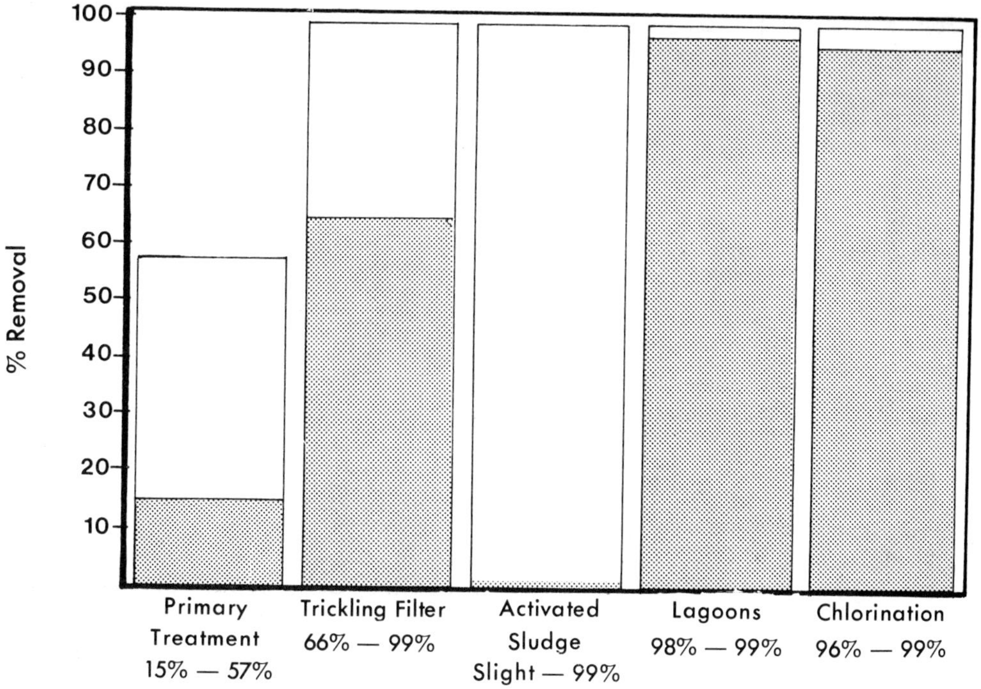

Figure 3. Processes differ in their removal of bacteria.

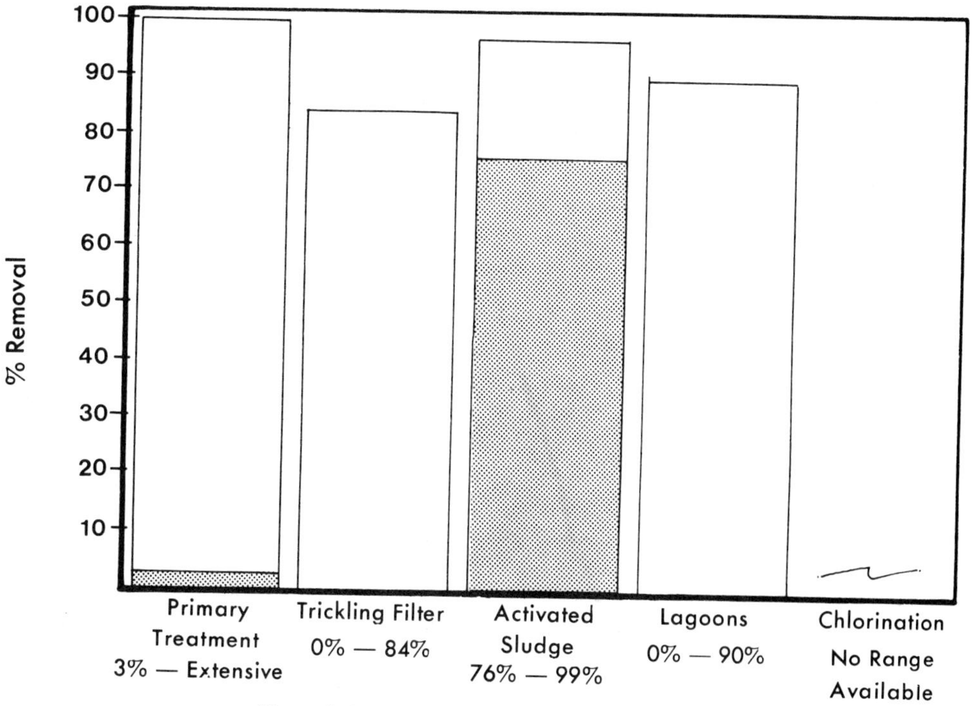

Figure 4. Less certainty exists on virus removal.

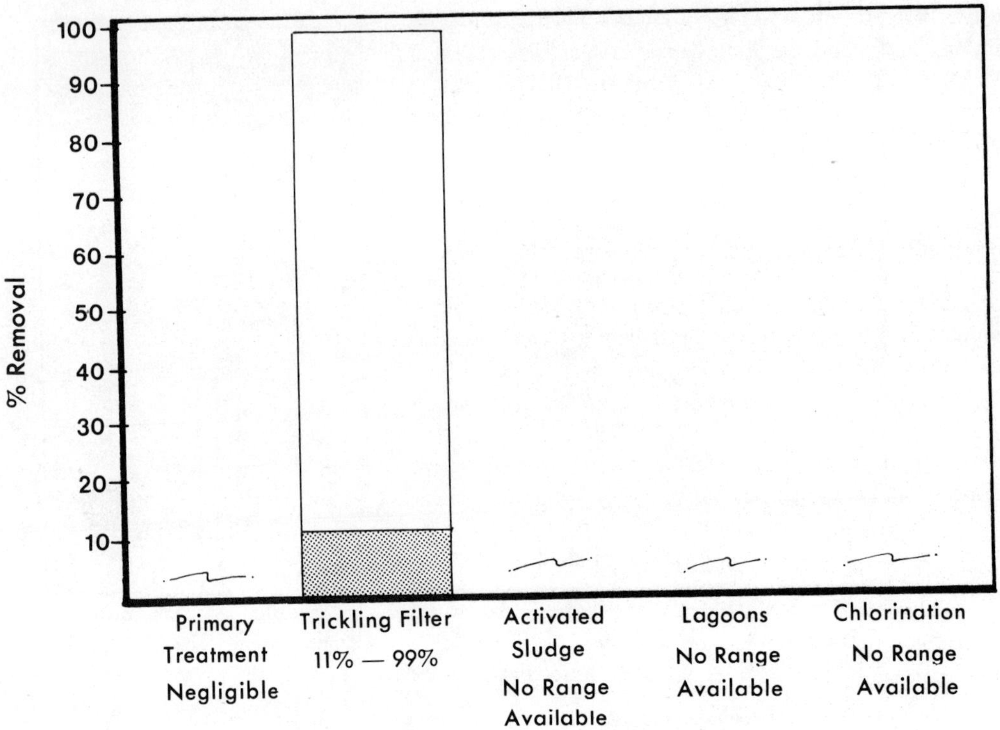

Figure 5. Protozoan removal data are very scarce.

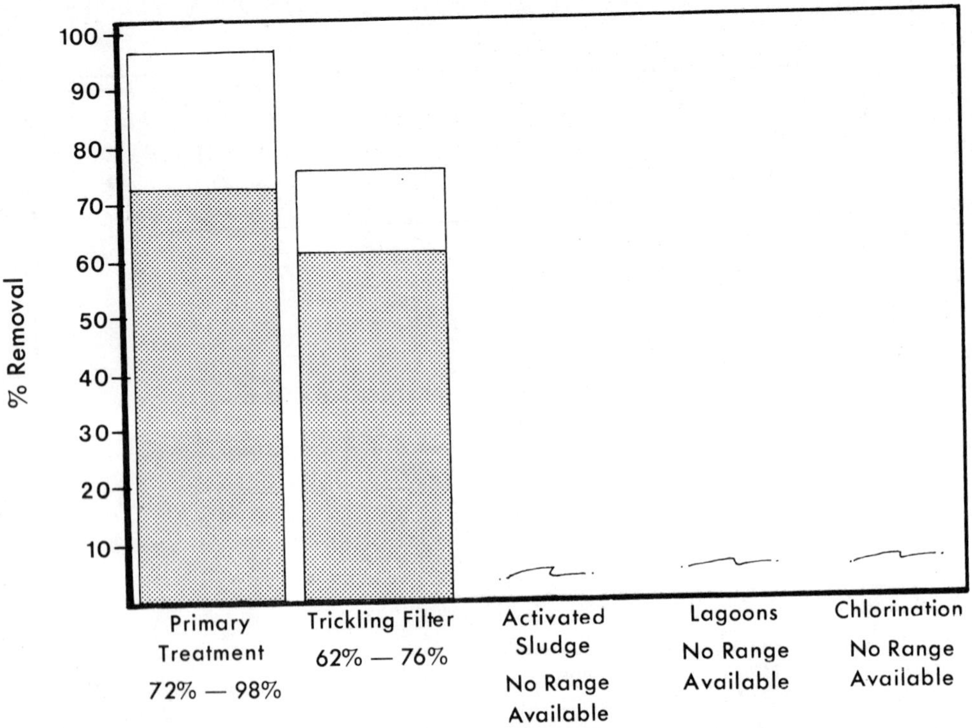

Figure 6. Primary treatment appears more efficient for helminth removal.

a literature review comprising studies by the following authors, all cited in the bibliography: Bryan, Cody and Tischer, Cram, Kabler, Fitzgerald and Rohlick, Foster and Engelbrecht, McKinney, Newton *et al.*, Parsons *et al.*, Rudolfs *et al.*, and Wang and Dunlop. Several species are represented in the graphs for each organism, therefore, the figures show the broadest range of values reported in any category.

Chlorination Does not Destroy Pathogens Completely

Chlorination cannot be relied upon to provide complete destruction of pathogenic bacteria and viruses. It is highly effective in achieving large reductions of bacterial pathogens, but it is much less effective against cysts and enteric viruses. For example, a chlorine dose of 2 mg/l killed 99.9% of the coliform bacteria in 60 minutes in wastewater, whereas it took a chlorine dose of 20 mg/l to achieve the same kill of poliovirus. (Hoadley and Goyal, 1976). This level of chlorination is not economically feasible and is undesirable due to formation of chloramines and organic complexes. Another factor which complicates transfer of laboratory data obtained with pure viruses to wastewater in the field situation is that polioviruses and other viruses are much more resistant to chlorine if organic matter is present (Trask, *et al.*, 1945). Table 7 describes other factors influencing the effectiveness of chlorination.

Chlorination greatly reduces bacterial pathogens but is less effective against cysts and enteric viruses.

High levels of chlorination are undesirable for land treatment because of high cost and unsatisfactory results.

Chlorine disinfection works by preventing normal respiration by the organisms. It should be remembered that bacteria can multiply very quickly when environmental conditions become favorable. Thus, a high kill figure may not necessarily mean that no disease threat remains. Effectiveness of bacterial removal should be measured as residual bacteria counts rather than percentage kill.

A high kill of bacteria by chlorination does not eliminate the threat of bacterial disease.

Table 7. Several Factors Affect the Efficiency of Chlorine in Destroying Wastewater Pathogens.

- Increased exposure time increases efficiency
- Increased concentration of chlorine increases efficiency
- Decreased organic content increases efficiency
- Decreased ammonia concentration increases efficiency
- Increased temperature increases efficiency
- Decreased pH increases efficiency
- Increased residual chlorine increases efficiency
- Decreased suspended solids increase efficiency

Infective Doses must be Considered

The planner or engineer must consider the infective dose—the number of organisms necessary to cause disease in man or animals—to evaluate the disease potential of the system. Infective doses for most bacterial and protozoan pathogens are relatively high, but the infective dose for viruses is low.

For instance, ingestion of 10^8 enteropathogenic *E. coli* or *V. Cholera*, 10^3 to 10^5 *Salmonella*, and 10^1 10^2 *Shigella* organisms are necessary to cause infection in man. (Hoadley and Goyal, 1976) The infective dose of a protozoan, such as *E. histolytica*, is believed to be as high as 20 cysts (Foster and Engelbrecht, 1973). However, a single virus may produce disease in a human host.

We can conclude that the potential for bacterial disease transmission after disinfection of treated wastewater would appear to be quite small, but the low infective dose of viruses gives importance to even relatively low concentrations of these agents in water.

Because a single virus can cause disease, even relatively low concentrations of viral pathogens in wastewater are a cause for concern.

AEROSOL PATHOGENS CAN BE REDUCED

Aerosols containing bacterial and viral pathogens may be infectious upon inhalation. The body of information at present indicates that at least some potential health effects are related to the production of wastewater aerosols. However, these potential deleterious effects have yet to be fully established.

Bacterial, viral pathogens in aerosols can be inhaled by humans, causing infection.

The following practices are recommended to reduce or eliminate these effects:

- Creating buffer zones.
- Controlling sprinkling operations to minimize the production of fine droplets.
- Eliminating sprinkling during high winds.
- Restricting sprinkling to daylight hours.
- Decreasing nozzle pressure.
- Aiming the nozzles downward.
- Planting hedgerows on buffer zones.

Aerosols are defined as particles in the size range of 0.01 to 50 μm that are suspended in air. When an airborne water droplet is created, the water evaporates rapidly under average atmospheric conditions, resulting in a nucleus of the originally dissolved solids plus the microorganisms contained in the original droplet. This high rate of evaporation results in the die-off of many of the original organisms, but resistant organisms that remain may persist for a long time.

High rate of evaporation of aerosols results in die-off of many pathogenic organisms.

During sprinkler irrigation about 0.3% aerosolization of the liquid was reported by Sorber, *et al.*, (1975) using fluorescein dye tracers. Some pathogens found in wastewater have a lower infective dose in aerosol form than when ingested directly. The threat of aerosol-borne disease is often discussed. It must be remembered, however, that large concentrations of aerosols are also generated from conventional secondary treatment plants, particularly those employing activated sludge processes or trickling filters. In fact, most of the information available today on wastewater aerosols concerns their generation by conventional wastewater facilities. Airborne coliform bacteria have been recovered at night as far as 0.8 mile (1.3 km) from a large trickling filter plant (Hickey and Reist, 1975).

It may take fewer pathogens to cause infection when inhaled in aerosols than when ingested directly.

There have been no reports of disease transmission from these conventional plants. This fact seems to support the belief that aerosol transmission of disease is not likely from a properly designed and well operated land application system using sprinkler irrigation.

Aerosols are generated from conventional treatment plants as well as spray irrigation. There have been no reports of disease transmitted from these plants.

Pathogen Survival Time in Aerosols Depends on Several Factors

How long do pathogenic bacteria and viruses live in aerosols? Under favorable conditions, bacterial aerosols may remain viable for several hours. These bacterial aerosols remain viable and travel further with increased wind velocity and relative humidity, decreased temperature, and darkness. A primary cause of bacterial aerosol destruction seems to be rapid desiccation.

High humidity, increased wind speed, low temperature, darkness enhance survival and travel of bacterial aerosols.

Viruses in aerosols also seem to remain viable for long periods of time. A study by Walker (1970) revealed that polioviruses in aerosols remained active for more than 23 hours when the relative humidity was high. Further study is needed to definitely ascertain the threat of airborne viruses. Factors that have been found to affect the survival and dispersion of bacteria and viruses in such aerosols are summarized in Table 8.

A model to describe aerosol movement of pathogens has been developed by Sorber, *et al.* (1974) and was given a preliminary field test in 1975 (Sorber, *et al.*, 1975). The field study enabled these workers to reach six conclusions:

1. Under spray irrigation conditions studied, relatively high concentrations of bacterial aerosols were transmitted for considerable distances.
2. The aerosols carrying the bacteria were in a size range which can be inhaled by humans.
3. Sunlight had little effect on bacterial decay after the bacteria were exposed to initial aerosol shock and desiccation, for the aerosol ages studied.
4. The concentration of organisms at the source and atmospheric stability are the two most important factors influencing downwind aerosol concentrations.
5. Low wind speed and darkness are the stable atmospheric conditions which lead to the highest bacterial aerosol concentrations downwind.
6. Effective disinfection was found capable of reducing aerosols to background levels in reasonable distances.

In addition, the authors provided two inferences relating to the decrease of viable particles. The authors concluded that a distance in excess of 1800 meters is required to reduce the

Table 8. Factors that Affect the Survival and Dispersion of Bacteria and Viruses in Wastewater Aerosols. (Metcalf and Eddy, 1977)

Factor	Remarks
Relative humidity	Bacteria and most enteric viruses survive longer at high relative humidities, such as those occurring during the night. High relative humidity delays droplet evaporation and retards organism die-off.
Wind speed	Low wind speeds reduce biological aerosol transmission.
Sunlight	Sunlight, through ultraviolet radiation, is deleterious to microorganisms. The greatest concentration of biological aerosols from wastewater occurs at night.
Temperature	Increased temperature can also reduce the viability of microbial aerosols mainly by accentuating the effects of relative humidity. Pronounced temperature effects do not appear until a temperature of 80°F (26.7°C) is reached.
Open air	It has been observed that bacteria and viruses are inactivated more rapidly when aerosolized and when the captive aerosols are exposed to the open air than when held in the laboratory.

concentration of viable particles to within 5 organisms/m^3 air of the background concentration. This figure was extrapolated from the test data, which was collected at distances up to about 150 meters from the spray point. No correlation was attempted between background levels of bacteria and infectious doses for pathogenic organisms which would relate to the desirability of achieving this sort of reduction in atmospheric concentration.

The bacterial populations from the chlorinated wastewater were found to be much less affected by aerosol shock and age than those in the unchlorinated wastewater. That is, although the bacterial concentrations were much less at the source in the chlorinated samples, the percent reduction of viable organisms was such that reasonable levels of atmospheric bacteria could be achieved within relatively short distances by atmospheric die-off without disinfection. The organisms most susceptible to disinfection by chlorine may also be those most easily killed by desiccation in aerosols.

Organisms most susceptible to destruction by chlorine may also be those most easily destroyed by desiccation in aerosols.

Few statistics are Available on Spread of Aerosols

Another quantitative study was done on the spread of aerosols from land application systems. In this study, Merz (1957–now out of print) investigated the hazards associated with sprinkling treated wastewater onto a golf course. He concluded that the hazards were limited to direct contact with unevaporated droplets.

In spraying treated wastewater, one investigator found that hazards were limited to direct contact with unevaporated drops.

Conclusions of other articles and unpublished studies on aerosol viability and drift from sprinkler systems follow. All of these, along with Merz, are referenced in Metcalf and Eddy (1977).

- Reploh and Handloser estimated that a viable aerosol can be carried 1,312 feet downwind by a 16 ft/s (5 m/s) wind. They recommended the planting of hedges as a safety measure. Type of wastewater: untreated

- Bringmann and Trolldenier estimated that coliform organisms can remain viable 1,312 feet downwind from the source under a 23 ft/s (7 m/s) wind under conditions of darkness and 100% humidity. Type of wastewater: settled but not disinfected.
- Sepp found coliform bacteria in aerosols 10 feet downwind in a dense brushy area; up to 200 feet downwind in a sparsely vegetated area. Type of wastewater: ponded and chlorinated secondary effluent
- Shtarkas and Krasilshchikov found bacteria 2,133 feet downwind. Type of wastewater: water from a small stream contaminated with untreated domestic wastewater
- Bausum *et al.* (1976) found bacterial viruses 2,067 feet from the wetted zone at a sprinkler irrigation site.

ISRAELI STUDY CALLS FOR MORE EPIDEMIOLOGICAL RESEARCH

Katzenelson, *et al.* (1976) reported the first epidemiological evidence of a disease risk associated with wastewater sprinkler irrigation. In a study of 77 kibbutzim (collective farm settlements) that used wastewater for irrigation and 130 kibbutzim that did not use wastewater for any purpose, the team found that the incidence of disease was 2 to 4 times higher in those settlements that used sewage for spray irrigation. In Israel, domestic wastewater effluents have been used extensively for agricultural irrigation for more than 20 years, mainly by sprinkler systems. The level of enteric microorganisms in these partially treated nondisinfected oxidation pond effluents (mean detention time of 3 to 7 days) often approaches that of raw domestic sewage.

> *Incidence of sewage-associated disease at Israeli settlements using spray irrigation of wastewater was 2 to 4 times higher than at settlements that did not use wastewater at all, one study reported.*

Diseases that occurred more often included shigellosis, salmonellosis, infectious hepatitis, typhoid fever, and influenza. The incidence of diseases not associated with sewage, such as strep infection and tuberculosis, was not significantly different between the two types of settlements. Neither was there any difference in disease outbreaks during the winter between the two settlements, a time when irrigation is not practiced.

The investigators noted that the pathogens from wastewater might reach the kibbutz populations by an alternate pathway, such as on the bodies and clothes of irrigation workers who live in the community and return from the fields at mealtime and at the end of the day. Indeed, very poor sanitation has been observed at many of the sites, and it is extremely likely that pathogens are being transported directly from sewage in the fields to dining facilities and other communal areas on the clothes of farm workers. The pathway of disease spread from wastewater is not clearly defined in this study. More detailed and controlled epidemiological research is needed on disease transmission from waste treatment sites that use irrigation.

> *Pathway of disease transmission in this Israeli study is not clear; more epidemiological studies on disease transmission from waste treatment sites are needed.*

BIBLIOGRAPHY

Arkin, E. W., W. H. Benton, and W. F. Hill. 1971. Enteric viruses in ground and surface water: a review of their occurrence and survival. *In* Virus and water quality: occurrence and control. 13th Water Quality Conference, 59, University of Illinois. 222 p.

Alexander, M., *et al.* 1972. Accumulation of nitrate. Nat'l Academy of Sciences, Washington, D.C. 106 p.

Anon. 1970. "Mercury poisoning leaves tragic mark on N. M. family." The Minneapolis Tribune, 16 Aug.: 14A.

Batey, T., C. Berrgman, and C. Line. 1972. The disposal of copper-enriched pig-manure slurry on grassland. *J. Brit. Grassland Soc.* **27**: 139–143.

Bausum, H. T., S. A. Schaub, and C. A. Sorber. 1976. Viral and bacterial aerosols at a wastewater spray irrigation site. *In:* Abstract of the annual meeting, American Society of Microbiologists. p. 193.

Bernarde, M. A. 1973. Land disposal and sewage effluent: appraisal of health effects of pathogenic organisms. *J. Amer. Water Works Assn.* **65**(6):432–440.

Bryan, F. L. 1974. Diseases transmitted by foods contaminated by wastewater. Presented at the Conference on the Use of Wastewater in the Production of Food and Fiber, Oklahoma City, Okla.

Buras, N. 1976. Concentration of enteric viruses in wastewater. *Water Resources.* **10**(4):295–298.

Burge, W. D. 1974. Pathogen considerations. *In* National program staff, factors involved in land application of agriculture and municipal wastes. (DRAFT) Agricultural Research Service. Soil, water and air sciences. USDA. Beltsville, Md. 20705. pp. 37–50.

Butler, R. G., C. T. Orlob, and P. H. McGauhey. 1954. Underground movement of bacterial and chemical pollutants. *Jour. AWWA* **46**:97–111.

Carlson, G. F., *et al.* 1968. Virus inactivation on clay particles in natural waters. *Jour. WPCF* **40**(2):R89–R106.

Clark, C. S., *et al.* 1976. Disease risks of occupational exposure to sewage. *J. of the Environ. Eng. Div., Proc. of the A.S.C.E.* **102**(EE2):375–388.

Cody, R. M. and R. G. Tischer. 1965. Isolation and frequency of occurrence of Salmonella and Shigella in stabilization ponds. *J. Water Pollut. Contr. Fed.* **37**(10):1399–1403.

Cohen, J., *et al.* 1971. Epidemiological aspects of cholera: El Tor outbreak in a non-epidemic area. *Lancet.* **2**:86.

Committee on Environmental Quality Management of the Sanitary Engineering Division. 1970. Engineering evaluation of the virus hazard in water. *J. Sanitary Eng. Division. A.S.C.E.*, Vol. 96, SA1, pp. 111–161.

Cram, E. B. 1943. The effect of various treatment processes on the survival of helminth ova and protozoan cysts in sewage. *Sewage Works J.*, **15**:1119.

Curtis, H. 1968. Biology. Worth Publishers, Inc., New York. 854 p.

Davis, B. D., R. Dulbecco, H. M. Eisen, H. S. Ginsberg, W. B. Wood, Jr. 1967. Microbiology. Harper & Row, New York 520 p.

Drewry, W. A. and R. Eliassen. 1968. Virus movement in groundwater. *Jour. WPCF* **40**(8):R257–R271.

Duboise, S. M., B. E. Moore, and B. P. Sagik. 1976. Poliovirus survival and movement in a sandy forest soil. *Appl. Environmental Microbiology* **31**(4):543–546.

Dunlop, S. G. 1968. Survival of pathogens and related disease hazards. *In* Municipal sewage effluent for irrigation. C. W. Wilson, and F. E. Beckett, (eds.). Louisiana Tech. Alumni Foundation, Tech. Station, Ruston, LA.

Fitzgerald, G. P. and G. A. Rohlich. 1959. An evaluation of stabilization pond literature, cited by Kabler, P., *In* Removal of pathogenic microorganisms by sewage treatment processes. *Sewage Ind. Wastes* **31**:12, 1373–1382.

Foster, D. H. and R. S. Engelbrecht. 1973. Microbial hazards of disposing of wastewater on soil. *In* W. E. Sopper and L. T. Kardos, eds. Recycling treated municipal wastewater and sludge through forest and cropland. The Pennsylvania State University Press, University Park, Penn. pp. 247–270.

Gallagher, R. 1969. Diseases that plague modern man. Oceana Publications, Dobbs Ferry, N.Y. 230 p.

Gerba, C. P., C. Wallis, and J. L. Melnick. 1975. Fate of wastewater bacteria and viruses in soil. *Journal of the Irrigation and Drainage Division, Proc. of the A.S.C.E.* **101**(IR3):157–174.

Gilbert, R. G. *et al.* 1976. Wastewater renovation and reuse: virus removal by soil filtration. *Science* **192**:1004–1005.

Goldshmid, J., D. Zohar, Y. Argaman, and Y. Kott. 1972. Effect of dissolved salts on the filtration of coliform bacteria in sand dunes. (Presented at the 6th International Water Pollution Research Conference).

Hickey, J. L. S. and P. C. Reist. 1975. Health significance of airborne microorganisms from wastewater treatment processes. *J. Water Pollut. Contr. Fed.* **47**(12):2741–2773.

Hickman, Cleveland P. 1966. Integrated principles of zoology. The C. V. Mosby Company, St. Louis. 965 p.

Hoadley, A. S. and S. M. Goyal. 1976. Public health implications of the application of wastewater to land. In: Land treatment and disposal of municipal and industrial wastewater. Sanks, R. L. and T. Asano (eds.) Ann Arbor, Ann Arbor Science. pp. 102–132.

Iwanszuk, J. and W. Dozanska. 1959. Effects of chlorination on the survival of *Ascaris suis* eggs in sewage. *Water Pollut. Abstr.*, **32**:2074.

Jawetz, E., J. L. Melnick, and E. A. Adelberg. 1974. Review of medical microbiology. Lange Medical Publications. Los Altos, CA 528 p.

Jensen, K. A. and K. E. Jensen. 1942. Occurrence of tubercle bacilli in sewage and experiments on sterilization of tubercle bacilli-containing sewage with chlorine. *Acta Tuberc. Pnevmol. Scan.* **16**:217.

Kabler, P. 1959. Removal of pathogenic microorganisms by sewage treatment processes. *Sewage Ind. Wastes.* **31**(12):1373–1382.

Katzenelson, E., I. Borum, and H. I. Shuval. 1976. The risk of communicable disease infection associated with wastewater irrigation in agricultural settlements. *Science.* **194**:944–946.

Kelley, S. and W. W. Sanderson. 1959. The effect of sewage treatment on viruses. *Sewage Ind. Wastes.* **31**(6): 683–689.

Kingsbury, J. M. 1964. Poisonous plants of the U.S. and Canada. Hall, NJ. 626 p.

Kradel, D. C., G. LaMotte, H. Rothenbacher, P. J. Glantz, V. Pidcoe. 1975. Comparison of the occurrence of potentially pathogenic organisms in wild animals ranging over municipal wastewater irrigated and non-irrigated areas. *In* Wood, G. W., P. J. Glantz, and D. C. Kradel (eds.). 1975 Faunal response to spray irrigation of chlorinated sewage effluent. Inst. for Res. on Land and Water Resources. The Pennsylvania State Univ. Research Publ. #87. 89 p.

Krishnaswami, S. K. 1971. Health aspects of land disposal of municipal wastewater effluents. *Can. J. Public Health.* **62**:36.

Krone, R. B., G. T. Orlob, and C. Hodgkinson. 1958. Movement of coliform bacteria through porous media. *Sewage & Industrial Wastes.* **30**:1–13.

Krone, R. B. 1968. The movement of disease producing organisms through soil. (Presented at the Symposium on the Use of Municipal Sewage Effluent for Irrigation. Louisiana Polytechnic Institute, Ruston.)

Kruse, C. W., Y. C. Hsu, A. C. Griffiths, and R. Stringer. 1970. Halogen action of bacteria, viruses, and protozoa. Proc. National Specialty Conf. on Disinfection, 113. A.S.C.E., New York.

Lance, J. C., C. P. Gerba, and J. L. Melnick. 1976. Virus movement in soils columns flooded with secondary sewage effluent. *Applied Environmental Microbiology.* **32**(4):520–526.

Liu, O. C., H. R. Seraichekas, E. W. Akin, D. A. Brashear, E. L. Katy, and W. F. Hill, Jr. 1971. Relative resistance of 20 human viruses to free chlorine in Potomic water. *In* Virus and water quality, occurrence and control. 13th Water Quality Conf. Univ. of Illinois. 222 p.

McGauhey, P. H. and R. B. Krone. 1967. Soil mantle as a wastewater treatment system. Sanitary Eng. Res. Laboratory Report No. 67–11. Berkeley, Univ. of California. 201 p.

McKinney, R. E., H. E. Langely, and H. D. Tomlinson. 1959. Survival of *Salmonella typhosa* during anaerobic digestion. *Sewage Ind. Wastes.* **30**:1469–1477.

Merrell, J. C., Jr. and P. C. Ward. 1968. Virus control at the Santee, Calif. project. *J. Amer. Water Works Assoc.* **60**:145–153.

Metcalf and Eddy, Inc. 1977. Process design manual for land treatment of municipal wastewater. U.S. EPA, U.S. Army Corps of Engineers, U.S. Dept. of Agriculture. EPA 625/1-77-008. EPA Technology Transfer, Washington, D.C.

Meyer, R. C., F. C. Hinds, H. R. Isacson, and T. D. Hinesly. 1971. Porcine enterovirus survival and anaerobic sludge digestion. Proc. Intl. Symposium on Livestock Wastes, 183, A.S.A.E. St. Joseph, Michigan.

Miller, R. H. 1973. Some microbiological aspects of recycling sewage sludges and waste effluents on land. *In* Proc. of a joint conf. on recycling municipal sludges and effluents on land. EPA/USDA/Nat'l. Ass. of State Universities and Land Grant Colleges. Champaign, Illinois. July 9–13, pp. 79–90.

Miller, R. H. 1973a. The microbiology of sewage sludge decomposition in soil. Unpublished report. Dept. of Agronomy, The Ohio State University.

Mirzoev, G. G. 1968. Extent of survival of dysentery bacilli at low temperatures and self-disinfection of soil and water in the far north. *Hygiene and Sanitation.* **31**:437–439.

Napolitano, P. J. and D. R. Rowe. 1966. Microbial content of air near sewage treatment plants. *Water Sewage Works.* **11**(12):480–483.

Nelson, N., *et al.* 1971. Hazards of mercury. *Environ. Res.* **4**(1):1–69. Academic Press, New York.

Newfield, R. D. and E. R. Hermann. 1975. Heavy metal removal by acclimated activated sludge. *J. Water Pollut. Contr. Fed.* **47**(2):310–329.

Newton, W. L., A. J. Bennett, and W. B. Figgatt. 1949. Observations of the effects of various sewage treatment processes upon eggs of *Taenia saginata. Amer. J. Hygiene.* **49**(2):166–175.

New York State Health Commission. 1932. Public health in New York State. Albany.

Nordberg, G. F. 1974. Health hazards of environmental cadmium pollution. *Ambio* **3**(2):55–66.

Norman, U. H. and P. W. Kablen. 1953. Bacteriological study of irrigated vegetables. *Sewage Ind. Wastes.* **25**(5):605–609.

Parizek, R. R., L. T. Kardos, W. E. Sopper, E. A. Myers, D. E. Davis, M. A. Farrell, and J. B. Nesbitt. 1967. Wastewater renovation and conservation. Penn State Studies No. 23. The Pennsylvania State University, University Park, Penn. 71 p.

Parsons, D., C. Brownlee, D. Wetter, A. Maurer, E. Haughton, L. Kornder, M. Slezak. 1975. Health aspects of sewage effluent irrigation. Pollu. Cont. Branch, British Columbia Water Res. Service, Dept. of Lands, Forests, and Water Res., Parliament Bldgs. Victoria, British Columbia. 75 p.

Reed, S. C., *et al.* 1972. Wastewater management by disposal on the land. Special Report 171. Cold Regions Res. & Eng. Lab., Hanover, New Hampshire. 183 p.

Rudolfs, W., L. L. Falk, and R. A. Ragotzkie. 1950. Literature on the occurrence and survival of enteric, pathogenic, and related organisms in soil, water, sewage, and sludges, and on vegetation: II. Animal parasites. *Sewage Ind. Wastes.* **22**(11):1417–1427.

Rudolfs, W. L., L. L. Falk, and R. A. Ragotzkie. 1951. Contamination of vegetables grown in polluted soil, II. Field and laboratory studies on *Entamoeba* cysts. *Sewage Works J.*, **23**:478.

Schaub, S. A., *et al.* 1975. Land application of wastewater: The fate of viruses, bacteria, and heavy metals at a rapid infiltration site. TR7504, U.S. Army Medical Bioengineering Research and Development Laboratory, Fort Detrick, Frederick, Md.

Sorber, C. A. 1973. Protection of public health. Land disposal of municipal effluents and sludges. EPA-902-9-73-001. U.S. Environ. Protection Agency, New York.

Sorber, C. A., S. A. Schaub, and H. T. Bausum. 1974. An assessment of a potential virus hazard associated with spray irrigation of domestic wastewaters. *In* J. F. Malina, Jr. and B. P. Sagik. eds. Virus survival in water and wastewater systems. Center for Res. in Water Resources, University of Texas, Austin.

Sorber, C. A., H. T. Bausum, and S. A. Schaub. 1975. Bacterial aerosols created by spray irrigation of wastewater. *In* Wastewater resource manual. Sprinkler Irrigation Association, Silver Spring Maryland. Section 1B9:33–43.

Sproul, O. J. 1975. Virus movement into groundwater from septic tank systems. *In* W. J. Jewell and R. Swan (eds.). Water pollution control in low density areas. Proc. of a rural environmental eng. conf. Published for the Univ. of Vermont by the Univ. Press of New England, Hanover, N.H.

Tannock, G. W. and J. M. B. Smith. 1971. Studies on the survival of *Salmonella typhimurium* and *S. bovis-morbificans* on pasture and in water. *Aust. Veg. J.* **47**:557–559.

Taras, M. J., A. E. Greenberg, R. D. Hoak, and M. C. Rand. 1971. Standard methods for the examination of water and wastewater. Amer. Public Health Assn. Washington, D.C. 874 p.

Taylor, T. J. and M. R. Burrows. 1971. The survival of *Escherichia coli* and *Salmonella dublin* in slurry on pasture and the infectivity of *S. dublin* for grazing calves. *Brit. Vet. J.* **127**:536–543.

Trask, J. D., J. L. Melnick, and H. A. Wenner. 1945. Chlorination of human, monkey-adapted, and mouse strains of poliomyelitis virus. *American Journal of Hygiene* **41**:30–40.

Walker, B. 1970. Viruses respond to environmental exposure. *J. Environ. Health.* **32**(5):532–550.

Wallis, C., M. Henderson, and J. L. Melnick. 1972. Enterovirus concentration on cellulose membranes. *Applied Microbiology* **23**(3):476–480.

Wang, W. L. and S. G. Dunlop. 1959. Animal parasites in sewage and irrigation water. *Sewage Ind. Waste* **26**(8):1020–1032.

Weber, W. J., Jr. 1972. Physicochemical processes for water quality control. Wiley-Interscience, New York 640 p.

Wellings, F. M., A. L. Lewis, and C. W. Mountain. 1974. Virus survival following wastewater spray irrigation of sandy soils. *In:* Virus survival in water and wastewater systems. Malina, J. F., Jr. and B. P. Sagik (eds.) Austin, Center for Research in Water Resources. pp. 253–260.

Wellings, F. M., A. L. Lewis, and C. W. Mountain. 1975. Pathogenic viruses may thwart land disposal. *Water Wastes Eng.* Mar. Vol. 12.

APPENDIX A

VARIOUS GOVERNMENT, AGENCY GUIDELINES FOR USE OF RECLAIMED WASTEWATER

GUIDELINES FOR WASTEWATER REUSE VARY

To illustrate how two regional governments have responded to the risk of disease transmission from wastewater, we have included the California guidelines for use of reclaimed water and the

British Columbia Water Resources Service guidelines on sewage effluent irrigation. The British Columbia Water Resources Service recommends those guidelines established by the World Health Organization.

In the third section of Appendix A are guidelines on wastewater reuse set up by the National Technical Advisory Committee on Water Quality and the World Health Organization.

In the final section are suggestions for crop use of wastewater made by leading researchers in the field.

SUMMARY AND CONCLUSIONS FROM

HEALTH ASPECTS OF SEWAGE EFFLUENT IRRIGATION

by

D. Parsons, C. Brownlee, D. Wetter,
A. Maurer, E. Haughton, L. Kornder,
and M. Slezak

April, 1975

British Columbia Water Resources Service

1. Treated sewage effluents represent a valuable source of water and nutrients which may be reused to advantage in land irrigation systems. The degree of risk to public health which is associated with effluent irrigation is related to the microbial characteristics of the untreated effluent, the nature of the treatment process, the nature of the crop being irrigated, and the characteristics and location of the irrigated area.
2. In many parts of North America, regulatory agencies have established bacteriological guidelines for agricultural irrigation waters. These guidelines generally allow the use of water having from 1,000 to 10,000 MPN total coliforms/100 ml. Some states have also established guidelines for the agricultural use of sewage effluent.

   In California, for example, primary effluents are considered suitable for the surface or spray irrigation of fodder, fiber, and seed crops, and for the surface irrigation of produce eaten cooked. Effluents used to irrigate dairy pastures and to spray irrigate produce eaten raw should have MPN coliform levels of $< 23/100$ ml.

   Raw or untreated sewage cannot be used for irrigation purposes in Oregon. An adequately disinfected secondary effluent is required for the surface or spray irrigation of alfalfa, grass, timber land, or other fodder crops, but coliform levels of up to 1000/100 ml are permitted.
3. A meeting of experts convened by the World Health Organization concluded that primary treatment would be sufficient to permit reuse through irrigation of crops not for direct human consumption. Secondary treatment and most probably disinfection and filtration were considered necessary if the effluent were to be used for irrigation of produce for human consumption.
4. A review of literature has shown that outbreaks of communicable disease have been caused by sewage discharges and irrigating produce which is eaten raw with sewage or sewage-polluted water. In all cases, night soil, untreated sewage, or primary sludge has been the source of contamination. Most outbreaks have been associated with the ingestion of contaminated water, the consumption of shellfish raised in contaminated water, or the consumption of fruits and vegetables, particularly leafy vegetables, irrigated with sewage or sewage-polluted water and eaten raw.
5. Raw sewage will likely contain each of the intestinal bacteria, viruses, worms, and protozoa that infect the population from which the sewage is derived. The numbers of each micro-

organism present in the sewage depends upon geographical and cultural factors and upon the extent of sewage dilution during transport.

6. Effluents can be applied to land by subsurface discharge, by flooding, by overland flow, by ridge and furrow irrigation, or by spray irrigation systems. The spray irrigation of effluent will result in the formation of aerosols, and these may contain pathogenic microorganisms. Coliform organisms have been detected within the mist zone of spray irrigation systems.
7. Although there are numerous diseases which may be transmitted through sewage, relatively few are of significance in British Columbia. The incidence of typhoid, paratyphoid, cholera and poliomyelitis is low in British Columbia and throughout North America. Tuberculosis and brucellosis have been eradicated from dairy cattle in Canada. Schistosomiasis, strongyloidiasis, and hookworm infections are diseases of the tropics. The numbers of the pathogenic organisms which cause any of these diseases or infections should therefore be extremely low in sewage effluents in British Columbia. It is unlikely that the use of treated sewage effluents for irrigation will alter the incidence of these diseases in this province.
8. There are some pathogenic microorganisms which can be transmitted via sewage and which may be present in raw sewage in British Columbia. These include the causative agents of salmonellosis, shigellosis, infectious hepatitis, taeniasis, enterobiasis, ascariasis, amoebiasis, and giardiasis.

   Because of the possible presence of these organisms in raw sewage, the use of untreated or of primary treated sewage effluent for irrigation purposes constitutes a potential hazard to public health and should not be allowed. Decisions concerning the degree of treatment required before irrigation, system design, buffer zones, and site or crop selection should be made in order to protect the public from these or similar organisms.
9. Primary treatment will remove many of the ova of parasites from raw sewage, but few of the bacteria or viruses. Trickling filters will remove an additional 60% to 70% of the parasite ova and 70% to 90% of pathogenic bacteria, but very few viruses. Slow sand filters, however, can provide virtually complete removal of cysts and ova and greatly reduce viral concentrations.

   Activated sludge systems remove more than 90% of the bacteria and viruses remaining after primary treatment. Aeration has little effect on parasites, but settling during treatment in an activated sludge system should significantly reduce the numbers of parasites present in the effluent.

   Greater than 90% removal of bacteria and viruses is possible with lagoons or stabilization ponds which provide several months retention. Settling during storage in such ponds should also remove the ova and cysts of intestinal parasites. Storage for one month or longer also exceeds the normal life span of many pathogens or parasites.

   Effluent chlorination has little effect on parasite ova and amoeba cysts. 95 to 99+% removal of most bacteria from secondary effluents can be obtained through chlorination if residual of 1 mg/l is maintained for one hour. Such chlorination of highly clarified effluents may also provide freedom from viruses.
10. There is no single sewage treatment process which will remove all pathogenic microorganisms. Through combinations of treatment processes however, it is possible to produce effluents from which all, or almost all, of each pathogen group has been removed. In order to produce an effluent suitable for irrigation of a wide variety of crops and landscapes, it will probably be necessary to combine an aerobic, biological treatment process with clarification (through settling or filtration), disinfection, and storage.

(Detailed recommendations concerning effluent treatment and irrigation for specific purposes are presented in the following chapter, not reproduced here.)

Table 1. Treatment Processes Suggested by the World Health Organization for Wastewater Reuse.

	Irrigation			Recreation	
	Crops not for Direct Human Consumption	Crops Eaten Cooked Fish Culture	Crops Eaten Raw	No Contact	Contact
Health criteria (see below for explanation of symbols)	1 + 4	2 + 4 or 3 + 4	3 + 4	2	3 + 5
Primary treatment	●●●	●●●	●●●	●●●	●●●
Secondary treatment		●●●	●●●	●●●	●●●
Sand filtration or equivalent polishing methods		●	●		●●●
Disinfection		●	●●●	●	●●●

Health criteria:
1 Freedom from gross solids; significant removal of parasite eggs.
2 As 1, plus significant removal of bacteria.
3 Not more than 100 coliform organisms per 100 ml in 80% of samples.
4 No chemicals that lead to undesirable residues in crops or fish.
5 No chemicals that lead to irrigation of mucous membranes and skin.

In order to meet the given health criteria, processes marked ● ● ● will be essential. In addition, one or more processes marked ● ● will also be essential, and further processes marked ● may sometimes be required.

STATE OF CALIFORNIA DEPARTMENT OF HEALTH GUIDELINES FOR USE OF RECLAIMED WATER

Berkeley, CA 1974

STATE OF CALIFORNIA DEPARTMENT OF HEALTH

GUIDELINES FOR USE OF RECLAIMED WATER FOR SPRAY IRRIGATION OF CROPS

1. Reclaimed water shall meet the Regional Water Quality Control Board requirements and the quality requirements established by the State of California Department of Health for health protection.
2. The discharge shall be confined to the area designated and approved for disposal and reuse. Irrigation should be controlled to minimize ponding of wastewater and runoff should be contained and properly disposed.
3. Maximum attainable separation of reclaimed water lines and domestic water lines shall be practiced. Domestic and reclaimed water transmission and distribution mains shall conform to the "Separation and Construction Criteria" (see attached).

a. The use area facilities must comply with the "Regulations Relating to Cross-Connections," Title 17, Chapter V, Section 7583–7622, inclusive, California Administrative Code.
b. Plans and specifications of the existing and proposed reclaimed water system and domestic water system shall be submitted to State and/or local health agencies for review and approval.

4. All reclaimed water valves and outlets should be appropriately tagged to warn the public that the water is not safe for drinking or direct contact.
5. All piping, valves, and outlets should be color-coded or otherwise marked to differentiate reclaimed water from domestic or other water.
   a. Where feasible, differential piping materials should be used to facilitate water system identification.
6. All reclaimed water valves and outlets should be of a type that can only be operated by authorized personnel.
7. Adequate means of notification shall be provided to inform the public that reclaimed water is being used. Conspicuous warning signs with proper wording of sufficient size to be clearly read shall be posted at adequate intervals around the use area.
8. Irrigation should be done so as to prevent or minimize contact by the public with the sprayed material and precautions should be taken to insure that reclaimed water will not be sprayed on walkways, passing vehicles, buildings, domestic water facilities, or areas not under control of the user.
   a. The irrigated areas should be fenced where primary effluent is used.
   b. Windblown spray from the irrigation area should not reach areas accessible to the public.
   c. Irrigated areas must be kept completely separated from domestic water wells and reservoirs. A minimum of 500 feet should be provided.
9. Adequate measures should be taken to prevent the breeding of flies, mosquitoes and other vectors of public health significance during the process of reuse.
10. Operation of the use area facilities should not create odors, slimes, or unsightly deposits of sewage origin.
11. Adequate time should be provided between the last irrigation and harvesting to allow the crops and soil to dry.
    a. Animals, especially milking animals, should not be allowed to graze on land irrigated with reclaimed water until it is thoroughly dry.
12. There should be no subsequent planting of produce on lands irrigated with primary effluent.

STATE OF CALIFORNIA DEPARTMENT OF HEALTH

GUIDELINES FOR USE OF RECLAIMED WATER FOR SURFACE IRRIGATION OF CROPS

1. Reclaimed water shall meet the Regional Water Quality Control Board requirements and the quality requirements established by the State of California Department of Health for health protection.
2. The discharge shall be confined to the area designated and approved for disposal and reuse. Irrigation should be controlled to minimize ponding of wastewater and runoff should be contained and properly disposed.
3. Maximum attainable separation of reclaimed water lines and domestic water lines shall be practiced. Domestic and reclaimed water transmission and distribution mains shall conform to the "Separation and Construction Criteria" (see attached).
   a. The use area facilities must comply with the "Regulations Relating to Cross-Connections," Title 17, Chapter V, Sections 7583–7622, inclusive, California Administrative Code.
   b. Plans and specifications of the existing and proposed reclaimed water system and domestic water system shall be submitted to State and/or local health agencies for review and approval.

4. All reclaimed water valves and outlets should be appropriately tagged to warn the public that the water is not safe for drinking or direct contact.
5. All piping, valves, and outlets should be color-coded or otherwise marked to differentiate reclaimed water from domestic or other water.
   a. Where feasible, differential piping materials should be used to facilitate water system identification.
6. All reclaimed water valves and outlets should be of a type that can only be operated by authorized personnel.
7. Adequate means of notification shall be provided to inform the public that reclaimed water is being used. Conspicuous warning signs with proper wording of sufficient size to be clearly read shall be posted at adequate intervals around the use area.
8. The public shall be effectively excluded from contact with the reclaimed water used for irrigation.
   a. The irrigated areas should be fenced where primary effluent is used.
   b. Irrigated areas must be kept completely separated from domestic water wells and reservoirs. A minimum of 500 feet should be provided.
9. Adequate measures should be taken to prevent the breeding of flies, mosquitoes and other vectors of public health significance during the process of reuse.
10. Operation of the use area facilities should not create odors, slimes, or unsightly deposits of sewage origin.
11. Adequate time should be provided between the last irrigation and harvesting to allow the crops and soil to dry.
    a. Animals, especially milking animals, should not be allowed to graze on land irrigated with reclaimed water until it is thoroughly dry.
12. There should be no subsequent planting of produce on lands irrigated with primary effluent.
13. Adequate measures shall be taken to prevent any direct contact between the edible portion of the crops and the reclaimed water.

STATE OF CALIFORNIA DEPARTMENT OF HEALTH

GUIDELINES FOR USE OF RECLAIMED WATER FOR LANDSCAPE IRRIGATION

1. Reclaimed water shall meet the Regional Water Quality Control Board requirements and the quality requirements established by the State of California Department of Health for health protection.
2. The discharge shall be confined to the area designated and approved for disposal and reuse. Irrigation should be controlled to minimize ponding of wastewater and runoff should be contained and properly disposed.
3. Maximum attainable separation of reclaimed water lines and domestic water lines shall be practiced. Domestic and reclaimed water transmission and distribution mains shall conform to the "Separation and Construction Criteria" (see attached).
   a. The use area facilities must comply with the "Regulations Relating to Cross-Connections," Title 17, Chapter V, Sections 7583–7622, inclusive, California Administrative Code.
   b. Plans and specifications of the existing and proposed reclaimed water system and domestic water system shall be submitted to State and/or local health agencies for review and approval.
4. All reclaimed water valves, outlets and/or sprinkler heads should be appropriately tagged to warn the public that the water is not safe for drinking or direct contact.
5. All piping, valves, and outlets should be color-coded or otherwise marked to differentiate reclaimed water from domestic or other water.
   a. Where feasible, differential piping materials should be used to facilitate water system identification.

6. All reclaimed water valves, outlets, and sprinkler heads should be of a type that can only be operated by authorized personnel.
   a. Where hose bibbs are present on domestic and reclaimed water lines, differential sizes should be established to preclude the interchange of hoses.
7. Adequate means of notification shall be provided to inform the public that reclaimed water is being used. Such notification should include the posting of conspicuous warning signs with proper wording of sufficient size to be clearly read. At golf courses, notices should also be printed on score cards and at all water hazards containing reclaimed water.
8. Tank trucks used for carrying or spraying reclaimed water should be appropriately identified to indicate such.
9. Irrigation should be done so as to prevent or minimize contact by the public with the sprayed material and precautions should be taken to insure that reclaimed water will not be sprayed on walkways, passing vehicles, buildings, picnic tables, domestic water facilities, or areas not under control of the user.
   a. Irrigation should be practiced during periods when the grounds will have maximum opportunity to dry before use by the public unless provisions are made to exclude the public from areas during and after spraying with reclaimed water.
   b. Windblown spray from the irrigation area should not reach areas accessible to the public.
   c. Irrigated areas must be kept completely separated from domestic water wells and reservoirs. A minimum of 500 feet should be provided.
   d. Drinking water fountains should be protected from direct or windblown reclaimed water spray.
10. Adequate measures should be taken to prevent the breeding of flies, mosquitoes and other vectors of public health significance during the process of reuse.
11. Operation of the use area facilities should not create odors, slimes, or unsightly deposits of sewage origin in places accessible to the public.

STATE OF CALIFORNIA DEPARTMENT OF HEALTH

GUIDELINES FOR USE OF RECLAIMED WATER FOR IMPOUNDMENTS

1. Reclaimed water shall meet the Regional Water Quality Control Board requirements and the quality requirements established by the State of California Department of Health for health protection.
2. The discharge shall be confined to the area designated and approved for disposal and reuse. Runoff should be contained and properly disposed.
3. Maximum attainable separation of reclaimed water lines and domestic water lines shall be practiced. Domestic and reclaimed water transmission and distribution mains shall conform to the "Separation and Construction Criteria" (see attached).
   a. The use area facilities must comply with the "Regulations Relating to Cross-Connections," Title 17, Chapter V, Sections 7583–7622, inclusive, California Administrative Code.
   b. Plans and specifications of the existing and proposed reclaimed water system and domestic water system shall be submitted to State and/or local health agencies for review and approval.
4. At restricted recreational impoundments and landscape impoundments all valves and outlets should be appropriately tagged to warn the public that the water is not safe for drinking or bathing.
5. At non-restricted recreational impoundments all valves and outlets should be appropriately tagged to warn the public that the water is reclaimed from sewage and is not safe for drinking.
6. All piping, valves, and outlets should be color-coded or otherwise marked to differentiate reclaimed water from domestic or other water.

a. Where feasible, differential piping materials should be used to facilitate water system identification.
7. All reclaimed water valves and outlets should be of a type that can only be operated by authorized personnel.
8. Adequate means of notification shall be provided to inform the public that reclaimed water is being used. Such notification should include the posting of conspicuous warning signs with proper wording of sufficient size to be clearly read.
9. Adequate measures shall be taken to prevent body-contact activities, such as wading or swimming, at restricted recreational impoundments containing reclaimed water.
10. Adequate measures shall be taken to prevent direct public contact with reclaimed water at landscape impoundments.
11. Restricted and non-restricted recreational impoundments shall be maintained under the continuous supervision of qualified personnel during periods of use.
12. Impoundments containing reclaimed water must be kept completely separated from domestic water wells and reservoirs. A minimum of 500 feet should be provided.
13. Adequate measures should be taken to prevent the breeding of flies, mosquitoes and other vectors of public health significance during the process of reuse.
14. Operation of the use area facilities should not create odors, slimes, or unsightly deposits of sewage origin in places accessible to the public.

STATE OF CALIFORNIA DEPARTMENT OF HEALTH

GUIDELINES FOR WORKER PROTECTION AT WATER RECLAMATION USE AREAS

1. Employees should be made aware of the potential health hazards involved with contact or ingestion of reclaimed water.
2. Employees should be subjected to periodic medical examinations for intestinal diseases and to adequate immunization shots.
3. Adequate first aid kits should be available on location, and all cuts and abrasions should be treated promptly to prevent infection. A doctor should be consulted where infection is likely.
4. Precautionary measures should be taken to minimize direct contact of employees with reclaimed water.
   a. Employees should not be subjected to reclaimed water sprays.
   b. For work involving more than a casual contact with reclaimed water, employees should be provided with protective clothing.
   c. At crop irrigation sites, the crops and soil should be allowed to dry before harvesting by employees.
5. Provisions should be made for a supply of safe drinking water for employees. Where bottled water is used for drinking purposes, the water should be in contamination-proof containers and protected from contact with reclaimed water or dust.
   a. The water should be of a source approved by the local health authority.
6. Toilet and washing facilities should be provided.
7. Precautions should be taken to avoid contamination of food taken to areas irrigated with reclaimed water, and food should not be taken to areas still wet with reclaimed water.
8. Adequate means of notification shall be provided to inform the employees that reclaimed water is being used. Such notification should include the posting of conspicuous warning signs with proper wording of sufficient size to be clearly read.
   a. In some locations, especially at crop irrigation use areas, it is advisable to have the signs in Spanish as well as English.
9. All reclaimed water valves, outlets, and/or sprinkler heads should be appropriately tagged

SEPARATION AND CONSTRUCTION CRITERIA DOMESTIC AND RECLAIMED WASTEWATER TRANSMISSION AND DISTRIBUTION MAINS

Basic Separation		Water Main Involved		Reclaimed Wastewater Main Construction Minimum Separation if Basic Separation is not Feasible		
Parallel Construction	Perpendicular Construction	Reclaimed Wastewater	Domestic Water	Parallel Construction	Perpendicular Construction Reclaimed wastewater main *above* domestic water main	Perpendicular Construction Reclaimed wastewater main *below* domestic water main Clearance less than three (3) feet
25 ft[a]	3 ft[b]	Pressure	Gravity[c]	No Exception	Minimum pipe class 2 × wwp; Steel casing 25 ft both sides of crossings	Minimum pipe class 2 × wwp; Steel casing 25 ft both sides of crossing
25 ft	3 ft	Gravity	Gravity	VCP, AC, CIP, or equal, class 150; 15 ft minimum separation; Mechanical compression joints	Steel casing 25 ft both sides of crossing	VCP, AC, CIP, or equal, class 150; Mechanical compression joints 25 ft both sides of crossing
10 ft	3 ft	Pressure	Pressure	Minimum pipe class 2 × wwp; 4 ft minimum separation; no common trench	Minimum pipe class 2 × wwp; Mechanical compression joints 4 ft both sides of crossing	Minimum pipe class 2 × wwp; Mechanical compression joints 4 ft both sides of crossing
10 ft	3 ft	Gravity	Pressure	VCP; Mechanical compression joints 4 ft minimum separation	Concrete encasement or steel casing 4 ft both sides of crossing	VCP, AC, CIP; Mechanical compression joints 4 ft both sides of crossing

[a] All distances measured from pipeline O.D.
[b] Domestic water main 3 ft above reclaimed wastewater main.
[c] Less than 5 psi.

to warn employees that the water is not safe for drinking or direct contact (direct contact is allowed at non-restricted recreational impoundments).

10. All piping, valves and outlets should be color-coded or otherwise marked to differentiate reclaimed water from domestic or other water.
    a. Where feasible, differential piping materials should be used to facilitate water system identification.
11. All reclaimed water valves, outlets and sprinkler heads should be of a type that can only be operated by authorized personnel.
    a. Where hose bibbs are present on domestic and reclaimed water lines, differential sizes should be established to preclude the interchange of hoses.

RECOMMENDATIONS FROM NATIONAL TECHNICAL ADVISORY COMMITTEE ON WATER QUALITY

Federal Water Pollution Control Administration, 1968

For waters intended for agricultural use, the monthly average coliform bacteria counts should not exceed 5,000/100 ml and the fecal coliform concentration should not exceed 1,000/100 ml

For recreational water, the coliforms should not exceed 1,000/100 ml.

For drinking water, the coliforms should not exceed 1/100 ml.

LIMITATIONS ON USING WASTEWATER ON AGRICULTURAL CROPS

A Summary of Relevant Research

Limitations on crop use put forth by Rudolfs, Frank, and Ragotzkie (1950); Krishnaswami (1971); and Geldreich and Bordner (1971) can be summarized as follows:

1. Crops that are eaten after they are cooked, or industrial crops that are eaten after satisfactory processing, may be irrigated with treated wastewater.
2. Oxidized and disinfected wastewater effluent may be used to irrigate fruit and vegetable crops. Vegetables should not be sprinkler irrigated for 4 weeks prior to harvest. Similarly, application on pasture and hay should stop 2 weeks before pasturing or harvesting. (This provides a drying period for farm equipment access.)
3. Reclaimed water used for the surface or spray irrigation of fodder, fiber, and seed crops shall have a level of quality no less than that of primary effluent.

Module II-6

CLIMATE AND WASTEWATER STORAGE

SUMMARY

Five components of the hydrologic cycle are of utmost importance in land application systems. These are precipitation, infiltration/percolation, evapotranspiration, runoff, and groundwater. An understanding of each of these is necessary prior to the evaluation and design of land application systems. Hydrologic events which occur during the application season must be evaluated separately from those which occur during the nonapplication or storage season.

The application season for the site under investigation must be defined. Undoubtedly, it will include the growing season, but how far may it stretch into the spring and fall? This is dependent upon the waste characteristics, soil-plant treatment medium, and the groundwater quality required. Precipitation during the application season affects the number of application days. This interruption in the irrigation scheduling should be estimated and built into the system. Extreme rainfall events must be contained. Water sinks such as evapotranspiration, infiltration/percolation, and runoff may all be estimated. These capacities must be maximized within the limitations of the system under investigation. The soils, vegatative cover, and effluent quality are integral components in estimating these limitations.

The nonapplication season defines the storage period. The storage period is defined in terms of climatic constraints which the designer imposes on the treatment system. These constraints vary according to the location of the site and may include temperature, precipitation, snow cover, soil moisture, etc. These factors can be analyzed by computer programs developed by the National Weather Service. The programs are summarized in this module, and Appendix A is a reprint of an Environmental Protection Agency publication describing the programs and their use in detail.

Hydrologic events which affect cropping systems are not of concern during the storage period. The storage volume required is dependent upon the flow of wastewater during the storage season, rainfall into the reservoir, evaporation from the reservoir, and seepage from the reservoir. These may all be estimated with reasonable accuracy.

CONTENTS

GLOSSARY

available moisture—The portion of soil water that can be readily absorbed by plant roots. Usually defined as water held in the soil against a pressure of up to 15 bars (atm).

dew point—The temperature at which air becomes saturated when cooled without addition of moisture or change of pressure.

evapotranspiration—The combined loss of water from an area by evaporation from soil, water and plant surfaces and transpiration (release of vapor) from plants.

field capacity—The moisture content which a soil will hold against the force of gravity after it has drained for 48 hours following saturation; the amount of water retained by the soil micropores.

groundwater—Water that fills all the unblocked pores of underlying material below the water table, which is the upper limit of saturation.

growing season—The period of the year when climatic conditions are favorable for plant growth, common to a given area.

infiltration—The downward entry of water into the soil.

leachate—Water which enters the groundwater from a land application site.

percolation–The downward movement of water through soil.
precipitation–Any solid or liquid water particles which fall from the atmosphere and reach the ground. Also the amount, expressed in terms of depth of liquid water, which has fallen at a given point over a specified period of time.
regolith–The unconsolidated mantle of earth above the bedrock resulting from weathering of the underlying rock and/or deposition of material transported by gravity, wind, water or ice.
soil moisture tension–The equivalent negative pressure in the soil water, equal to the pressure which must be applied to the soil water to bring it to hydraulic equilibrium, through a porous permeable membrane, with a pool of water of the same composition.

OBJECTIVES

Upon completion of this module, the reader should be able to:

1. Discuss each component of the hydrologic cycle which has an impact on land application systems. Which components are subject to control by design variables?
2. Define the application season and nonapplication period for a given site.
3. Obtain the applicable computer program for storage requirements from the National Climatic Center, choose proper input variables, and use the output in design of storage, for a given land application site.
4. Differentiate between climatic and hydrologic factors of importance during the application and nonapplication periods, and perform water balance calculations for each period.

INTRODUCTION: THE HYDROLOGIC CYCLE

The hydrologic cycle, Figure 1, is an accumulation of many climatic and hydrogeologic phenomena which describes the cyclic nature of water movement in the natural environment. Solar energy induces the cycle through evaporation and transpiration of water from aquatic and terrestrial surfaces. Air masses collect this water vapor and transport it large distances. When the temperature and pressure of the air mass reflect the dew point, precipitation occurs. Once reaching the earth, the precipitation is either intercepted by vegetation, infiltrated into the soil profile, or accumulated in runoff to surface water bodies whereupon it is recycled.

**Several water movement mechanisms of the hydrologic cycle affect the water balance at land application sites. These factors can influence irrigation schedules, application rates, drainage design, storage design, crop selection, and other aspects of the system.*

Six components of the hydrologic cycle are reviewed, each with respect to land application systems. These include: (1) precipitation, (2) infiltration/percolation, (3) transpiration, (4) evaporation, (5) runoff, and (6) groundwater.

Precipitation includes all forms of moisture falling from clouds such as rain, snow, sleet or hail. Each season has an established precipitation pattern which we expect year after year. In the evaluation and design of land application systems, these precipitation patterns must be known in order that we may (1) impose our artificial components upon the hydrologic cycle

*This and other italicized summaries are intended to highlight key ideas, provide a basis for later review or to aid in skimming sections that are relatively familiar. They can be ignored in a complete reading of the text.

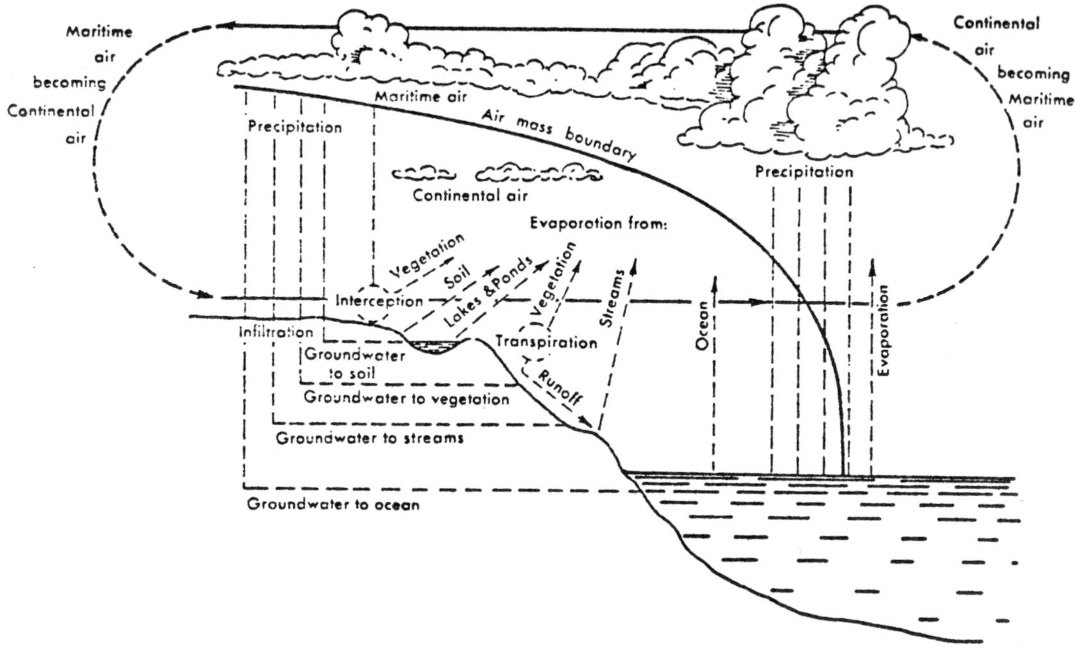

Figure 1. The hydrologic cycle ("Climate and Man," USDA Yearbook of Agriculture, 1941).

with little, if any, affect upon the surrounding ecosystem, and (2) avoid hydrologic events which may cause the land application system to fail.

Precipitation seasonal patterns may be deduced from published climatological data. They will affect application scheduling, drainage and storage design.

The *infiltration and subsequent percolation* of water into and through the soil profile is the major replenishment of groundwater. The variables associated with this penetration and transport of water are many, however only a few may be changed so as to enhance the infiltration capacity of the soil. These include establishment of vegetative cover and installation of graded gravel filters and subsurface drains. Penetration of water into agricultural soils generally proceeds at rates varying between 0.05 and 5 in./hr. Application volumes and rates, e.g., 4 in./wk at $\frac{1}{4}$ in./hr, utilized in land application systems fall well within the general range of soil permeabilities found in irrigated agriculture. Therefore, unless unusually heavy and prolonged precipitation is encountered, hydraulic conductivity will not be a limiting factor where irrigated crops are maintained.

Infiltration/percolation rates have a bearing on process selection. For crop irrigation systems, infiltration rates are generally much in excess of application rates; infiltration/percolation capacity can be enhanced by vegetative cover, gravel filters, and underdrainage.

Evapotranspiration (evaporation and transpiration) from soil and plant surfaces varies with the soil moisture conditions, vegetative cover, and solar energy available at the site. The soil moisture condition will be increased considerably during the application season. This increase in soil moisture will also increase the evapotranspiration rate of the land application surface.

Evapotranspiration is composed of evaporation from soil and plant surfaces—a function of solar and wind energy—and transpiration—a function of plant uptake and soil moisture.

Runoff of wastewater and precipitation from land application sites must be controlled for two obvious reasons: (1) contact time between wastewater and the treatment system must be maintained to ensure removal of pollutants, and (2) extreme rainfalls must be collected and routed at nonerosive flows to protect the wastewater application site from excessive soil erosion. Once collected and treated, this water may be released to stream flow.

Runoff control affects contact time of applied wastewater and possible soil erosion. Collected runoff may be returned to the head end of the treatment system.

Groundwater within the regolith replenishes water utilized through evapotranspiration, stream flow, and evaporation of large water bodies. The groundwater beneath land application sites has been protected from polluted leachates by PL 92-500, the Water Pollution Control Act Amendments of 1972. Prior to reaching groundwater levels, the leachate must meet groundwater quality criteria as outlined in PL 93-523, the Safe Drinking Water Act. Some flexibility on permissable water quality will follow from the states' interpretations of these laws. Module I-9, "Legal Aspects," contains a full discussion of these points.

Groundwater quality criteria for land application sites is set out in Federal legislation.

Perhaps the two most comprehensive data sources describing the details of the hydrologic cycle for all areas of the U.S. are the U.S. Geological Survey and the National Weather Service. The USGS is most well known for the extensive information it provides on surface water flow. Of major concern to those working on land application systems is the extensive network of wells and groundwater quality information which is also reported by USGS. This information is useful in establishing the background quality of groundwater in regions considered as potential land treatment sites. The National Oceanic and Atmospheric Administration (NOAA), National Weather Service (NWS) maintains approximately 3,000 weather stations across the United States. These report a variety of weather data which is accumulated and published as "Climatological Data." Precipitation data may be found in this publication as well as average monthly temperatures for use in determining evapotranspiration of the land application site. This data is valuable in determining specific storage requirements required for varying climates. Soils and cover crop information may be obtained from USDA Soil Conservation Service soil surveys, Module I-6, "Site Evaluation" and Module II-7, "Crop Selection and Management Alternatives."

The most comprehensive data sources are the U.S. Geological Survey for surface and groundwater quantity and quality, the National Weather Service for climatological data.

HYDROLOGIC CONSIDERATIONS IN LAND APPLICATION SYSTEMS

Hydrologic considerations are of paramount importance to the design, operation, and maintenance of land application systems. Wastewater is applied at a rate designed to optimize the renovation capacity of the soil/plant medium and to maximize the utilization of the available nutrients within the wastes. Unscheduled runoff events due to high intensity storms will occur over land application systems. Protection of the land from excessive erosion as well as collection of the runoff must be included in the overall design. Module II-10, "Drainage for Land Application Sites," discusses these factors.

Water balance considerations are of primary concern to the design engineer. Without water, nutrient transport into and through the soil profile would be impossible. Conversely, too rapid movement may generate groundwater pollution. Water management to optimize nutrient availability in the root zone and at the same time provide for water transport at the design effluent quality is difficult with the monitoring and application equipment available today. A thorough discussion of crop water requirements is contained in Module II-7, "Crop Selection and Management Alternatives." What kind of hydraulic load will the system sustain over how long a period of time and still attain the expected effluent quality? There are many variables within the soil/plant treatment medium which must be considered before a question such as this can be answered.

In developing a land application system, the system boundaries need to be defined. Once this is accomplished a hydraulic balance must be attempted to consider all avenues of inflow and outflow through the system boundaries. Figure 2 briefly outlines the considerations typically faced in the design of land application systems.

The hydrologic phenomena that must be considered in land application systems vary with the seasons. During the application season, such things as precipitation, evapotranspiration, runoff, infiltration/percolation, groundwater quality, and wastewater application must be taken into account in the vicinity of the application site. However, during the non-application season all wastewater is routed to storage. Hydrologic considerations for storage design include wastewater flow, precipitation, evaporation, seepage, and groundwater quality.

Hydraulic considerations differ for periods of wastewater application and storage. These periods are determined by growing season and climatic considerations.

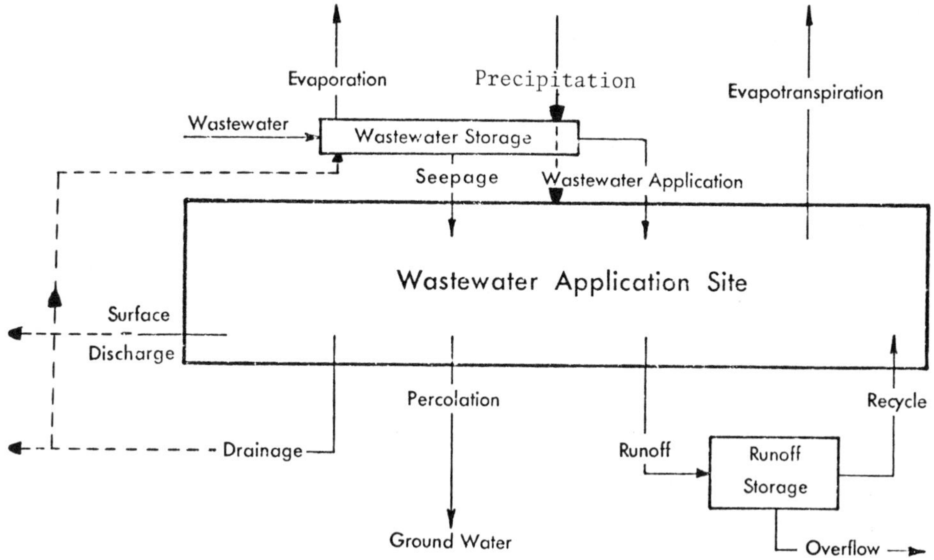

Figure 2. Hydrologic considerations in land application systems.

Application Season vs. Nonapplication Periods

In determining the application design for wastewater renovation, the growing season for cover crops is of major concern where they are used for pollutant control. The growing season, as reported by USDA in Agricultural Handbook No. 283 (USDA, SRS, 1972), is dependent upon the specific crop as well as the latitude of the site. This agricultural handbook estimates the usual planting and harvesting dates of field and seed crops across the United States. The growing season, during which nutrient uptake rates are greatest at irrigation sites, is determined by climate (primarily temperature). Therefore, climatic records can be used to estimate storage requirements for land application sites by setting certain criteria for days favorable for waste application, and designating days not meeting these criteria as unfavorable or "nonapplication" days.

Computer Programs Available for Storage Volume Computation

The number of consecutive nonapplication days due to climatic constraints may be determined through the use of computer programs developed by the National Weather Service which are fully described in Appendix A. This provides a rational basis for the design of the storage requirements for any specific situation. Two programs (EPA-1 and -3) are suited to areas constrained by cold periods, and one program (EPA-2) is geared to use in wet areas. Each program uses certain variable constraints (temperature, rainfall, snow cover, etc. for cold area programs; rainfall, soil moisture condition, percent saturation, etc. for wet areas) to define favorable and unfavorable application days which are tabulated for the entire year. In the case of the cold area programs, data is given on the length and severity of cold seasons through use of the cumulative degree day tabulation. Therefore, the designer may use his discretion in basing storage needs on cold period length rather than on the unfavorable day analysis. The specific values of the various constraints within the program may be adjusted to meet differing design limitations. About 25 years of weather data are required at a station to obtain meaningful results from the programs.

> *Computer programs are available from the NWS, National Climatic Center that compute storage requirements based on climatic factors applicable in different parts of the U.S. NWS computer programs present data on duration and intensity of cold weather along with various other climatic variables. The programs are useful design tools.*

The time between the established storage period by NWS procedures and the growing season as compiled by the USDA may be several weeks to 2 months. During this time, nutrient uptake by the cover crop will be negligible. In selecting design parameters for irrigation water requirements, a mean daily temperature of 45°F has been used as the beginning and ending date of the growing season for pasture grasses (USDA, SCS, 1967). If nutrient uptake by the cover crop is to be relied upon, then the growing season limitations must be respected in the design procedure. The element of prime concern is nitrogen, since chemical precipitation of phosphorus is efficient even at low temperatures. It is also likely that problems will be minimized if the applied wastewater contains nitrogen mainly in the ammonium form. During the colder periods, adsorption of the ammonium ion will hold this form of nitrogen. As the temperature increases, nitrification may occur before plants start to grow. However, proper selection of an early growing grass crop will enable maximum nitrogen control. See Module II-1, "Nitrogen Considerations," for further details.

There can be a significant lag time between the end of storage period and beginning of crop growing season. Decisions on application or storage strategy during this period depend on factors such as ammonium adsorption and establishment of early-growing grass species on irrigation sites to control nitrogen leaching.

Wastewater Application Season

Evapotranspiration. As mentioned earlier, evapotranspiration (ET) is the volume of water comprised of transpiration due to plant respiration and evaporation from the soil and plant surfaces. Water losses through ET can be estimated through the use of the Blaney-Criddle Formula (see Appendix B and Module II-7, "Crop Selection and Management Alternatives"). In estimating the consumptive use (or evapotranspiration), the factors taken into consideration are temperature, length of day, available moisture, and the crop growth stage. Each factor varies with respect to the next resulting in a very complex expression.

Evapotranspiration rates vary with soil moisture, cover crop, and energy available. ET can be estimated by the Blaney-Criddle and other formulas.

The rate of evapotranspiration is dependent upon the soil moisture conditions (soil type, available moisture, soil-moisture tension), the cover crop (type of crop and stage of growth), and the energy available for respiration and evaporation (temperature and daylight hours).

With regard to the *soil moisture* condition, the available moisture and soil moisture tension vary as shown in Figure 3 for different soil types. As the soil-moisture tension increases, the

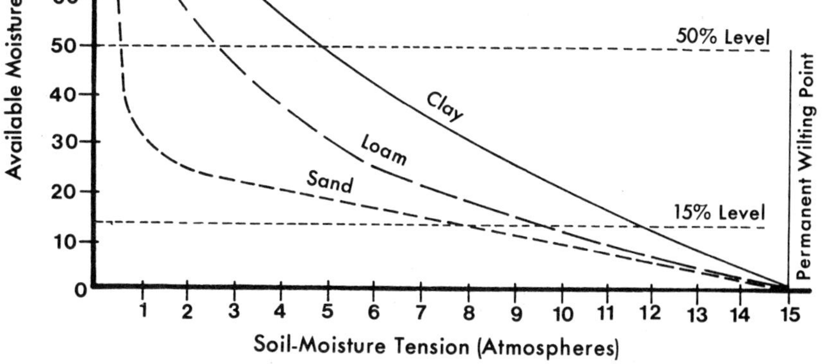

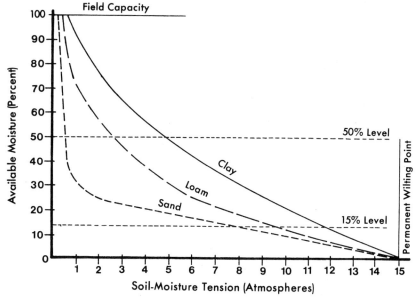

Figure 3. Moisture-release curves for three soils (Thorne and Raney, 1956).

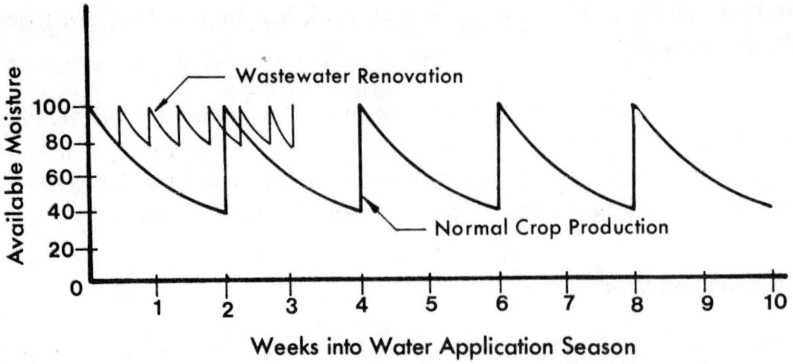

Figure 4. Irrigation scheduling at shorter intervals will maintain a higher percentage of available moisture in the soil profile for evapotranspiration.

availability of moisture for evapotranspiration decreases. In utilizing these facts in land application systems, it would behoove the design engineer to maintain as little soil-moisture tension as possible in order that the maximum amount of moisture may be available for evapotranspiration. This may come into effect in the scheduling of the irrigation period as shown in Figure 4.

In the irrigated areas of the United States, irrigation scheduling is a topic of great importance among farmers, soil-crop specialists, and irrigation engineers. The farmers and soil-crop specialists rely upon delivery of water to the irrigated crops between specified limits, i.e., field capacity and approximately 50% of field capacity. In this manner, water is delivered to the crop prior to the occurrence of plant stresses which would decrease productivity. This approach for water replenishment was developed around a shortage of water. However, in land application systems, the object is to maximize the flow of water, therefore irrigation scheduling techniques which maximize this water sink (evapotranspiration) should be developed. Scheduling water delivery when approximately 20% of the available moisture is depleted (80% of field capacity) is a simple way of maximizing the evapotranspiration potential of the site.

> *Scheduling wastewater application at soil moisture not less than 80% of field capacity will maximize uptake due to ET. Normal irrigation practice (to conserve water) is to apply at about 50% field capacity.*

Certain management schemes in the humid northeast and southeast will probably include wastewater applications when soil water is at field capacity. By applying the water at a slow rate the macropore space can remain free of water and provide the aeration needed for a healthy soil.

Continuous high moisture conditions within the root zone may jeopardize many cropping systems, especially if anaerobic conditions occur. Reed canarygrass has come to the forefront in land application systems because of its high moisture tolerance. Additional data on water tolerance of other crops are discussed in Module II-7, "Crop Selection and Management Alternatives."

The *type of crop cover* selected for wastewater renovation will depend on many variables. A decision must be made between annual and perennial crops. Perennial crops provide year-round soil cover, protecting the land application site from excessive erosion. Annual crops can provide high peak nutrient uptake during certain periods, as discussed below. Many successful systems employ a combination of annual and perennial crops.

Other things being equal (soil moisture and available energy), evapotranspiration from the cover crop is largely dependent upon the stage of growth and is more apparent with annual crops than perennials. Figure 5 outlines three stages: (1) emergence and development of a complete vegetative cover which generally proceeds over $\frac{1}{2}$ of the growth period; (2) a period of maximum vegetative cover where evapotranspiration is also at a maximum; and (3) crop fruiting and maturing where evapotranspiration begins to drop off.

The *energy available for evapotranspiration* is not a design variable for any given site, but is a function of local climatic conditions. Air temperature and number of daylight hours are two easily measured and widely reported variables which may be used to compute the total energy for ET. This is explained at length in the discussion of the Blaney-Criddle Formula, Appendix B. Other formulas for ET calculation, and a general discussion of their usefulness, are included in Module II-7, "Crop Selection and Management Alternatives."

In summary, evapotranspiration rates are dependent upon the soil moisture condition, cover crop, and energy available for evaporation and transpiration. The soil moisture condition and the cover crop may be manipulated to maximize the evapotranspiration from the energy available at the site.

Seasonal consumptive use (ET) averages between 1.5 and 4.0 acre-ft/acre-yr depending upon the many factors previously outlined. The higher values would be expected from the irrigated areas of the Southwest, whereas the lower would be more representative of the cooler Northeast.

Precipitation. Precipitation over the land application site has a relatively minor negative effect on the volume of water that may pass through the system, and it would have a slight positive effect through dilution of nutrients and mineral salts. However, precipitation which induces erosive flows within the land application site may be quite damaging to the physical system. The

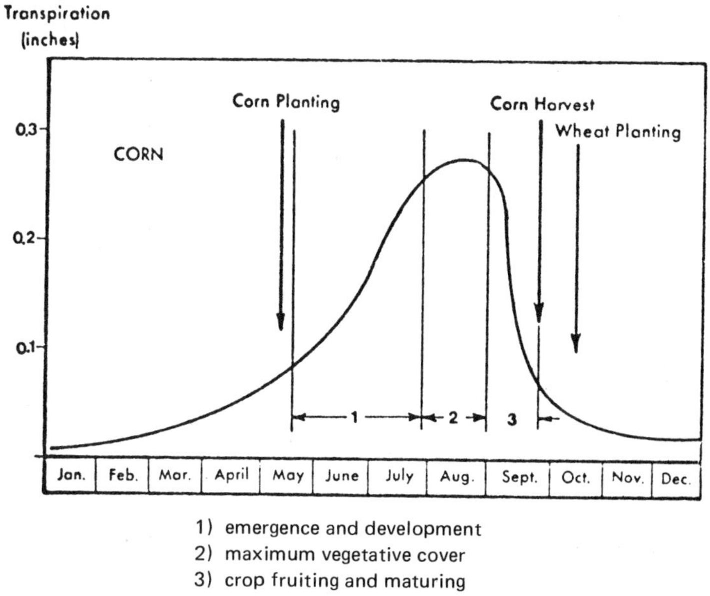

Figure 5. Evapotranspiration is a function of climate, soil moisture, and crop characteristics. These results are for irrigated crops under similar climatic conditions in Ohio; observed differences are due to crop (adapted from Soil Conservation Service, 1964).

site must be guarded against these extreme rainfall runoff events. The most common controls against runoff include diversion structures, terraces, and waterways which collect and deliver the runoff water to storage at nonerosive flows.

The most important factor concerning precipitation is control of flow from unusually heavy storms.

The *average monthly precipitation* is an accumulation of normal rainfall events which occur over an hour or two or possibly whole days. These few days of rain could upset the irrigation scheduling if they occurred so as to postpone the application day. The number of potential days postponed should be evaluated and included in the irrigation scheduling. The infiltration/percolation capacity of the soil profile must be matched against typical rainfall events over the application season to estimate the number of nonapplication days expected. However, where runoff from the site is not expected and where the vegetative cover permits, rainfall will dilute any additional wastewater application.

Climatological data published by the National Weather Service gives daily precipitation as well as average monthly and yearly precipitation. Whiting (1975) reviews the climatic data available from the National Weather Service.

Extreme rainfall events have also been monitored by Hershfield (1961). In these cases, the intensity as well as the frequency and duration are also monitored. This type of data is used to estimate requirements for protection of the system from excessive erosion and component failure, i.e., terraces, dams, pipes, channels, diversion ditches, etc., during high rainfall-runoff events.

The frequency for the chances of a storm occurring in 2, 5, 10, 25, 50, and 100 years and having a duration of 6, 12, and 24 hours are detailed on maps of the United States. Figure 6 is one map taken from Hershfield (1961). This map indicates 25-year, 24-hour frequency and duration storms only. This data can serve as a basis to design an acceptable level at which runoff from a site would be deemed to be unavoidable by the regulatory agency. Thus runoff would be contained from a 6-inch precipitation event that occurred over a 24-hour period. But any amount exceeding 6 inches could be acceptable runoff. More precise maps for the western states are available from the Soil Conservation Service within those states. Figure 7 is a detailed map of Colorado, which is included to indicate the extreme variability of rainfall in the western U.S., making broader generalizations useless.

These kinds of precipitation data help to make possible design estimation of average yearly and monthly rainfall, and rainfall-runoff potential. This is considered as basic information to this type of engineering design.

Runoff. Runoff from land application sites may be part of the wastewater renovation process as in overland flow systems, or it may occur as the result of an extreme rainfall event. In either case, it must be controlled.

In design, additional storage facilities as well as land application area must be provided for the expected runoff events. Where combined sewer and urban drainage systems are used, the land treatment system must be capable of storing and applying not only sewage but also nonpoint source urban runoff. Runoff controls from the land application site must also be provided to preclude nonpoint source runoff of nutrients, pathogens, sediments, etc. and to protect the capital investment in the land application system from excessive erosion. Erosion control measures may include such things as terraces, diversions, grassed waterways, drop structures, detention reservoirs, etc. All of these soil and water conservation measures are used either to store ex-

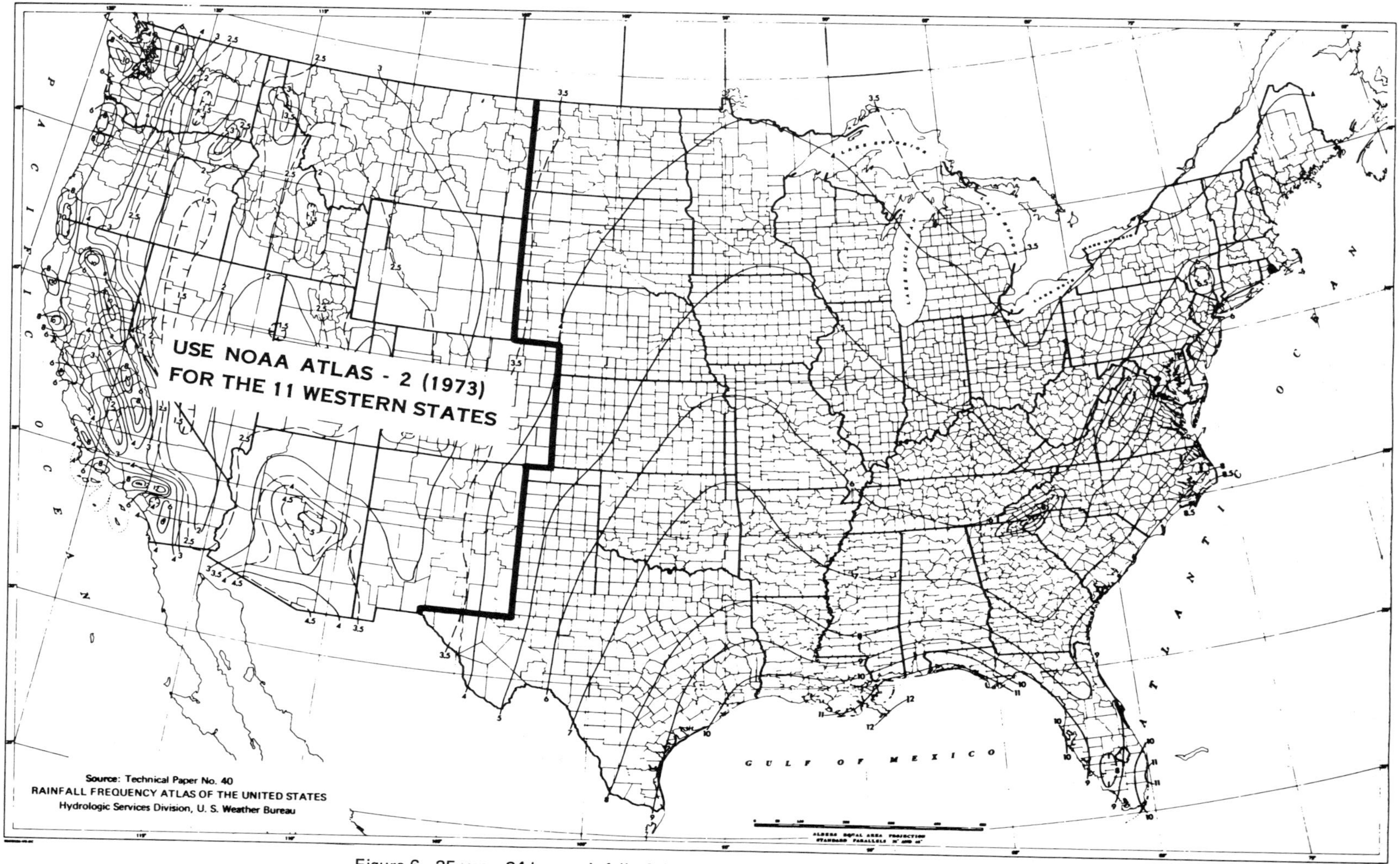

Figure 6. 25-year, 24-hour rainfall of the Eastern United States (Hershfield, 1961).

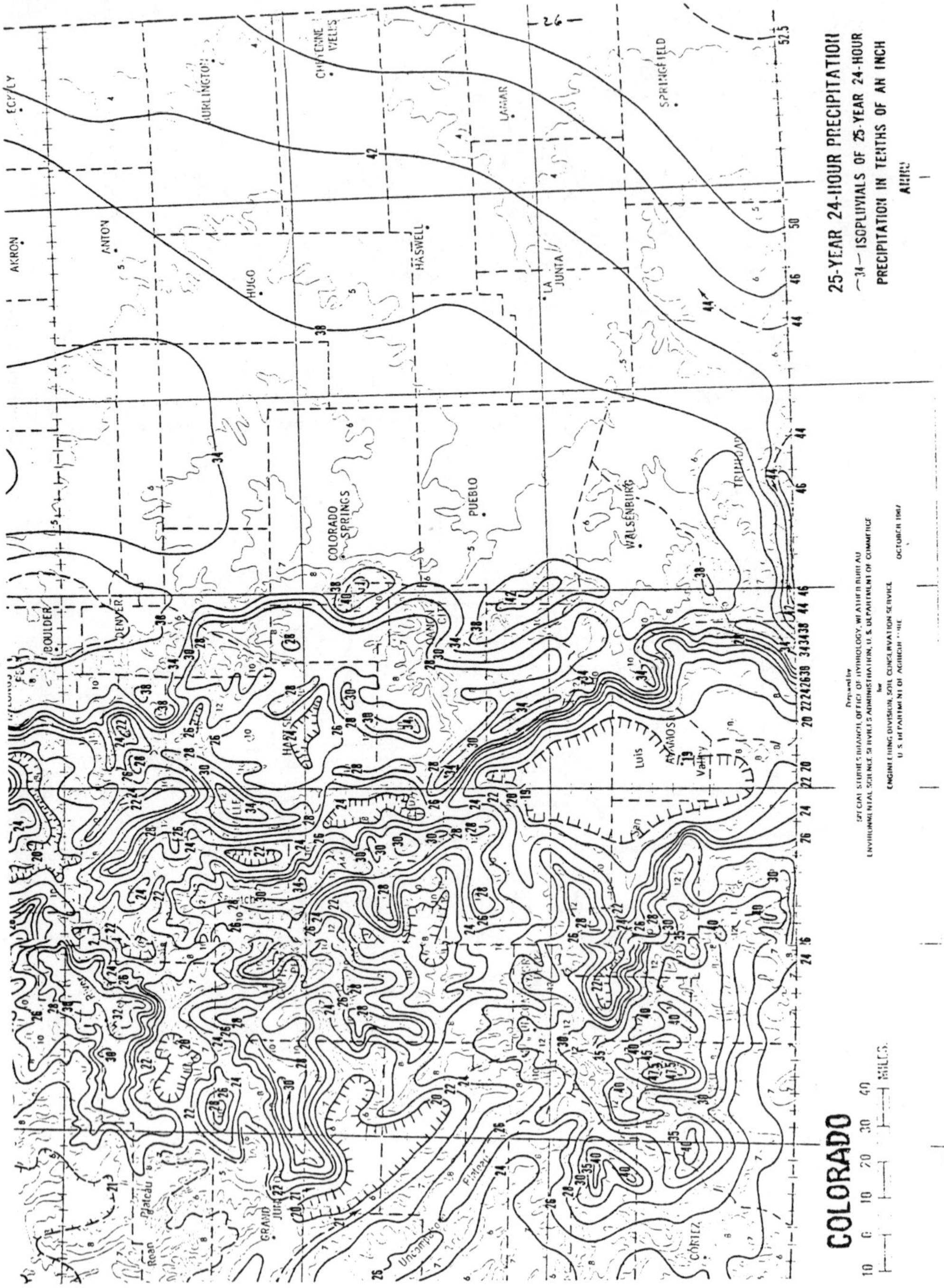

Figure 7. 25-year 24-hour rainfall of Colorado (SCS, 1971).

cessive water in place, transport water at nonerosive flows, or store water after being safely transported.

Runoff can produce erosion and unwanted discharge of nutrients and pollutants to surrounding land or watercourses. Typical drainage structures are used to control runoff; these must be sized to accommodate a design storm flow.

The 25-year 24-hour storm is used as the precipitation event from which all runoff must be contained for open beef cattle feedlots. Generally the nutrient and microorganism level on an active feedlot surface would be much higher than that of a land application system. The nature and the source of the wastewater being applied is an important factor in selecting runoff control criteria. The unchlorinated effluent from preliminary treatment would certainly have more stringent controls than that from a pea processing plant.

Infiltration/Percolation. The infiltration and percolation of wastewater into and through the soil profile may be a major wastewater sink in land application systems. The variables associated with the transport of water through the regolith are many. For purposes of this discussion, it is assumed that no extraordinary avenues of escape are available which will allow the wastewater to short circuit the soil treatment medium. Careful investigation of the treatment site must be undertaken in order to preclude this from happening. Potential problems which may arise from short circuiting and prevalent regolith configurations which enhance its occurrence are discussed in Module I-6, "Site Evaluation."

Infiltration and percolation of wastewater into the soil is important in most land treatment systems. Site evaluation must assure that this mechanism is not short-circuited by subsurface faults.

Permeability values for many soils are listed within the local soil survey or irrigation guide developed by the USDA Soil Conservation Service. These values are generally obtained from a simple clean water infiltrometer test although an explanation of the procedure used is generally provided in the accompanying text. Table 1 briefly outlines the Soil Conservation Service classification system from which the permeability class in the soil survey is obtained.

Generalized tables of infiltration ranges based upon textural class are also available, as given in Table 2. The variability in the approach and comparative values found in such tabulations emphasize the need for on site evaluation of this design parameter.

Table 1. Permeability Classes for Saturated Soil (USDA-SCS, 1971).

Permeability Class	Inches Per Hour
Very slow	less than 0.06
Slow	0.06 to 0.20
Moderately slow	0.20 to 0.60
Moderate	0.60 to 2.0
Moderately rapid	2.0 to 6.0
Rapid	6.0 to 20
Very rapid	more than 20

Table 2. Typical Ranges of Infiltration Rates and Available Soil Moisture by Soil Textural Class (Powell, 1975).

	Available Soil Moisture Storage[a]		Good Condition Base Soil Basic Infiltration Rates[b]		
			Slope		
	Range	Average	0–3%	3–9%	9+%
	--------in/ft------		------------in/ft------------		
Very coarse textured sands and fine sands	0.50–1.00	0.75	1.0+	0.7+	0.5+
Coarse textured loamy sands and loamy fine sands	0.75–1.25	1.00	0.7–1.5	0.50–1.00	0.40–0.70
Moderately coarse textured sandy loams and fine sandy loams	1.25–1.75	1.50	0.5–1.0	0.40–0.70	0.30–0.50
Medium textured very fine sandy loams, loams, and silt loams	1.50–2.30	2.00	0.3–0.7	0.20–0.50	0.15–0.30
Moderately fine textured sandy clay loams and silty clay loams	1.75–2.50	2.20	0.2–0.4	0.15–0.25	0.10–0.15
Fine textured sandy clays, silty clays and clays	1.60–2.50	2.30	0.1–0.2	0.10–0.15	<0.10

[a]Storage between field capacity ($\frac{1}{10}$ to $\frac{1}{3}$ atm) and wilting point (15 atm).
[b]For good vegetative cover these rates may increase by 25 to 50%. For poor surface soil conditions these rates may decrease by as much as 50%.

It must be pointed out that infiltrometer tests and other infiltration/percolation estimates are based upon the flow of clear water. Secondary treated effluent nearly matches this as far as hydraulic conductivity is concerned. Even raw sewage which has been exposed to preliminary treatment has less than 0.5% solids. Clogging of the soil profile is not likely with these types of effluents in systems other than rapid infiltration/percolation. But where sludges are separated from the wastewater stream and concentrated, their solids content is appreciably increased. Heavy applications of this type of material may cause soil clogging, but scheduled rest periods have kept this problem at a minimum.

Hydraulic conductivity, measured by soil infiltration and permeability, should not be limiting except in high rate rapid infiltration systems. Liquid sludge application systems may require evaluation to determine if clogging may be a problem.

When comparing the soil permeability (Table 1) with typical wastewater application rates (2 to 60+ in./wk), it is easy to see that the hydraulic conductivity of the soil will rarely be a limiting factor except for rapid infiltration/percolation systems. Once the wastewater has infiltrated the soil, the percolate below the treatment medium must be acceptable quality, as defined by applicable laws or standards. In most cases at 5 feet into the soil profile, land application systems will have to produce water quality that will meet drinking water standards.

Groundwater. As mentioned previously, groundwater quality is protected from degradation by national and local laws. The quality of groundwater must be known prior to wastewater application for several reasons: (1) degradation of groundwaters beyond specified limits will not be allowed, therefore some pollutants may have to be removed entirely prior to wastewater application to the land, i.e., nitrates leached to nitrate laden groundwaters, (2) proof of the source of groundwater contamination, if and when it occurs, must be established for remedial as well as legal reasons, and (3) knowledge of the groundwater quality prior to operation of the land treatment system will assist in making management decisions.

Groundwater quality at a proposed site may affect pretreatment requirements, establishes the necessary baseline to measure system performance, and assists in planning site management.

A system of monitoring the quality and movement of groundwater must be established to provide the background information needed for quality assurance and design base. Monitoring of the groundwater is discussed in Module II-11, "Monitoring at Land Application Sites."

Wastewater Application. Some of the soil and vegetative responses to land application of wastewater are predictable. This area of concern has been researched from the cropping systems standpoint as well as the wastewater renovation standpoint.

In many *soils*, the bulk density increases appreciably when large volumes of water are applied over short periods of time. This increase in bulk density is at the expense of soil macroporosity which, in turn, results in substantial decreases in infiltration capacity, aeration, and plant productivity (Keller, 1970). Low application rates, less than 0.2 in./hr, over longer periods of time allow the water to pass through the soil profile without exhausting the air from the macropore space. In wastewater distribution systems, this would also allow more time for evaporation to extract additional water from the spray.

Heavy water application can affect soil macroporosity. Vegetation may be established because of its water tolerance or allowed to adapt naturally.

Some types of *vegetation* have an established tolerance for high moisture conditions. Where prolonged moisture conditions prevail, vegetative cover will change and adapt to the new soil moisture extremes and where additional water is imposed upon the hydrologic cycle, a vegetative change should be expected. Water tolerant forage crops such as reed canarygrass are popular for that property. Vegetative change may be realized through purposeful planning and design for re-vegetation or this may be left to dominant species take-over at the site. In any case, a vegetation change should be expected and investigated in planning and design of land application systems.

Nonapplication Season

Wastewater Storage. The storage capacity of effluent required in land application systems depends upon the incoming wastewater rate and the number of consecutive unfavorable application days over the year. An extensive study of climate and weather variations has enabled general predictions of the storage requirements for all portions of the U.S. to be estimated by use of the NWS computer programs discussed earlier (see Appendix A). A summary of this study is shown in Figure 8. For New York State, the average number of nonapplication days for which storage would normally be required is around 120 days. All additions or deletions from the storage system must also be considered. These include such things as precipitation, evaporation and seepage into or out of the reservoir.

The *flow of wastewater* from municipalities varies considerably not only during the day, but also over the seasons. These variations are due to many activities, depending largely upon the climatic conditions of the area, the seasonal influx of people, etc. Sewers collecting raw sewage are often far from watertight. The quality control during construction and the ground movement after construction is such that inevitably cracks occur along cemented pipe joints. During the seasons where a high water table is to be expected, seepage into the sewers adds to the base flow. Of course this base flow fluctuates during the year. The seasonal fluctuation of population such as students, tourists, etc. also places an extreme variation on the waste treatment facility. The volume of wastewater derived from municipalities varies between 70 to 150 gallons

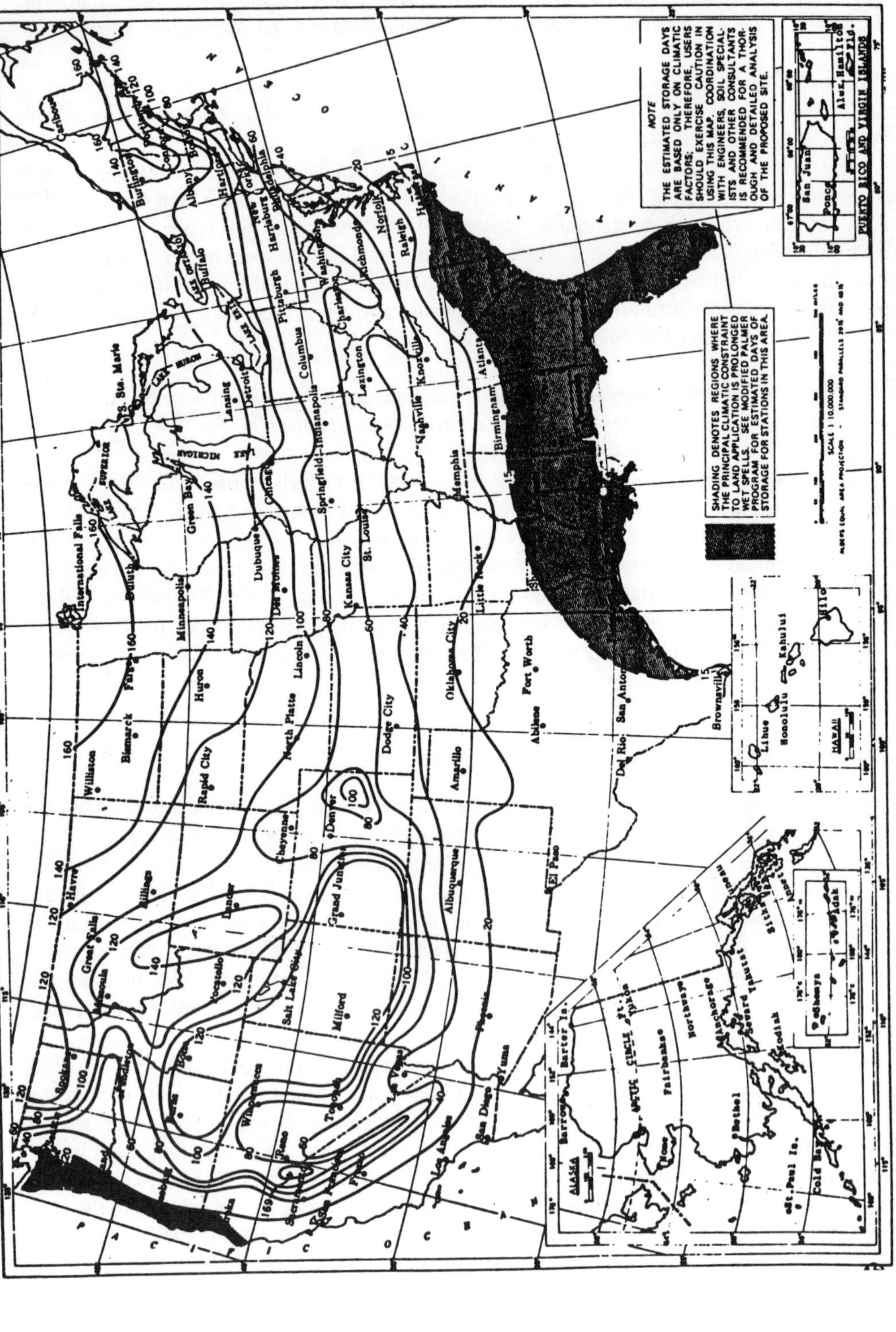

Figure 8. Storage days required as estimated from the use of the EPA-1 computer program as described in Appendix A.

per capita day. This depends largely upon the population density, commercial and industrial discharges, and the seasonal water table which encourages seepage into the system.

Storage design factors include length of nonapplication season, wastewater flow, precipitation, evaporation, and seepage.

The *precipitation* which falls directly into the storage facility must be applied to the land during the wastewater application season. The design values selected for this parameter should be taken from local climatological data from the nearest NWS reporting station. This accumulation could account for appreciable amounts of water especially in areas of the country where precipitation greatly exceeds evaporation.

Evaporation from the storage reservoir occurs throughout the year, varying from 20 in./yr in Maine to 86 in./yr in Arizona. Evaporation losses from reservoirs could be a substantial water sink, however total dissolved solids accumulate in the remaining wastewater. Two factors which minimize the effect of evaporation from storage are (1) in cold areas, storage ponds are full during low evaporation periods, and (2) in high evaporation areas like Arizona, storage needs are very small.

Seepage of wastewater from the storage reservoir will no doubt be kept at a minimum due to laws and regulations. Impermeable reservoir linings are available but expensive. Where a water table is reasonably close to the reservoir bottom, interceptor drains such as those at Muskegon, Michigan may be used to capture the wastewater for recycle to the reservoir. Whether or not the bottom of a reservoir handling organics seals itself is an unanswered question. For all practical purposes, tight soils in reservoir bottoms will seal, sands and gravels may never seal. In most systems it is assumed that the net transfer of water out of the storage facility due to seepage will be zero. In cases where seepage cannot be avoided, recovery of the seepage and recycle to storage is one acceptable control alternative.

BIBLIOGRAPHY

Hershfield, D. M. 1961. Rainfall frequency atlas of the United States. U.S. Dept. of Commerce, Weather Bureau. Technical Paper No. 40. Washington, D.C. 115 p.

Keller, J. 1970. Sprinkler intensity and soil tilth. pp. 118–125. *In* Transactions of the ASAE, Vol. 13, No. 1. American Society of Agricultural Engineers. St. Joseph, Mich.

Powell, G. M. 1975. Design factors, part II. Prepared for the design seminar for land treatment of municipal wastewater effluents, New York City, U.S. EPA, Technology Transfer.

Soil Conservation Service. 1964. National engineering handbook. Section 15, Irrigation. p. 1–39. Chapter 1, Soil-plant-water relationships.

Soil Conservation Service. 1967. Engineering Division. Irrigation water requirements. Technical Release No. 21, U.S. Govt. Printing Office, Washington, D.C.

Soil Conservation Service. Soil Survey Staff. 1971. Guide for interpreting engineering uses of soils. USDA. U.S. Govt. Printing Office, Washington, D.C. 87 p.

Thorne, M. D. and W. A. Raney. 1956. Soil moisture evaluation. USDA, ARS 41-6. 14 p.

USDA-SRS. 1972. Usual planting and harvesting dates–field and seed crops by states in principal producing areas. Agricultural Handbook No. 283. U.S. Dept. of Agriculture, Statistical Reporting Service, U.S. Govt. Printing Office, Washington, D.C. 84 p.

Whiting, D. M. 1975. Use of climatic data in design of soils treatment systems. EPA-660/2-75-018. (Interagency agreement with National Climatic Center). National Environmental Research Center. Corvallis, Ore.

Whiting, D. M. 1976. Use of climatic data in estimating storage days for soils treatment systems. EPA 660/2-76-250. (Interagency agreement with National Climatic Center). Robert S. Kerr Environmental Research Laboratory, Ada, Okla.

Appendix A

USE OF CLIMATIC DATA IN ESTIMATING STORAGE DAYS FOR SOILS TREATMENT SYSTEMS

By

Dick M. Whiting
National Climatic Center
Environmental Data Service
National Oceanic and Atmospheric Administration
Asheville, North Carolina

Interagency Agreement
EPA-IAG-D5-F694

Project Officer

Richard E. Thomas
Wastewater Management Branch
Robert S. Kerr Environmental Research Laboratory
Ada, Oklahoma

ROBERT S. KERR ENVIRONMENTAL RESEARCH LABORATORY
OFFICE OF RESEARCH AND DEVELOPMENT
U.S. ENVIRONMENTAL PROTECTION AGENCY
ADA, OKLAHOMA

DISCLAIMER

ABSTRACT

The number of days each year that Soils Treatment Systems may be inoperative because of unfavorable weather conditions can be estimated from analysis of daily climatological data. In cold regions each winter season (Nov.–Apr.) is examined for a 20- to 25-year period. Each day is defined as favorable, partly favorable or unfavorable using a set of thresholds for temperature, precipitation and snow depth. A daily accounting procedure adjusts the increase or decrease in storage on the basis of the type of day and the drawdown rate. The maximum storage days each year are summarized in a final table which also presents the mean, the standard deviation, the unbiased third moment about the mean, the coefficient of skewness and storage days for recurrence intervals of 5, 10, 25, and 50 years.

A separate program is used for stations in wet regions where the primary constraint to land application is saturated soil. The daily mean temperature and precipitation are examined for a 20- to 25-year period using the modified Palmer program. This program estimates the condition of both the underlying and surface soil layers in an attempt to identify periods when saturation occurs and runoff would result from additional precipitation or application. The longest consecutive period of unfavorable days each year is summarized in a table that also shows estimated storage days at the 5, 10, 25, and 50th percentiles.

Chronological listings of the actual data and computations for the entire period of record can also be furnished. Thresholds of most of the parameters can be changed to suit individual systems. The program can also be modified to examine the condition of the surface layer during the growing season as an estimate of irrigation needs.

This report is submitted in fulfillment of Interagency Agreement EPA-IAG-D5-F694 by the National Climatic Center, Asheville, North Carolina. Work was completed on June 30, 1976.

CONTENTS

CONVERSION TABLES ENGLISH TO METRIC UNITS

Length:

1 inch (in.) = 25.4 millimeter (mm)
= 2.54 centimeter (cm)

1 foot (ft) = 30.48 centimeter (cm)
= 0.3048 meter (m)

1 yard (yd) = 91.44 centimeter (cm)
= 0.9144 meter (m)
1 statute mile (stat. mi.) = 1609.344 meter (m)
= 1.609344 kilometer (km)

Speed:
1 mile per hour (mi hr^{-1}, mph) = 0.868391 knot (kt)
= 0.44704 m sec^{-1}
= 1.609344 km hr^{-1}

Density, Specific Volume:
1 pound per cubic foot (lb ft^{-3}) = 0.0160185 grams per cubic centimeter (g cm^{-3})

Pressure:
1 standard atmosphere (14.7 lb $in.^{-2}$) = 760 millimeters of mercury (mm Hg)
1 millibar (mb.) = 0.750062 millimeters of mercury (mm Hg)

Temperature:
Celsius (C) = $\frac{5}{9}$ (F - 32) where F is temperature in degrees Fahrenheit
Absolute (A) or Kelvin (K) = C + 273.16

ACKNOWLEDGMENTS

The support and assistance of many individuals within the several divisions of the National Climatic Center (NCC) is acknowledged with sincere thanks. The writer is grateful to Mr. Frank Quinlan, Chief, Climatological Analysis Division and to Dr. Nathaniel Guttman, Chief, Statistical Climatology Branch, NCC, for their suggestions and words of encouragement. Others in the Center who provided special assistance include: Mrs. Irma Lewis, Chief, Data Translation Branch, Mr. Ray Barr, Chief, Programming Section, Mr. Coy Johnson, Programmer, Mrs. June Radford and Miss Dottie Goodman, Statistical Climatology Branch, and personnel in the Audiovisual Services and Photographic Laboratory of the National Climatic Center.

Appreciation is also extended to the Robert S. Kerr Environmental Research Laboratory, Ada, Oklahoma for its support of this project, and especially to Mr. Richard E. Thomas, Project Officer.

Special recognition is extended to Dr. J. R. Mather, Wayne C. Palmer and others for their work in developing many of the techniques used in these programs. The concept of assessing the water balance in the general manner described was developed by the late C. W. Thornthwaite.

SECTION I CONCLUSIONS

Engineers, planners, and designers of Soils Treatment Systems may need an estimate of the number of days when land application would be undesirable due to certain weather conditions. If storage must be planned during extended periods when application would result in surface runoff, the programs developed can be used to estimate the duration of such conditions. In addition, the modified Palmer program (EPA-2) might be useful in estimating irrigation needs during the growing season.

The programs developed by the National Climatic Center, through support of the Environmental Protection Agency, provide estimates of storage requirements using 20 to 30 years of daily climatological data. Two of them (EPA-1 and EPA-3) are designed for use at stations in cold regions, while another (EPA-2) is best suited to those in wet regions. The EPA-3 program is similar to EPA-1 but examines daily maximum and minimum temperatures instead of the daily mean temperature. In this program an unfavorable (UNF) day is defined as one with precipitation of one-half inch or more, snow depth one inch or more, or a maximum temperature less than 40°F. A favorable (FAV) day is one with less than one-half inch precipitation, snow depth less than one inch, a maximum temperature of 40°F or more and a minimum temperature of 25°F or more. Partly favorable (L) days have less than one-half inch precipitation, less than one inch of snow on the ground, maximum temperatures of 40°F or more, but with minimum temperatures less than 25°F. On FAV days the decrease in storage is equal to the daily flow (Q) minus the drawdown rate (DD). On 'L' days, the increase (gain) in storage is equal to Q-(DD/2), while on UNF days the increase in storage is equal to Q.

A final summary of EPA-3 includes the yearly storage values from which the mean, standard deviation and other statistics are derived. Finally, the estimated storage days are computed by the Pearson Type III method for recurrence intervals of 5, 10, 25, and 50 years.

The programs represent a rather simple approach to a complex problem. They deal almost exclusively with climatological factors rather than biologic or hydrologic ones. In spite of the great variability in soils, climate and waste characteristics, reasonable estimates of storage requirements due to climate can be obtained from daily climatological records.

Estimates derived from these programs should be examined carefully to determine if adjustments are necessary because of local conditions. No program can account for all possible variables such as the quality and concentration of the effluent, depth to water table, soil type and condition, loading rates and others.

Hill [1] points out that the Thornthwaite method for estimating evapotranspiration from the mean daily temperature and length of day is generally conceded to be a poor estimator for daily amounts. However, it appears that in identifying wet spells neither evapotranspiration nor the available water capacity of the soil significantly affects the results; that is, when rainy spells occur, the amount of rainfall is sufficient to saturate the soil regardless of its assigned capacity.

Estimates from EPA-2 based on 25 years of record for stations in wet regions do not always fit a smooth pattern. Estimates range from 11 to 35 days in Louisiana, 12 to 33 days in Mississippi and 21 to 165 days in Oregon. The greatest number of storage days due to saturated soil occurs along the coast of Washington and Oregon.

A modification of EPA-2 examines the runoff and the condition of the surface and underlying soil in defining unfavorable days. Additional modification might be advisable in order to classify days as unfavorable unless the computed soil moisture is lower than the proposed loading rate. There are times when runoff is not indicated yet the underlying layer (SU) is saturated and the surface layer (SS) is within a few hundredths of an inch of being saturated. The program can be altered to class such days as unfavorable. This feature is identified on the listing by a percent following the EPA-2 indicator; for example, EPA-2(10%) denotes a set of data processed using not only the runoff feature in defining unfavorable days, but includes those days when the sum of the surface (SS) and underlying (SU) layers is within the indicated percent (10) of saturation. Of course, saturation is defined as the assigned available water capacity (AWC) for that location. This feature has the effect of extending the wet spells between a series of days with runoff. It can also be useful in estimating irrigation needs during the growing season since the moisture in the surface layer is indicative of the amount available in the root zone. It should also prove useful in estimating, along with other factors, the maximum loading rate during certain times of the year.

Finally, the variability in the amount of storage needed from year-to-year can be significant at some stations. For example, standard deviations of 15 to 20 are common at stations having a mean of 45 storage days. The yearly storage values can be used as input to the Gumbel extreme value analysis which gives estimates of storage days for return periods up to 100 years with confidence bands of ±34%.

SECTION II RECOMMENDATIONS

The programs defined as PO3B75:NCC-EPA-1 or EPA-3 generally should be considered at sites where the normal January temperature is colder than 40°F (4.4°C), while SOILMT:NCC-EPA-2 should be used in areas where the normal annual precipitation is 50.00 inches (1266 mm) or more. The selection of these arbitrary thresholds was made after processing a number of stations on both programs. The transition zones are not clearly defined and the 40°F temperature and the 50 inches of precipitation are only rough approximations. It may be desirable in some cases to process the data both ways, especially in the Pacific Northwest.

Each program was developed to accept variable thresholds which can be changed to account for differences in soils, systems, etc.; however, the basic programs remain intact. Sharp differences in temperature, precipitation, and soil can occur over very short distances, hence each site should be examined on an individual basis. Consultation with hydrologists, geologists, soil scientists, and other specialists is highly recommended.

SECTION III INTRODUCTION

A need was expressed in 1974 by the Robert S. Kerr Environmental Research Laboratory, Environmental Protection Agency, to find a way of using climatic data to estimate storage days for Soils Treatment Systems. Some thought was given to using soil temperature and soil moisture measurements, but this was not possible because of the paucity of data. The programs developed at the National Climatic Center (NCC) require serially complete, long-term digitized data. This type of daily climatological data is available for many locations in the NCC's tape library.

Once the data base had been selected, it was necessary to determine just what constituted climatic constraints to land application. It soon became apparent that a constraint for one system and location often was not a constraint elsewhere. In addition, while cold and snow might be definite constraints in some regions, prolonged rainy spells would be the major deterrent in others. Add to these variables the different types of soils, systems, types of effluents, etc. and one can sense the misgivings with which this project was started.

There were one or two hopeful signs. First, if the user can specify the types of weather unfavorable for a particular operation, those days can be identified. Second, Palmer's program [2] to identify droughts might be modified to define days when the soil was saturated and runoff would occur. The development of these programs provided a method of estimating storage days in cold and in wet regions. Only experience and observation will reveal how well these programs work. No field tests or controlled studies have been made; however, continued alterations are being made to the programs.

Each program can accept variable input, i.e. different thresholds, in order to be of maximum use. The basic programs remain standard, while the adjustment factors can be changed to account for the variations in soils, systems and climate. As a first approximation these programs are being used with a set of threshold values which are applicable for wastewater irrigation systems. Thresholds applicable for high-rate infiltration systems and overland-flow systems will be developed later.

The U.S. Army Corps of Engineers has done a vast amount of research in connection with soil trafficability. Much of their investigative work [3] is geared toward developing programs that can predict the state of the soil at a given time and place. Other efforts have been centered on the prediction of stream flow and with the hydrologic quality of soils [4]. Nothing in the literature examined dealt with the problem of identifying the frequency or duration of periods when the ground was saturated, although a number of studies [5-12] examined the water balance during the growing season.

The soil type and its condition more than any other factors generally determine the amount of runoff. The Soil Conservation Service (SCS) has completed surveys in many areas and has published detailed information for various soil series that may be obtained at local SCS offices [13].

Although the available water capacity (AWC) may not be a major factor in the programs discussed, it is an important soil property for other reasons. It is often given in reports as estimated amounts of water, in inches per inch of soil for specific soil layers. The following adjective ratings refer to the sum total of available water from the surface to bedrock, or to a depth of 60 inches.

Very low	<3	inches
Low	3 to 6	inches
Medium	6 to 9	inches
High	9 to 12	inches
Very high	>12	inches

The four major hydrologic soil groups, as described by the Soil Conservation Service, are presented here for information.

A. (Low runoff potential). Soils having high infiltration rates even when thoroughly wetted. These consist chiefly of deep, well to excessively drained sands or gravels. These soils have a high rate of water transmission in that water readily passes through them.

B. Soils having moderate infiltration rates when thoroughly wetted. These consist chiefly of moderately deep to deep, moderately well to well drained soils with moderately fine to moderately coarse textures. These soils have a moderate rate of water transmission.

C. Soils having slow infiltration rates when thoroughly wetted. These consist chiefly of soils with a layer that impedes downward movement of water or soils with moderately fine to fine texture. These soils have a slow rate of water transmission.

D. (High runoff potential). Soils having very slow infiltration rates when thoroughly wetted. These consist chiefly of clay soils with a high swelling potential, soils with a permanent high water table, soils with a claypan or clay layer at or near the surface, and shallow soils over nearly impervious material. These soils have a very slow rate of water transmission.

The programs developed at the NCC appear to follow actual physical processes rather well, at least as far as climatology is concerned. The depletion rate in EPA-2 was introduced to offset a unique condition in Palmer's original program. It can best be explained by saying that his weekly and monthly accounting systems had no provision to carry-over runoff. In the daily accounting, it became necessary to gradually deplete the excess defined as runoff. Since only a percentage of the surplus water runs off on any day [13], it was necessary to assign some estimated amount to account for gravitational water that is made available as runoff in the following days. The amount lost each day depends on, among other things, soil type, structure, and depth of the soil layer.

It became clear that any attempt to introduce all of the variables affecting storage was either not possible, or not justified on the grounds of time and money. An empirical approach seemed to offer a reasonable escape from this dilemma. Stations with different soil groups (clay, loam, and sand) were processed using assigned depletion rates (.50, .75, and 1.00 in./day, respectively). The only guide used in selecting these numbers came from experience and observation. For example, if the indicated amounts fall on saturated soil it will very likely take at least one day for that particular soil to dry enough so that application is possible. The programs discussed in this report use limited input data in an attempt to attain certain objectives. Most of the development discussed in the report focuses on irrigation-type systems whose operation is influenced much more by precipitation than by either overland-flow systems or high-rate infiltration systems.

One of the obvious limitations in this report is in the area of hydrology. None of the programs discussed in this report provide information dealing with the intensity, duration or amount of precipitation from individual storms. The Office of Hydrology, National Weather Service has prepared precipitation frequency analyses for all locations in the United States [14, 15, 16]. Publications are available with generalized charts from which the magnitude of 10-year 24-hour and 25-year 24-hour precipitation amounts can be obtained for any location. Information about the areal variability of rainfall, recurrence intervals, as well as information on evaporation should be requested from that office. Other offices to be contacted for related information include the U.S. Geological Survey, the U.S. Army Corps of Engineers and the U.S. Department of Agriculture's Soil Conservation Service.

SECTION IV ESTIMATING STORAGE CAPACITY IN COLD REGIONS

If a Soils Treatment System is designed to operate during the winter months in regions where the normal January temperature is colder than about 40°F (4.4°C), storage needs can be estimated by converting days defined as unfavorable into days of storage using either the EPA-1 or EPA-3 program. Temperature, snow depth, and precipitation are examined each day during the months of November through April for a period of 20 to 25 years. The exact period depends on the completeness and length of the digital record for the particular station.

The EPA-1 program [17] defines each day as favorable or unfavorable for operation on the basis of the assigned thresholds. While these thresholds can be varied, almost all stations processed on EPA-1 were assigned thresholds of: mean daily temperature 32°F, 1.00 inch or more of snow cover and 0.50 inch of precipitation as the threshold between favorable and unfavorable days. The maximum storage capacity each winter is estimated by a daily accounting system, where one day's flow is added to storage on days classed as unfavorable, while on favorable days the accumulated storage is reduced by one-half the daily flow. This amount can be changed from station-to-station to match the expected drawdown rate. The maximum storage estimated during each winter season is printed out at the end of the season. A final summary is presented showing the estimated storage days for each of the winter seasons along with percentiles at the .05, .10, .25, and .50 levels. A detailed chronological listing of the daily input can also be furnished that gives one an opportunity to examine the accounting process from day-to-day. Cumulative degree days to base 32°F are also listed. A freezing index is computed for each season and is defined as the difference, in degree days, between the highest and lowest values on a cumulative curve. This index can be considered a measure of the intensity of the cold period as explained in an earlier publication [17]. In the event the system is to be shut down completely during the winter, the duration of the freezing period is usually a better measure of storage needs than the estimate obtained from the favorable-unfavorable day computations.

When a high percentage of the annual precipitation falls in the form of rain or drizzle and the mean daily temperature seldom falls below 32°F (0°C) during the winter months, EPA-1 may not give reasonable estimates of storage. Olympia, Washington is an example of such a station. The problem arises partly from the fact that precipitation can occur almost every day for weeks between the months of November and April. The amounts are often less than 0.50 inch, while at the same time the mean daily temperatures are above 32°F. Although such days are defined as favorable by EPA-1, it is obvious from a review of the daily listings that land application would be undesirable during much of this time because of saturated soil and EPA-2 should be used.

The built-in flexibility of each program is considered one of the principal features since it permits each user to determine the thresholds for his particular situation. One modification to EPA-1 eliminated the snow depth and precipitation constraints entirely. This was done to use the long-term experience at an existing high-rate infiltration system to get an estimate of the temperature threshold for high-rate infiltration systems. If a waste treatment system in a cold region has been operating for many years without being seriously hampered by weather conditions, it follows that similar systems should be able to operate in a similar climate. The mean daily temperature threshold was reduced to a point where the lowest storage estimate at the control station was reduced to a minimum as a best approximation of a reasonable threshold value.

The station selected for this special processing was Lake George, New York, which has operated on a 12-month basis for 40 years despite its relatively cold climate. Spier Falls, New York, about 15 miles to the south, was selected as the weather station with the most suitable climatological record. The temperature threshold was lowered from 32° to 10°F, and each day below this threshold was considered unfavorable. Storage days were estimated using the same daily accounting system as before. These values ranged from 21 days during the most unfavorable season, to one day during the most favorable season.

With this background for the Lake George area, runs were then made using the same 10°F threshold for La Crosse, Wisconsin, and Greenville, Maine, again without regard to snow depth or precipitation. Table 1 gives the results of these runs, along with comparative values based on standard thresholds.

Another modification suggested by Parmalee [18] and Griffes [19] resulted in the development of the EPA-3 program. The proposed changes to EPA-1 included the use of the daily maximum and minimum temperatures instead of the daily mean temperature as the criteria for an unfavorable day. In addition, the daily accounting system was changed slightly and the summary table includes several additional statistics. The storage gain on UNF days is equal to the daily flow (Q). On L days, the gain is equal to the daily flow (Q) minus one-half the assigned drawdown rate, while on FAV days the decrease in storage is equal to the daily flow minus the drawdown rate (DD). The summary table shows the mean number of storage days, the standard deviation of the samples, the unbiased third moment about the mean and the coefficient of skewness. Finally, storage days are computed using the Pearson Type III method. Detailed information about the EPA-3 program can be found in Appendix A.

Table 1. Estimated Annual Storage Days Based on:
(a) Mean Daily Temperature ≤10°F, and
(b) Standard Thresholds of ≤32°F, Snow Depth ≥1.00 Inch and Precipitation ≥0.50 Inch

Program: P03B75—NCC—EPA—1 (Modified)

	Spier Falls, N. Y.			Greenville, Maine			La Crosse, Wis.		
	Mean Jan. Temp. 20.4 F			Mean Jan. Temp. 12.5 F			Mean Jan. Temp. 16.1 F		
	(a)	YEAR	(b)	(a)	YEAR	(b)	(a)	YEAR	(b)
	21.0	1969	132.5	25.0	1970	164.5	13.5	1964	123.5
	14.0	1962	121.5	30.0	1958	149.5	18.5	1950	133.5
	18.0	1960	107.0	29.5	1971	172.0	19.0	1958	123.0
	10.0	1967	108.5	20.0	1964	152.0	20.5	1970	119.5
	8.5	1970	137.0	30.5	1960	152.0	20.0	1962	98.0
	3.0	1958	125.0	20.5	1962	158.0	8.0	1955	133.0
	9.5	1963	116.0	30.5	1967	122.0	7.5	1949	113.5
	7.0	1961	110.5	13.5	1968	176.0	12.5	1961	119.0
	8.0	1964	107.5	13.5	1955	160.5	15.0	1971	120.5
	3.0	1968	131.5	24.0	1961	146.5	12.5	1968	109.5
	5.0	1966	104.5	10.5	1949	147.0	15.0	1969	108.5
	2.5	1971	138.5	12.0	1966	153.5	13.0	1948	104.0
	4.0	1955	130.0	15.5	1963	145.5	12.0	1951	119.0
	4.5	1965	96.0	18.0	1969	150.0	6.0	1966	107.0
	6.0	1954	92.5	18.5	1956	139.0	11.5	1956	98.0
	3.0	1959	114.5	12.5	1951	156.5	8.0	1954	112.5
	5.5	1956	93.5	13.0	1954	146.0	5.0	1959	115.5
	3.5	1951	129.5	6.5	1965	164.5	17.0	1965	80.0
	3.5	1958	104.5	8.5	1959	157.5	12.0	1972	97.0
	4.5	1953	73.0	15.5	1953	136.0	3.0	1952	111.5
	2.0	1950	97.5	6.0	1950	132.5	7.5	1957	95.5
	1.0	1952	73.0	7.0	1952	131.5	13.0	1963	90.0
				9.0	1957	152.5	10.0	1960	70.0
							7.0	1953	84.5
							11.0	1967	76.0
Max	21		139	31		176	21		134
10%	17		135	30		169	19		127

SECTION V ESTIMATING STORAGE CAPACITY IN WET REGIONS

The principal climatic constraint to land application at locations along the Gulf States and the Pacific Northwest Coastal Region is prolonged wet spells, rather than cold or snow. Program EPA-2 examines daily climatological data during a 20 to 25 year period in an attempt to identify days when the soil is saturated. If land application is undesirable during periods when surface runoff could occur, the longest consecutive period defined as "wet" would be an estimation of storage needs. Palmer [2] developed a program that examines weekly (and monthly) temperature and precipitation in order to estimate the duration and magnitude of abnormal moisture deficiency. His analysis yields successive index values on weekly (or monthly) periods for State Climatic Divisions and is currently used to identify meteorological drought.

In the original program, positive index values indicate that the moisture supply from current or antecedent rainfall exceeded the amount required to sustain the evapotranspiration, runoff and moisture storage which could be considered as normal and appropriate for the climate of the area. High positive values mean that fields are too wet to work, or that rains have actually caused flooding. Rains in excess of the estimated water use produce positive values of recharge until the soils reach field capacity. Excess water then shows up as runoff. His classification system is given in Table 2.

The underlying concept of Palmer's work is that the amount of precipitation required for the near-normal operation of the agricultural economy of an area during some stated period is dependent on the average climate of the area and on the prevailing meteorological conditions, both during and preceding the period in question. Although his goal was to identify droughts, he found it necessary to identify the entire range of conditions from dry to wet, as shown above. His program appears to work well in identifying droughts and thus became an obvious choice in an attempt to identify prolonged wet spells.

A soil's ability to hold water is dependent upon the thickness of its various horizons, its texture, bulk density, percentage of coarse fragments and the organic content in the profile. The available water capacity (AWC) is commonly defined as the amount of water available to plant roots from the surface to bedrock, or to the unconsolidated nonconforming material such as sand or gravel, or to an arbitrary depth in the case of deep soils [13].

The effect of changing the AWC in the EPA-2 program does not change the number of storage days significantly. This apparently is caused by several factors, perhaps the most obvious is that when wet spells occur, the rainfall is more than sufficient to saturate the soil regardless of its assigned AWC. The surface layer of the soil is assumed to be roughly equivalent to the plow layer. At field capacity it is expected to hold one inch of the available moisture. This is the

Table 2. Drought Classification by Palmer

Index	Degree
$\geq$ 4.00	Extremely wet
3.00 to 3.99	Very wet
2.00 to 2.99	Moderately wet
1.00 to 1.99	Slightly wet
.50 to .99	Incipient wet spell
.49 to -.49	Near normal
-.50 to -.99	Incipient drought
-1.00 to -1.99	Mild drought
-2.00 to -2.99	Moderate drought
-3.00 to -3.99	Severe drought
$\leq$ -4.00	Extreme drought

layer onto which the rain falls and from which evaporation takes place. Therefore, in the moisture accounting it is assumed that evapotranspiration takes place at the potential rate from this surface layer until all the available moisture in the layer has been removed. Only then can moisture be removed from the underlying layer of soil. Likewise, it is assumed that there is no recharge to the underlying portion of the root zone until the surface layer has been brought to field capacity. The available capacity of the soil in the lower layer depends on the depth of the effective root zone and on the soil characteristics in the area under study. It is further assumed that the loss from the underlying layer depends on initial moisture content as well as on the computed potential evapotranspiration (PE) and the available water capacity (AWC) of the soil system.

The distribution of the estimated annual storage days for 25 years from EPA-2 may be a good indicator of how the program is working. One knows from experience that soils are neither always wet nor always dry.

Furthermore, the occurrence of extremely long wet periods is a rarity in most places. The summary of the annual values for 25 years should reveal an almost predictable distribution in terms of range and maximum values. To be more explicit, a large number of years with no storage days, or of estimates over 100 days due to saturated soil, would be suspect almost everywhere except deserts and swamps.

Palmer's original program dealt specifically with weekly and monthly time periods and although he considered the antecedent conditions, it was not necessary for him to be concerned with the amount of runoff. In the daily accounting, the assigned depletion rate serves to approximate the actual drying rate of the soil. It is essential to the program, but it can be changed to suit different conditions. This artificial method of gradually reducing the excess daily runoff is simply a means of accounting for the long-term effect of extremely heavy rainfall. It is a way of describing the continuing restrictive factors of daily amounts over and above that which would just cause saturation. The program defines a day as unfavorable whenever the amount of moisture in the surface soil (SS), plus that in the underlying soil (SU) and the accumulated daily runoff depleted by a constant (SRO) is equal to or greater than the assigned AWC.

About 75 stations have been processed on the original EPA-2 program that defines a day with runoff as unfavorable. Examination of the daily listings for a number of stations revealed some interesting conditions. Days following those with runoff frequently showed a moisture content in the surface layer very close to saturation. Since the surface layer had not dried out enough to accept the intended amount of application, these days should also be classed as unfavorable and the string of storage days continued. The program was modified to check the amount of moisture in the surface (SS) layer as well as the runoff and to class either condition as unfavorable. The amount of moisture in the surface layer at saturation is always 1.00 inch, with the remainder of the AWC assigned to the underlying layer (SU). Thus, if one selects 0.10 inch as the daily application rate it may be necessary to consider days with SS equal to or greater than 0.90 inch as unfavorable along with those when saturation is indicated. When this feature is used, the program indicator is followed by the percent of SS being considered. For example, in the case stated above, the indicator would be EPA-2(10%); if the application rate was 0.20 inch, the indicator would be EPA-2(20%), etc. This simply says the amount of moisture needed before the surface layer is saturated is the difference between the amount of moisture in that layer and 1.00 inch. This feature may also be useful to planners in estimating irrigation requirements in some areas. Additional changes would have to be made to include a check on the amount of moisture in both SS and SU. The general concept would be to establish minimum moisture needs and then analyze the daily data in terms of the frequency, duration and even total moisture deficit by months, or other time periods.

Examination of the daily listings from EPA-2 shows that most of the wet periods occur during the winter months when evapotranspiration is lowest. The heavy rains of summer do not often result in significant wet spells for a number of reasons. The soil is more likely to be at less than capacity and therefore, in need of recharge. Also, evapotranspiration rates are highest during the growing season. Detailed information about the EPA-2 program is given in Appendix A.

SECTION VI DISCUSSION

Daily weather reports from the principal observing stations usually include measurements of a number of elements, such as temperature, precipitation, snowfall, snow depth, winds, relative humidity, sunshine, sky cover, etc. These observations have been reported regularly from a network of about 300 stations in the United States for many years, but have been placed on magnetic tape only since about 1948. This network is staffed by meteorologists and most of the stations are located at airports.

Another important source of weather reports is the cooperative network of approximately 10,000 to 12,000 stations. These stations are operated by dedicated volunteers from all walks of life who are making a significant contribution to climatology. Daily reports from this network generally consist of temperature, precipitation, snowfall, snow depth and remarks about unusual weather events. Some networks report only precipitation and river stages, while others include soil temperature and soil moisture measurements. State universities and other interested organizations have made a special effort to place these climatological observations on magnetic tape prior to 1948. Several hundred stations now have digitized record as far back as 1895, or earlier.

In general, observing networks have been established primarily to provide specific weather information to a particular segment of the economy. Some are multi-purpose, of course, and most overlap to some degree. Aviation, hydrology and agriculture are a few examples that need unique and different measurements of the environment. Weather forecasting probably has the broadest base of all in its use of weather observations on a current, or real-time basis. The operational use of weather data is only part of the total picture. Most weather records forwarded to the NCC are edited, sorted and placed in magnetic tape files for future processing. While statistical analysis of climatological data can be very useful, it also has limitations.

One of the first steps in applying climatological data to a specific problem is to insure the presence of an adequate base in digital form (magnetic tape). Consideration should be given to the quality, completeness and length of record, as well as having a large number of stations well distributed throughout the country. The tape decks meeting the above criteria at the NCC are TD-30 and TD-486.

Inventories should be reviewed for completeness before the data can be processed by the programs discussed in this report. If the requested record is inadequate, a substitute station must be used, since a significant number of missing elements or days can bias the results. This means that climatological data from stations 50 to 75 miles from the proposed site may sometimes be used. While each program furnishes selected information about weather conditions at a station during the past quarter of a century, the lack of refinement is obvious. Thresholds to be assigned should be determined by specialists on the site since numerous factors complicate the relationship between rainfall, temperature and wet periods.

Although there is reason to believe that the output from these programs can provide realistic estimates of storage requirements, the limitations are so great in some areas that any measure of confidence may be misleading. At best, these programs offer a simple approximation of storage based on the examination of a few elements. Palmer, and others, developed the basic computer program that provides a realistic picture of moisture excesses and deficiencies many years ago. As more complete and longer records become available, more refined methods of estimating storage will be developed. Future studies will undoubtedly include such elements as solar radiation, wind, moisture and temperature measurements at selected levels above and below the ground, evaporation, precipitation, and others. The present practice at some observing sites of recording certain weather elements only during the growing season should be expanded to include all months. In any case, it is hoped that some of the material presented in this report will stimulate others to examine the data base and respond with improved methods and techniques for estimating storage days.

Finally, daily observations from the cooperative stations reflect weather conditions that existed during the previous *24-hour period.* These periods do not coincide with calendar days, since most cooperative observers take their observations in the morning (7 A.M.), or in the evening (5 P.M.). It is entirely possible for one or more of the reported values to have occurred on the previous day. Since there is no real solution to this dilemma, except by having all stations report on a midnight-to-midnight basis, it is the practice in most climatological data processing programs to ignore differences in the hours of observation. For example, it is impossible to tell from the digital records, or indeed, even from some of the original records from the 7 A.M. to 5 P.M. reporting stations, on which date the reported rain actually fell.

SECTION VII REFERENCES

1. Hill, Jerry D., "The Use of a 2-Layer Model to Estimate Soil Moisture Conditions in Kentucky," *Monthly Weather Review*, **102**(10), Oct. 1974.
2. Palmer, Wayne C., "Meteorological Drought," Research Paper No. 45, U.S. Department of Commerce, Weather Bureau, Washington, D.C., Feb. 1965, 58 p.
3. "Report of Conference on Soil Trafficability Prediction," U.S. Army Engineer Waterways Experiment Station, Corps of Engineers, Vicksburg, Miss., Nov. 1966, 125 p.
4. Smith, M. H. and M. P. Meyer, "Automation of a Model for Predicting Soil Moisture and Soil Strength" (SMSP Model), Miscellaneous Paper M-73-1, U.S. Army Engineer Waterways Experiment Station, Mobility and Environmental Systems Laboratory, Vicksburg, Miss., Jan. 1973, 120 p.
5. Thornthwaite, C. W. and J. R. Mather, "The Water Balance," Publications in Climatology, Vol. VIII, Number 1, Drexel Institute of Technology, Laboratory of Climatology, Centerton, N.J., 1955, 104 p.
6. Mather, John R., "The Climatic Water Balance," Publications in Climatology, Volume XIV, Number 3, C. W. Thornthwaite Associates, Laboratory of Climatology, Centerton, N. J., 1961, pp. 249-264.
7. Barger, G. L., Robert Shaw, and R. F. Dale, "Chances of Receiving Selected Amounts of Precipitation in the North Central Region of the U.S.," Agricultural and Home Economics Experiment Station, Iowa State University of Science and Technology, Ames, Iowa, July 1959, 277 p.
8. Barger, G. L., Robert Shaw, and R. F. Dale, "The Gamma Distribution from 2- and 3-Week Precipitation Totals in the North Central Region of the U.S.," Agricultural and Home Economics Experiment Station, Iowa State University of Science and Technology, Ames, Iowa, Dec. 1959, 183 p.
9. Barger, Gerald L., "Weather Planning for Direct Seeding of Southeastern Pines," reprinted from "Direct Seeding in the South," 1959, 12 p.
10. Pengra, Ray F., "Seasonal Variations of Soil Moisture in South Dakota," Agricultural Economics Pamphlet No. 99, Agricultural Experiment Station, S.D. State College, Brookings, S.D., Feb. 1959.
11. Pierce, L. T., "A Practical Method of Determining Evapotranspiration from Temperature and Rainfall," paper presented at the American Society of Agricultural Engineers, Chicago, Ill., Dec. 1956, 6 p.
12. van Bavel, C. H. M., "Agricultural Drought in North Carolina," Technical Bulletin No. 122, North Carolina Agricultural Experiment Station, North Carolina State College, Raleigh, N.C., June 1956, 35 p.
13. "Guide for Sediment Control on Construction Sites," U.S. Department of Agriculture, Soil Conservation Service, Raleigh, N.C., Mar. 1973, 122 p.
14. "Precipitation-Frequency Atlas of the Western United States," NOAA-Atlas-2, U.S. Department of Commerce, National Oceanic and Atmospheric Administration, National Weather Service, Hydrology, Silver Spring, Md. Vols. I-XI, 1973, 50 p.
15. Technical Paper #40, Hydrology, National Weather Service, National Oceanic and Atmospheric Administration, Silver Spring, Md., May 1961, 115 p.
16. Technical Paper #49, Hydrology, National Weather Service, National Oceanic and Atmospheric Administration, Silver Spring, Md., May 1964, 29 p.
17. "Use of Climatic Data in Design of Soils Treatment Systems," U.S. Environmental Protection Agency Office of Research and Development, EPA-660/2-75-018, Corvallis, Ore., June 1975, 67 p.
18. Parmalee, Donald M., Consultant, Alloway, New Jersey, Camp Edge Road, personal correspondence, 1975.
19. Griffes, Douglas A., Metcalf & Eddy Engineers, Palo Alto, Calif., personal correspondence, 1976.
20. Thom, H. C. S., "New Distributions of Extreme Winds in the United States." In: Journal of the Structural Division, *Proc. Am. Soc. Civil Eng.*," **94**, No. ST 7, July 1968, 12 p.
21. "Glossary of Meteorology," edited by Ralph E. Huschke, American Meteorological Society, Boston, Mass., 1959, 638 p.

22. Holzworth, George C., "A Climatological Analysis of Pasquill Stability Categories based on 'STAR' summaries," Environmental Protection Agency, Environmental Sciences Research Laboratory, Research Triangle Park, N.C., Apr. 1976, 51 p.
23. "Evaluation of Land Application Systems," U.S. Environmental Protection Agency, Technical Bulletin, EPA-430/9-75-001, Office of Water Program Operations, Washington, D.C., Mar. 1975, 182 p.
24. "Factors Involved in Land Application of Agricultural and Municipal Wastes," U.S. Department of Agriculture, Agricultural Research Service, Beltsville, Md., July 1974, 200 p.
25. Pound, Charles E. and Ronald W. Crites, Metcalf and Eddy Engineers, "Design Seminar for Land Treatment of Municipal Wastewater Effluents," Design Factors, Part I, Palo Alto, Calif., Apr. 1975.
26. Sorber, C. A., "Protection of Public Health," Proceedings of the Conference on Land Disposal of Municipal Effluents and Sludges," Rutgers University, New Brunswick, N.J., Mar. 1973.
27. Cry, George W., "Effects of Tropical Cyclone Rainfall on the Distribution of Precipitation Over the Eastern and Southern United States," Environmental Science Services Administration, U.S. Department of Commerce, Washington, D.C., June 1967, 195 p.
28. "Environmental Guide for the U.S. Gulf Coast," National Oceanic and Atmospheric Administration, Environmental Data Service, National Climatic Center, Asheville, N.C., Nov. 1972, 177 p.
29. "Environmental Guide for Seven U.S. Ports and Harbor Approaches," National Oceanic and Atmospheric Administration, Environmental Data Service, National Climatic Center, Asheville, N.C., Feb. 1972, 166 p.
30. W. B. Langbein, "Water Yield and Reservoir Storage in the United States," Geological Survey Circular 409, U.S. Department of the Interior, Geological Survey, 1959.
31. A. F. Meyer, "Evaporation from Lakes and Reservoirs," Minnesota Resources Commission, St. Paul, Minn., 1942.
32. R. E. Horton, "Evaporation Maps of the United States," *Trans. Am. Geophysical Union*, Vol. 24, Part 2, Apr. 1943, pp. 750–751.
33. "Water-Loss Investigations: Vol. 1–Lake Hefner Studies," Geological Survey Professional Paper No. 269, U.S. Geological Survey, 1954.
34. "Water-Loss Investigations: Lake Mead Studies," Geological Survey Professional Paper No. 298, U.S. Geological Survey, 1958.
35. M. A. Kohler, T. J. Nordenson, and W. E. Fox, "Evaporation from Pans and Lakes," Research Paper No. 38, U.S. Weather Bureau, 1955.

SECTION VIII APPENDIXES

APPENDIX A THE COMPUTER PROGRAMS

LIST OF FIGURES AND TABLES

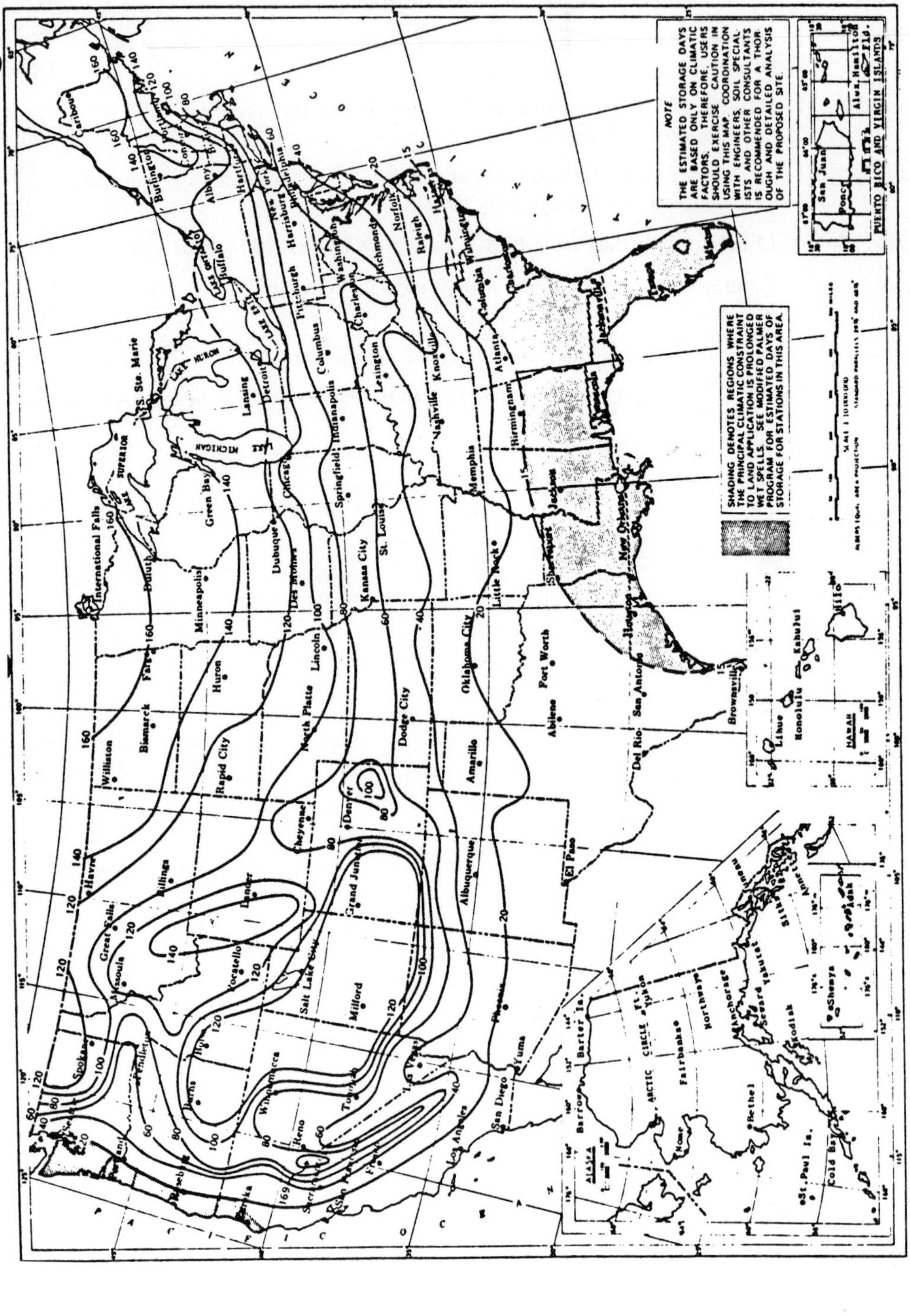

Figure 1. Estimated maximum annual storage days from EPA-1 program.

Table 3. Selected Information for Stations, EPA-1 Program.

STATION	LENGTH OF FREEZE PERIOD (10 %)	ESTIMATED STORAGE DAYS (10 %)	FREEZE INDEX (10 %) (BASE 32° F)	LENGTH OF MAXIMUM FREEZE PERIOD (DAYS)	ESTIMATED MAXIMUM STORAGE DAYS	NORMAL JANUARY TEMPERATURE	GROWING SEASON (DAYS)	YEARS
BIRMINGHAM, AL	008	012	0061	013	015	44.2	241	24
MADISON, AL	018	019	0075	020	021	41.2	215	22
MOBILE, AL	004	005	0019	004	006	51.2	298	20
MUSCLE SHOALS, AL	014	018	0089	016	021	41.3	230	25
ONEONTA, AL	012	017	0085	014	020	41.9	219	24
ST. BERNARD, AL	016	018	0100	020	021	40.9	204	20
SCOTTSBORO, AL	012	016	0095	017	020	41.1	210	25
SELMA, AL	004	005	0037	004	006	49.1	262	23
THOMASVILLE, AL	006	008	0041	007	009	47.4	242	20
ASH FORK, AZ	025	024	0115	045	040	36.8	155	22
GANADO, AZ	110	074	0633	114	078	27.5	129	21
CORNING, AR	023	022	0150	024	022	37.7	207	20
DUMAS, AR	011	013	0075	014	021	43.7	223	22
FAYETTEVILLE, AR	029	018	0145	033	031	37.1	181	27
GILBERT, AR	021	020	0154	025	023	37.3	179	20
LITTLE ROCK, AR	018	016	0091	020	019	39.5	238	25
NASHVILLE EXP STN, AR	014	012	0087	017	015	41.9	219	23
NEWPORT, AR	020	016	0094	023	022	40.2	215	21
WALDRON, AR	016	012	0080	017	013	40.8	199	25
CEDARVILLE, CA	066	071	0437	075	072	29.6	131	20
LOS ANGELES, CA	001	004	0001	001	004	54.5	350	25
SALINAS, CA	001	004	0001	001	004	50.0	262	20
SAN FRANCISCO, CA	001	005	0001	001	007	48.3	350	25
SANTA CRUZ, CA	001	006	0001	001	007	48.8	279	24
SUSANVILLE, CA	060	059	0418	068	061	29.9	121	22
TAHOE CITY, CA	123	164	0758	130	169	28.2	075	22
BOULDER, CO	052	049	0346	056	058	33.0	148	24
DENVER, CO	082	056	0563	100	065	29.9	150	24
DURANGO, CO	110	099	0642	112	116	25.9	106	22
FORT COLLINS, CO	105	067	0587	122	071	26.8	140	23
LAMAR, CO	065	046	0531	069	053	29.3	160	22
STRATTON, CO	088	066	0571	094	102	29.7	153	23
TRINIDAD, CO	065	047	0450	093	065	31.0	156	22
BRIDGEPORT, CT	083	064	0410	084	068	30.2	135	23
WILMINGTON, DE	062	053	0312	081	061	32.0	191	25
APALACHICOLA, FL	001	003	0006	002	003	53.7	322	24
AVON PARK, FL	001	002	0001	001	005	62.9	344	27
DAYTONA BEACH, FL	001	003	0001	001	005	58.4	326	17
FT. MYERS, FL	001	003	0001	001	005	63.5	365	24
HOMESTEAD, FL	001	003	0001	001	006	65.3	344	24
JACKSONVILLE, FL	002	003	0006	003	004	54.6	313	19
TAMPA, FL	001	004	0001	001	005	60.4	349	21
WEST PALM BEACH, FL	001	004	0001	001	004	65.5	365	21

Table 3. Selected Information for Stations, EPA-1 Program. (*Continued*)

STATION	LENGTH OF FREEZE PERIOD (10 %)	ESTIMATED STORAGE DAYS (10 %)	FREEZE INDEX (10 %) (BASE 32°F)	LENGTH OF MAXIMUM FREEZE PERIOD (DAYS)	ESTIMATED MAXIMUM STORAGE DAYS	NORMAL JANUARY TEMPERATURE	GROWING SEASON (DAYS)	YEARS
ATHENS, GA	008	010	0044	010	016	44.5	221	24
AUGUSTA, GA	004	005	0026	005	007	45.8	219	23
ATLANTA, GA	012	011	0055	014	016	42.4	236	24
BLAIRSVILLE EXP STA,GA	035	031	0154	053	048	37.8	172	25
MACON, GA	004	005	0030	004	007	47.8	240	23
BOISE, ID	085	077	0476	094	087	29.0	159	24
CAMBRIDGE, ID	115	101	1256	122	120	22.5	117	24
LEWISTON, ID	050	048	0505	063	059	31.2	179	25
POCATELLO, ID	116	109	1252	125	125	23.2	142	25
CAIRO,IL	022	028	0179	035	034	36.3	230	25
CHICAGO,IL/MIDWAY	093	081	0954	129	086	24.3	192	25
HOOPESTON,IL	095	072	0746	119	076	26.2	170	25
MOLINE,IL	110	092	1106	135	095	21.5	174	25
MT.VERNON,IL	076	042	0342	080	046	32.1	191	24
PANA,IL	076	061	0519	086	065	28.9	186	25
PEORIA,IL	095	083	0906	129	088	23.8	181	25
BERNE, IN	089	074	0661	101	086	27.1	160	25
COLUMBUS, IN	077	061	0472	087	062	30.1	167	25
CRAWFORDSVILLE, IN	090	073	0714	101	086	27.2	163	25
EVANSVILLE, IN	060	045	0343	084	049	32.6	216	25
SOUTH BEND, IN	104	101	0926	132	109	24.0	165	25
BELLE PLAINE, IA	128	113	1387	135	126	19.8	155	25
DES MOINES, IA	128	106	1416	136	111	19.4	175	25
HAMPTON, IA	140	126	1788	146	136	16.3	144	23
KEOSAUQUA, IA	091	084	0875	119	090	24.4	168	27
LOGAN, IA	114	105	1152	128	107	20.6	162	25
SHENANDOAH, IA	101	077	0822	119	095	23.8	167	25
DODGE CITY, KS	066	048	0381	094	053	30.8	184	23
GOODLAND, KS	093	066	0614	121	073	27.6	157	23
HAYS, KS	090	064	0593	129	065	27.8	167	23
HEALY, KS	073	062	0463	094	067	29.9	164	23
HERINGTON, KS	084	052	0445	092	064	29.1	185	23
INDEPENDENCE, KS	034	026	0224	041	032	33.9	196	23
JUNCTION CITY, KS	059	049	0396	069	058	28.7	175	23
BOWLING GREEN, KY	032	039	0217	036	040	35.6	204	25
LEXINGTON, KY	073	048	0310	084	052	32.9	198	25
PADUCAH, KY	041	037	0228	047	038	36.0	209	25
ARCADIA, LA	007	012	0060	009	012	45.6	246	22
LAFAYETTE, LA	003	004	0015	004	005	51.9	275	25
LAKE PROVIDENCE, LA	007	012	0061	011	016	46.0	252	25
LEESVILLE, LA	004	005	0036	005	006	49.7	243	21
MONROE, LA	006	009	0051	006	010	46.7	253	25
SHREVEPORT, LA	005	006	0047	006	009	47.2	272	25
WINNFIELD, LA	005	006	0043	005	008	48.4	240	21

Table 3. Selected Information for Stations, EPA-1 Program. *(Continued)*

STATION	LENGTH OF FREEZE PERIOD (10 %)	ESTIMATED STORAGE DAYS (10 %)	FREEZE INDEX (10 %) (BASE 32°F)	LENGTH OF MAXIMUM FREEZE PERIOD (DAYS)	ESTIMATED MAXIMUM STORAGE DAYS	NORMAL JANUARY TEMPERATURE	GROWING SEASON (DAYS)	YEARS
BAR HARBOR, ME	111	128	0976	120	140	23.7	156	23
EASTPORT, ME	124	122	1012	127	130	22.6	174	23
GREENVILLE, ME	158	169	2268	159	172	12.5	110	23
BALTIMORE, MD	028	053	0183	033	055	36.1	200	22
ADAMS, MA	126	117	1266	132	123	21.0	127	25
BOSTON, MA	084	080	0418	089	081	29.2	217	25
HAVERHILL, MA	094	102	0640	112	104	26.8	181	25
WORCESTER, MA	122	118	0953	133	126	23.6	148	25
MUSKEGON, MI	124	116	1040	127	119	24.0	161	23
INTERNATIONAL FALLS,MN	161	168	3506	164	172	01.9	095	23
MINNEAPOLIS, MN	141	143	2174	146	143	12.2	166	23
PARK RAPIDS, MN	159	155	3186	164	159	05.0	122	24
ABERDEEN, MS	008	011	0051	013	013	44.7	228	24
CLARKSDALE, MS	017	016	0083	020	021	43.1	236	25
COLUMBIA, MS	003	006	0020	004	009	50.4	239	24
MERIDIAN, MS	006	008	0033	011	015	46.9	246	24
PONTOTOC, MS	010	014	0077	013	019	43.2	223	24
STONEVILLE EXP STA, MS	011	013	0077	012	017	44.2	229	23
ST. LOUIS, MO	075	044	0367	085	051	31.3	206	25
SALISBURY, MO	082	054	0470	090	071	27.8	186	25
SPRINGFIELD, MO	040	039	0274	047	043	32.9	201	25
TARKIO, MO	098	084	0869	119	090	24.0	170	22
WARSAW, MO	043	037	0296	050	038	32.5	185	23
WEST PLAINS, MO	038	033	0232	040	035	33.8	178	25
BILLINGS, MT	128	100	1342	139	102	21.9	132	25
BOZEMAN, MT	134	144	1348	137	152	20.8	107	23
DILLON, MT	138	128	1647	139	132	20.2	099	25
GREAT FALLS, MT	128	091	1557	132	102	20.5	135	25
MILES CITY, MT	134	123	2356	138	140	15.4	150	25
MISSOULA, MT	133	121	1309	137	128	20.8	095	25
GRAND ISLAND, NE	121	098	1173	132	103	22.3	151	25
IMPERIAL, NE	093	082	0765	121	092	26.9	150	25
SCOTTSBLUFF, NE	117	075	0804	134	081	24.9	135	25
VALENTINE, NE	128	104	1401	136	119	20.4	146	25
ADAVEN, NV	085	086	0425	096	102	30.6	119	24
ELKO, NV	124	117	1230	125	124	23.2	089	25
ELY, NV	133	120	1362	134	137	23.6	126	25
LOVELOCK, NV	083	061	0459	087	080	28.9	135	25
RENO, NV	058	050	0374	064	067	31.9	141	25

Table 3. Selected Information for Stations, EPA-1 Program. *(Continued)*

STATION	LENGTH OF FREEZE PERIOD (10 %)	ESTIMATED STORAGE DAYS (10 %)	FREEZE INDEX (10 %) (BASE 32°F)	LENGTH OF MAXIMUM FREEZE PERIOD (DAYS)	ESTIMATED MAXIMUM STORAGE DAYS	NORMAL JANUARY TEMPERATURE	GROWING SEASON (DAYS)	YEARS
LEBANON, NH	136	135	1569	137	147	18.1	120	25
ATLANTIC CITY, NJ	040	032	0138	045	036	32.7	225	25
NEWARK, NJ	070	054	0302	082	059	31.4	219	25
TRENTON, NJ	064	053	0280	083	056	32.1	218	25
ALBUQUERQUE, NM	034	024	0159	057	026	35.2	196	24
ARTESIA, NM	013	013	0160	013	013	40.8	220	21
CIMARRON, NM	074	034	0296	099	074	32.2	150	25
CLAYTON, NM	049	035	0235	056	042	33.1	166	25
CLOVIS, NM	018	015	0149	018	020	37.2	198	22
SANTA FE, NM	095	065	0493	095	065	29.8	165	17
BUFFALO, NY	119	103	0948	126	108	23.7	179	25
ELMIRA, NY	120	104	1021	125	113	25.0	156	25
POUGHKEEPSIE, NY	089	086	0760	108	092	26.5	177	25
ROCHESTER, NY	124	115	0970	125	121	24.0	176	23
SPIER FALLS, NY	125	135	1405	138	139	20.4	150	23
SYRACUSE, NY	117	115	0975	125	118	23.6	168	25
WATERTOWN, NY	127	126	1467	132	128	18.8	151	25
CANTON, NC	023	035	0171	052	041	38.0	195	25
CHARLOTTE, NC	010	011	0052	010	014	42.1	239	25
ELIZABETH CITY, NC	010	013	0036	011	016	42.3	226	25
FAYETTEVILLE, NC	008	010	0045	013	012	43.0	222	25
GREENSBORO, NC	016	024	0086	023	030	38.7	211	25
HICKORY, NC	013	020	0070	020	025	39.1	210	24
LAURINBURG, NC	006	008	0028	009	012	43.9	212	25
MAYSVILLE, NC	009	011	0040	013	012	45.0	241	23
RALEIGH, NC	013	017	0061	021	021	40.5	237	25
SALISBURY, NC	010	014	0061	016	015	40.9	205	25
WELDON, NC	015	017	0082	021	029	40.5	193	23
WILMINGTON, NC	004	007	0023	006	007	46.4	262	25
WILSON, NC	008	011	0044	011	015	42.5	222	24
BISMARCK, ND	146	140	3103	148	144	08.2	136	24
DEVILS LAKE, ND	158	156	3533	165	168	04.2	127	24
WILLISTON, ND	144	141	2975	147	146	08.3	132	24
AKRON CANTON, OH	103	087	0761	109	092	26.3	173	25
COLUMBUS, OH	084	071	0578	087	076	28.4	196	25
DAYTON, OH	084	069	0599	092	071	28.1	165	25
FINDLAY, OH	106	088	0772	130	095	25.8	160	25
ADA, OK	017	013	0085	020	014	40.9	220	22
BARTLESVILLE, OK	022	020	0170	025	020	35.4	210	23
HOBART, OK	020	018	0169	026	020	37.3	210	23
MADILL, OK	013	010	0069	018	012	42.1	227	23
OKLAHOMA CITY, OK	022	018	0166	026	020	36.8	223	22
PAULS VALLEY, OK	013	014	0091	016	016	39.7	207	23
POTEAU, OK	016	012	0082	018	014	41.1	212	23

Table 3. Selected Information for Stations, EPA-1 Program. *(Continued)*

STATION	LENGTH OF FREEZE PERIOD (10 %)	ESTIMATED STORAGE DAYS (10 %)	FREEZE INDEX (10 %) (BASE 32°F)	LENGTH OF MAXIMUM FREEZE PERIOD (DAYS)	ESTIMATED MAXIMUM STORAGE DAYS	NORMAL JANUARY TEMPERATURE	GROWING SEASON (DAYS)	YEARS
BURNS, OR	108	102	0774	122	119	25.2	111	25
EUGENE, OR	010	026	0119	027	032	39.4	199	25
KLAMATH FALLS, OR	087	089	0400	103	091	29.7	126	25
MEDFORD, OR	018	020	0085	025	022	36.6	184	25
PENDLETON, OR	037	042	0526	062	052	32.0	196	25
PORTLAND, OR	014	027	0094	023	037	38.1	211	25
REDMOND, OR	044	049	0379	075	071	30.2	075	23
ROSEBURG, OR	006	010	0056	008	015	40.9	219	28
SALEM, OR	012	023	0101	028	029	38.8	192	25
ERIE, PA	113	098	0741	114	109	25.1	200	23
PHILADEPHIA, PA	068	057	0322	083	064	32.3	190	25
PITTSBURGH, PA	069	049	0421	085	053	28.1	187	25
STATE COLLEGE, PA	104	094	0709	113	106	27.0	166	25
WILLIAMSPORT, PA	093	088	0726	098	095	27.2	164	25
PROVIDENCE, RI	089	076	0526	090	080	28.4	197	24
CHARLESTON, SC	003	004	0018	004	006	48.6	294	24
COLUMBIA, SC	005	006	0025	005	007	45.4	252	25
CONWAY, SC	004	006	0024	005	008	46.5	240	26
FLORENCE, SC	005	008	0024	006	009	45.6	241	25
RAINBOW LAKE, SC	007	012	0041	009	012	42.1	210	25
SUMMERVILLE, SC	004	005	0021	004	006	47.4	252	25
ABERDEEN, SD	145	138	2690	148	142	09.5	137	25
BROOKINGS, SD	144	131	2365	147	136	12.0	137	24
PIERRE, SD	136	126	2205	142	136	15.6	155	25
RAPID CITY, SD	128	099	1291	136	100	21.9	150	25
CHATTANOOGA, TN	012	020	0097	020	022	40.2	229	25
CROSSVILLE, TN	054	052	0322	085	055	34.5	176	20
KINGSPORT, TN	022	024	0146	022	028	36.4	190	27
KNOXVILLE, TN	021	022	0125	035	031	40.6	220	25
MEMPHIS, TN	019	017	0108	022	025	40.5	237	25
NASHVILLE, TN	022	028	0150	032	031	38.3	224	25
AMARILLO, TX	021	023	0205	031	027	36.0	191	25
CORPUS CHRISTI, TX	001	003	0003	003	005	56.3	306	21
DALHART, TX	038	027	0195	051	032	33.8	185	24
DALLAS, TX	005	008	0052	007	010	45.4	239	25
EL PASO, TX	005	005	0048	006	006	43.6	243	24
LUBBOCK, TX	016	014	0118	018	020	39.1	204	24
LUFKIN, TX	004	005	0034	006	006	48.8	231	24
MIDLAND, TX	006	007	0058	007	010	43.6	232	24
WICHITA FALLS, TX	016	015	0090	018	017	41.5	216	24
BLANDING, UT	098	091	0623	109	110	27.7	148	25
HANKSVILLE, UT	082	073	0934	090	089	26.1	156	24
LOGAN, UT	130	109	0931	131	113	24.0	165	21
MILFORD, UT	112	090	1007	129	101	25.7	128	25
NEPHI, UT	092	075	0675	109	089	26.0	137	23
SALT LAKE CITY, UT	098	089	0765	107	115	28.0	202	25

Table 3. Selected Information for Stations, EPA-1 Program. (*Continued*)

STATION	LENGTH OF FREEZE PERIOD (10 %)	ESTIMATED STORAGE DAYS (10 %)	FREEZE INDEX (10 %) (BASE 32°F)	LENGTH OF MAXIMUM FREEZE PERIOD (DAYS)	ESTIMATED MAXIMUM STORAGE DAYS	NORMAL JANUARY TEMPERATURE	GROWING SEASON (DAYS)	YEARS
BURLINGTON, VT	138	134	1800	143	136	16.8	148	25
BLACKSTONE, VA	016	025	0100	026	031	38.3	181	24
HOT SPRINGS, VA	080	066	0338	101	067	31.6	135	22
NORFOLK, VA	013	015	0059	017	022	40.5	219	25
WASHINGTON, DC/NATIONAL	039	034	0122	062	047	35.6	200	25
LONGVIEW, WA	021	029	0104	028	049	38.2	182	24
OLYMPIA, WA	034	035	0227	034	045	37.2	344	23
SEATTLE, WA	022	033	0081	034	036	38.2	233	25
SPOKANE, WA	105	100	0953	124	106	25.4	169	25
SUNNYSIDE, WA	058	050	0536	102	059	30.5	158	25
VANCOUVER, WA	028	028	0192	033	034	38.4	233	23
WALLA WALLA, WA	036	042	0476	054	051	33.4	202	25
WENATCHEE, WA	093	087	0849	110	122	26.6	188	22
BLUESTONE DAM, WV	065	052	0322	084	058	31.1	150	25
CHARLESTON, WV	065	044	0272	084	049	34.5	193	25
MORGANTOWN, WV	072	060	0486	085	070	31.5	165	25
ASHLAND, WI	147	148	2353	149	149	12.1	109	22
EAU CLAIRE, WI	144	141	2339	148	147	11.7	151	24
GREEN BAY, WI	140	135	1932	148	139	15.4	161	25
LACROSSE, WI	139	127	1925	146	134	16.1	161	25
MADISON, WI	134	119	1700	146	125	16.8	177	25
RHINELANDER, WI	146	149	2344	147	156	12.3	085	21
WEYERHAUSER, WI	145	145	2289	150	148	11.3	125	21
AFTON, WY	149	144	2163	159	156	14.3	018	25
CASPER, WY	136	095	1140	138	101	23.2	130	25
GILLETTE, WY	134	108	1229	125	113	21.7	129	22
ROCK SPRINGS, WY	145	136	1718	151	142	19.2	060	25
WHEATLAND, WY	099	058	0536	127	066	28.9	102	25

Station	Normal January Temp. (°F)	Normal Annual Precip. (in.)	EPA-1	EPA-2	General Period
Alabama					
Mobile	51	67	6	15	1949-73
Selma	49	52	6	20	1949-73
Thomasville	45	56	9	25	1953-74
Arkansas					
Dumas	44	50	21	19	1951-71
Little Rock	40	49	19	12	1949-72
California					
Los Angeles	55	12	4	5	1952-69
San Francisco	48	20	7	13	1949-72
Florida					
Avon Park	63	55	5	13	1945-72
Daytona Beach	58	50	5	8	1955-73
Tampa	60	49	5	16	1953-73
Georgia					
Augusta	46	43	7	10	1949-72
Macon	48	44	7	12	1949-72
Louisiana					
Lafayette	52	57	5	12	1948-73
Lake Providence	46	53	16	19	1951-72
Leesville	50	54	6	35	1950-73
Monroe	47	50	10	12	1949-72
Shreveport	47	45	9	11	1949-72
Winnfield	48	55	8	16	1950-73
Mississippi					
Aberdeen	45	53	13	24	1951-72
Clarksdale	43	49	21	18	1949-72
Columbia	50	61	9	33	1947-73
Meridian	47	52	15	14	1948-73
Pontotoc	43	55	19	20	1949-71
Stoneville	44	50	17	17	1951-72

Station	Normal January Temp. (°F)	Normal Annual Precip. (in.)	EPA-1	EPA-2	General Period
North Carolina					
Charlotte	42	42	14	12	1949-73
Raleigh	41	43	21	14	1949-73
Weldon	41	43	29	11	1926-50
Wilmington	46	54	7	11	1949-73
Oregon					
Eugene	39	43	32	35	1948-73
Medford	37	21	22	21	1948-72
Roseburg	41	34	15	22	1935-64
Salem	39	41	29	36	1948-73
South Carolina					
Charleston	49	52	6	24	1948-73
Columbia	45	46	7	15	1949-73
Conway	47	52	8	9	1945-73
Tennessee					
Crossville	35	57	55	24	1953-73
Texas					
Amarillo	36	20	27	11	1949-71
Corpus Christi	56	29	5	13	1951-72
Dallas	45	36	10	15	1948-72
El Paso	44	8	6	0	1948-72
Wichita Falls	42	27	17	8	1950-72
Washington					
Longview	38	46	49	60	1948-73
Olympia	37	51	45	65*	1949-73
Vancouver	38	40	34	31	1924-50
Walla Walla	33	16	51	14	1950-73

*** This station shows 145 storage days when processed under EPA-2 (10%)**

Figure 2. Comparison between maximum annual storage days estimated from EPA-1 and EPA-2 programs.

LIST OF SYMBOLS

STA: Station
YR: Year (20 to 30 years of record)
MO: Month 01 = January, 02 = February, etc.
DA: 01 – 31
MN: Daily minimum temperature (°F)
MX: Daily maximum temperature (°F)
T: Daily mean temperature (°F)
P: Daily precipitation (inches), Trace = 0.00
SP: Available moisture in the soil at the start of a day. Saturation is assumed when SP = AWC (available water capacity)
SS: Amount of available moisture in the surface soil at the end of a day.
SU: Amount of available moisture in the underlying soil at the end of a day.
PE: Daily potential evapotranspiration (Thornthwaite)
PL: Daily potential moisture loss
PR: Potential recharge; at the start of a day this is the number of inches required to bring the soil to field capacity
R: Daily recharge; net gain in the surface and underlying soil
L: Daily moisture loss from the surface and underlying soil
ET: Daily evapotranspiration
DR: Depletion rate
RO: Daily runoff
ARO: Accumulated runoff. This is the sum of the previous days SRO and the current day's runoff; ARO = SRO′ + RO
SRO: Accumulated daily runoff minus the depletion rate; SRO = ARO – DR
AWC: Available water capacity of the soil. At saturation, AWC = SS + SU
UNF: "X" denotes an unfavorable day for land application because of possible surface runoff

(If SRO + SS + SU is equal to or greater than the available water capacity, the day is unfavorable)

Note: Symbols with a prime (′) indicate conditions at the start of a day. Subscripts s and u indicate "surface" and "underlying", respectively.

Constants which must be supplied for the station being analyzed are:

- I, the heat index
- b, a coefficient which depends on the heat index
- g, the tangent of the station's latitude
- W, the available water capacity of the soil minus 1.0 inch in inches
- ϕ, Daily solar declination, in radians

To start the analysis two values must be assigned: SS′, in inches (to hundreths) of soil moisture available in the surface soil at the start, and SU′, in inches (to hundreths) of soil moisture available in the underlying soil at the start.

Figure 3. Explanation of symbols used in the listings from EPA-2.

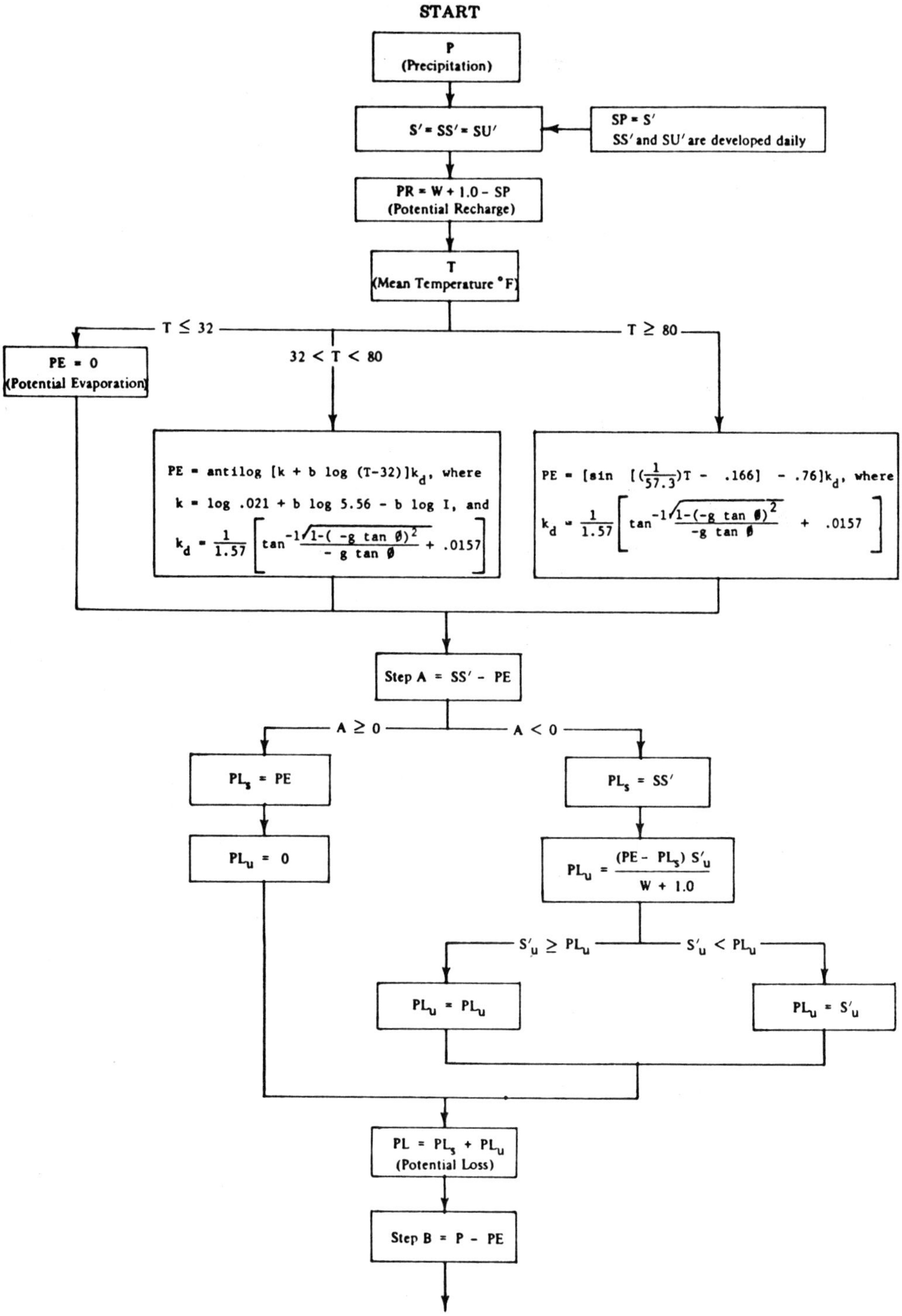

Figure 4. Flow diagram for the modified Palmer program (EPA-2).

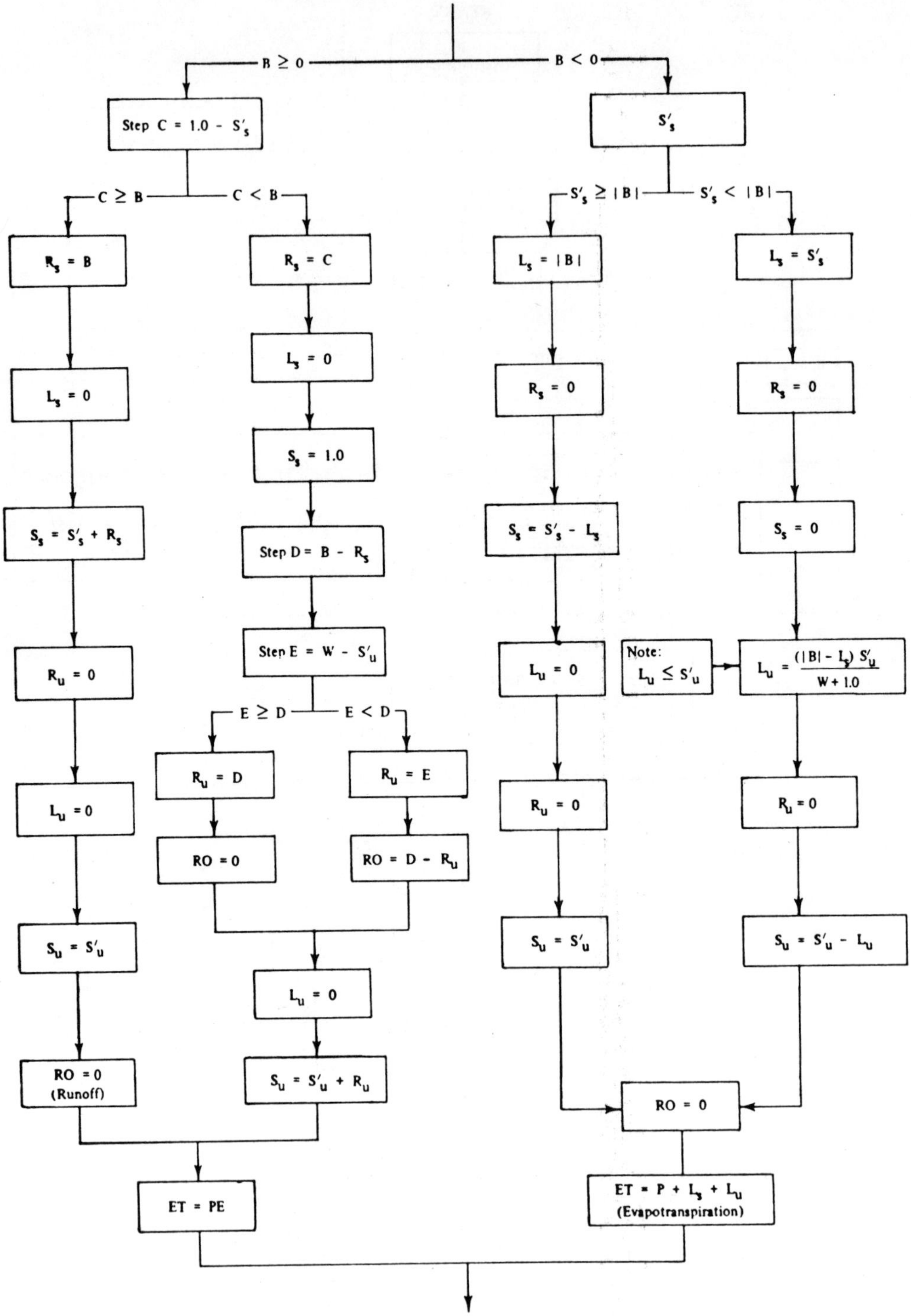

Figure 4. Flow diagram for modified Palmer program (EPA-2)—*continued.*

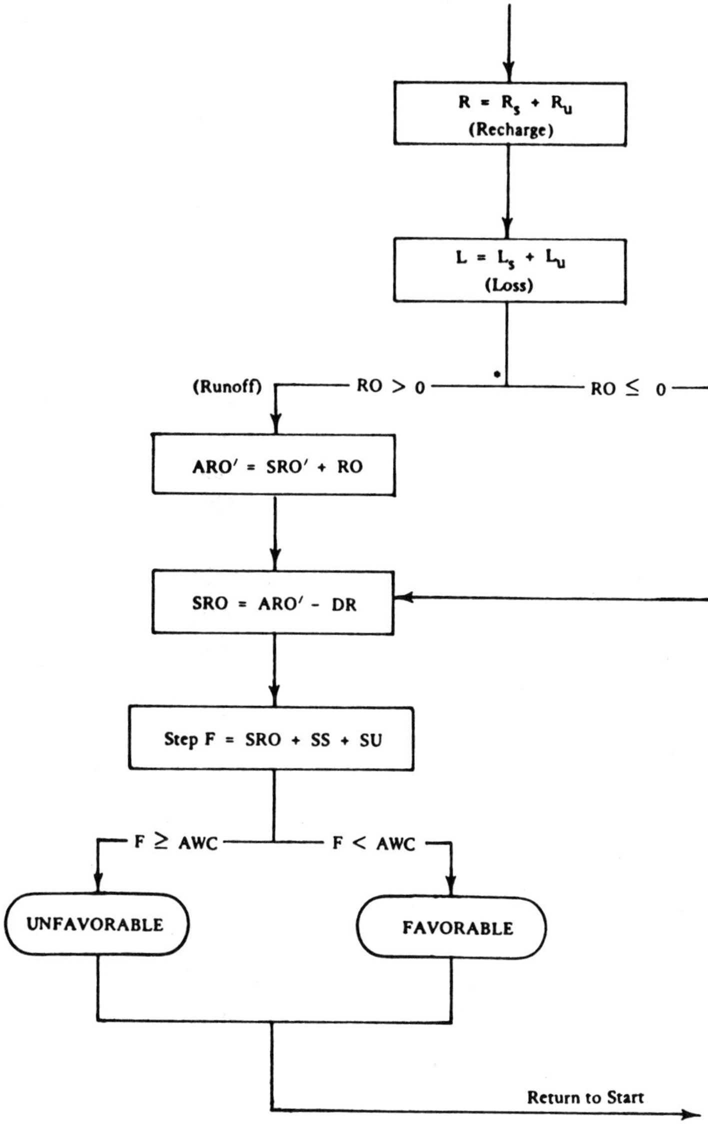

• The original Palmer program that computes the drought severity indices continues from this point and runs about as long as the portion shown here. Most of the constants for EPA–2 are stored on a separate tape at NCC; however, some values must be extracted from the earlier Palmer tabulations. This means that although there are no restrictions on the use or dissemination of the programs, a considerable amount of time and effort are required to collect the necessary material. In addition, a 20 to 25-year period of record can be processed at the Center for less than the cost of the program. Finally, the daily data must be purchased along with the program for each station of interest.

Figure 4. Flow diagram for modified Palmer program (EPA-2)—*continued.*

STATION	DATE	MN	MX	T	P	SP	SS	SU	PE	PL	PR	R	L	ET	RO	SRO	UNF
223627	58 828	60	89	74.5	0.00	3.07	0.00	2.92	0.15	0.15	2.93	0.00	0.15	0.15	0.00	0.00	
223627	58 829	63	90	76.5	0.00	2.92	0.00	2.83	0.17	0.08	3.08	0.00	0.08	0.08	0.00	0.00	
223627	58 830	65	90	77.5	0.00	2.83	0.00	2.75	0.17	0.08	3.17	0.00	0.08	0.08	0.00	0.00	
223627	58 831	70	94	82.0	0.00	2.75	0.00	2.66	0.21	0.10	3.25	0.00	0.10	0.10	0.00	0.00	
223627	58 9 1	71	94	82.5	0.00	2.66	0.00	2.56	0.21	0.09	3.34	0.00	0.09	0.09	0.00	0.00	
223627	58 9 2	69	95	82.0	0.00	2.56	0.00	2.47	0.21	0.09	3.44	0.00	0.09	0.09	0.00	0.00	
223627	58 9 3	70	93	81.5	0.00	2.47	0.00	2.39	0.20	0.08	3.53	0.00	0.08	0.08	0.00	0.00	
223627	58 9 4	70	89	79.5	0.00	2.39	0.00	2.32	0.19	0.07	3.61	0.00	0.07	0.07	0.00	0.00	
223627	58 9 5	71	85	78.0	0.00	2.32	0.00	2.25	0.17	0.07	3.68	0.00	0.07	0.07	0.00	0.00	
223627	58 9 6	69	93	81.0	0.00	2.25	0.00	2.17	0.20	0.07	3.75	0.00	0.07	0.07	0.00	0.00	
223627	58 9 7	70	92	81.0	0.00	2.17	0.00	2.10	0.20	0.07	3.83	0.00	0.07	0.07	0.00	0.00	
223627	58 9 8	67	90	78.5	0.00	2.10	0.00	2.04	0.18	0.06	3.90	0.00	0.06	0.06	0.00	0.00	
223627	58 9 9	61	89	75.0	0.00	2.04	0.00	1.99	0.15	0.05	3.96	0.00	0.05	0.05	0.00	0.00	
223627	58 910	63	92	77.5	0.24	1.99	0.07	1.99	0.17	0.06	4.01	0.07	0.00	0.17	0.00	0.00	
223627	58 911	70	79	74.5	1.56	2.06	1.00	2.47	0.15	0.10	3.94	0.07	0.00	0.15	0.00	0.00	
223627	58 912	68	79	73.5	0.00	3.47	0.86	2.47	0.14	0.14	2.53	0.00	0.14	0.14	0.00	0.00	
223627	58 913	66	79	72.5	0.00	3.33	0.73	2.47	0.13	0.13	2.67	0.00	0.13	0.13	0.00	0.00	
223627	58 914	65	84	74.5	0.00	3.20	0.58	2.47	0.15	0.15	2.80	0.00	0.15	0.15	0.00	0.00	
223627	58 915	72	88	80.0	0.48	3.05	0.87	2.47	0.19	0.19	2.95	0.29	0.00	0.19	0.00	0.00	
223627	58 916	72	90	81.0	1.07	3.34	1.00	3.22	0.19	0.19	2.68	0.29	0.00	0.19	0.00	0.00	
223627	58 917	69	79	74.0	0.75	4.22	1.00	3.83	0.14	0.14	1.78	0.29	0.00	0.14	0.00	0.00	
223627	58 918	67	79	73.0	0.00	4.83	0.87	3.83	0.13	0.13	1.17	0.00	0.13	0.13	0.00	0.00	
223627	58 919	69	83	76.0	2.67	4.69	1.00	5.00	0.15	0.15	1.31	0.00	0.00	0.15	1.21	0.46	X
223627	58 920	72	76	74.0	8.07	6.00	1.00	5.00	0.14	0.14	0.00	0.00	0.00	0.14	7.93	7.64	X
223627	58 921	71	79	75.0	2.48	6.00	1.00	5.00	0.15	0.15	0.00	0.00	0.00	0.15	2.33	9.22	X
223627	58 922	69	84	76.5	0.00	6.00	0.84	5.00	0.16	0.16	0.00	0.00	0.16	0.16	0.00	8.47	X
223627	58 923	70	86	78.0	0.16	5.84	0.84	5.00	0.17	0.17	0.16	0.00	0.01	0.17	0.00	7.72	X
223627	58 924	73	88	80.5	0.00	5.84	0.65	5.00	0.19	0.19	0.16	0.00	0.19	0.19	0.00	6.97	X
223627	58 925	74	92	83.0	0.00	5.65	0.45	5.00	0.20	0.20	0.35	0.00	0.20	0.20	0.00	6.22	X
223627	58 926	73	92	82.5	0.00	5.45	0.25	5.00	0.20	0.20	0.55	0.00	0.20	0.20	0.00	5.47	X
223627	58 927	64	80	72.0	0.00	5.25	0.13	5.00	0.12	0.12	0.75	0.00	0.12	0.12	0.00	4.72	X
223627	58 928	55	75	65.0	0.00	5.13	0.04	5.00	0.09	0.09	0.87	0.00	0.09	0.09	0.00	3.97	X
223627	58 929	51	80	65.5	0.00	5.04	0.00	4.96	0.09	0.08	0.96	0.00	0.08	0.08	0.00	3.22	X
223627	58 930	56	76	66.0	2.17	4.96	1.00	5.00	0.09	0.07	1.04	0.00	0.00	0.09	1.04	3.52	X
223627	5810 1	51	69	60.0	0.32	6.00	1.00	5.00	0.06	0.06	0.00	0.00	0.00	0.06	0.26	3.02	X
223627	5810 2	48	70	59.0	0.00	6.00	0.94	5.00	0.06	0.06	0.00	0.00	0.06	0.06	0.00	2.27	X
223627	5810 3	53	63	58.0	0.12	5.94	1.00	5.00	0.05	0.05	0.06	0.00	0.00	0.05	0.01	1.53	X
223627	5810 4	57	75	66.0	0.00	6.00	0.91	5.00	0.07	0.09	0.00	0.00	0.09	0.09	0.00	0.78	X
223627	5810 5	57	78	67.5	0.00	5.91	0.81	5.00	0.10	0.10	0.09	0.00	0.10	0.10	0.00	0.03	016
223627	5810 6	54	84	69.0	0.00	5.81	0.71	5.00	0.10	0.10	0.19	0.00	0.10	0.10	0.00	0.00	
223627	5810 7	60	81	70.5	0.00	5.71	0.60	5.00	0.11	0.11	0.29	0.00	0.11	0.11	0.00	0.00	
223627	5810 8	64	84	74.0	0.00	5.60	0.46	5.00	0.13	0.13	0.40	0.00	0.13	0.13	0.00	0.00	
223627	5810 9	69	86	77.5	0.06	5.46	0.37	5.00	0.16	0.16	0.54	0.00	0.10	0.16	0.00	0.00	
223627	581010	58	76	67.0	0.00	5.37	0.28	5.00	0.09	0.09	0.63	0.00	0.09	0.09	0.00	0.00	
223627	581011	53	72	62.5	0.00	5.28	0.21	5.00	0.07	0.07	0.72	0.00	0.07	0.07	0.00	0.00	
223627	581012	52	77	64.5	0.00	5.21	0.13	5.00	0.08	0.08	0.79	0.00	0.08	0.08	0.00	0.00	
223627	581013	49	81	65.0	0.00	5.13	0.04	5.00	0.08	0.08	0.87	0.00	0.08	0.08	0.00	0.00	
223627	581014	51	79	65.0	0.00	5.04	0.00	4.97	0.08	0.08	0.96	0.00	0.08	0.08	0.00	0.00	
223627	581015	55	79	67.0	0.00	4.97	0.00	4.89	0.09	0.08	1.03	0.00	0.08	0.08	0.00	0.00	
223627	581016	60	80	70.0	0.00	4.89	0.00	4.81	0.11	0.09	1.11	0.00	0.09	0.09	0.00	0.00	
223627	581017	59	81	70.0	0.00	4.81	0.00	4.72	0.11	0.09	1.19	0.00	0.09	0.09	0.00	0.00	
223627	581018	57	82	69.5	0.00	4.72	0.00	4.64	0.10	0.08	1.28	0.00	0.08	0.08	0.00	0.00	
223627	581019	52	77	64.5	0.00	4.64	0.00	4.58	0.08	0.06	1.36	0.00	0.06	0.06	0.00	0.00	
223627	581020	45	80	62.5	0.00	4.58	0.00	4.53	0.07	0.05	1.42	0.00	0.05	0.05	0.00	0.00	
223627	581021	47	82	64.5	0.00	4.53	0.00	4.47	0.08	0.06	1.47	0.00	0.06	0.06	0.00	0.00	

STATION 223627 GREENWOOD, MS
AWC= 6.0 DEPLETION RATE=0.75

BEG.DATE	END.DATE	#DAYS
490102	490110	9
500105	500113	9
511213	511222	10
521230	530102	4
530429	530506	8
540428	540505	8
550321	550328	8
560201	560212	12
571112	571124	13
580919	581004	16
590417	590421	5
600301	600305	5
611208	611219	12
621224	621229	6
631213	631224	12
640422	640427	6
650208	650212	5
660209	660216	8
671214	671221	8
680103	680114	12
691229	700102	5
700105	700111	7
710508	710514	7
720101	720107	7
730314	730325	12
MAX FOR TOTAL PERIOD OF RECORD		
580919	581004	16

PERCENTILES	DAYS
0.05	14.7
0.10	12.1
0.25	11.5
0.50	8.0

Figure 5. Daily listing from EPA-2 for Greenwood, MS, Aug.-Oct. 1958, with summary table.

STATION	DATE	MN	MX	T	P	SP	SS	SU	PE	PL	PR	R	L	ET	RO	SRO	UNF
456114	4910 4	46	55	50.5	0.68	3.96	0.73	3.85	0.06	0.06	8.04	0.62	0.00	0.06	0.00	0.00	
456114	4910 5	44	60	52.0	0.18	4.58	0.84	3.85	0.06	0.06	7.42	0.12	0.00	0.06	0.00	0.00	
456114	4910 6	40	52	46.0	0.34	4.69	1.00	3.99	0.04	0.04	7.31	0.12	0.00	0.04	0.00	0.00	
456114	4910 7	37	61	49.0	0.00	4.99	0.95	3.99	0.05	0.05	7.01	0.00	0.05	0.05	0.00	0.00	
456114	4910 8	32	55	43.5	0.01	4.94	0.92	3.99	0.03	0.03	7.06	0.00	0.02	0.03	0.00	0.00	
456114	4910 9	46	60	53.0	0.43	4.92	1.00	4.28	0.07	0.07	7.08	0.00	0.00	0.07	0.00	0.00	
456114	491010	42	57	49.5	0.19	5.28	1.00	4.42	0.05	0.05	6.72	0.00	0.00	0.05	0.00	0.00	
456114	491011	45	60	52.5	0.31	5.42	1.00	4.66	0.06	0.06	6.58	0.00	0.00	0.06	0.00	0.00	
456114	491012	36	63	49.5	0.00	5.66	0.95	4.66	0.05	0.05	6.34	0.00	0.05	0.05	0.00	0.00	
456114	491013	32	58	45.0	0.00	5.61	0.91	4.66	0.04	0.04	6.39	0.00	0.04	0.04	0.00	0.00	
456114	491014	29	66	47.5	0.00	5.57	0.86	4.66	0.05	0.05	6.43	0.00	0.05	0.05	0.00	0.00	
456114	491015	39	61	50.0	0.04	5.53	0.85	4.66	0.05	0.05	6.47	0.00	0.01	0.05	0.00	0.00	
456114	491016	32	58	45.0	0.00	5.51	0.81	4.66	0.04	0.04	6.49	0.00	0.04	0.04	0.00	0.00	
456114	491017	27	53	40.0	0.02	5.48	0.81	4.66	0.02	0.02	6.52	0.00	0.00	0.02	0.00	0.00	
456114	491018	27	50	38.5	0.00	5.47	0.79	4.66	0.02	0.02	6.53	0.00	0.02	0.02	0.00	0.00	
456114	491019	22	52	37.0	0.00	5.46	0.78	4.66	0.01	0.01	6.54	0.00	0.01	0.01	0.00	0.00	
456114	491020	23	55	39.0	0.00	5.44	0.76	4.66	0.02	0.02	6.56	0.00	0.02	0.02	0.00	0.00	
456114	491021	29	58	43.5	0.00	5.43	0.73	4.66	0.03	0.03	6.57	0.00	0.03	0.03	0.00	0.00	
456114	491022	29	53	41.0	0.00	5.40	0.71	4.66	0.02	0.02	6.60	0.00	0.02	0.02	0.00	0.00	
456114	491023	35	49	42.0	0.00	5.37	0.68	4.66	0.03	0.03	6.63	0.00	0.03	0.03	0.00	0.00	
456114	491024	37	60	48.5	0.02	5.34	0.66	4.66	0.05	0.05	6.66	0.00	0.03	0.05	0.00	0.00	
456114	491025	36	51	43.5	0.01	5.32	0.63	4.66	0.03	0.03	6.68	0.00	0.02	0.03	0.00	0.00	
456114	491026	43	53	48.0	0.25	5.30	0.84	4.66	0.04	0.04	6.70	0.21	0.00	0.04	0.00	0.00	
456114	491027	47	57	52.0	1.24	5.50	1.00	5.69	0.06	0.06	6.50	0.21	0.00	0.06	0.00	0.00	
456114	491028	45	60	52.5	1.36	6.69	1.00	6.99	0.06	0.06	5.31	0.21	0.00	0.06	0.00	0.00	
456114	491029	40	57	48.5	0.01	7.99	0.96	6.99	0.05	0.05	4.01	0.00	0.04	0.05	0.00	0.00	
456114	491030	36	63	49.5	0.00	7.95	0.92	6.99	0.05	0.05	4.05	0.00	0.05	0.05	0.00	0.00	
456114	491031	36	74	55.0	0.00	7.90	0.85	6.99	0.07	0.07	4.10	0.00	0.07	0.07	0.00	0.00	
456114	4911 1	35	65	50.0	0.00	7.84	0.80	6.99	0.05	0.05	4.16	0.00	0.05	0.05	0.00	0.00	
456114	4911 2	34	72	53.0	0.00	7.79	0.74	6.99	0.06	0.06	4.21	0.00	0.06	0.06	0.00	0.00	
456114	4911 3	33	71	52.0	0.00	7.73	0.69	6.99	0.05	0.05	4.27	0.00	0.05	0.05	0.00	0.00	
456114	4911 4	38	74	56.0	0.00	7.68	0.62	6.99	0.07	0.07	4.32	0.00	0.07	0.07	0.00	0.00	
456114	4911 5	35	64	49.5	0.00	7.61	0.57	6.99	0.05	0.05	4.39	0.00	0.05	0.05	0.00	0.00	
456114	4911 6	36	52	44.0	0.00	7.56	0.54	6.99	0.03	0.03	4.44	0.00	0.03	0.03	0.00	0.00	
456114	4911 7	38	55	46.5	0.19	7.53	0.70	6.99	0.04	0.04	4.47	0.15	0.00	0.04	0.00	0.00	
456114	4911 8	45	51	48.0	0.25	7.68	0.91	6.99	0.04	0.04	4.32	0.21	0.00	0.04	0.00	0.00	
456114	4911 9	40	48	44.0	0.14	7.89	1.00	7.00	0.03	0.03	4.11	0.21	0.00	0.03	0.00	0.00	
456114	491110	42	50	46.0	0.38	8.00	1.00	7.35	0.04	0.04	4.00	0.21	0.00	0.04	0.00	0.00	
456114	491111	46	53	49.5	1.16	8.35	1.00	8.46	0.05	0.05	3.65	0.21	0.00	0.05	0.00	0.00	
456114	491112	45	49	47.0	2.89	9.46	1.00	11.00	0.04	0.04	2.54	0.21	0.00	0.04	0.31	0.00	X
456114	491113	46	58	52.0	0.11	12.00	1.00	11.00	0.05	0.05	0.00	0.21	0.00	0.05	0.06	0.00	X
456114	491114	45	56	50.5	0.19	12.00	1.00	11.00	0.05	0.05	0.00	0.21	0.00	0.05	0.14	0.00	X
456114	491115	40	54	47.0	0.00	12.00	0.96	11.00	0.04	0.04	0.00	0.00	0.04	0.04	0.00	0.00	003
456114	491116	40	64	52.0	0.00	11.96	0.91	11.00	0.05	0.05	0.04	0.00	0.05	0.05	0.00	0.00	
456114	491117	37	55	46.0	0.04	11.91	0.92	11.00	0.03	0.03	0.09	0.01	0.00	0.03	0.00	0.00	
456114	491118	45	55	50.0	0.04	11.92	0.91	11.00	0.05	0.05	0.08	0.00	0.01	0.05	0.00	0.00	
456114	491119	46	51	48.5	0.00	11.91	0.87	11.00	0.04	0.04	0.09	0.00	0.04	0.04	0.00	0.00	
456114	491120	38	56	47.0	0.00	11.87	0.83	11.00	0.04	0.04	0.13	0.00	0.04	0.04	0.00	0.00	
456114	491121	36	50	43.0	0.05	11.83	0.86	11.00	0.03	0.03	0.17	0.02	0.00	0.03	0.00	0.00	
456114	491122	37	51	44.0	0.88	11.86	1.00	11.00	0.03	0.03	0.16	0.02	0.00	0.03	0.71	0.00	X
456114	491123	48	52	50.0	0.63	12.00	1.00	11.00	0.04	0.04	0.00	0.02	0.00	0.04	0.59	0.00	X
456114	491124	50	60	55.0	0.36	12.00	1.00	11.00	0.06	0.06	0.00	0.02	0.00	0.06	0.30	0.00	X
456114	491125	48	59	53.5	0.46	12.00	1.00	11.00	0.05	0.05	0.00	0.02	0.00	0.05	0.41	0.00	X
456114	491126	50	58	54.0	2.50	12.00	1.00	11.00	0.05	0.05	0.00	0.02	0.00	0.05	2.45	1.70	X
456114	491127	43	56	49.5	1.00	12.00	1.00	11.00	0.04	0.04	0.00	0.02	0.00	0.04	0.96	1.90	X

STATION 456114 OLYMPIA, WA
AWC=12.0 DEPLETION RATE=0.75

BEG.DATE	END.DATE	#DAYS
491225	500228	65
501119	501209	21
510109	510125	17
520118	520126	9
530106	530207	33
540113	540130	18
551111	551202	21
560210	560309	29
570114	570211	29
581217	581230	14
591209	591220	12
600113	600129	17
611212	611229	18
621119	621205	17
631112	631124	13
641206	650110	36
651223	660114	23
661127	661220	24
670313	670326	14
681213	681224	12
690103	690212	41
701226	710130	36
711202	720101	31
720222	720313	21
730101	730120	20
MAX FOR TOTAL PERIOD OF RECORD		
491225	500228	65

PERCENTILES	DAYS
0.05	57.8
0.10	38.0
0.25	30.0
0.50	21.0

Figure 6. Daily listing from EPA-2 for Olympia, WA, Oct. 1949—Nov. 1949, with summary table.

STATION	DATE	MN	MX	T	P	SP	SS	SU	PE	PL	PR	R	L	ET	RO	SRO	UNF
456114	491128	41	52	46.5	0.54	12.00	1.00	11.00	0.03	0.03	0.00	0.02	0.00	0.03	0.51	1.66	X
456114	491129	40	47	43.5	0.34	12.00	1.00	11.00	0.03	0.03	0.00	0.02	0.00	0.03	0.31	1.22	X
456114	491130	42	51	46.5	0.18	12.00	1.00	11.00	0.03	0.03	0.00	0.02	0.00	0.03	0.15	0.62	X
456114	4912 1	47	53	50.0	0.72	12.00	1.00	11.00	0.04	0.04	0.00	0.02	0.00	0.04	0.68	0.55	X
456114	4912 2	33	49	41.0	0.44	12.00	1.00	11.00	0.02	0.02	0.00	0.02	0.00	0.02	0.42	0.22	X
456114	4912 3	30	41	35.5	0.01	12.00	1.00	11.00	0.01	0.01	0.00	0.02	0.00	0.01	0.00	0.00	X
456114	4912 4	29	51	40.0	0.02	12.00	1.00	11.00	0.02	0.02	0.00	0.02	0.00	0.02	0.00	0.00	X
456114	4912 5	35	48	41.5	0.46	12.00	1.00	11.00	0.02	0.02	0.00	0.02	0.00	0.02	0.44	0.00	X
456114	4912 6	29	45	37.0	0.00	12.00	0.99	11.00	0.01	0.01	0.00	0.00	0.01	0.01	0.00	0.00	014
456114	4912 7	26	46	36.0	0.00	11.99	0.98	11.00	0.01	0.01	0.01	0.00	0.01	0.01	0.00	0.00	
456114	4912 8	25	44	34.5	0.02	11.98	1.00	11.00	0.00	0.00	0.02	0.02	0.00	0.00	0.00	0.00	
456114	4912 9	27	46	36.5	0.00	12.00	0.99	11.00	0.01	0.01	0.00	0.00	0.01	0.01	0.00	0.00	
456114	491210	24	43	33.5	0.00	11.99	0.99	11.00	0.00	0.00	0.01	0.00	0.00	0.00	0.00	0.00	
456114	491211	26	34	30.0	0.05	11.99	1.00	11.00	0.00	0.00	0.01	0.00	0.00	0.00	0.04	0.00	X
456114	491212	32	38	35.0	0.19	12.00	1.00	11.00	0.01	0.01	0.00	0.00	0.00	0.01	0.18	0.00	X
456114	491213	36	43	39.5	0.06	12.00	1.00	11.00	0.02	0.02	0.00	0.00	0.00	0.02	0.04	0.00	X
456114	491214	37	47	42.0	0.00	12.00	0.98	11.00	0.02	0.02	0.00	0.00	0.02	0.02	0.00	0.00	003
456114	491215	36	43	39.5	0.07	11.98	1.00	11.00	0.02	0.02	0.02	0.00	0.00	0.02	0.03	0.00	X
456114	491216	32	40	36.0	0.20	12.00	1.00	11.00	0.01	0.01	0.00	0.00	0.00	0.01	0.19	0.00	X
456114	491217	32	44	38.0	0.69	12.00	1.00	11.00	0.01	0.01	0.00	0.00	0.00	0.01	0.68	0.00	X
456114	491218	26	37	31.5	0.24	12.00	1.00	11.00	0.00	0.00	0.00	0.00	0.00	0.00	0.24	0.00	X
456114	491219	17	34	25.5	0.00	12.00	1.00	11.00	0.00	0.00	0.00	0.00	0.00	0.00	0.00	0.00	X
456114	491220	22	37	29.5	0.13	12.00	1.00	11.00	0.00	0.00	0.00	0.00	0.00	0.00	0.13	0.00	X
456114	491221	36	44	40.0	0.13	12.00	1.00	11.00	0.02	0.02	0.00	0.00	0.00	0.02	0.11	0.00	X
456114	491222	42	49	45.5	0.31	12.00	1.00	11.00	0.03	0.03	0.00	0.00	0.00	0.03	0.28	0.00	X
456114	491223	40	47	43.5	1.27	12.00	1.00	11.00	0.02	0.02	0.00	0.00	0.00	0.02	1.25	0.50	X
456114	491224	33	49	41.0	0.00	12.00	0.98	11.00	0.02	0.02	0.00	0.00	0.02	0.02	0.00	0.00	009
456114	491225	37	49	43.0	0.66	11.98	1.00	11.00	0.02	0.02	0.02	0.00	0.00	0.02	0.62	0.00	X
456114	491226	39	43	41.0	0.78	12.00	1.00	11.00	0.02	0.02	0.00	0.00	0.00	0.02	0.76	0.01	X
456114	491227	42	48	45.0	2.94	12.00	1.00	11.00	0.03	0.03	0.00	0.00	0.00	0.03	2.91	2.17	X
456114	491228	46	52	49.0	0.20	12.00	1.00	11.00	0.04	0.04	0.00	0.00	0.00	0.04	0.16	1.58	X
456114	491229	31	51	41.0	0.42	12.00	1.00	11.00	0.02	0.02	0.00	0.00	0.00	0.02	0.40	1.24	X
456114	491230	31	36	33.5	0.54	12.00	1.00	11.00	0.00	0.00	0.00	0.00	0.00	0.00	0.54	1.02	X
456114	491231	30	42	36.0	0.18	12.00	1.00	11.00	0.01	0.01	0.00	0.00	0.00	0.01	0.17	0.45	X
456114	50 1 1	20	33	26.5	0.10	12.00	1.00	11.00	0.00	0.00	0.00	0.00	0.00	0.00	0.10	0.00	X
456114	50 1 2	6	25	15.5	0.06	12.00	1.00	11.00	0.00	0.00	0.00	0.00	0.00	0.00	0.06	0.00	X
456114	50 1 3	5	28	16.5	0.11	12.00	1.00	11.00	0.00	0.00	0.00	0.00	0.00	0.00	0.11	0.00	X
456114	50 1 4	23	32	27.5	0.00	12.00	1.00	11.00	0.00	0.00	0.00	0.00	0.00	0.00	0.00	0.00	X
456114	50 1 5	24	37	30.5	0.04	12.00	1.00	11.00	0.00	0.00	0.00	0.00	0.00	0.00	0.04	0.00	X
456114	50 1 6	34	37	35.5	0.86	12.00	1.00	11.00	0.01	0.01	0.00	0.00	0.00	0.01	0.85	0.10	X
456114	50 1 7	32	36	34.0	0.46	12.00	1.00	11.00	0.00	0.00	0.00	0.00	0.00	0.00	0.46	0.00	X
456114	50 1 8	30	34	32.0	0.04	12.00	1.00	11.00	0.00	0.00	0.00	0.00	0.00	0.00	0.04	0.00	X
456114	50 1 9	28	37	32.5	0.26	12.00	1.00	11.00	0.00	0.00	0.00	0.00	0.00	0.00	0.26	0.00	X
456114	50 110	31	39	35.0	0.76	12.00	1.00	11.00	0.01	0.01	0.00	0.00	0.00	0.01	0.75	0.00	X
456114	50 111	27	36	31.5	0.09	12.00	1.00	11.00	0.00	0.00	0.00	0.00	0.00	0.00	0.09	0.00	X
456114	50 112	18	32	25.0	0.11	12.00	1.00	11.00	0.00	0.00	0.00	0.00	0.00	0.00	0.11	0.00	X
456114	50 113	12	19	15.5	1.21	12.00	1.00	11.00	0.00	0.00	0.00	0.00	0.00	0.00	1.21	0.46	X
456114	50 114	6	21	13.5	0.07	12.00	1.00	11.00	0.00	0.00	0.00	0.00	0.00	0.00	0.07	0.00	X
456114	50 115	18	25	21.5	0.19	12.00	1.00	11.00	0.00	0.00	0.00	0.00	0.00	0.00	0.19	0.00	X
456114	50 116	16	36	26.0	0.01	12.00	1.00	11.00	0.00	0.00	0.00	0.00	0.00	0.00	0.01	0.00	X
456114	50 117	8	26	17.0	0.00	12.00	1.00	11.00	0.00	0.00	0.00	0.00	0.00	0.00	0.00	0.00	X
456114	50 118	13	26	19.5	0.38	12.00	1.00	11.00	0.00	0.00	0.00	0.00	0.00	0.00	0.38	0.00	X
456114	50 119	24	41	32.5	1.89	12.00	1.00	11.00	0.00	0.00	0.00	0.00	0.00	0.00	1.89	1.14	X
456114	50 120	38	46	42.0	1.07	12.00	1.00	11.00	0.02	0.02	0.00	0.00	0.00	0.02	1.05	1.44	X
456114	50 121	42	48	45.0	0.73	12.00	1.00	11.00	0.03	0.03	0.00	0.00	0.00	0.03	0.70	1.39	X

Figure 7. Daily listing from EPA-2 for Olympia, WA, Nov. 1949—Jan. 1950.

STATION	DATE	MN	MX	T	P	SP	SS	SU	PE	PL	PR	R	L	ET	RO	SRO	UNF
456114	50 122	36	45	40.5	0.37	12.00	1.00	11.00	0.02	0.02	0.00	0.00	0.00	0.02	0.35	0.99	X
456114	50 123	23	37	30.0	0.17	12.00	1.00	11.00	0.00	0.00	0.00	0.00	0.00	0.00	0.17	0.41	X
456114	50 124	12	27	19.5	0.02	12.00	1.00	11.00	0.00	0.00	0.00	0.00	0.00	0.00	0.02	0.00	X
456114	50 125	2	31	16.5	0.06	12.00	1.00	11.00	0.00	0.00	0.00	0.00	0.00	0.00	0.06	0.00	X
456114	50 126	28	37	32.5	0.20	12.00	1.00	11.00	0.00	0.00	0.00	0.00	0.00	0.00	0.20	0.00	X
456114	50 127	22	30	26.0	0.00	12.00	1.00	11.00	0.00	0.00	0.00	0.00	0.00	0.00	0.00	0.00	X
456114	50 128	8	31	19.5	0.02	12.00	1.00	11.00	0.00	0.00	0.00	0.00	0.00	0.00	0.02	0.00	X
456114	50 129	5	26	15.5	0.00	12.00	1.00	11.00	0.00	0.00	0.00	0.00	0.00	0.00	0.00	0.00	X
456114	50 130	6	27	16.5	0.00	12.00	1.00	11.00	0.00	0.00	0.00	0.00	0.00	0.00	0.00	0.00	X
456114	50 131	1	18	9.5	0.03	12.00	1.00	11.00	0.00	0.00	0.00	0.00	0.00	0.00	0.03	0.00	X
456114	50 2 2	4	36	20.0	0.00	12.00	1.00	11.00	0.00	0.00	0.00	0.00	0.00	0.00	0.00	0.00	X
456114	50 2 3	1	28	14.5	0.08	12.00	1.00	11.00	0.00	0.00	0.00	0.00	0.00	0.00	0.08	0.00	X
456114	50 2 4	27	40	33.5	0.15	12.00	1.00	11.00	0.00	0.00	0.00	0.00	0.00	0.00	0.15	0.00	X
456114	50 2 5	33	43	38.0	0.51	12.00	1.00	11.00	0.01	0.01	0.00	0.00	0.00	0.01	0.50	0.00	X
456114	50 2 6	33	42	37.5	0.55	12.00	1.00	11.00	0.01	0.01	0.00	0.00	0.00	0.01	0.54	0.00	X
456114	50 2 7	34	44	39.0	0.22	12.00	1.00	11.00	0.02	0.02	0.00	0.00	0.00	0.02	0.20	0.00	X
456114	50 2 8	34	46	40.0	0.92	12.00	1.00	11.00	0.02	0.02	0.00	0.00	0.00	0.02	0.90	0.15	X
456114	50 2 9	34	46	40.0	0.14	12.00	1.00	11.00	0.02	0.02	0.00	0.00	0.00	0.02	0.12	0.00	X
456114	50 210	33	43	38.0	0.20	12.00	1.00	11.00	0.01	0.01	0.00	0.00	0.00	0.01	0.19	0.00	X
456114	50 211	37	44	40.5	0.11	12.00	1.00	11.00	0.02	0.02	0.00	0.00	0.00	0.02	0.09	0.00	X
456114	50 212	38	46	42.0	0.95	12.00	1.00	11.00	0.03	0.03	0.00	0.00	0.00	0.03	0.92	0.17	X
456114	50 213	44	53	48.5	0.28	12.00	1.00	11.00	0.04	0.04	0.00	0.00	0.00	0.04	0.24	0.00	X
456114	50 214	42	52	47.0	0.07	12.00	1.00	11.00	0.04	0.04	0.00	0.00	0.00	0.04	0.03	0.00	X
456114	50 215	44	53	48.5	0.15	12.00	1.00	11.00	0.05	0.05	0.00	0.00	0.00	0.05	0.10	0.00	X
456114	50 216	35	53	44.0	0.05	12.00	1.00	11.00	0.03	0.03	0.00	0.00	0.00	0.03	0.02	0.00	X
456114	50 217	35	52	43.5	0.04	12.00	1.00	11.00	0.03	0.03	0.00	0.00	0.00	0.03	0.01	0.00	X
456114	50 218	37	57	47.0	0.19	12.00	1.00	11.00	0.04	0.04	0.00	0.00	0.00	0.04	0.15	0.00	X
456114	50 219	33	46	39.5	0.03	12.00	1.00	11.00	0.02	0.02	0.00	0.00	0.00	0.02	0.01	0.00	X
456114	50 220	35	48	41.5	0.06	12.00	1.00	11.00	0.03	0.03	0.00	0.00	0.00	0.03	0.03	0.00	X
456114	50 221	37	46	41.5	0.06	12.00	1.00	11.00	0.03	0.03	0.00	0.00	0.00	0.03	0.03	0.00	X
456114	50 222	40	52	46.0	0.13	12.00	1.00	11.00	0.04	0.04	0.00	0.00	0.00	0.04	0.09	0.00	X
456114	50 223	39	50	44.5	2.37	12.00	1.00	11.00	0.03	0.03	0.00	0.00	0.00	0.03	2.34	1.59	X
456114	50 224	40	50	45.0	0.50	12.00	1.00	11.00	0.04	0.04	0.00	0.00	0.00	0.04	0.46	1.30	X
456114	50 225	40	56	48.0	1.37	12.00	1.00	11.00	0.05	0.05	0.00	0.00	0.00	0.05	1.32	1.87	X
456114	50 226	36	53	44.5	0.53	12.00	1.00	11.00	0.04	0.04	0.00	0.00	0.00	0.04	0.49	1.62	X
456114	50 227	28	51	39.5	0.00	12.00	0.98	11.00	0.02	0.02	0.00	0.00	0.02	0.02	0.00	0.87	X
456114	50 228	26	52	39.0	0.00	11.98	0.96	11.00	0.02	0.02	0.02	0.00	0.02	0.02	0.00	0.12	X
456114	50 3 1	27	57	42.0	0.00	11.96	0.93	11.00	0.03	0.03	0.04	0.00	0.03	0.03	0.00	0.00	065
456114	50 3 2	39	50	44.5	0.62	11.93	1.00	11.00	0.04	0.04	0.07	0.00	0.00	0.04	0.52	0.00	X
456114	50 3 3	48	55	51.5	2.00	12.00	1.00	11.00	0.06	0.06	0.00	0.00	0.00	0.06	1.94	1.19	X
456114	50 3 4	41	49	45.0	1.12	12.00	1.00	11.00	0.04	0.04	0.00	0.00	0.00	0.04	1.08	1.52	X
456114	50 3 5	38	49	43.5	0.25	12.00	1.00	11.00	0.03	0.03	0.00	0.00	0.00	0.03	0.22	0.99	X
456114	50 3 6	31	48	39.5	0.01	12.00	0.99	11.00	0.02	0.02	0.00	0.00	0.01	0.02	0.00	0.24	X
456114	50 3 7	26	45	35.5	0.00	11.99	0.98	11.00	0.01	0.01	0.01	0.00	0.01	0.01	0.00	0.00	005
456114	50 3 8	34	48	41.0	0.11	11.98	1.00	11.00	0.03	0.03	0.02	0.00	0.00	0.03	0.07	0.00	X
456114	50 3 9	32	44	38.0	0.22	12.00	1.00	11.00	0.02	0.02	0.00	0.00	0.00	0.02	0.20	0.00	X
456114	50 310	30	45	37.5	0.14	12.00	1.00	11.00	0.01	0.01	0.00	0.00	0.00	0.01	0.13	0.00	X
456114	50 311	26	47	36.5	0.04	12.00	1.00	11.00	0.01	0.01	0.00	0.00	0.00	0.01	0.03	0.00	X
456114	50 312	33	45	39.0	0.02	12.00	1.00	11.00	0.02	0.02	0.00	0.00	0.00	0.02	0.00	0.00	X
456114	50 313	35	47	41.0	0.05	12.00	1.00	11.00	0.03	0.03	0.00	0.00	0.00	0.03	0.02	0.00	X
456114	50 314	37	48	42.5	0.00	12.00	0.97	11.00	0.03	0.03	0.00	0.00	0.03	0.03	0.00	0.00	006
456114	50 315	36	49	42.5	0.28	11.97	1.00	11.00	0.03	0.03	0.03	0.00	0.00	0.03	0.22	0.00	X
456114	50 316	43	50	46.5	0.97	12.00	1.00	11.00	0.05	0.05	0.00	0.00	0.00	0.05	0.92	0.17	X
456114	50 317	40	53	46.5	0.75	12.00	1.00	11.00	0.05	0.05	0.00	0.00	0.00	0.05	0.70	0.13	X
456114	50 318	40	47	43.5	0.90	12.00	1.00	11.00	0.04	0.04	0.00	0.00	0.00	0.04	0.86	0.24	X

Figure 8. Daily listing from EPA-2 for Olympia, WA, Jan. 1950–Mar. 1950.

STATION	AWC	YEARS	MO/YR	MAX	5%	10%	25%	50%
ALABAMA								
BAY MINETTE,AL	6	1949-73	MAR	62	13	13	13	9	7
BREWTON,AL	6	1949-73	MAY	70	17	16	11	8	7
CLANTON,AL	6	1949-73	DEC	61	24	20	11	9	7
MOBILE,AL	6	1949-73	APR	55	15	14	11	10	7
SELMA,AL	6	1949-73	DEC	61	20	18	11	8	6
THOMASVILLE,AL	6	1953-74	DEC	61	25	23	13	10	8
ARKANSAS								
DUMAS,AR	6	1951-71	APR	58	19	19	14	10	7
LITTLE ROCK,AR	6	1949-72	MAY	68	12	12	12	9	7
CALIFORNIA								
LOS ANGELES,CA	7	1952-69	FEB	69	5	5	3	1	0
SAN FRANCISCO,CA	7	1949-72	JAN	69	13	13	11	5	4
FLORIDA								
AVON PARK,FL	6	1945-72	JUN	68	13	12	9	7	4
BELLE GLADE,FL	6	1951-73	OCT	52	11	10	8	8	6
DAYTONA BEACH,FL	6	1955-73	SEP	64	8	8	8	6	4
TAMPA,FL	6	1953-73	JUL	60	16	15	9	6	3
GEORGIA								
AUGUSTA,GA	6	1949-72	FEB	71	10	10	9	6	4
MACON,GA	6	1949-72	JAN	64	12	11	9	7	5
SAVANNAH,GA	6	1949-72	SEP	50	17	16	11	7	4
LOUISIANA								
HOUMA,LA	6	1950-73	MAY	59	17	16	11	8	6
LAFAYETTE,LA	6	1948-73	JAN	59	12	12	10	8	6
LAKE PROVIDENCE,LA	6	1951-72	DEC	67	19	18	14	12	7
LEESVILLE,LA	6	1950-73	APR	53	35	31	16	8	6
MONROE,LA	6	1949-72	DEC	61	12	12	12	8	7
NEW ORLEANS,LA	6	1954-73	MAY	59	16	16	9	7	6
ST. JOSEPH,LA	6	1956-73	DEC	61	12	11	11	9	6
SCHRIEVER,LA	6	1948-73	MAR	48	16	15	13	8	7
SHREVEPORT,LA	6	1949-72	JAN	68	11	11	9	7	6
WINNFIELD,LA	6	1950-73	MAY	53	16	16	14	11	6
MISSISSIPPI								
ABERDEEN,MS	6	1951-72	DEC	61	24	23	13	9	7
BILOXI,MS	6	1951-73	SEP	57	14	13	10	8	7
CANTON,MS	6	1948-73	DEC	61	16	15	11	8	6
CLARKSDALE,MS	6	1949-72	DEC	72	18	16	11	10	8
COLUMBIA,MS	6	1947-73	FEB	61	33	27	16	10	8

STATION	AWC	YEARS	MO/YR	MAX	5%	10%	25%	50%
MISSISSIPPI								
GREENWOOD,MS	6	1949-73	SEP	58	16	15	12	12	8
JACKSON,MS	6	1943-61	DEC	61	12	12	10	8	7
MERIDIAN,MS	6	1948-73	FEB	61	14	13	11	8	7
PONTOTOC,MS	6	1949-71	DEC	67	20	19	14	12	9
POPLARVILLE,MS	6	1945-73	DEC	61	25	22	13	8	6
STONEVILLE,MS	6	1951-72	APR	58	17	17	15	12	7
VICKSBURG,MS	6	1942-62	DEC	61	17	17	14	10	7
NORTH CAROLINA								
CHARLOTTE,NC	6	1949-73	JAN	73	12	12	11	8	6
RALEIGH,NC	6	1949-73	JAN	63	14	13	12	7	6
WELDON,NC	6	1926-50	JAN	30	11	11	10	7	5
WILMINGTON,NC	6	1949-73	JUL	50	11	10	9	8	6
OREGON								
EUGENE,OR	12	1948-73	JAN	49	35	34	31	21	15
MEDFORD,OR	12	1948-72	DEC	64	21	19	11	5	3
ROSEBURG,OR	12	1935-64	JAN	50	22	20	18	12	9
SALEM,OR	12	1948-73	DEC	49	36	34	25	17	13
SOUTH CAROLINA								
CHARLESTON,SC	6	1948-73	JUN	73	24	21	11	6	5
COLUMBIA,SC	6	1949-73	JAN	63	15	14	9	7	5
CONWAY,SC	6	1945-73	JAN	70	9	9	9	7	6
TENNESSEE								
CROSSVILLE,TN	6	1953-73	JAN	66	24	24	22	20	17
TEXAS								
ABILENE,TX	6	1950-70	JAN	61	6	6	6	3	0
AMARILLO,TX	6	1949-71	FEB	60	11	10	8	2	0
BROWNSVILLE,TX	6	1954-72	SEP	67	12	11	6	2	0
CORPUS CHRISTI,TX	6	1951-72	SEP	67	13	11	5	3	1
DALLAS,TX	6	1948-72	APR	66	15	15	12	7	4
HOUSTON,TX	6	1948-72	OCT	49	14	13	9	7	4
WICHITA FALLS,TX	6	1950-72	APR	57	8	8	6	3	3
WASHINGTON								
LONGVIEW,WA	12	1948-73	DEC	49	60	53	35	21	17
OLYMPIA,WA	12	1949-73	DEC	49	65	58	38	30	21
SEATTLE,WA	7	1949-73	DEC	49	47	40	24	14	11
SEQUIM	12	1948-73	MAR	56	6	5	3	0	0
VANCOUVER,WA	12	1924-50	JAN	50	31	28	19	17	11
WALLA WALLA,WA	7	1950-73	JAN	69	14	14	10	6	0

Figure 9. Storage days estimated from EPA-2 program.

STATION 1 583 BAY MINETTE, AL
AWC= 6.0 DEPLETION RATE=0.75

BEG.DATE	END.DATE	#DAYS
490322	490329	8
500831	500906	7
510317	510324	8
520215	520224	10
531204	531216	13
541231	550101	2
550413	550425	13
560311	560318	8
570405	570412	8
581231	590102	3
590519	590525	7
600402	600407	6
611210	611218	9
620331	620412	13
630721	630727	7
640108	640112	5
650930	651009	10
661231	670106	7
671210	671213	4
681231	690105	6
690816	690823	8
700106	700111	6
710905	710908	4
721221	721225	5
730329	730401	4
MAX FOR TOTAL PERIOD OF RECORD		
620331	620412	13

PERCENTILES	DAYS
0.05	13.0
0.10	13.0
0.25	8.5
0.50	7.0

STATION 168295 SCHRIEVER, LA
AWC= 6.0 DEPLETION RATE=0.75

BEG.DATE	END.DATE	#DAYS
480301	480317	16
490328	490409	13
500402	500407	6
510328	510402	6
520331	520406	7
530716	530720	5
540729	540803	6
550206	550209	4
560930	561008	9
570604	570610	7
580911	580917	7
590530	590612	14
601224	601226	3
610907	610914	8
620109	620112	4
631109	631116	8
641003	641009	7
651218	651223	6
660210	660217	8
570205	670212	8
681230	681231	2
690409	690419	11
701027	701030	4
710907	710911	5
721220	721224	5
730416	730423	8
MAX FOR TOTAL PERIOD OF RECORD		
480301	480317	16

PERCENTILES	DAYS
0.05	15.3
0.10	13.3
0.25	8.0
0.50	7.0

STATION 319191 WELDON, NC
AWC= 6.0 DEPLETION RATE=0.75

BEG.DATE	END.DATE	#DAYS
260311	260314	4
270110	270116	7
280128	280201	5
291223	291225	3
300116	300126	11
310112	310116	5
321211	321220	10
330208	330212	5
340226	340301	4
351223	360101	10
360127	360202	7
371209	371214	6
380618	380622	5
390825	390902	9
400124	400201	9
410307	410309	3
421221	421223	3
430214	430216	3
441127	441201	5
451229	460102	5
460411	460413	3
470118	470121	4
480208	480213	6
490511	490512	2
501229	501230	3
MAX FOR TOTAL PERIOD OF RECORD		
300116	300126	11

PERCENTILES	DAYS
0.05	10.7
0.10	10.0
0.25	7.0
0.50	5.0

STATION 445120 LYNCHBURG, VA
AWC= 9.0 DEPLETION RATE=0.75

BEG.DATE	END.DATE	#DAYS
480124	480202	10
490103	490106	4
500127	500202	7
511214	511221	8
520125	520131	7
530102	530106	5
540110	540118	9
550201	550207	7
560313	560316	4
570205	570212	8
580206	580220	15
591218	591220	3
600302	600314	13
611227	620103	8
621209	621215	7
630118	630129	11
640209	640213	5
650130	650205	7
661223	670101	10
671228	680117	21
680123	680126	4
690120	690123	4
700117	700124	7
710113	710205	24
720201	720210	10
730126	730203	9
MAX FOR TOTAL PERIOD OF RECORD		
710113	710205	24

PERCENTILES	DAYS
0.05	23.0
0.10	16.8
0.25	10.0
0.50	7.5

Figure 10. Summary tables showing annual estimated storage days for four stations (EPA-2).

SUMMARY TABLE FROM NCC—EPA—3 PROGRAM

A summary of each winter season examined shows the Freeze Index, its duration and the maximum storage days computed from the daily listings. Additional information includes the mean number of storage days, the standard deviation, unbiased third moment about the mean, the coefficient of skewness and a list of the thresholds used for each of the elements examined. The thresholds may be changed to fit individual systems or other special conditions. Storage days are also computed for recurrence intervals of 5, 10, 25 & 50--years using the Pearson Type III method. The procedure used to define each day and to compute the coefficient of skewness is shown below;

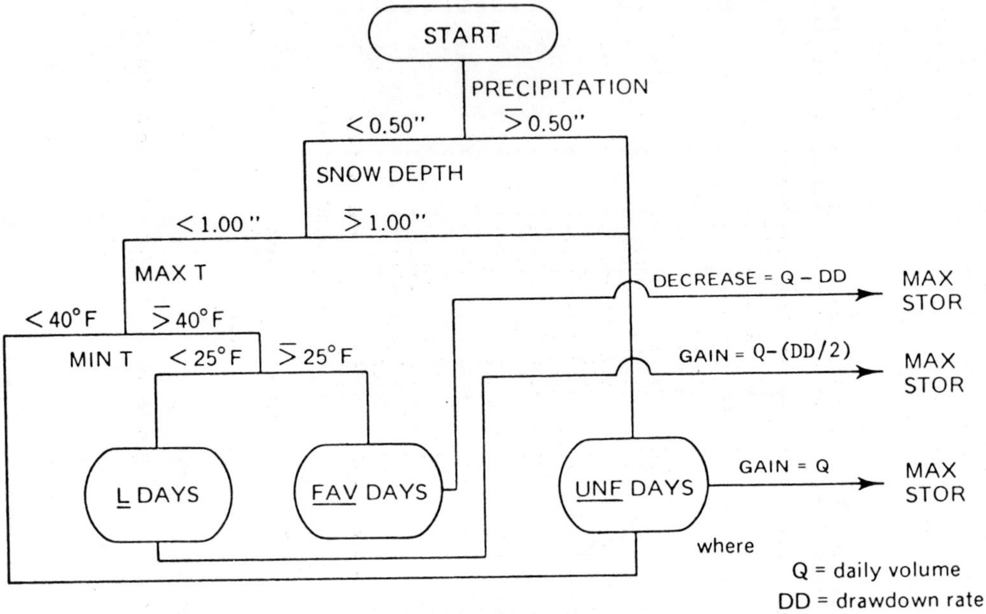

1. Using the values in the "MAX STOR" column, compute the mean storage days ($\bar{x}$) and the standard deviation (σ).

2. Cube the difference between each storage value and the mean $(STOR-\bar{x})^3$ and sum algebraically.

3. Compute the unbiased third moment about the mean: $a = \dfrac{N\Sigma(STOR-\bar{x})^3}{(N-1)(N-2)}$

4. Compute the coefficient of skewness: $C_s = \dfrac{a}{\sigma^3}$

5. Enter table with C_s and recurrence interval to find k.

6. Storage days are then equal to $\bar{x} + k\sigma$.

Figure 11. Outline of the procedure used in developing the summary table for EPA-3 shown in Figure 12.

USE OF CLIMATIC DATA IN DESIGN OF SOILS TREATMENT SYSTEMS TABLE 1 EPA-3 05/13/76
STA # 485390 LANDER, WY POR: 481101-740430

HIGH TO LOW FREEZE INDEX

INDX	DATE BEGIN	DATE END	DUR	MAX FA	MAX UNF	MAX STOR	MAX (STOR-MN)	MAX (STOR-MN)3
2154	721110	730410	151	7	149	169.00	64.24	265104
1880	481103	490314	131	16	109	128.00	23.24	12552
1704	611101	620316	135	14	103	123.50	18.74	6581
1397	671101	680218	109	10	105	129.75	24.99	15606
1357	551110	560315	125	19	46	109.00	4.24	76
1286	591103	600317	134	39	98	126.50	21.74	10275
1230	631206	640328	112	19	102	121.00	16.24	4283
1156	511113	520325	132	28	48	106.75	1.99	8
1075	541127	550328	121	27	59	111.00	6.24	243
1035	711101	720215	106	21	112	114.75	9.99	997
1010	581114	590315	121	10	47	123.75	18.99	6848
1005	681123	690315	112	12	29	104.25	-.51	0
985	621219	630317	88	14	42	73.50	-31.26	-30547
883	561113	570207	86	14	58	101.25	-3.51	-43
879	641110	650328	138	17	24	88.00	-16.76	-4708
863	491202	500214	74	35	48	73.00	-31.76	-32036
836	731118	740225	99	15	31	90.75	-14.01	-2750
822	651211	660307	86	23	45	87.25	-17.51	-5369
812	601102	610308	126	10	78	115.25	10.49	1154
752	521115	530306	111	11	40	102.00	-2.76	-21
733	701209	710307	88	16	44	81.50	-23.26	-12584
698	571101	580320	139	13	50	115.50	10.74	1239
664	501107	510319	132	9	18	77.75	-27.01	-19705
560	661130	670308	98	15	25	82.00	-22.76	-11790
495	691116	700404	139	13	25	90.25	-14.51	-3055
413	531117	540331	134	17	14	78.50	-26.26	-18109

STORAGE DATA N = 26 MEAN = 104.76 SD = 22.45 A = 7984 (STOR-MN)3 = 184249 CS = .70

AVERAGE INDEX 1026

PERCENTILES	INDX	DUR	STOR
MAX	2154	151	169.00
5%	2044	146	153.30
10%	1757	139	128.55
25%	1241	134	121.50
50%	934	121	105.45

RECURRENCE INTV

STOR05 = 122.50
STOR10 = 134.69
STOR25 = 148.93
STOR50 = 158.81

THRESHOLDS : 32 F, 1 DEPTH, .50 PRECIP, 1.50 RATE < 25 F, MIN < 40 F, MAX

Figure 12. Summary of 26 winter seasons at Lander, WY from the EPA-3 program (see Figs. 13–18 for the daily listings).

USE OF CLIMATIC DATA IN DESIGN OF SOILS TREATMENT SYSTEMS EPA-3 05/13/76
STA # 485390 LANDER, WY PORI 481101-740430

YR	MO	DA	MAX	MIN	MEAN	SNOW DPTH	PPPP	FOG	DD	CDD	FA	DUR FA	UNF	DUR UNF	MAX STOR
48	11	01	60	30	45				13	13	X				
48	11	02	57	33	45				13	26	X				
48	11	03	59	31	45				13	39	X				
48	11	04	38	24	31				-1	38		3	X		1.00
48	11	05	49	15	32				0	38			L	1	1.25
48	11	06	47	17	32				0	38			L		1.50
48	11	07	30	13	22	5	.41		-10	28		2	X		2.50
48	11	08	30	4	17	4			-15	13			X		3.50
48	11	09	39	9	24	2			-8	5			X		4.50
48	11	10	37	21	29	2	.01		-3	2			X		5.50
48	11	11	35	17	26	2			-6	-4			X		6.50
48	11	12	46	21	34	1			2	-2			X		7.50
48	11	13	53	28	41	T			9	7	X			6	7.00
48	11	14	58	28	43	T			11	18	X				6.50
48	11	15	53	26	40				8	26	X				6.00
48	11	16	48	21	35				3	29			L		6.25
48	11	17	44	18	31	T	.01		-1	28			L		6.50
48	11	18	33	15	24				-8	20		5	X		7.50
48	11	19	27	10	19		.16		-13	7			X		8.50
48	11	20	34	23	29	7	.45		-3	4			X		9.50
48	11	21	34	10	22	9	.18		-10	-6			X		10.50
48	11	22	35	4	20	8			-12	-18			X		11.50
48	11	23	39	14	27	6			-5	-23			X		12.50
48	11	24	45	21	33	4			1	-22			X		13.50
48	11	25	35	14	25	3			-7	-29			X		14.50
48	11	26	31	12	22	3	.01		-10	-39			X		15.50
48	11	27	25	5	15	3			-17	-56			X		16.50
48	11	28	39	4	22	3			-10	-66			X		17.50
48	11	29	37	11	24	2			-8	-74			X		18.50
48	11	30	29	5	17	2			-15	-89			X		19.50

Figure 13. Listing of daily data from EPA-3 program, Lander, WY, Nov. 1948.

USE OF CLIMATIC DATA IN DESIGN OF SOILS TREATMENT SYSTEMS EPA-3 05/13/76
STA # 485390 LANDER, WY PORI 481101-740430

YR	MO	DA	MAX	MIN	MEAN	SNOW DPTH	PPPP	FOG	DD	CDD	FA	DUR FA	UNF	DUR UNF	MAX STOR
48	12	01	33	16	25	2			-7	-96			X		20.50
48	12	02	52	17	35	1			3	-93			X		21.50
48	12	03	48	38	43	1			11	-82			X		22.50
48	12	04	40	7	24	2	.24		-8	-90			X		23.50
48	12	05	26	0	13	3			-19	-109			X		24.50
48	12	06	31	8	20	3			-12	-121			X		25.50
48	12	07	24	5	15	3			-17	-138			X		26.50
48	12	08	32	9	21	2			-11	-149			X		27.50
48	12	09	35	3	19	2			-13	-162			X		28.50
48	12	10	40	19	30	2			-2	-164			X		29.50
48	12	11	40	22	31	2			-1	-165			X		30.50
48	12	12	43	29	36	1			4	-161			X		31.50
48	12	13	34	21	26	1			-4	-165			X		32.50
48	12	14	23	16	20	1	.01		-12	-177			X		33.50
48	12	15	26	10	18	1			-14	-191			X		34.50
48	12	16	23	5	14	1			-18	-209			X		35.50
48	12	17	21	3	12	1			-20	-229			X		36.50
48	12	18	27	7	17	1			-15	-244			X		37.50
48	12	19	34	14	24	1			-8	-252			X		38.50
48	12	20	42	17	30	1			-2	-254			X		39.50
48	12	21	25	6	16	1			-16	-270			X		40.50
48	12	22	22	1	12	1			-20	-290			X		41.50
48	12	23	17	11	14	2	.32		-18	-308			X		42.50
48	12	24	18	0	9	6	.02	F	-23	-331			X		43.50
48	12	25	3	-14	-6	6		F	-38	-369			X		44.50
48	12	26	5	-17	-6	6			-38	-407			X		45.50
48	12	27	15	-12	2	5			-30	-437			X		46.50
48	12	28	32	7	20	5			-12	-449			X		47.50
48	12	29	19	3	11	4			-21	-470			X		48.50
48	12	30	27	1	14	3			-18	-488			X		49.50
48	12	31	23	1	12	3			-20	-508			X		50.50

Figure 14. Listing of daily data from EPA-3 program, Lander, WY, Dec. 1948.

USE OF CLIMATIC DATA IN DESIGN OF SOILS TREATMENT SYSTEMS EPA-3 05/13/76
STA # 485390 LANDER, WY POR: 481101-740430

YR	MO	DA	MAX	MIN	MEAN	SNOW DPTH	PPPP	FOG	DD	CDD	FA	DUR FA	UNF	DUR UNF	MAX STOR
49	01	01	23	3	13	3			-19	-527			X		51.50
49	01	02	21	7	14	6	.10		-18	-545			X		52.50
49	01	03	8	1	5	7	.21	F	-27	-572			X		53.50
49	01	04	16	1	9	9	.05		-23	-595			X		54.50
49	01	05	25	1	13	8			-19	-614			X		55.50
49	01	06	22	-2	10	8			-22	-636			X		56.50
49	01	07	23	3	13	7			-19	-655			X		57.50
49	01	08	30	-1	15	7	.07	F	-17	-672			X		58.50
49	01	09	-1	-15	-8	9	.11		-40	-712			X		59.50
49	01	10	-1	-15	-8	10	.09		-40	-752			X		60.50
49	01	11	7	-8	-1	10	.07		-33	-785			X		61.50
49	01	12	5	-14	-5	9		F	-37	-822			X		62.50
49	01	13	13	0	7	9		F	-25	-847			X		63.50
49	01	14	35	3	19	9		F	-13	-860			X		64.50
49	01	15	25	12	19	9	.04		-13	-873			X		65.50
49	01	16	18	-5	7	9			-25	-898			X		66.50
49	01	17	6	-12	-3	9			-35	-933			X		67.50
49	01	18	18	-5	7	9	.02		-25	-958			X		68.50
49	01	19	5	-12	-4	9			-36	-994			X		69.50
49	01	20	-5	-21	-13	9			-45	-1039			X		70.50
49	01	21	7	-19	-6	9			-38	-1077			X		71.50
49	01	22	14	0	7	10	.24		-25	-1102			X		72.50
49	01	23	1	-23	-11	17	.26		-43	-1145			X		73.50
49	01	24	-12	-31	-22	17			-54	-1199			X		74.50
49	01	25	-8	-31	-20	17			-52	-1251			X		75.50
49	01	26	2	-25	-12	17			-44	-1295			X		76.50
49	01	27	10	-16	-3	16	.22		-35	-1330			X		77.50
49	01	28	13	-7	3	19	.14		-29	-1359			X		78.50
49	01	29	2	-15	-7	17			-39	-1398			X		79.50
49	01	30	14	-8	3	15			-29	-1427			X		80.50
49	01	31	13	-7	3	15	.03		-29	-1456			X		81.50

Figure 15. Listing of daily data from EPA-3 program, Lander, WY, Jan. 1949.

USE OF CLIMATIC DATA IN DESIGN OF SOILS TREATMENT SYSTEMS — EPA-3 — 05/13/76
STA # 485390 LANDER, WY — PORI 481101-740430

YR	MO	DA	MAX	MIN	MEAN	SNOW DPTH	PRPP	FOG	DD	CDD	FA	DUR FA	UNF	DUR UNF	MAX STOR
49	02	01	-2	-17	-10	15			-42	-1498			X		82.50
49	02	02	-1	-19	-10	15			-42	-1540			X		83.50
49	02	03	2	-19	-9	15			-41	-1581			X		84.50
49	02	04	25	-14	6	15			-26	-1607			X		85.50
49	02	05	21	2	12	15			-20	-1627			X		86.50
49	02	06	25	-6	10	15			-22	-1649			X		87.50
49	02	07	27	9	18	15	.01		-14	-1663			X		88.50
49	02	08	22	5	14	15			-18	-1681			X		89.50
49	02	09	30	2	16	14			-16	-1697			X		90.50
49	02	10	46	11	29	10			-3	-1700			X		91.50
49	02	11	38	6	22	10	.15		-10	-1710			X		92.50
49	02	12	6	-16	-5	12	.23		-37	-1747			X		93.50
49	02	13	7	-28	-11	12			-43	-1790			X		94.50
49	02	14	18	-8	5	11			-27	-1817			X		95.50
49	02	15	32	1	17	11			-15	-1832			X		96.50
49	02	16	44	8	26	10			-6	-1838			X		97.50
49	02	17	50	22	36	9			4	-1834			X		98.50
49	02	18	49	32	41	8			9	-1825			X		99.50
49	02	19	46	26	36	6			4	-1821			X		100.50
49	02	20	36	17	27	6			-5	-1826			X		101.50
49	02	21	42	21	32	5			0	-1826			X		102.50
49	02	22	47	20	34	5			2	-1824			X		103.50
49	02	23	45	25	35	3			3	-1821			X		104.50
49	02	24	37	15	26	2			-6	-1827			X		105.50
49	02	25	43	17	30	2			-2	-1829			X		106.50
49	02	26	39	23	31	2			-1	-1830			X		107.50
49	02	27	38	19	29	2			-3	-1833			X		108.50
49	02	28	40	17	29	2			-3	-1836			X		109.50

Figure 16. Listing of daily data from EPA-3 program, Lander, WY, Feb. 1949.

USE OF CLIMATIC DATA IN DESIGN OF SOILS TREATMENT SYSTEMS EPA-3 05/13/75
STA # 485390 LANDER, WY PORI 461101-740430

YR	MO	DA	MAX	MIN	MEAN	SNOW DPTH	PPPP	FOG	DD	CDD	FA	DUR FA	UNF	DUR UNF	MAX STLR
49	03	01	43	18	31	2			-1	-1837			X		110.50
49	03	06	49	25	37	1			5	-1832			X		111.50
49	03	07	46	25	36	1			4	-1828			X		112.50
49	03	08	36	27	32	1	.07		0	-1828			X		113.50
49	03	09	44	18	31	1			-1	-1829			X		114.50
49	03	10	36	18	27	1			-5	-1834			X		115.50
49	03	11	40	22	31	T			-1	-1835			L	109	115.75
49	03	12	39	25	32	T	.05		0	-1835		1	X		116.75
49	03	13	49	23	36	T			4	-1831			L	1	117.00
49	03	14	32	11	22	T			-10	-1841		1	X		118.00
49	03	15	53	13	33	T			1	-1840			L	1	118.25
49	03	16	39	25	32	T			0	-1840		1	X		119.25
49	03	17	51	21	36	T			4	-1836			L	1	119.50
49	03	18	54	27	41	T			9	-1827	X				119.00
49	03	19	62	30	46	T			14	-1813	X				118.50
49	03	20	57	30	44	T	.25		12	-1801	X				118.00
49	03	21	50	25	38				6	-1795	X				117.50
49	03	22	50	28	39				7	-1768	X				117.00
49	03	23	37	28	33	7	1.19		1	-1787		6	X		118.00
49	03	24	32	20	26	13	.10		-6	-1793			X		119.00
49	03	25	40	12	26	7			-6	-1799			X		120.00
49	03	26	41	22	32	4			0	-1799			X		121.00
49	03	27	42	20	31	3			-1	-1800			X		122.00
49	03	28	33	14	24	3	.04		-8	-1808			X		123.00
49	03	29	29	21	25	4	.17		-7	-1815			X		124.00
49	03	30	39	18	29	1			-3	-1818			X		125.00
49	03	31	39	23	31	1	.01		-1	-1819			X		126.00

Figure 17. Listing of daily data from EPA-3 program, Lander, WY, Mar. 1949.

USE OF CLIMATIC DATA IN DESIGN OF SOILS TREATMENT SYSTEMS EPA-3 05/13/76
STA # 485390 LANDER, WY PORI 481101-740430

YR	MO	DA	MAX	MIN	MEAN	SNOW DPTH	PPPP	FOG	DD	CDD	FA	DUR FA	UNF	DUR UNF	MAX STOR
49	04	01	41	23	32	1			0	-1819			X		127.00
49	04	02	46	22	34	1			2	-1817			X		128.00
49	04	03	51	32	42	T			10	-1807	X			11	127.50
49	04	04	57	31	44	T			12	-1795	X				127.00
49	04	05	60	32	46				14	-1781	X				126.50
49	04	06	62	31	47				15	-1766	X				126.00
49	04	07	67	35	51				19	-1747	X				125.50
49	04	08	58	37	48		.32		16	-1731	X				125.00
49	04	09	47	32	40				8	-1723	X				124.50
49	04	10	59	27	43				11	-1712	X				124.00
49	04	11	73	34	54				22	-1690	X				123.50
49	04	12	63	40	52				20	-1670	X				123.00
49	04	13	49	32	41	T	.15		9	-1661	X				122.50
49	04	14	39	29	34	T	.48	F	2	-1659		11	X		123.50
49	04	15	59	23	41				9	-1650			L	1	123.75
49	04	16	68	35	52				20	-1630	X				123.25
49	04	17	64	37	51				19	-1611	X				122.75
49	04	18	69	34	52				20	-1591	X				122.25
49	04	19	70	39	55				23	-1568	X				121.75
49	04	20	60	38	49				17	-1551	X				121.25
49	04	21	66	34	50				18	-1533	X				120.75
49	04	22	64	40	52				20	-1513	X				120.25
49	04	23	75	41	58				26	-1487	X				119.75
49	04	24	78	50	64				32	-1455	X				119.25
49	04	25	70	45	58				26	-1429	X				118.75
49	04	26	65	42	54				22	-1407	X				118.25
49	04	27	71	40	56				24	-1383	X				117.75
49	04	28	77	42	60				28	-1355	X				117.25
49	04	29	69	42	56		.04		24	-1331	X				116.75
49	04	30	55	35	45				13	-1318	X	16			116.25
FREEZE INDEX			1880		481103		490314	131				16		109	128.00

Figure 18. Listing of daily data from EPA-3 program, Lander, WY, Apr. 1949.

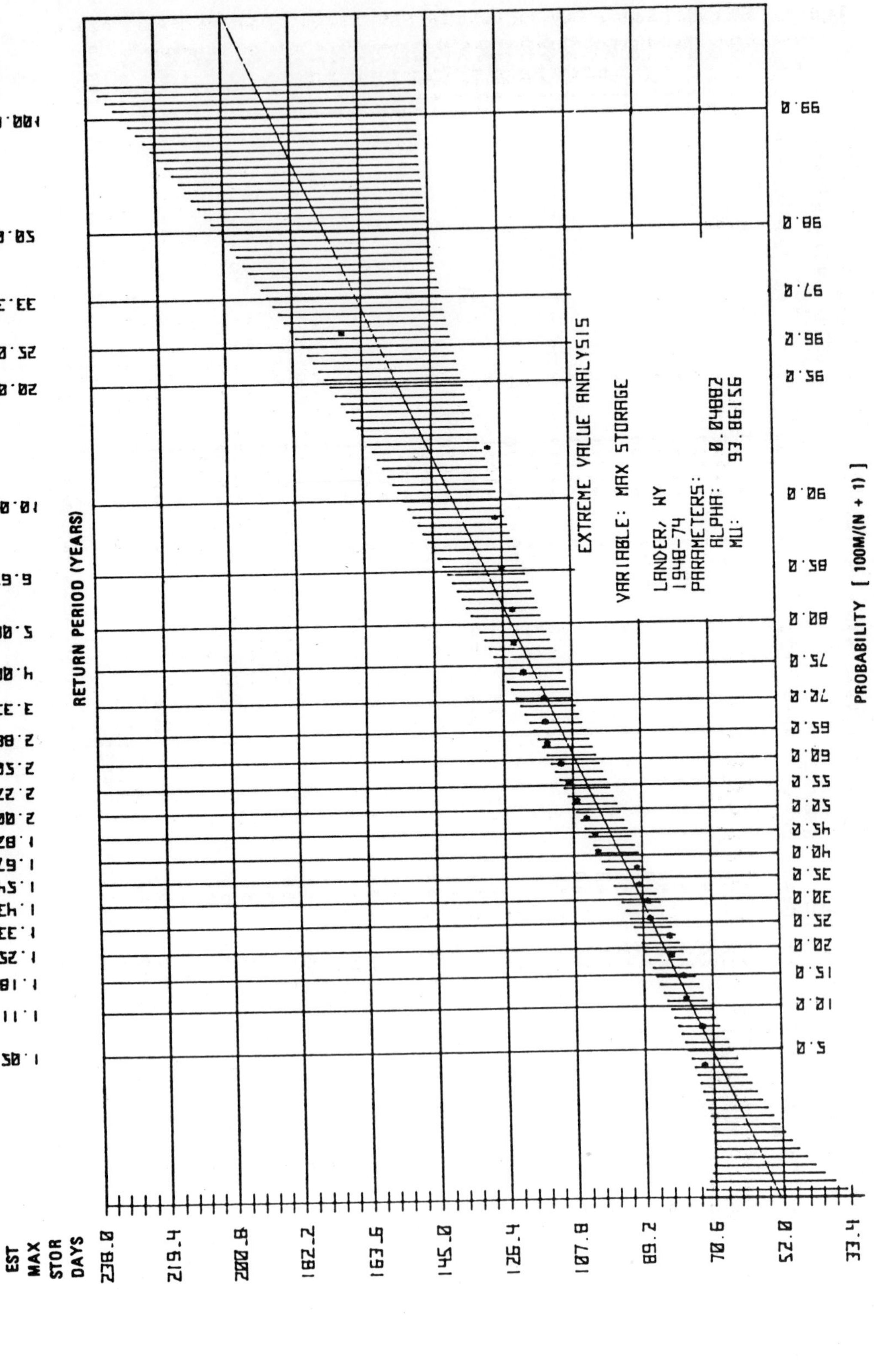

Figure 18A. Estimated maximum annual storage days for return periods to 100 years (Gumbel).

Table 4. Estimated Storage Days for Indicated Return Periods from the EPA-3 Program. Note the Effect of Changing the Drawdown Rate (DD) at Baltimore and the Period of Record at Riverton.

STATION	5-YR	10-YR	25-YR	50-YR	PERIOD	DD
BELLE PLAINE,IA	104.3	112.7	121.9	128.0	11/48-04/73	1.50
DES MOINES,IA	98.3	105.8	113.9	119.1	11/48-04/73	1.50
GRINNELL,IA	102.1	110.0	118.8	124.7	11/48-04/73	1.50
HAMPTON,IA	121.2	130.6	141.2	148.4	11/49-04/72	1.50
INDIANOLA,IA	85.5	93.9	103.1	109.2	11/48-04/73	1.50
KEOSAUQUA,IA	77.2	84.9	93.5	99.3	11/48-04/73	1.50
KNOXVILLE,IA	87.4	96.8	107.5	114.7	11/49-04/70	1.50
LOGAN,IA	93.1	105.6	110.6	116.5	11/47-04/73	1.50
NEWTON,IA	97.6	106.9	116.9	123.5	11/52-04/72	1.50
OSCEOLA,IA	79.1	87.3	96.3	102.3	11/48-04/67	1.50
OSKALOOSA,IA	90.8	99.1	108.3	114.4	11/46-04/73	1.50
SHENANDOAH,IA	75.3	83.3	92.0	97.7	11/48-04/73	1.50
WINTERSET,IA	96.5	105.4	115.0	121.2	11/53-04/73	1.50
BALTIMORE,MD	51.8	57.4	63.4	67.3	11/50-04/72	1.25
BALTIMORE,MD	48.8	54.8	61.4	65.7	11/50-04/72	1.33
BALTIMORE,MD	43.0	49.9	57.7	63.0	11/50-04/72	1.50
CROSSVILLE,TN	48.3	53.9	60.1	64.3	11/53-04/73	1.50
DIVERSON DAM,WY	80.7	88.8	98.0	104.7	11/48-04/74	1.50
LANDER,WY	122.5	134.7	148.9	158.8	11/48-04/74	1.50
PAVILLION,WY	85.3	92.4	100.1	105.2	11/48-04/74	1.50
RIVERTON,WY	100.2	105.7	111.6	115.5	11/47-04/74	1.50
RIVERTON,WY	102.1	110.3	119.5	125.8	11/22-04/47	1.50

APPENDIX B SUPPLEMENTAL INFORMATION

The tables in Figure 42 show the calendar dates in spring (in fall) after which (before which) threshold temperatures, or lower, may be expected for nine probability levels. The actual dates of the last occurrence in spring and the first occurrence in the fall are recorded and the number of days between these dates each year is computed. These periods are defined as the length of the growing season, the freeze-free period, or simply the number of days between the indicated threshold temperatures. The number of days from the first freeze in fall to the last freeze in spring is sometimes referred to as the dormant period.

This period is not the same as the Freeze Index described in the EPA-1 program which is computed by accumulating the daily differences between the *mean* temperature and 32°F. The freeze tables described in the previous paragraph are computed from daily *minimum* temperatures instead of the mean temperature. Another difference is that the freeze tables make no provision for the effect of warmer days that may follow the first low temperature in the fall, or those that precede the last one in the spring, while the Freeze Index does.

Hourly observations of surface wind direction and speed, including peak gusts, are taken routinely at most airport stations. Daily peak gusts are also reported at many of these stations.

A variety of wind summaries covering 5 to 10 years of record are readily available at the NCC and can be furnished for the cost of duplication. Caution should be exercised in using wind summaries, since many of them are prepared for special applications. Some of these include low ceiling-visibility-wind distributions or all weather wind graphs for airport master plans (Figure 19) and extreme wind probabilities for design purposes (Figures 20 and 21). Stations in coastal or mountain regions usually have quite different day and night wind patterns. The wind distribution at the proposed site may be influenced by local conditions and, therefore, quite different from that at the nearest reporting station. Special wind equipment might be installed at the proposed site if conditions warrant; however, analyses of wind direction and speed should be based on at least five or more years of record in order to obtain a representative distribution.

Thom [20] has chosen the annual extreme fastest mile wind speed as the vest available measure of wind for design purposes. He found that only airport or open country wind data could be used because of unknown surface friction conditions at city office exposures. Figures (20) and (21) are from Thom's paper and show the 10-year and 25-year mean recurrence intervals in miles per hour. The charts are based on 21 years of record; however, suitable adjustments must be made to the values where experience indicates the wind speeds in the figures are inadequate. These conditions can exist where unusual channeling occurs, such as the Santa Ana in California and the winds along the eastern slopes of the Rockies, especially in the region of Boulder and Colorado Springs.

Thunderstorms and tropical cyclones generally produce their high winds during the summer months, while the highest wind speeds in winter are usually caused by extratropical cyclones. Tropical cyclones are often accompanied by tornadoes and the effects of a well developed tropical storm can extend inland for several hundred miles. Tropical storms that move along the Atlantic coastline in a northeasterly direction can remain intense systems as far north as Maine. According to Thom, thunderstorms account for about one-third of the extreme wind speeds in this country.

Mountain and valley winds move along the axis of a valley [21], blowing uphill (valley wind) by day and downhill (mountain wind) by night. The valley wind sets in about one-half hour after sunrise and continues until about one-half hour before sunset, reaching its greatest strength at the time of maximum insolation along the slopes. On southerly slopes it may reach 14 mph, while on the north facing slopes it is barely noticeable. The mountain wind is due to nocturnal cooling and is somewhat weaker, reaching perhaps 9 mph on occasion, but it is usually stronger a few hundred feet above the ground than at the surface. Detailed information about mixing heights, stability and wind patterns in the upper air are also available [22]. Additional information of interest to planners may be found in the literature [23, 24, 25, 26, 27, 28, 29].

Note: The values in Figures 25–27 are roughly equivalent to the minimum temperatures which have an average return period of 100 years (1%), 33 years (3%) and 20 years (5%).

SUPPLEMENTAL INFORMATION

LIST OF FIGURES

Figure

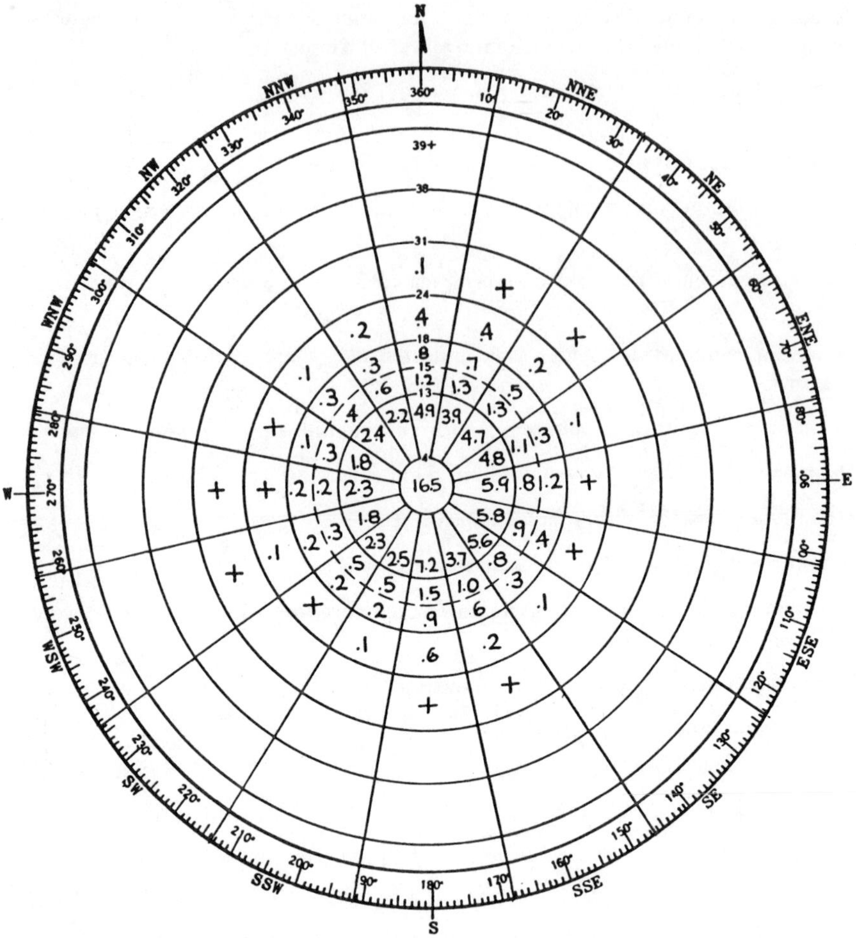

THE WIND ROSE IS A SCALED GRAPHICAL PRESENTATION OF SURFACE WIND DATA IN TERMS OF SPEED AND DIRECTION. THE RADIAL LINES OF THE DIAGRAM ARE POSITIONED SO THAT AREAS BETWEEN THEM ARE CENTERED ON THE DIRECTION FROM WHICH THE WINDS ARE REPORTED. THE CONCENTRIC CIRCLES REPRESENT LIMITS BETWEEN SPEED GROUPS SECTORS, I.E., 4, 13, 15, 18, 24, 31, 38, AND 39+ MILES PER HOUR. RADII FOR THESE GROUPS ARE ACCURATELY SCALED TO THE RESPECTIVE SPEEDS. THE SEGMENTS ENCLOSED BY RADIAL LINES AND CONCENTRIC CIRCLES ON THE DIAGRAM REPRESENT WIND SPEED-DIRECTION COMBINATIONS. THE DATA FROM A WIND SUMMARY ARE TRANSFERRED TO THE APPROPRIATE AREA ON THE DIAGRAM AS A PERCENTAGE OF THE TOTAL OBSERVATIONS EXAMINED.

16.5 % of the winds were less than 4 m.p.h.

based on 11,576 observations

+ indicates less than 0.05

Figure 19. Annual distribution of surface wind direction and speed, Boothville, LA, May 1971–Apr. 1975, 8 observations per day, all weather conditions.

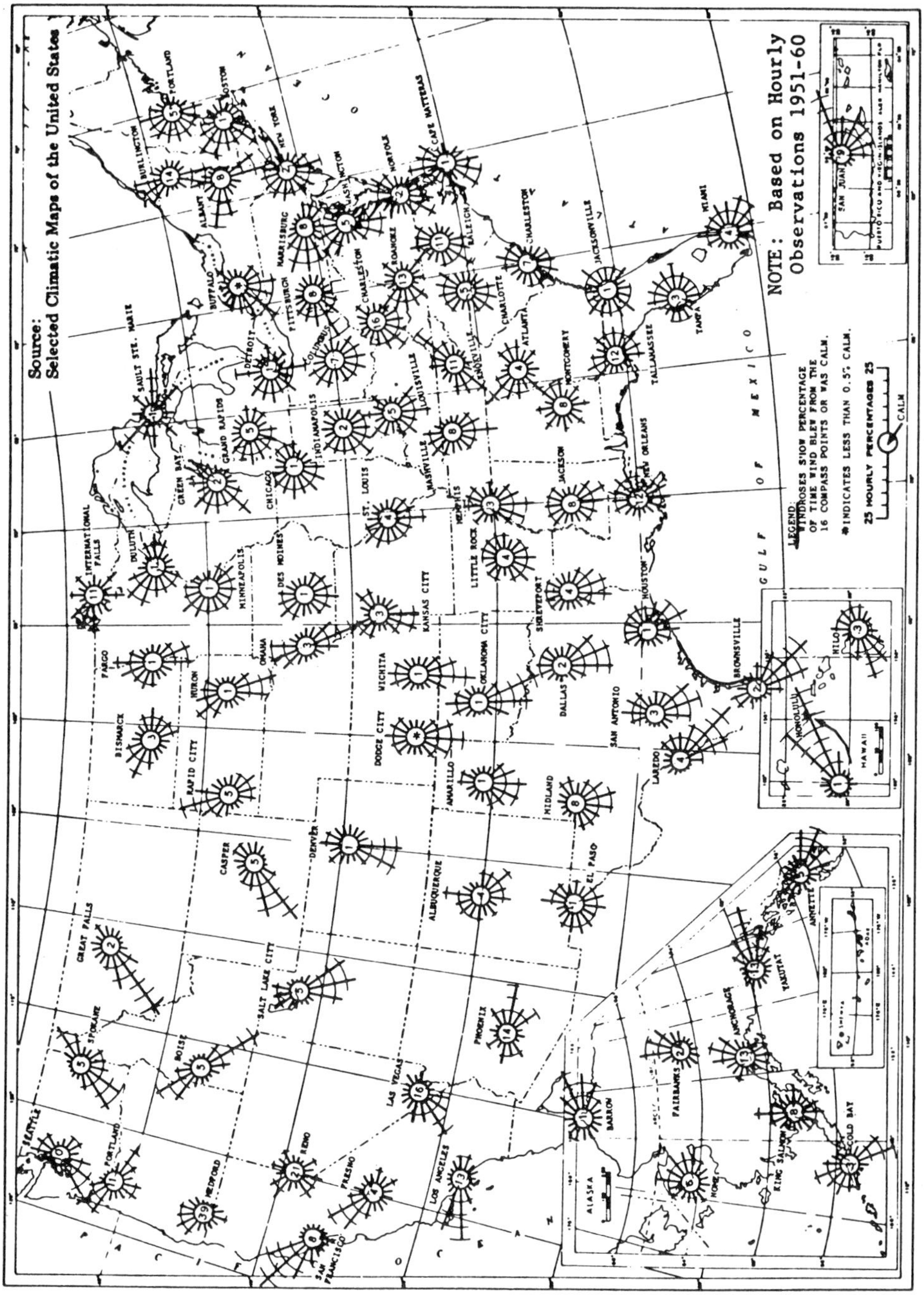

Figure 20. Surface wind roses, annual, 1951-60.

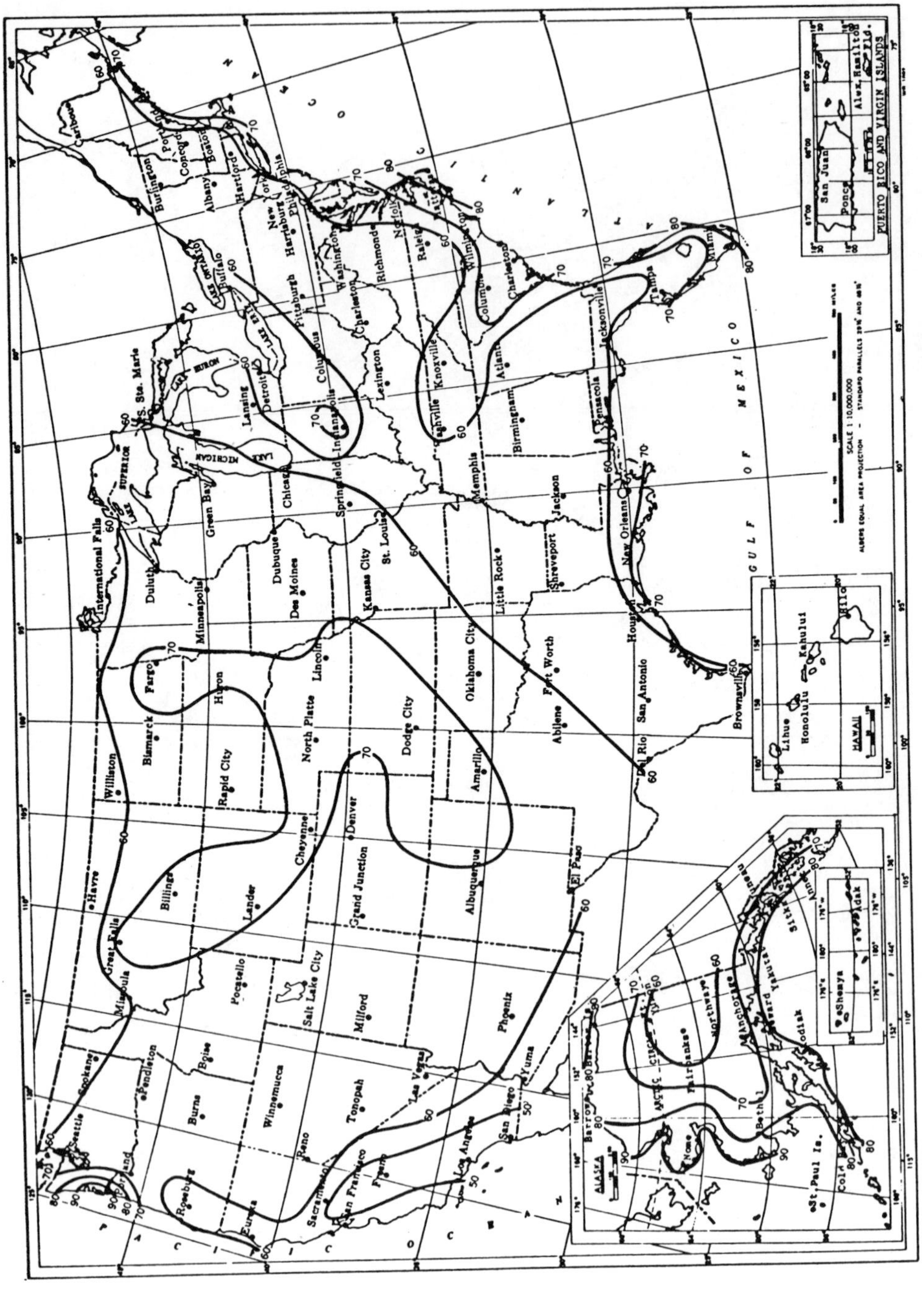

Figure 21. Isotach 0.10 quantiles, in miles per hour: annual extreme-mile 30 feet above ground, 10-year mean recurrence interval (after Thom).

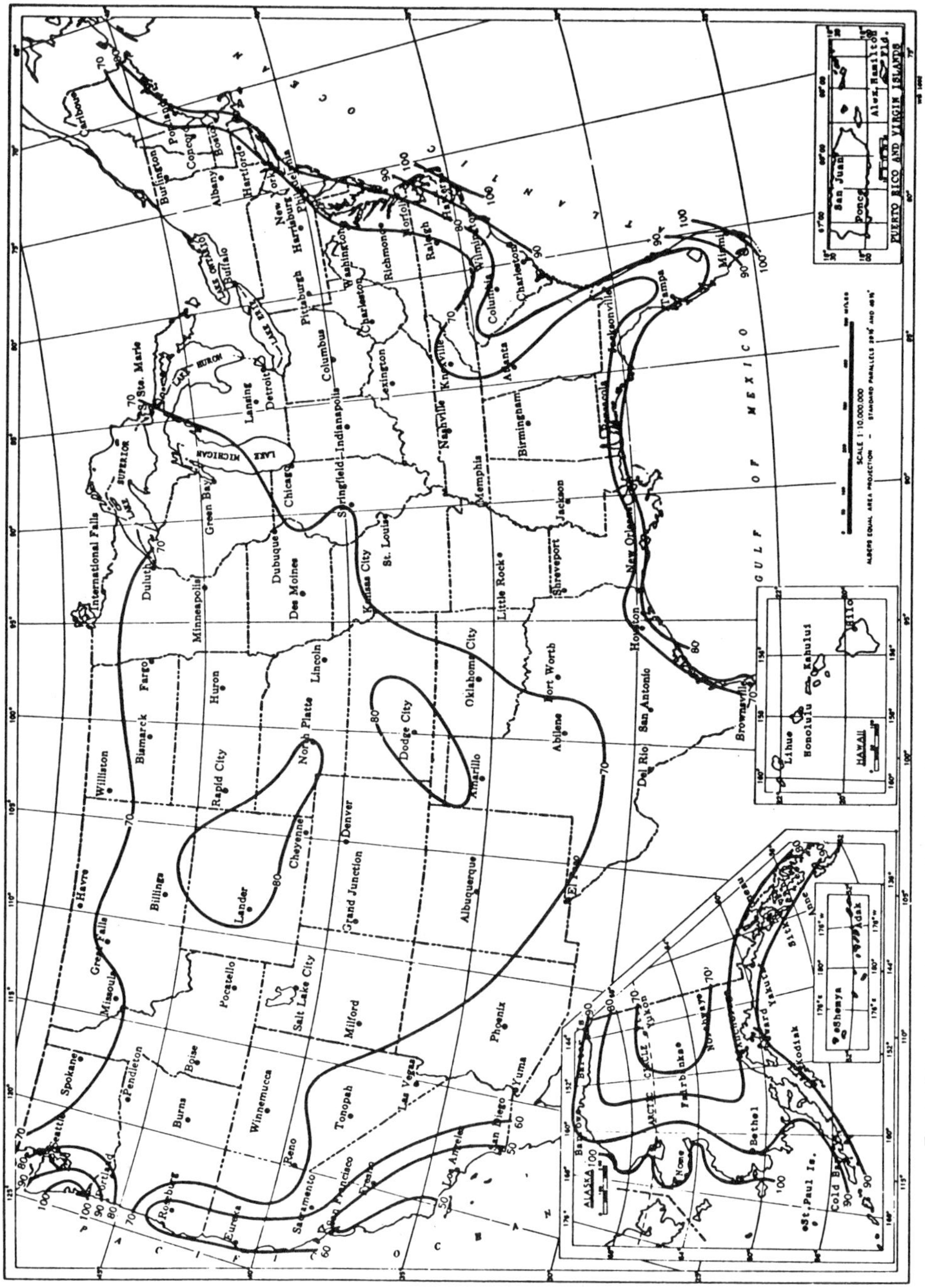

Figure 22. Isotach 0.04 quantiles, in miles per hour: annual extreme-mile 30 feet above ground, 25-year mean recurrence interval (after Thom).

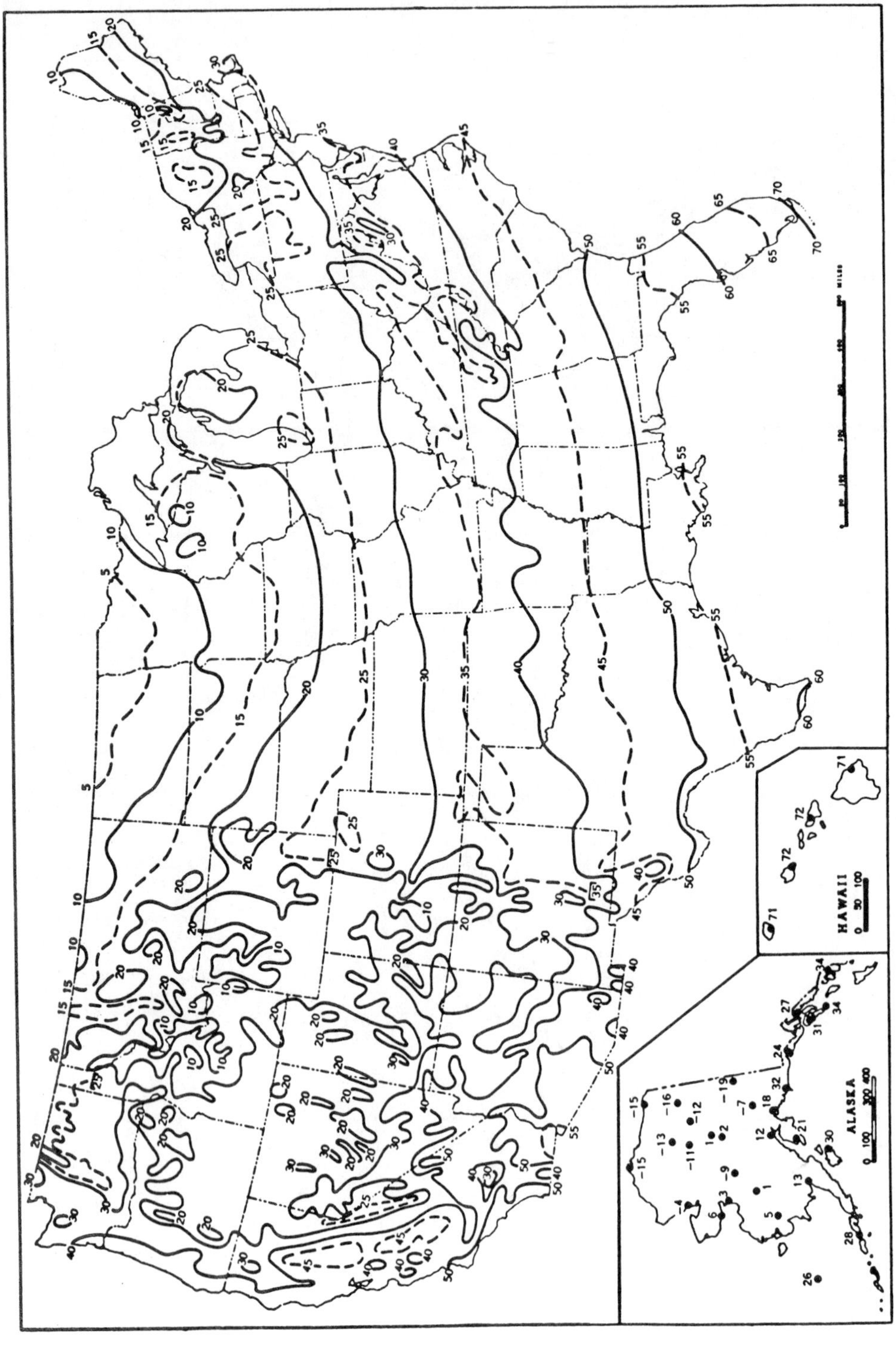

Figure 23. Normal daily average temperature (°F) for Jan. (1941–70).

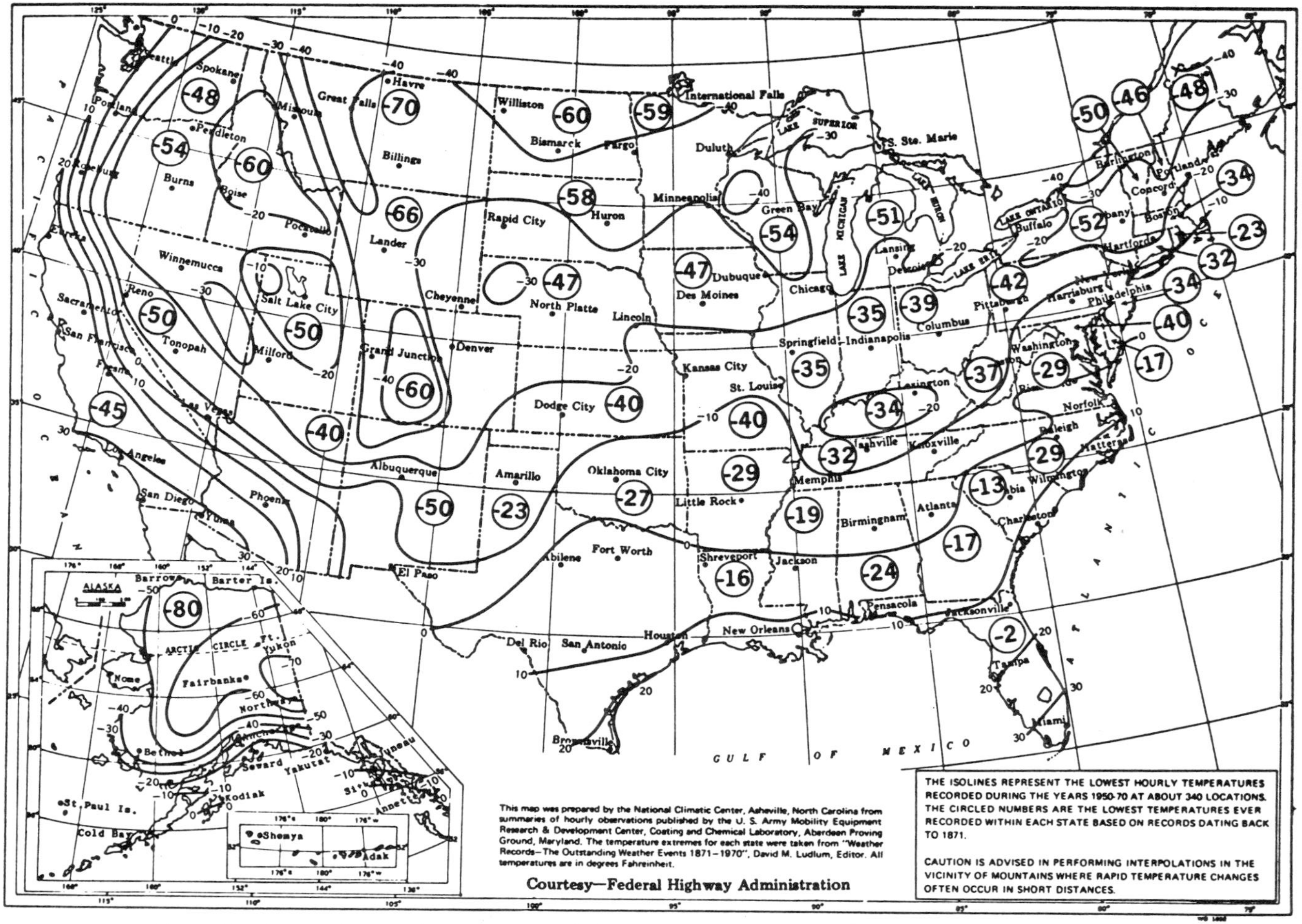

Figure 24. Extreme low hourly temperatures during the winter season with absolute minimums for each state.

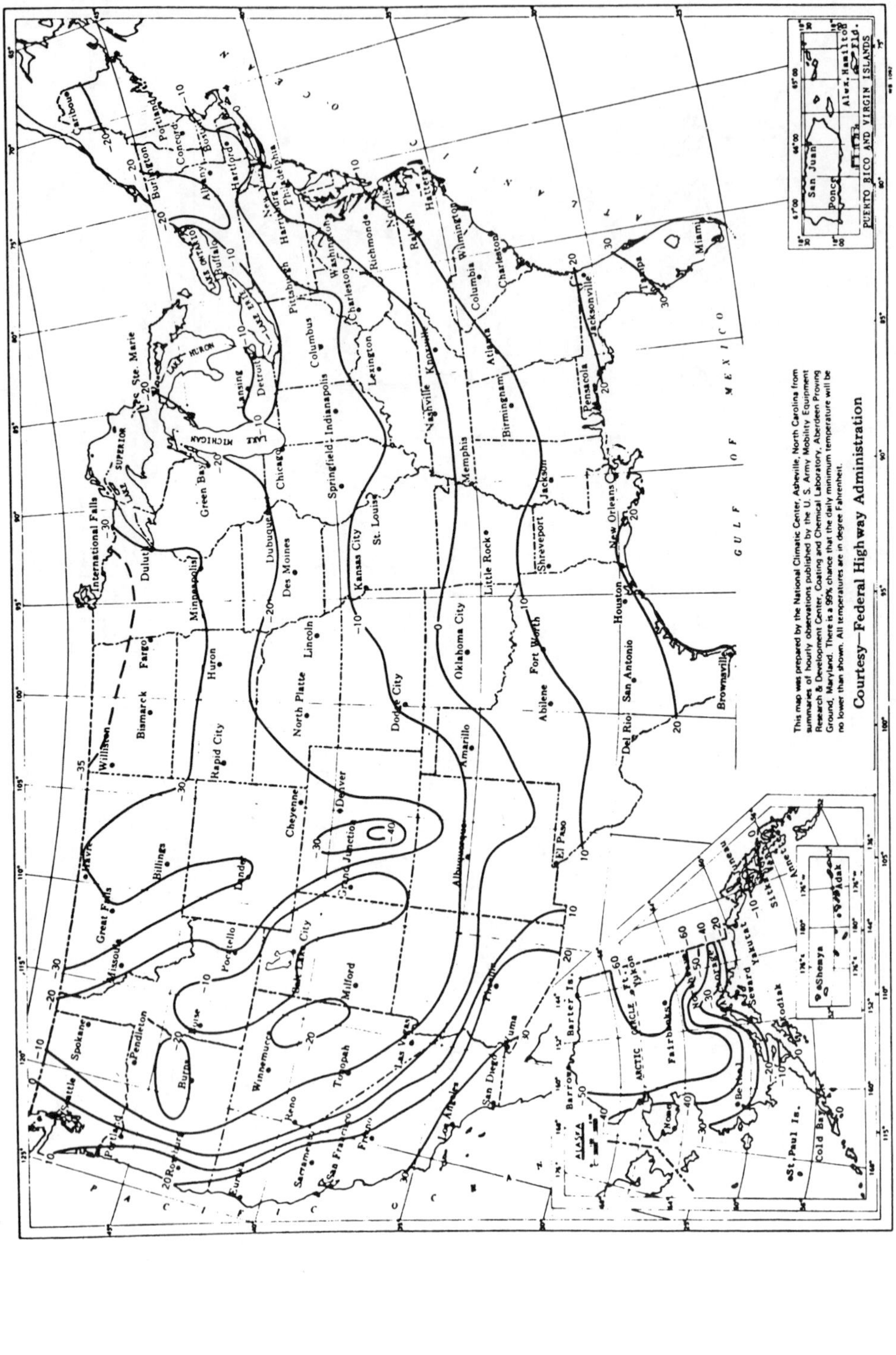

Figure 25. Minimum temperatures colder than indicated 1% of the hours during the winter season.

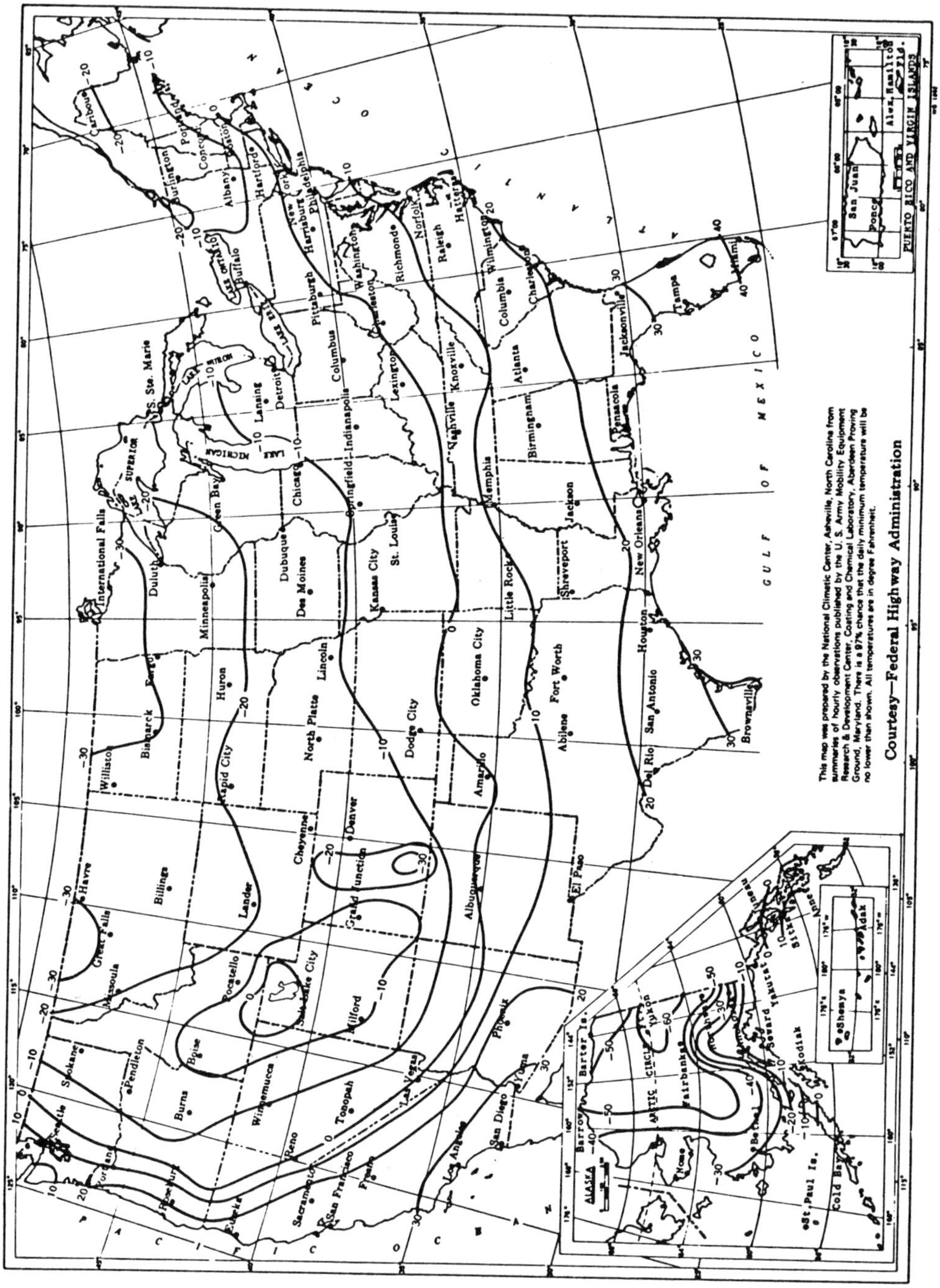

Figure 26. Minimum temperatures colder than indicated 3% of the hours during the winter season.

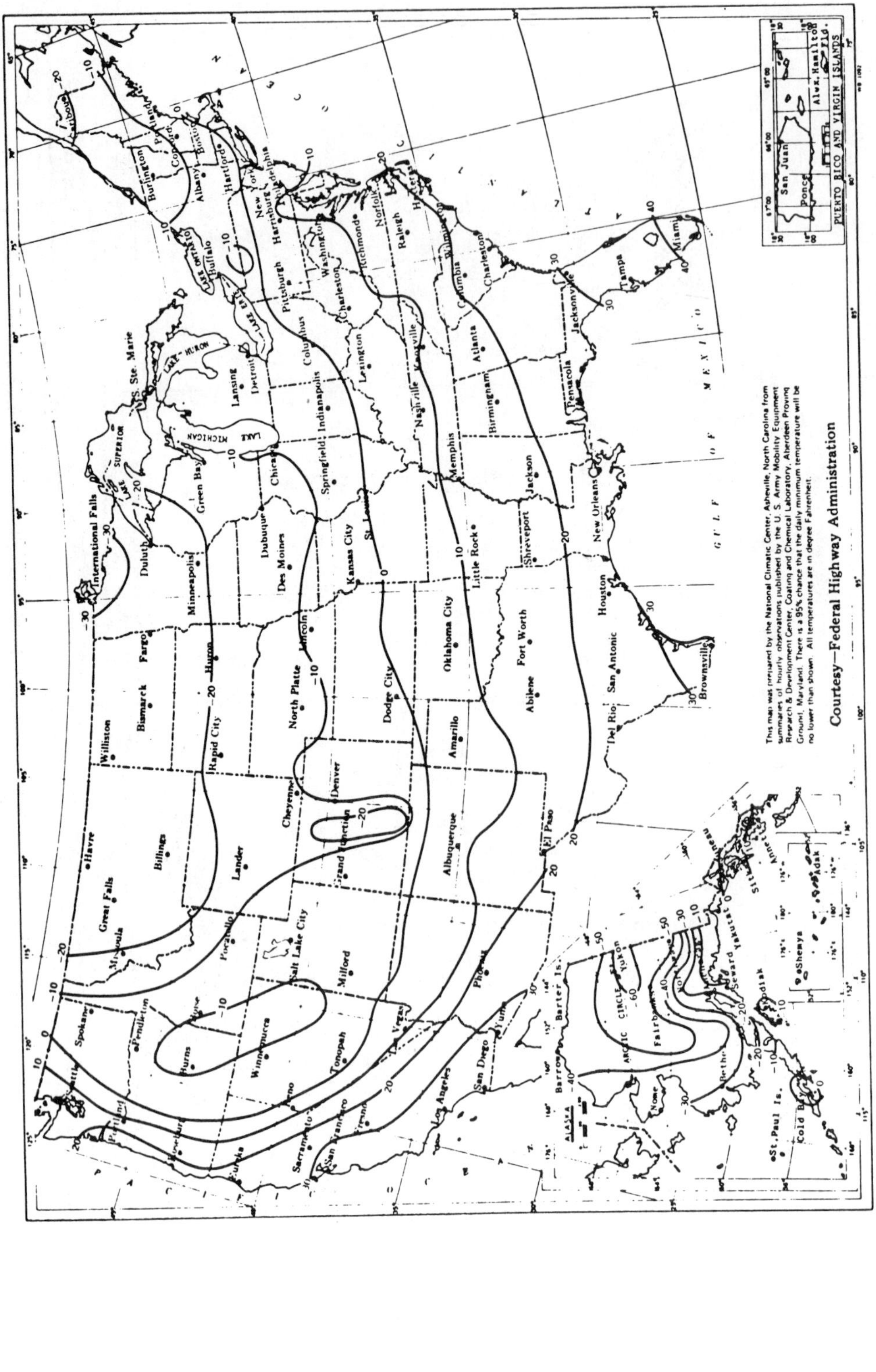

Figure 27. Minimum temperatures colder than indicated 5% of the hours during the winter season.

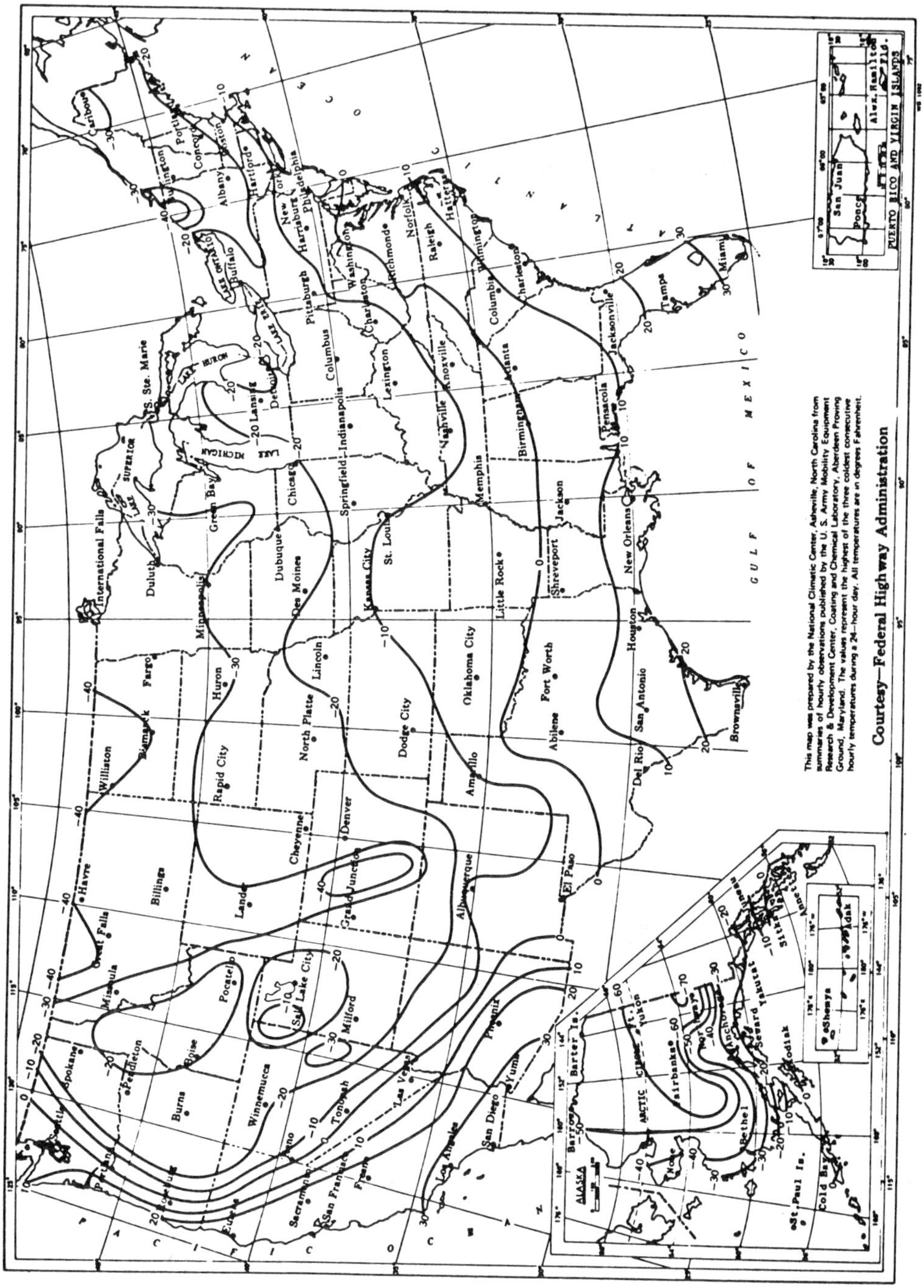

Figure 28. Consecutive 3-hour minimum temperature during the winter season.

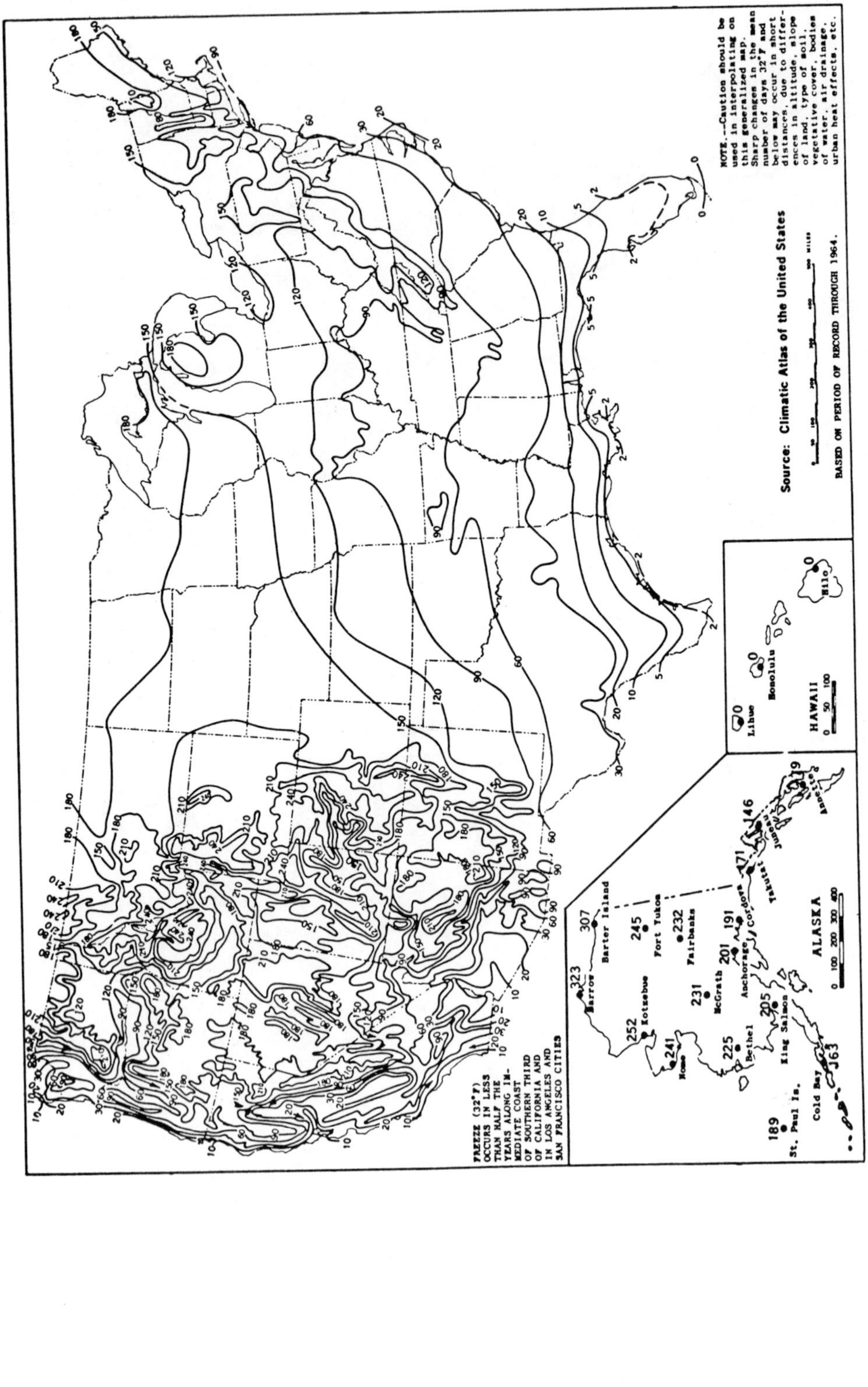

Figure 29. Mean annual number of days minimum temperature 32°F and below.

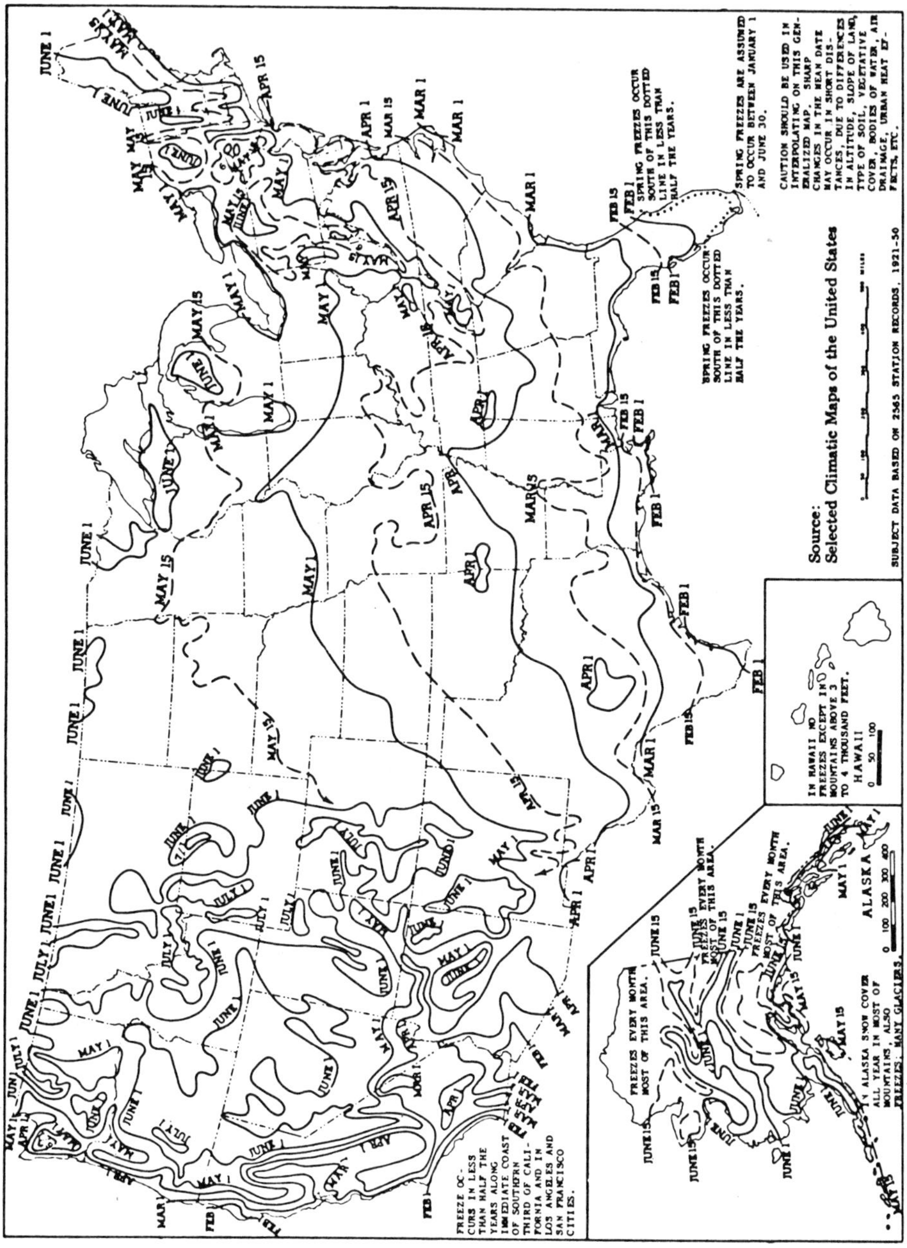

Figure 30. Mean date of last 32°F temperature in spring.

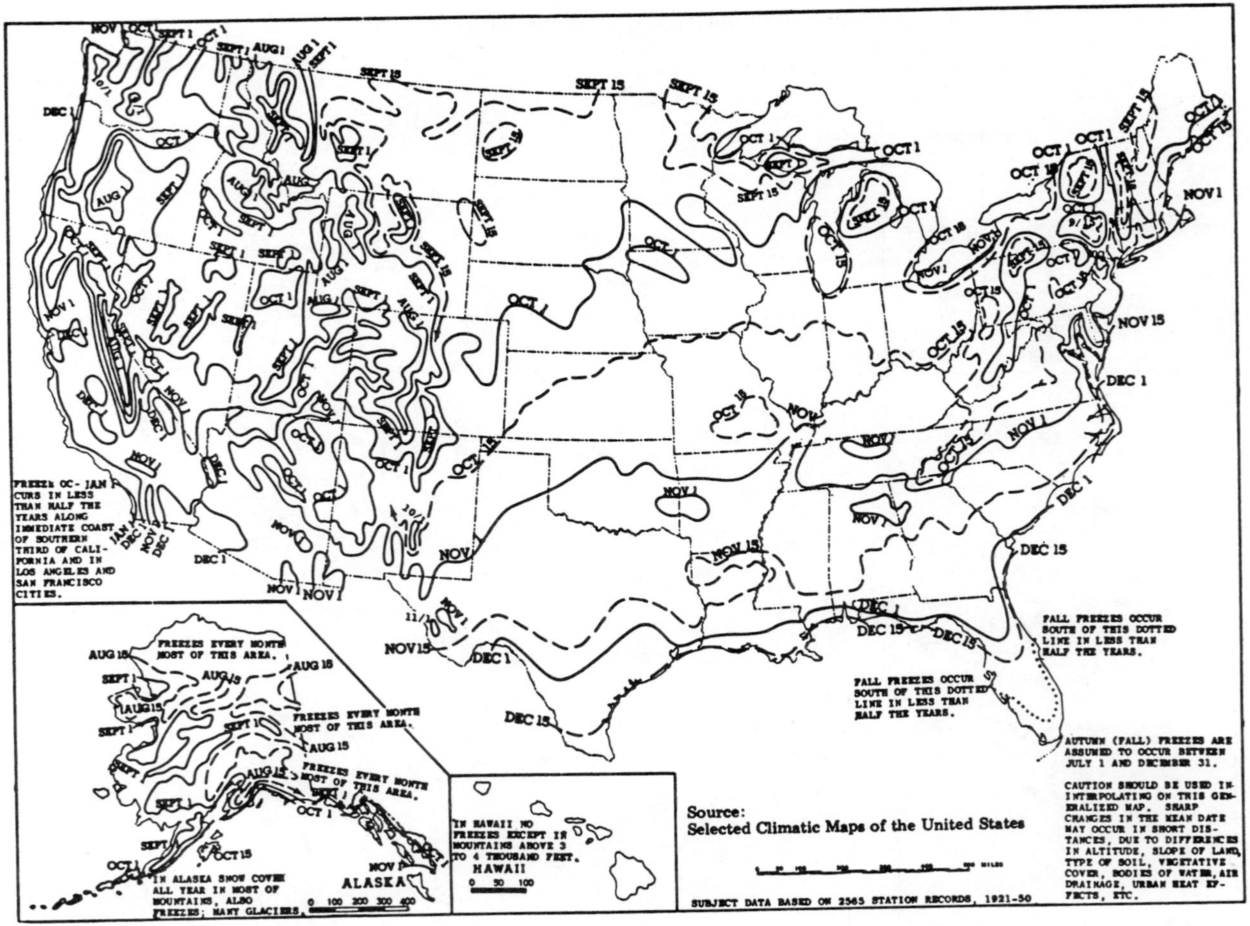

Figure 31. Mean date of first 32°F temperature in autumn.

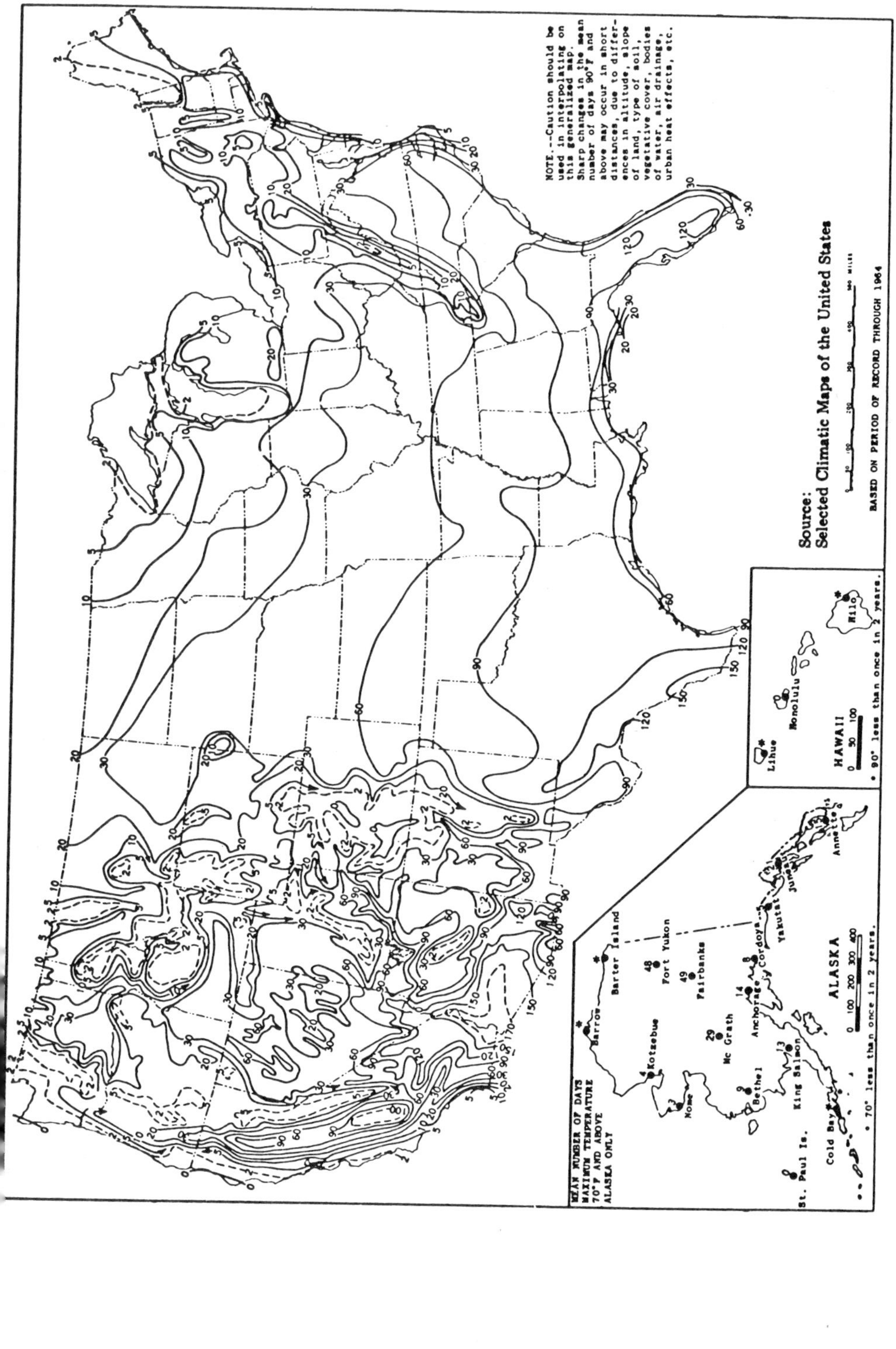

Figure 32. Mean annual number of days maximum temperature 90° F and above except 70° F and above in Alaska.

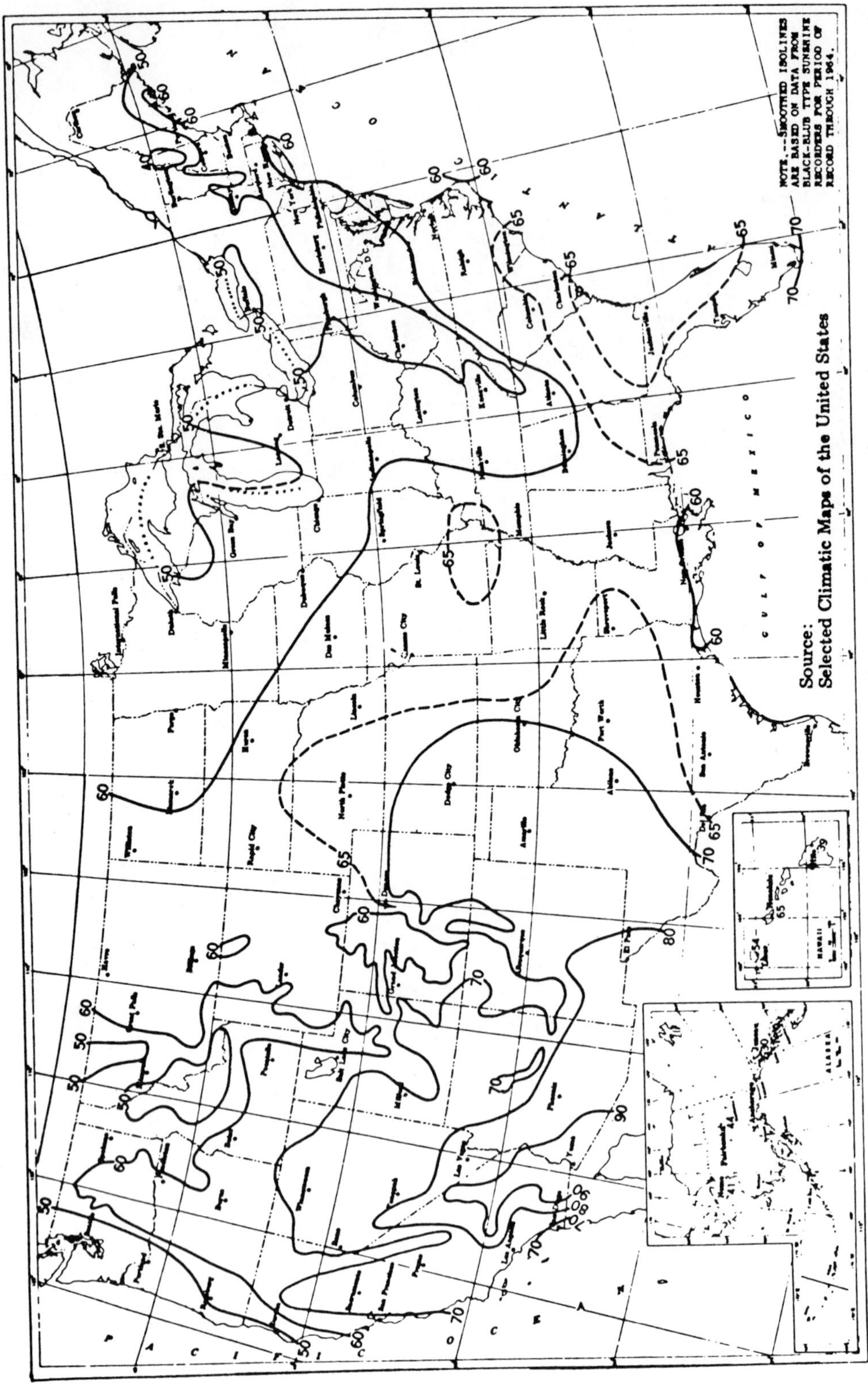

Figure 33. Mean annual percentage of possible sunshine.

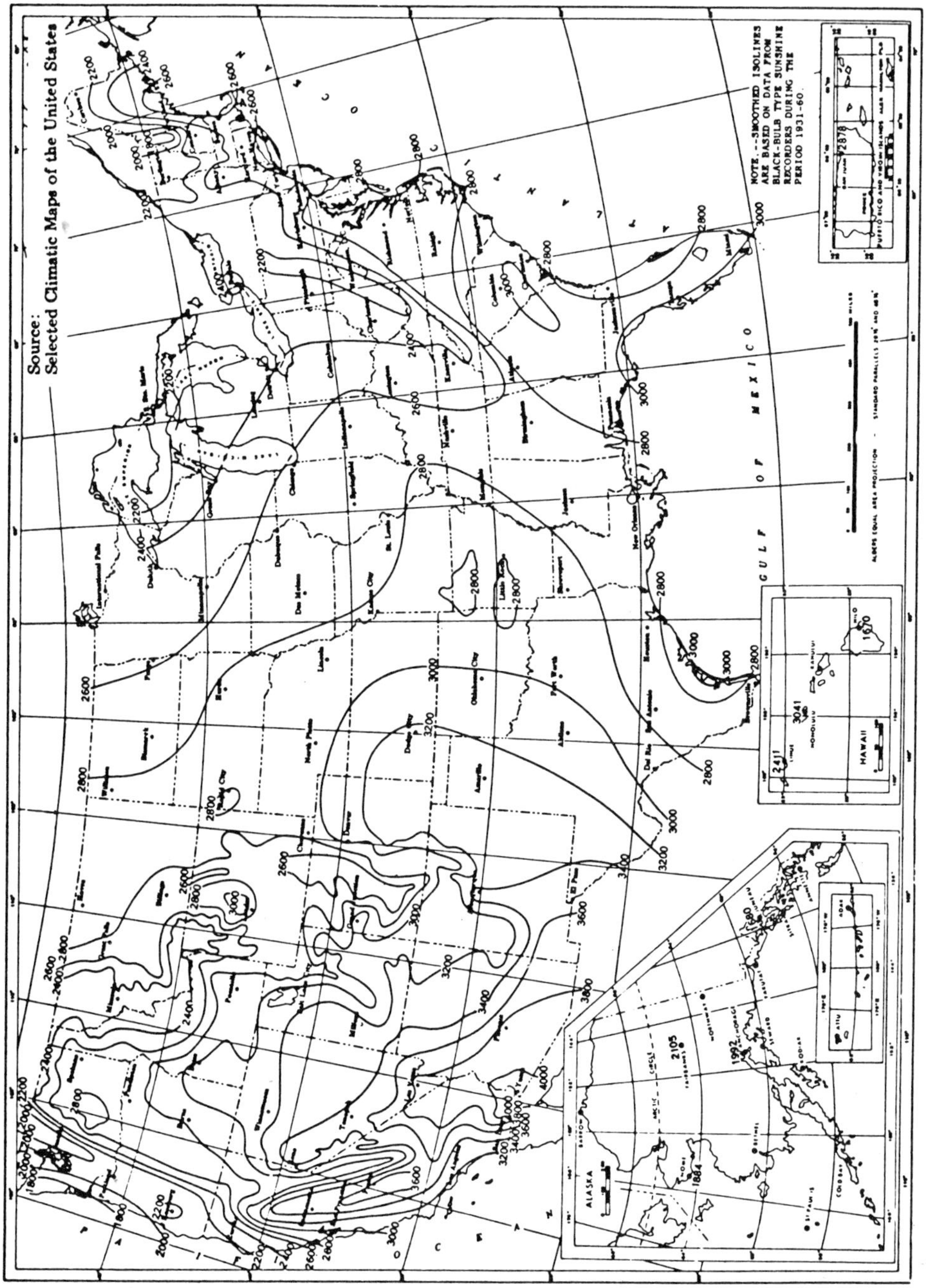

Figure 34. Mean annual total hours of sunshine.

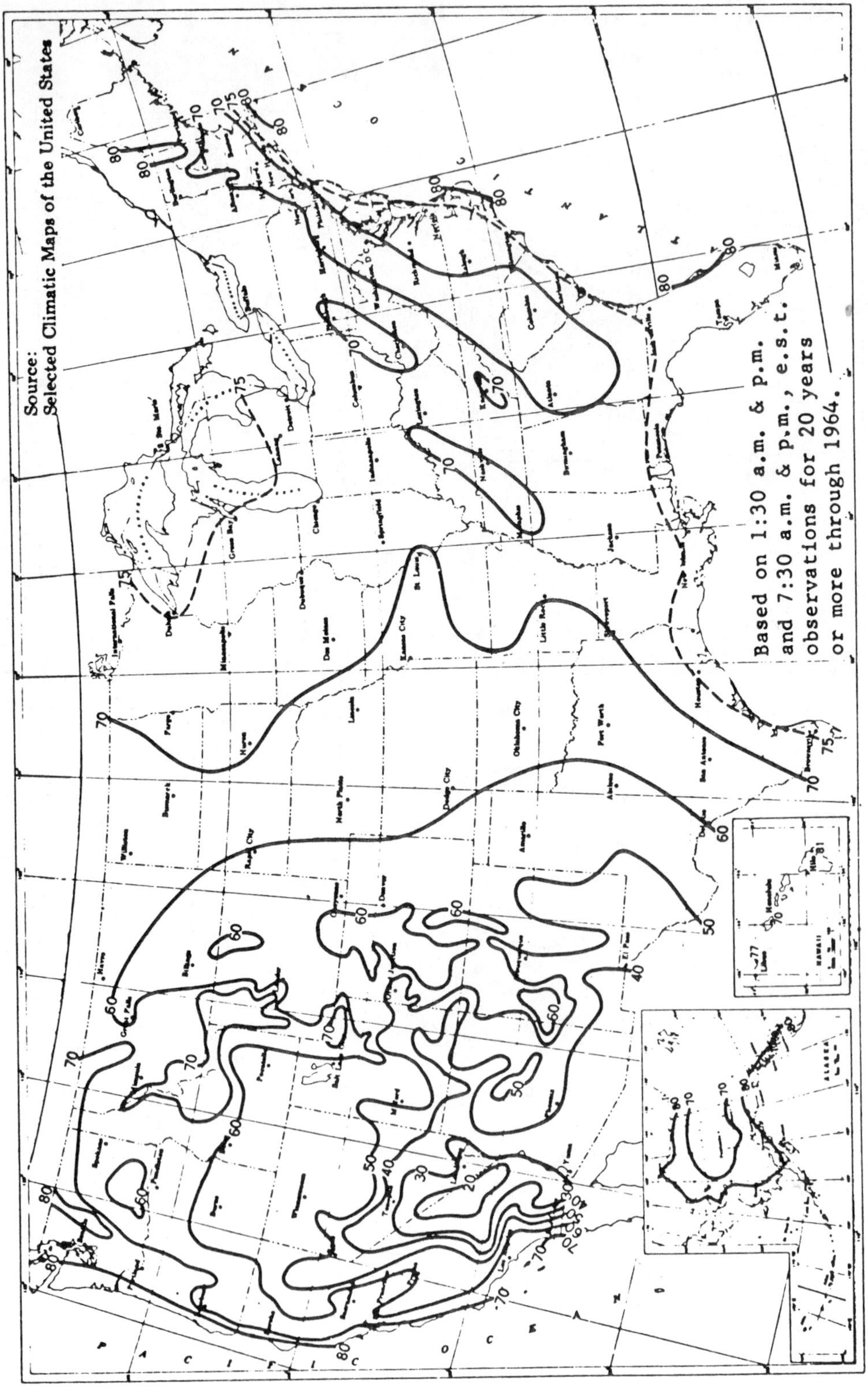

Figure 35. Mean annual relative humidity (%).

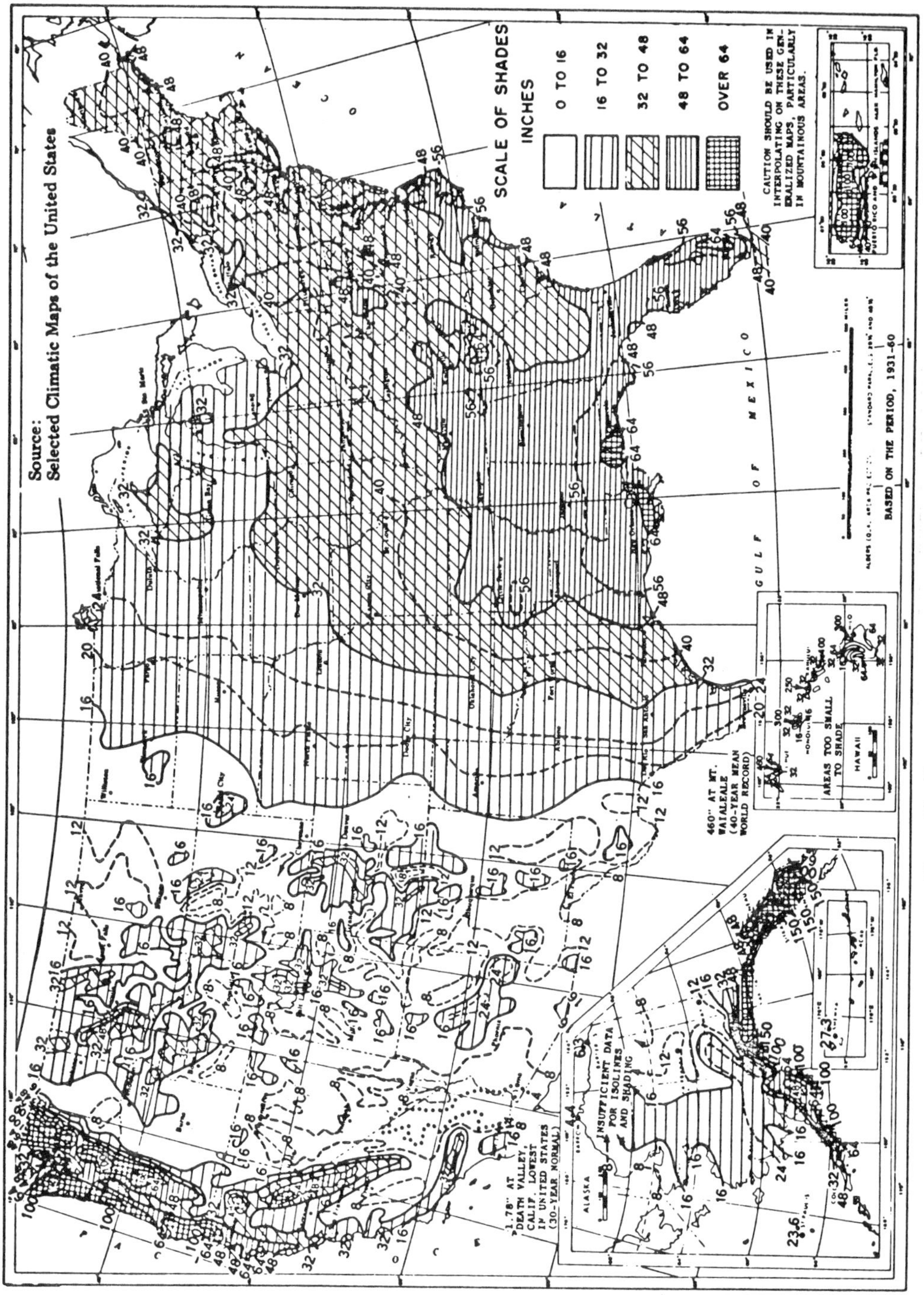

Figure 36. Normal annual total precipitation (inches).

STATION NAME	.01-.19IN	.20-.49IN	.50-.99IN	=>1.00IN	# YEARS
ALABAMA					
ANNISTON	59	25	10	2	18
BIRMINGHAM	74	28	10	3	23
DOTHAN	44	23	10	2	10
GADSDEN	58	28	10	2	16
HUNTSVILLE	74	29	10	3	14
MOBILE	75	27	14	6	14
MONTGOMERY	65	25	9	3	23
MUSCLE SHOALS	60	28	10	2	21
TUSCALOOSA	54	24	10	2	14
ARKANSAS					
EL DORADO	51	23	10	3	23
FAYETTEVILLE	40	27	8	3	4
FORT SMITH	64	20	7	2	24
HARRISON	56	21	7	1	23
HOT SPRINGS	50	24	10	3	17
JONESBORO	48	18	9	3	20
LITTLE ROCK	66	23	10	3	24
TEXARKANA	61	23	10	3	19
FLORIDA					
DAYTONA BEACH	71	25	11	5	25
FT. LAUDERDALE	54	23	13	8	20
FT. MYERS	61	23	14	8	13
GAINESVILLE	53	28	13	5	15
JACKSONVILLE	72	25	12	5	23
MELBOURNE	51	25	11	5	18
MIAMI	75	28	15	6	23
ORLANDO	69	26	13	5	25
PANAMA CITY	44	21	10	5	16
FLORIDA (CONTINUED)					
PENSACOLA	78	25	13	5	5
SARASOTA	45	23	11	5	12
TALLAHASSEE	66	31	16	7	14
TAMPA	62	24	13	5	17
W. PALM BEACH	77	30	15	6	24
GEORGIA					
ALBANY	52	23	8	2	22
ATHENS	72	28	8	2	14
ATLANTA	76	27	9	2	23
AUGUSTA	70	22	8	2	24
COLUMBUS	70	27	10	4	24
MACON	74	24	7	2	24
MOULTRIE	53	24	10	4	23
SAVANNAH	72	25	11	4	24
VALDOSTA	70	24	9	8	2
LOUISIANA					
ALEXANDRIA	47	27	12	4	13
BATON ROUGE	63	25	12	5	25
LAFAYETTE	42	26	12	6	20
LAKE CHARLES	56	24	12	5	25
MONROE	48	22	12	3	17
NEW ORLEANS	66	27	15	6	10
SHREVEPORT	66	21	9	3	17
MISSISSIPPI					
LAUREL-HATTIESBURG	52	27	13	5	22
MERIDIAN	59	28	11	3	24
OXFORD	54	26	9	3	21
TUPELO	48	25	10	2	17

Figure 37. Probable number of days per year that precipitation rates per hour can be expected.

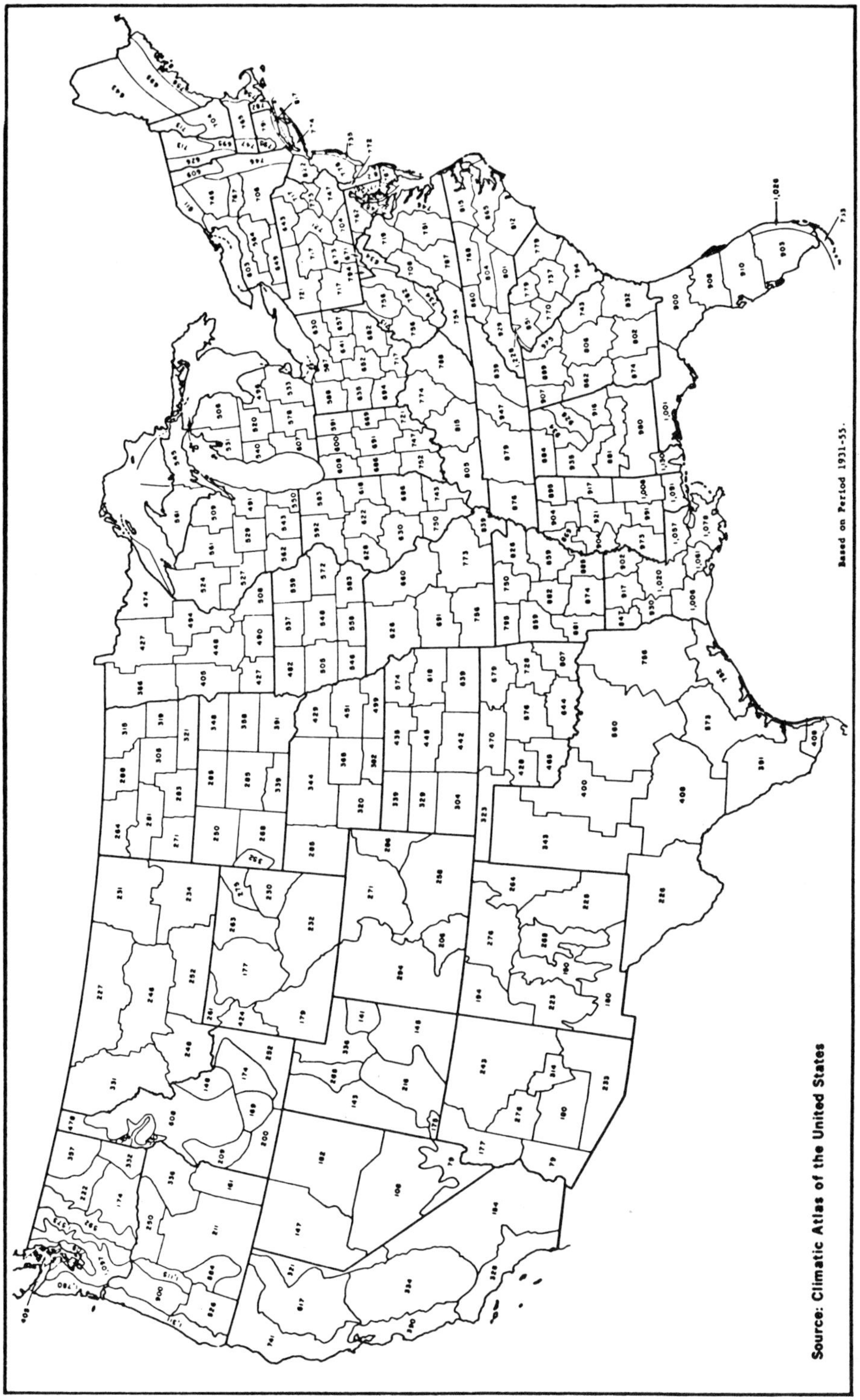
Based on Period 1931-55.
Source: Climatic Atlas of the United States

Figure 38. Mean annual precipitation in millions of gallons of water per square mile, by state climatic divisions.

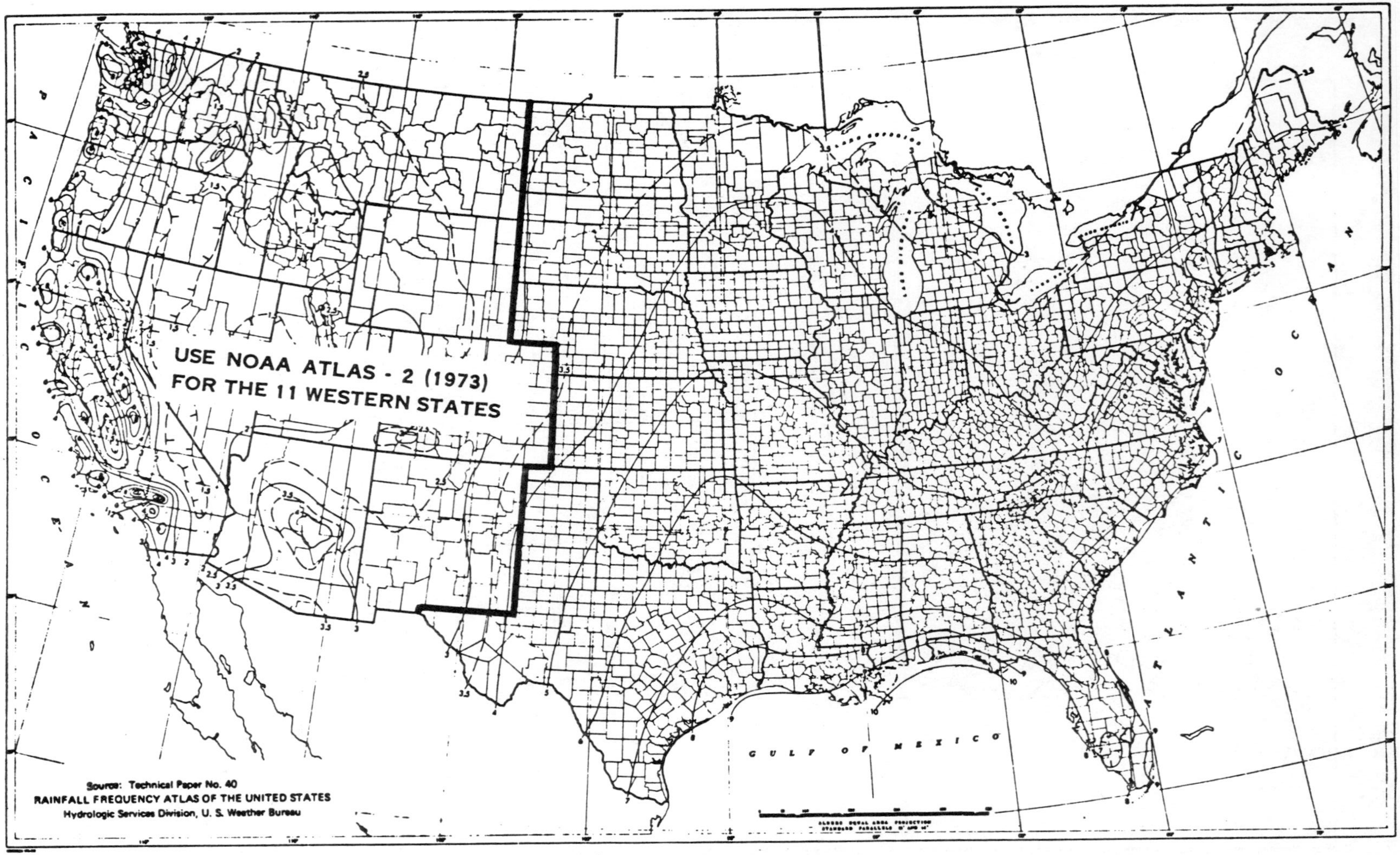

Figure 39. 10-year 24-hour rainfall (inches).

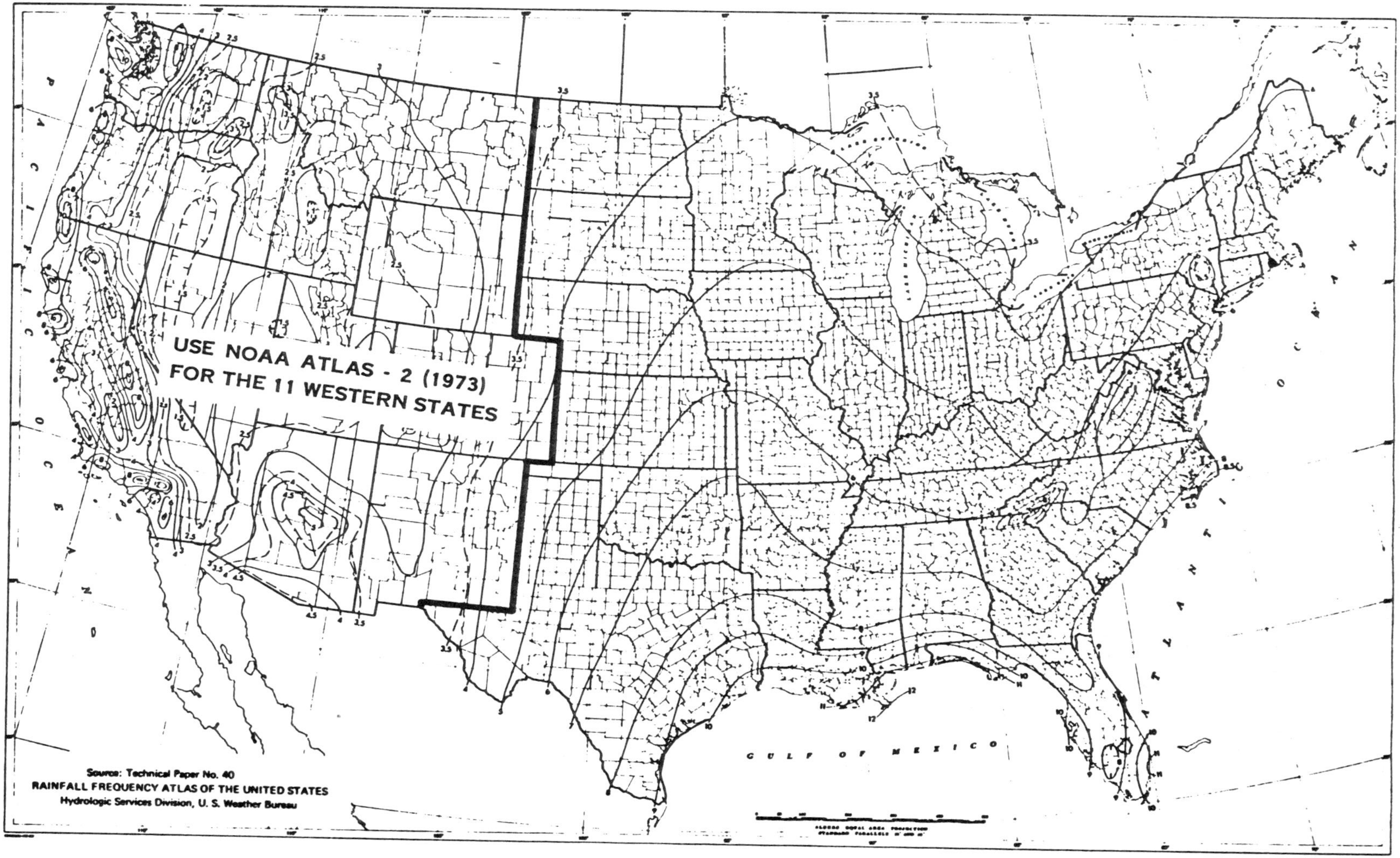

Figure 40. 25-year 24 hour rainfall (inches).

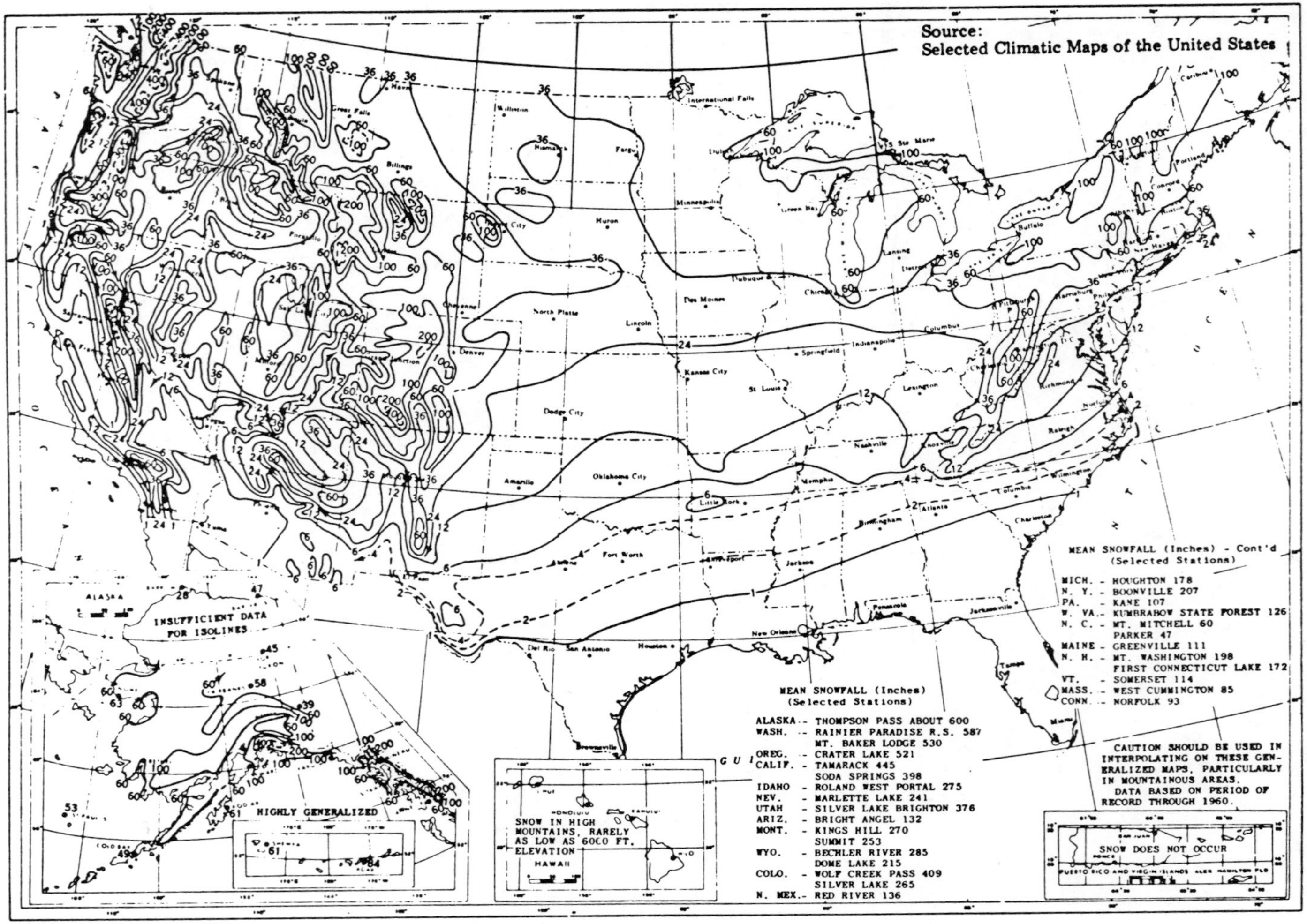

Figure 41. Mean annual total snowfall (inches).

CLIMATOGRAPHY OF THE UNITED STATES NO. 20

Climate of Greenwood FAA AP, Mississippi

DEPARTMENT OF COMMERCE • UNITED STATES OF AMERICA •

noaa NATIONAL OCEANIC AND ATMOSPHERIC ADMINISTRATION / ENVIRONMENTAL DATA SERVICE / NATIONAL CLIMATIC CENTER ASHEVILLE, N.C. APRIL 1975

Figure 42. Climatological summary for Greenwood, MS, from the series, "Climatography of the U.S. No. 20."

CLIMATOLOGICAL SUMMARY

LATITUDE N33 30
LONGITUDE W90 12

MEANS AND EXTREMES FOR PERIOD 1951-1973

GREENWOOD FAA AP, MS
ELEVATION 128

TEMPERATURE (°F)

MONTH	MEANS: DAILY MAXIMUM	MEANS: DAILY MINIMUM	MEANS: MONTHLY	EXTREMES: RECORD HIGHEST	YEAR	DAY	RECORD LOWEST	YEAR	DAY	MEAN NUMBER OF DAYS: MAX 90° AND ABOVE	MAX 32° AND BELOW	MIN 32° AND BELOW	MIN 0° AND BELOW
JAN	53.3	35.3	44.3	82	72	24	-1	62	12	0	2	14	0
FEB	57.4	37.8	47.6	81+	62	26	-4	51	3	0	1	9	0
MAR	64.8	44.2	54.5	88+	63	31	20	60	5	0	0	3	0
APR	75.6	54.0	64.8	92	72	15	32	73	11	0	0	0	0
MAY	83.5	61.7	72.6	100	51	31	43	54	4	7	0	0	0
JUN	90.5	69.2	79.8	104+	54	28	49	56	3	18	0	0	0
JULY	92.4	71.8	82.1	104+	54	1	53	67	15	23	0	0	0
AUG	91.7	70.5	81.1	105	51	30	55	56	23	22	0	0	0
SEPT	86.9	64.7	75.8	102	54	4	35	67	29	13	0	0	0
OCT	77.3	52.4	64.9	97+	54	3	27	52	30	2	0	1	0
NOV	64.6	42.6	53.6	89	72	1	19	55	29	0	0	6	0
DEC	56.0	37.5	46.8	81+	51	6	8	62	13	0	1	11	0
YEAR	74.5	53.5	64.0	105	AUG 51	30	-4	FEB 51	03	83	4	44	0

PRECIPITATION TOTALS (INCHES)

MONTH	MEAN	GREATEST MONTHLY	YEAR	GREATEST DAILY	YEAR	DAY	SNOW, SLEET: MEAN	MAXIMUM MONTHLY	YEAR	GREATEST DEPTH	YEAR	DAY	MEAN NUMBER OF DAYS: 10 or MORE	50 or MORE
JAN	4.72	12.12	51	3.76	73	21	.9	5.9	66	5.0	62	11	7	3
FEB	4.64	10.42	56	3.82	59	12	.6	8.0	60	8.0	51	01	7	3
MAR	5.83	16.84	73	6.83	55	21	.3	6.0	68	2.0	68	22	8	4
APR	5.61	10.91	64	4.66	53	29	.0						7	4
MAY	4.30	7.69	67	4.48	62	30	.0						7	3
JUN	2.86	7.58	61	3.24	66	14	.0						5	2
JULY	4.05	11.12	63	4.24	59	24	.0						6	3
AUG	2.92	7.69	64	3.56	64	16	.0						5	2
SEPT	3.00	19.65	58	8.07	58	20	.0						4	2
OCT	2.56	10.22	70	4.10	70	13	.0						4	2
NOV	4.34	13.20	57	3.61	57	13	.0						6	3
DEC	5.71	11.13	61	6.15	54	28	.3	4.0	63	2.0	63	25	7	4
YEAR	51.14	19.65	SEP 58	8.07	SEP 58	20	2.1	8.0	FEB 60	8.0	FEB 51	1	73	35

+ ALSO ON EARLIER DATES

FREEZE PROBABILITIES

PROBABILITY OF LATER DATE IN SPRING (MO/DA) THAN INDICATED

TEMP	.10	.20	.30	.40	.50	.60	.70	.80	.90
32	3/30	3/25	3/22	3/19	3/17	3/14	3/11	3/ 8	3/ 3
28	3/23	3/14	3/ 7	3/ 2	2/25	2/20	2/15	2/ 8	1/30
24	3/12	2/28	2/20	2/13	2/ 6	1/30	1/23	1/15	1/ 3
20	2/25	2/16	2/ 9	2/ 3	1/28	1/22	1/15	1/ 6	0/ 0
16	2/ 7	1/24	1/10	12/17	0/ 0	0/ 0	0/ 0	0/ 0	0/ 0

0/ 0 PROBABILITY OF OCCURRENCE OF THRESHOLD TEMP IS LESS THAN INDICATED PROBABILITY

PROBABILITY OF EARLIER DATE IN FALL (MO/DA) THAN INDICATED

TEMP	.10	.20	.30	.40	.50	.60	.70	.80	.90
32	10/23	10/28	11/ 1	11/ 4	11/ 8	11/11	11/14	11/18	11/23
28	10/31	11/ 8	11/13	11/18	11/23	11/27	12/ 2	12/ 8	12/15
24	11/10	11/18	11/24	11/29	12/ 4	12/ 8	12/13	12/19	12/27
20	11/26	12/ 8	12/19	12/26	1/ 3	1/12	1/22	2/ 9	0/ 0
16	12/17	1/ 4	1/23	0/ 0	0/ 0	0/ 0	0/ 0	0/ 0	0/ 0

0/ 0 PROBABILITY OF OCCURRENCE OF THRESHOLD TEMP IS LESS THAN INDICATED PROBABILITY

PROBABILITY OF LONGER THAN INDICATED FREEZE FREE PERIOD (DAYS)

TEMP	.10	.20	.30	.40	.50	.60	.70	.80	.90
32	252	246	242	239	235	232	228	224	218
28	302	291	283	276	270	263	257	249	237
24	337	324	319	307	305	292	285	276	263
20	>365	>365	>365	353	337	326	317	307	295
16	>365	>355	>365	>365	>365	>365	>365	>365	360

PRECIPITATION WITH PROBABILITY EQUAL OR LESS THAN

LVL	JAN	FEB	MAR	APR	MAY	JUN	JUL	AUG	SEP	OCT	NOV	DEC
0.05	1.20	1.75	1.93	2.41	1.26	0.28	[illegible]	0.55	0.18	0.00	0.85	1.89
0.10	1.63	2.14	2.45	2.91	1.45	0.40	[illegible]	0.81	0.36	0.38	1.24	2.41
0.20	2.31	2.81	3.27	3.62	2.16	0.87	[illegible]	1.24	0.73	1.03	1.87	3.20
0.30	2.91	3.32	3.95	[illegible]	2.79	1.27	2.94	1.63	1.14	1.45	2.45	3.88
0.40	3.51	3.81	4.41	4.74	[illegible]	1.70	[illegible]	[illegible]	1.60	1.84	3.04	4.52
0.50	4.13	4.32	5.28	5.29	3.84	2.18	[illegible]	2.46	2.14	2.25	3.47	5.18
0.60	4.84	4.86	[illegible]	5.88	[illegible]	2.76	[illegible]	2.95	2.79	2.69	4.39	5.92
0.70	5.64	5.49	4.86	[illegible]	[illegible]	3.67	5.56	3.53	3.63	3.22	5.23	6.72
0.80	6.08	6.24	7.92	[illegible]	[illegible]	4.45	[illegible]	[illegible]	4.81	3.84	6.34	7.76
0.90	8.47	7.53	9.79	[illegible]	7.40	[illegible]	[illegible]	5.59	6.80	4.95	8.24	9.56
0.95	9.91	8.65	11.12	[illegible]	8.42	7.86	[illegible]	6.69	8.78	5.85	9.83	10.91

MEDIAN PRECIPITATION AMOUNTS (0.50 PROBABILITY LEVEL) IN THIS TABLE DIFFER FROM THE MEANS SHOWN IN THE ABOVE TABLE BECAUSE OF THE METHOD USED IN MAKING THE COMPUTATIONS. THESE VALUES WERE DETERMINED FROM THE INCOMPLETE GAMMA DISTRIBUTION WHOSE CURVE HAS BEEN FOUND TO GIVE BEST FITS TO PRECIPITATION CLIMATOLOGICAL SERIES.

Figure 42a. Climatological summary for Greenwood, MS, from the series, "Climatography of the U.S. No. 20."

STATION: 22 3627 — MAX TEMP

YR	JAN	FEB	MAR	APR	MAY	JUN	JUL	AUG	SEP	OCT	NOV	DEC	ANNUAL
51	56.6	58.1	66.1	72.8	85.4	90.1	93.3	96.7	88.3	78.9	59.5	59.8	75.5
52	61.6	61.7	63.8	72.2	83.9	97.3	97.5	94.5	87.5	72.6	62.8	55.0	75.9
53	59.6	59.1	70.2	73.1	85.1	97.5	92.4	92.7	90.1	80.7	65.3	53.0	76.6
54	57.3	64.8	65.8	80.1	77.0	93.9	97.5	98.6	93.0	78.7	64.5	56.4	77.3
55	53.6	57.9	69.0	78.0	86.3	87.5	92.3	91.6	91.3	77.6	63.5	55.7	75.4
56	51.3	61.7	65.7	74.0	86.4	88.6	94.5	95.4	88.8	79.4	63.8	62.8	76.0
57	53.0	62.6	62.9	77.5	83.8	88.6	91.5	90.5	80.9	69.9	61.9	59.5M	73.6M
58	49.6	48.1	57.5	76.0	83.1	88.8	91.4	90.4	85.3	74.9	66.5	50.0	71.6
59	52.7	58.5	65.9	76.5	85.8	87.5	91.0	91.6	86.4	76.3	61.3	56.0	74.0
60	50.8	50.2	53.4	78.2	81.8	90.3	94.6	90.9	86.1	77.0	65.5	50.6	72.5
61	48.9	61.5	69.0	72.3	81.1	85.7	88.9	89.3	85.7	77.0	61.8	54.0	72.9
62	49.1	65.6	59.4	73.2	88.6	88.5	94.6	93.7	86.4	78.3	61.7	52.5	74.3
63	47.7	53.0	72.6	77.7	84.9	91.1	90.0	91.4	85.1	84.8	66.0	44.0	74.0
64	53.7	53.0	66.6	77.0	84.5	91.9	92.0	90.1	86.8	73.5	67.3	57.4	74.5
65	55.7	56.5	56.2	79.4	86.8	90.0	93.6	91.4	85.7	75.5	69.3	59.3	75.0
66	46.2	54.2	67.4	75.2	82.1	89.1	95.5	89.7	83.3	73.8	67.3	53.7	73.1
67	54.4	52.9	71.4	78.0	78.4	88.6	86.9	84.5	80.1	76.5	64.4	58.0	72.8
68	50.1M	49.9	63.4	73.6	79.5	89.7	88.3	88.7	82.7	76.4	62.8	56.3	71.6M
69	54.6	55.4	57.9	75.0	84.6	90.9	95.7	90.9	86.1	76.9	64.9	54.9	74.0
70	49.5	55.9	62.7	77.3	86.5	90.8	91.3	92.1	91.7	76.9	65.0	61.9	75.1
71	56.7	60.0	63.3	76.5	81.5	92.6	91.4	91.3	88.7	83.8	68.7	65.1	76.6
72	60.0	61.6	69.3	78.1	83.6	90.9	90.9	93.5	91.2	78.0	61.2	58.5	76.4
73	53.5	56.9	71.8	70.3M	79.2	91.0	90.8	88.9	87.2	80.9	70.1	55.2	74.7M
SUM	1226.2	1319.1	1491.3	1738.0	1919.9	2080.9	2125.9	2108.6	1998.6	1778.3	1485.1	1287.6	1713.4

STATION: 22 3627 — MIN TEMP

YR	JAN	FEB	MAR	APR	MAY	JUN	JUL	AUG	SEP	OCT	NOV	DEC	ANNUAL
51	37.0	38.6	47.6	51.1	59.2	70.2	73.2	72.8	64.8	52.9	38.1	41.1	53.9
52	44.6	43.8	43.9	49.6	60.3	72.1	71.9	71.8	60.4	39.6	40.7	36.5	52.9
53	39.9	39.3	49.9	49.1	64.5	73.8	72.7	69.2	60.4	50.5	37.9	33.9	53.6
54	38.4	41.3	43.6	57.6	57.6	69.6	73.4	73.1	64.4	56.1	39.9	35.2	54.0
55	34.9	38.4	46.4	57.2	64.0	64.4	72.7	70.5	65.6	48.8	40.0	37.1	53.3
56	32.2	42.9	43.0	52.0	65.0	68.1	71.5	69.9	59.8	54.6	40.4	44.1	53.6
57	37.3	46.6	43.2	55.0	63.0	70.9	71.6	69.5	65.2	51.2	45.9	40.2M	55.0M
58	31.6	30.5	41.8	53.2	63.2	69.9	72.7	70.3	67.6	52.3	44.5	32.8	52.5
59	32.5	39.6	42.0	52.1	65.5	68.9	71.3	72.3	66.2	57.2	38.1	38.6	53.7
60	38.1	33.9	36.8	54.8	58.9	69.0	72.1	72.2	66.3	55.0	41.3	33.2	52.6
61	29.6	41.1	49.9	49.6	58.9	66.5	70.2	69.1	64.7	50.0	44.9	37.9	52.7
62	30.8	44.3	41.5	51.6	64.9	69.5	73.2	71.9	65.9	57.2	43.2	34.0	54.0
63	28.3	30.5	47.8	57.3	62.7	70.2	71.2	71.5	63.7	54.5	44.5	27.6	52.5
64	33.7	33.6	43.2	57.6	63.6	70.5	71.5	71.2	64.9	48.9	46.3	39.4	53.7
65	36.8	36.6	39.8	57.5	65.1	69.1	72.4	71.3	67.0	51.1	49.8	38.4	54.6
66	31.7	38.6	44.3	56.3	60.8	67.1	73.7	69.1	63.7	48.4	46.1	37.7	53.0
67	37.1	34.0	47.6	58.1	59.6	68.9	69.1	67.2	59.7	48.1	39.2	40.4	52.4
68	35.5M	31.6	43.6	55.7	61.7	68.9	69.7	70.5	61.4	53.5	46.1	36.0	52.7M
69	38.0	38.6	39.6	56.5	62.1	68.8	73.2	68.4	63.1	54.6	40.0	35.7	53.2
70	32.0	34.9	43.6	58.0	61.5	68.6	70.3	71.5	67.8	51.6	41.2	40.0	53.4
71	37.2	37.2	41.8	51.2	58.1	70.4	71.7	71.1	68.9	58.3	42.6	48.1	54.7
72	40.4	39.4	45.7	54.3	60.9	67.4	70.0	69.9	69.3	55.8	42.4	38.8	54.5
73	35.1	34.1	50.0	49.5M	57.5	67.8	71.6	66.7	66.8	56.5	48.9	35.5	53.3M
SUM	812.5	869.6	1016.0	1242.1	1418.4	1590.6	1650.9	1621.0	1487.6	1204.5	980.0	862.2	1229.6

STATION: 22 3627 — AVERAGE TEMPERATURE

YR	JAN	FEB	MAR	APR	MAY	JUN	JUL	AUG	SEP	OCT	NOV	DEC	ANNUAL
51	46.8	48.4	56.8	62.0	72.3	80.2	83.3	84.8	76.6	65.9	48.8	50.5	64.7
52	53.0	52.8	53.7	60.8	72.1	84.7	86.7	83.2	76.0	56.1	51.8	45.8	64.6
53	49.8	49.2	60.1	61.1	74.8	85.7	82.6	81.0	75.3	65.6	51.6	43.5	65.0
54	47.9	53.1	54.7	68.8	67.3	81.8	85.5	85.9	78.7	66.4	52.2	45.8	65.7
55	44.3	48.2	57.7	67.6	75.2	76.0	82.5	81.1	78.5	63.2	51.8	46.4	64.4
56	41.8	52.3	54.4	63.0	75.7	78.4	83.0	82.7	74.3	66.9	52.1	53.5	64.8
57	45.2	54.6	53.1	66.3	73.4	79.8	81.6	80.0	73.1	60.6	53.9	49.9M	64.3M
58	40.6	39.3	49.7	63.6	73.2	79.4	82.1	80.4	76.5	63.6	55.5	41.4	62.1
59	42.6	49.1	54.0	63.3	75.7	78.2	81.2	82.0	76.3	66.8	49.7	47.3	63.9
60	44.5	42.1	45.1	66.5	70.4	79.7	83.4	81.6	76.2	66.0	53.4	41.9	62.6
61	39.3	51.3	59.5	61.0	70.0	76.1	79.6	79.2	75.2	63.5	53.4	46.0	62.8
62	40.0	55.0	50.5	62.3	76.8	79.0	83.9	82.8	76.2	67.8	52.5	43.3	64.2
63	38.0	41.8	60.2	67.5	73.8	80.7	80.6	81.5	74.4	69.7	55.3	35.8	63.3
64	43.7	43.3	54.9	67.2	74.0	81.2	81.8	80.7	75.9	61.2	56.8	48.4	64.1
65	46.3	46.6	48.0	68.5	76.0	79.6	83.0	81.4	76.4	63.3	59.6	48.9	64.8
66	39.0	46.4	55.9	64.8	71.5	78.1	84.6	79.4	73.5	61.1	56.7	45.7	63.1
67	45.8	43.5	59.5	68.1	69.0	78.8	78.0	75.9	69.9	62.3	51.8	49.2	62.7
68	42.8M	40.8	53.5	64.7	70.6	79.3	79.0	79.6	72.1	65.0	53.5	45.2	62.2M
69	46.3	47.0	48.8	65.8	73.4	79.9	84.5	79.7	74.6	65.8	52.5	45.3	63.6
70	40.8	45.4	53.2	67.7	74.0	79.7	80.8	81.8	79.8	64.3	53.1	51.0	64.3
71	47.0	48.6	52.6	63.9	69.8	81.5	81.6	81.2	78.8	71.1	55.7	56.6	65.7
72	50.2	50.5	57.5	66.2	72.3	79.2	80.5	81.7	80.3	66.9	51.8	48.7	65.5
73	44.3	45.5	60.9	59.9M	68.4	79.4	81.2	77.8	77.0	68.7	59.5	45.4	64.0M
SUM	1020.0	1094.8	1254.3	1490.6	1669.7	1836.4	1889.0	1865.4	1743.6	1491.8	1233.0	1075.5	1472.2

Figure 42b. Climatological summary for Greenwood, MS, from the series, "Climatography of the U.S. No. 20."

STATION: 22 3627 TOTAL PRECIPITATION

YR	JAN	FEB	MAR	APR	MAY	JUN	JUL	AUG	SEP	OCT	NOV	DEC	ANNUAL
51	12.12	3.66	8.28	4.27	.82	6.68	4.71	1.39	3.94	2.91	2.61	10.24	61.63
52	3.80	3.00	5.12	3.61	4.93	.04	1.34	4.18	1.06	T	3.80	4.51	35.39
53	4.01	7.70	6.16	9.33	5.95	.22	3.75	.73	1.10	.63	1.05	5.86	46.49
54	6.14	2.98	2.04	5.68	7.31	.56	2.09	1.19	.82	2.39	3.70	7.44	42.34
55	4.18	4.95	8.37	7.09	3.46	2.83	4.28	.89	.54	1.23	1.76	3.12	42.70
56	2.55	10.42	4.60	6.39	2.64	1.54	3.67	.70	.16	3.90	4.43	6.72	47.72
57	8.43	3.76	5.85	3.69	5.47	4.94	2.93	2.74	3.48	5.07	13.20	2.48	62.06
58	2.84	3.16	4.85	8.90	7.12	4.13	9.31	2.55	19.65	.91	1.76	.97	66.15
59	1.71	6.47	3.41	5.51	6.08	2.73	8.16	4.53	1.23	2.49	2.91	5.30	50.53
60	6.60	3.52	5.61	2.48	3.06	1.68	1.74	7.60	3.11	1.50	4.27	5.38	46.55
61	1.28	8.56	10.10	3.49	1.13	7.58	7.34	2.25	1.20	1.16	11.69	11.13	66.91
62	5.88	4.85	3.32	4.86	5.45	4.69	1.22	1.34	4.91	4.82	4.77	2.67	48.78
63	2.26	2.66	4.95	2.55	1.45	2.28	11.12	1.43	.46	.00	3.12	5.50	37.78
64	3.49	3.33	6.71	10.91	3.35	1.16	3.02	7.69	3.39	2.02	4.75	3.91	53.73
65	1.78	6.50	6.95	2.13	1.76	.88	2.02	2.78	2.92	1.64	1.04	1.91	32.31
66	5.17	7.95	1.06	4.33	2.85	4.64	1.86	1.17	4.77	1.85	1.30	5.24	42.19
67	1.52	4.90	2.54	4.00	7.69	2.89	2.30	3.38	.80	2.00	2.78	10.73	45.53
68	8.06	2.30	4.69	8.25	6.56	2.12	4.11	4.04	5.96	2.66	7.92	5.40	62.07
69	4.23	3.13	4.62	5.48	2.73	1.38	4.00	4.52	3.38	2.44	2.63	9.77	48.31
70	1.67	3.34	6.74	8.28	2.24	3.18	4.55	3.42	1.35	10.22	3.59	4.71	53.29
71	3.25	4.66	5.29	6.37	5.26	5.38	7.68	4.31	1.20	1.46	1.30	5.96	52.08
72	8.25	1.21	6.00	4.98	4.62	3.35	4.80	3.86	3.22	3.38	8.40	5.63	57.70
73	9.25	3.80	16.84	6.46	7.05	.83	10.93	.50	.44	4.13	6.99	6.85	74.07
SUM	108.47	106.81	134.06	129.04	98.98	65.71	106.93	67.19	69.09	58.81	99.77	131.43	1176.29

STATION: 22 3627 TOTAL SNOWFALL

SEASON	JUL	AUG	SEP	OCT	NOV	DEC	JAN	FEB	MAR	APR	MAY	JUN	SEASON
50-51							2.7	5.0	T	.0	.0	.0	
51-52	.0	.0	.0	.0	T	T	.0	.3	T	.0	.0	.0	.3
52-53	.0	.0	.0	.0	.0	.0	T	.0	T	.0	T	.0	.0
53-54	.0	.0	.0	.0	.0	T	T	.0	.0	.0	.0	.0	.0
54-55	.0	.0	.0	.0	.0	.0	T	.0	.0	.0	.0	.0	.0
55-56	.0	.0	.0	.0	.0	.0	T	.0	.0	.0	.0	.0	.0
56-57	.0	.0	.0	.0	.0	.0	.0	.0	T	.0	.0	.0	.0
57-58	.0	.0	.0	.0	.0	.0	1.0	T	.0	.0	.0	.0	1.0
58-59	.0	.0	.0	.0	.0	2.0	T	.0	.0	.0	.0	.0	2.0
59-60	.0	.0	.0	.0	.0	T	.0	8.0	T	.0	.0	.0	8.0
60-61	.0	.0	.0	.0	.0	T	T	T	.0	.0	.0	.0	.0
61-62	.0	.0	.0	.0	.0	.0	5.0	.0	.0	.0	.0	.0	5.0
62-63	.0	.0	.0	.0	.0	T	T	T	.0	.0	.0	.0	.0
63-64	.0	.0	.0	.0	.0	4.0	.5	.0	.0	.0	.0	.0	4.5
64-65	.0	.0	.0	.0	.0	T	T	.8	T	.0	.0	.0	.8
65-66	.0	.0	.0	.0	.0	.0	5.9	T	.0	.0	.0	.0	5.9
66-67	.0	.0	.0	.0	T	T	.4	T	T	.0	.0	.0	.4
67-68	.0	.0	.0	.0	T	1.0	1.0	.5	6.0	.0	.0	.0	8.5
68-69	.0	.0	.0	.0	.0	T	.0	.1	T	.0	.0	.0	.1
69-70	.0	.0	.0	.0	.0	.0	T	T	.0	.0	.0	.0	.0
70-71	.0	.0	.0	.0	.0	.0	4.0	T	T	T	.0	.0	4.0
71-72	.0	.0	.0	.0	T	.0	.0	T	.0	.0	.0	.0	.0
72-73	.0	.0	.0	.0	.0	.0	T	.0	.0	.0	.0	.0	.0
73-74	.0	.0	.0	.0	.0	T							
SUM	.0	.0	.0	.0	.0	7.0	20.5	14.7	6.0	.0	.0	.0	40.5

E AMOUNT IS WHOLLY OR PARTLY ESTIMATED.

T TRACE, AN AMOUNT TOO SMALL TO MEASURE.

M ONE OR MORE DAYS OF RECORD MISSING; IF AVERAGE VALUE IS ENTERED, LESS THAN 10 DAYS RECORD IS MISSING.

D WATER EQUIVALENT OF SNOWFALL WHOLLY OR PARTLY ESTIMATED.

MONTHLY NORMALS OF TEMPERATURE, PRECIPITATION AND HEATING AND COOLING DEGREE DAYS (1941-70)

	JAN	FEB	MAR	APR	MAY	JUN	JUL	AUG	SEP	OCT	NOV	DEC	ANN
TEMPERATURE	44.6	47.8	54.5	65.1	72.8	79.9	82.0	81.2	75.1	64.8	53.5	46.3	64.0
PRECIPITATION	4.87	5.06	5.83	5.15	4.31	3.31	4.14	2.99	3.24	2.56	4.58	5.28	51.32
HEATING DEGREE DAY	632	489	353	77	9	0	0	0	0	109	349	580	2598
COOLING DEGREE DAY	0	8	28	80	251	447	527	502	306	103	0	0	2252

Figure 42c. Climatological summary for Greenwood, MS, from the series, "Climatography of the U.S. No. 20."

EVAPORATION MAPS FOR THE UNITED STATES

Hydrologic Investigations Section, Hydrologic Services Division,
U.S. Weather Bureau, Washington, D.C.

Since evaporation inevitably extracts a portion of the gross water supply to a reservoir, the estimation of this loss is an important factor in reservoir design. In arid regions, the evaporation loss actually imposes a ceiling on the water supply obtainable through regulation. Speaking of storage on the main stem of the Colorado River, Langbein [30] states that "The gain in regulation to be achieved by increasing the present 29 million acre-feet to nearly 50 million acre-feet of capacity appears to be largely offset by a corresponding increase in evaporation."

In the final stages, the design of major storage projects requires detailed study of all data available, including observations made at the proposed reservoir sites. However, generalized estimates of free-water evaporation are invaluable in preliminary design studies of major projects, and are often fully adequate for the design of lesser projects. The maps presented herein have been prepared to serve these purposes, primarily, but they should be of value in other studies. For example, free-water evaporation [Figure 44] is a good index to potential evapotranspiration, or consumptive use, and the pan coefficient [Figure 45] is indicative of an aspect of climate. If solar radiation, wind, dew point, and air temperature are such that water in an exposed Class A pan is warmer than the air, the coefficient is greater than 0.7, and vice versa.

In 1942, A. F. Meyer [31] published a map comparable to that in Figure 44, and in the following year R. E. Horton [32] published a map of Class A pan evaporation similar to Figure 43. Subsequent to 1942, there has been a substantial increase in the Class A pan station network and significant progress in the development of techniques for estimating lake evaporation. However, the maps prepared by Horton and Meyer were carefully studied in the preparation of the new series—any pronounced differences are considered to be reasonably substantiated by data now available.

Figure 45 shows the ratio of annual lake evaporation to that from the Class A pan. It can be used to estimate free-water evaporation for any site for which representative pan data are available. Figure 46 has been included to assist in the extrapolation of seasonal pan evaporation data to annual values, as well as to provide an indication of the seasonal distribution of evaporation from a shallow free-water body. Figure 47 shows the variability of pan evaporation, year-to-year, and can be used to estimate the frequency distribution of annual lake evaporation. The correct interpretation and use of these figures are discussed later.

METHODS FOR COMPUTING EVAPORATION—The various methods for computing pan and lake evaporation are described in the Lake Hefner [33] and Lake Mead [34] Water-Loss Investigations Reports, and in Weather Bureau Research Paper No. 38 [35]. There are four generally accepted methods of computing lake evaporation: (a) water budget, (b) energy budget, (c) mass transfer, and (d) lake-to-pan relations. Very few reliable water-budget estimates are available because small errors in volume of inflow and outflow usually result in large errors in the residual evaporation value. The energy-budget approach requires such elaborate instrumentation that it is only feasible for special investigations. The mass-transfer method requires observations of lake surface-water temperature, dew point, and wind movement which are available for only a very few reservoirs. Methods (a), (b), and (c) are only applicable for existing lakes and reservoirs, and cannot be used in the design phase.

The few lake-evaporation determinations that have been made using water-budget, energy-budget, and mass-transfer methods were used in preparing Figures 44–45. However, from a practical point of view, the lake-evaporation map is based essentially on pan evaporation and related meteorological data collected at Class A evaporation and first-order synoptic stations.

DEVELOPMENT OF MAPS—The description of the development of the maps is given in Weather Bureau Technical Paper No. 37 "Evaporation Maps for the United States."

INTERPRETATION, USE, AND LIMITATIONS OF MAPS—Although the utility of the derived maps hinges largely on their reliability, it is virtually impossible to make any meaningful generalizations in this respect. In deriving Figures 43–45, all available pertinent data were utilized to the greatest extent feasible with present-day knowledge of the relationships involved. It can be reasonably assumed, therefore, that the maps provide the most accurate generalized estimates yet available. The reliability of the maps is obviously poorer in the areas of high relief than in the plains region, and the density of the observation network is an important factor throughout.

It is known that some of the data collected over the years are from sheltered sites which are not representative. Through subjective evaluation of the station descriptions and wind data, an attempt was made to derive pan evaporation and coefficient maps indicative of a representative exposure, reasonably free of obstructions to wind and sunshine. Variations in the data were smoothed to a considerable extent, and it is entirely possible that the true areal variation in evaporation exceeds that shown on the maps. For example, a pan or small reservoir located in a canyon of northerly orientation and partially shielded from the sun would experience considerably less evaporation than indicated by the maps.

The effect of topography has been taken into account only in a general way, except where the data provided definite indications. Thus, it will be noted that the isopleths tend to follow closely the topographic features in some portions of the maps while the resemblance is more casual in other areas. Both Class A pan and lake evaporation were assumed to decrease with elevation however, the decrease assumed for lake evaporation is less. With an increase in elevation, dew point and air temperatures tend to decrease, while wind movement usually increases. Solar radiation, on the other hand, increases upslope during cloudless days and may otherwise increase or decrease depending on the variation of cloudiness with elevation. There are but few reliable observations of the variation of all these factors up mountain slopes, but it is probable that the effect of these changes is less for lake evaporation than for pan evaporation.

There is good reason to expect that Figure 46, showing seasonal distribution of pan evaporation, is more reliable than any other map in the series. Fig. 47, on the other hand, is based on a sparse network, and time trends resulting from changes in site, exposure, etc., may have caused some bias in the derived values of standard deviation. Data which were obviously inconsistent were eliminated from the analysis, but any undetected inconsistencies result in values which tend to be too high. Even so, any bias in the final, smoothed isopleths should be small.

The use of Figures 43–47 is self-evident in most respects and need not be considered further here. Certain limitations and less obvious features are discussed in the following paragraphs.

Figures 43, 44, and 45. Unless the user has at hand pan-evaporation data not considered in the development of this series of maps, average annual lake evaporation can be taken directly from Figure 44. The value so determined will also suffice if pan-evaporation data collected at the site exceeds that given by Figure 43, application of the pan coefficient. (Figure 45) will probably provide a better estimate of lake evaporation than that given by Figure 44. If, on the other hand, observed pan evaporation is less than that given by Figure 43, a value of lake evaporation less than given by Figure 44 should be accepted only after it has been determined that the pan site is reasonably free of obstructions to wind and sunshine. This is to say that pan evaporation and the pan coefficient are both dependent upon exposure.

It should be emphasized that values of free-water evaporation given by Figure 44 (or Figures 43 and 45) assume that there is no net advection (heat content of inflow less outflow) over a long period of time. The mean annual advection is usually small and can be neglected, but this is not always the case. It was found at Lake Mead, for example, that advection results in a 5-inch increase in mean annual evaporation. If the advection term is appreciable, adjustment should be made as discussed in references [34] and [35].

Figure 46. The Class A pans are not in operation during the winter months over much of the country because of freezing weather. Figure 46 provides means of estimating average annual evaporation from that observed during the open season, May through October. When used in conjunction with Figure 43, it also provides a means of estimating average growing-season evaporation (Class A pan) which is so important in some studies.

Although the seasonal ratios of Figure 46 are based on Class A pan data, it is believed that they are equally applicable to free-water evaporation for shallow lakes. The ratios based on monthly computed lake evaporation for the first-order stations showed no significant deviation from those based on the pan values. It should be emphasized that the seasonal ratios can be applied to annual lake evaporation only in case of shallow lakes where energy storage can be ignored. In deep lakes, the energy storage becomes an important factor in determining seasonal or monthly evaporation. For example, at Lake Mead the maximum lake evaporation occurs in August, but maximum Class A pan evaporation is observed in June; for Lake Ontario, the maximum lake evaporation is in September, and maximum pan evaporation in July. Corrections can be made for changes in energy storage and heat advection into or out of the lake in the manner described in references [34] and [35].

The standard deviation of annual Class A pan evaporation can be obtained for any selected site directly from Figure 47. If the annual pan coefficient were constant, year-to-year, then the standard deviation of lake evaporation would be the product of that for pan evaporation and the pan coefficient. Because of variation in the annual pan coefficient, the standard deviation computed in this manner may be a few percent too low. Since the values given by Figure 47 are probably biased on the high side (discussed previously), the two possible errors tend to compensate.

Having obtained the mean and standard deviation, the frequency distribution of annual lake (or pan) evaporation can be derived, assuming the data are normally distributed. If it is further assumed that the annual evaporation totals occurring in successive years are independent, the frequency distribution of n-year evaporation can also be derived.

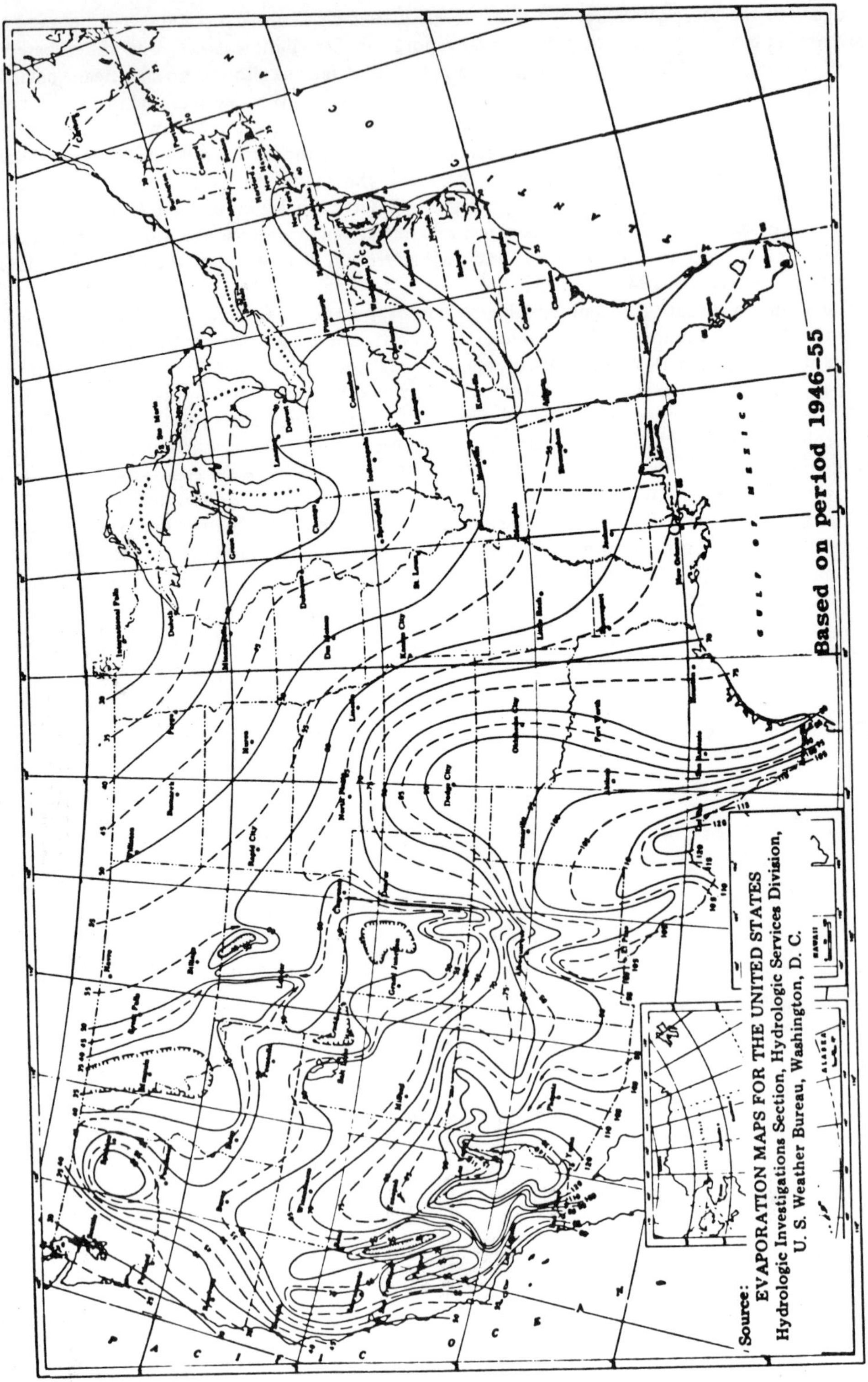

Figure 43. Mean annual class A pan evaporation (in inches).

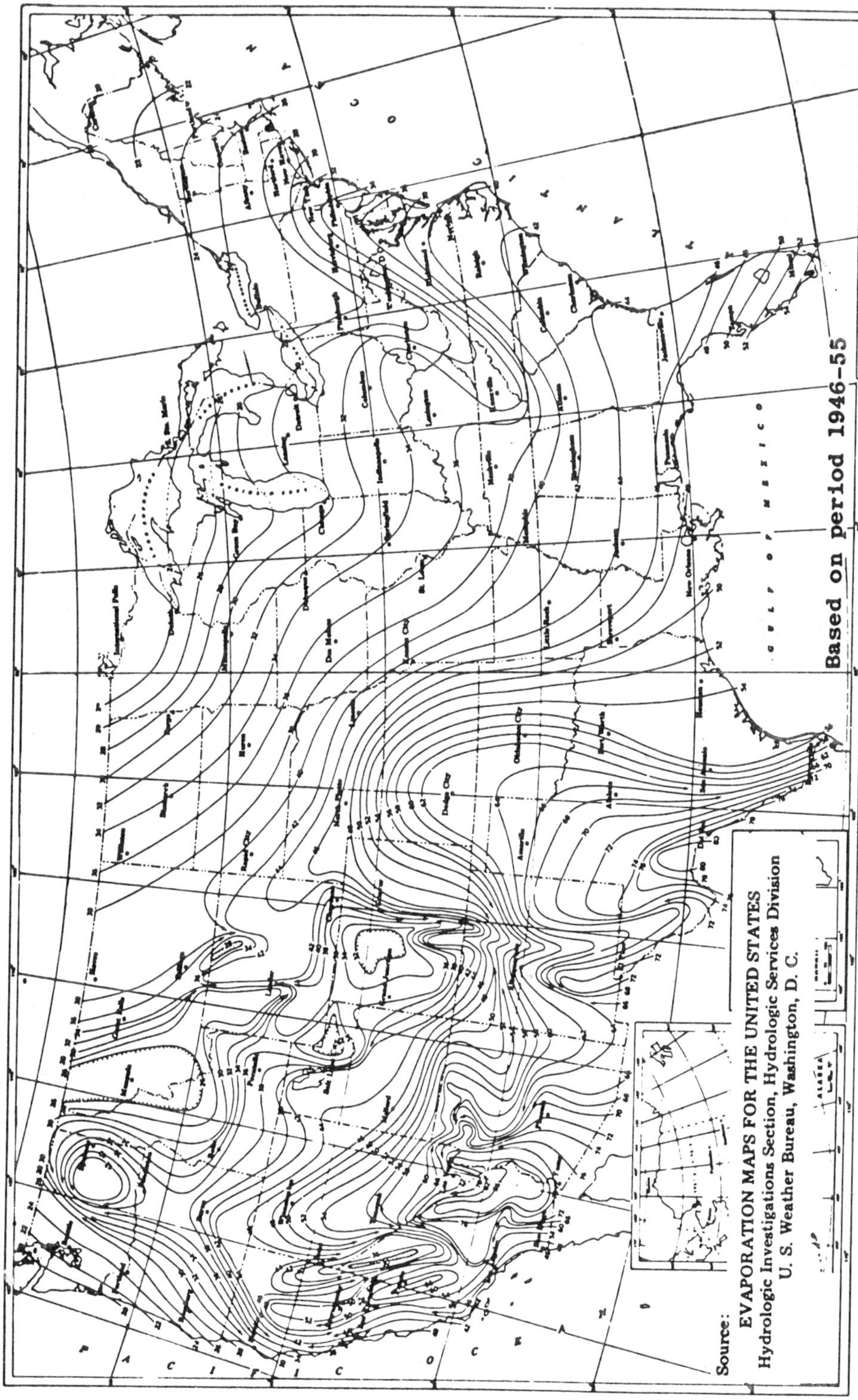

Figure 44. Mean annual lake evaporation (in inches).

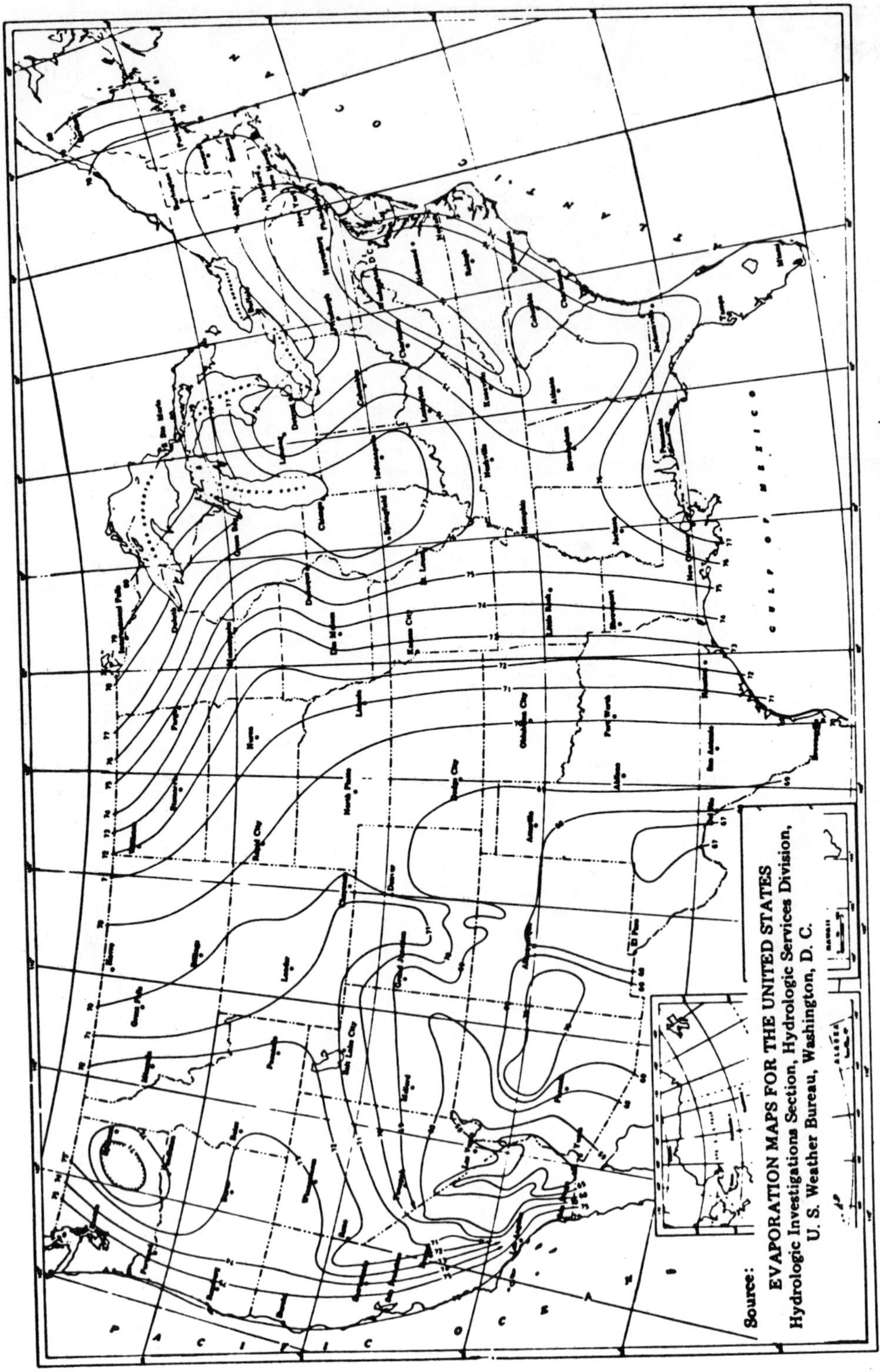

Figure 45. Mean annual class A pan coefficient (in percent).

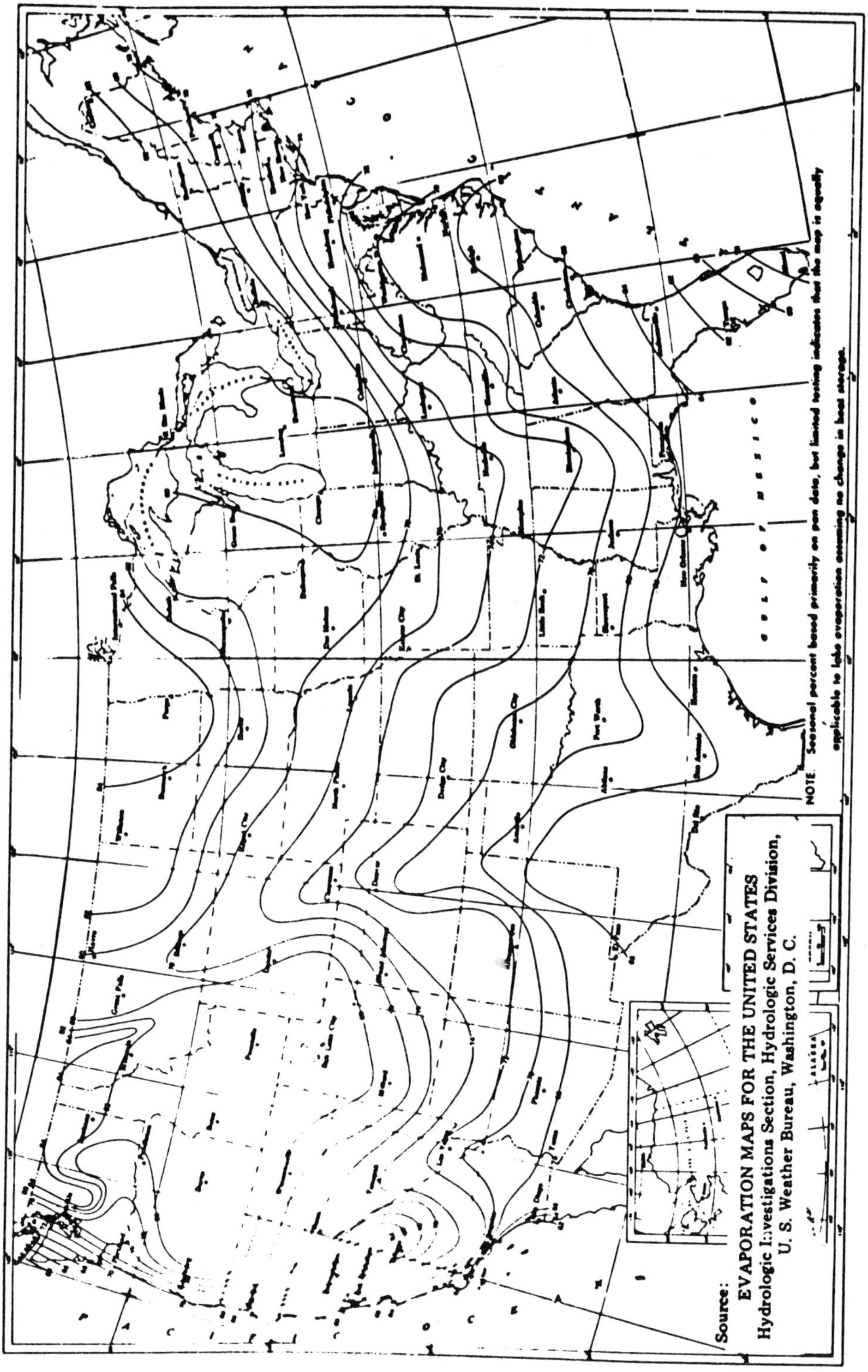

Figure 46. Mean May–October evaporation in percent of annual.

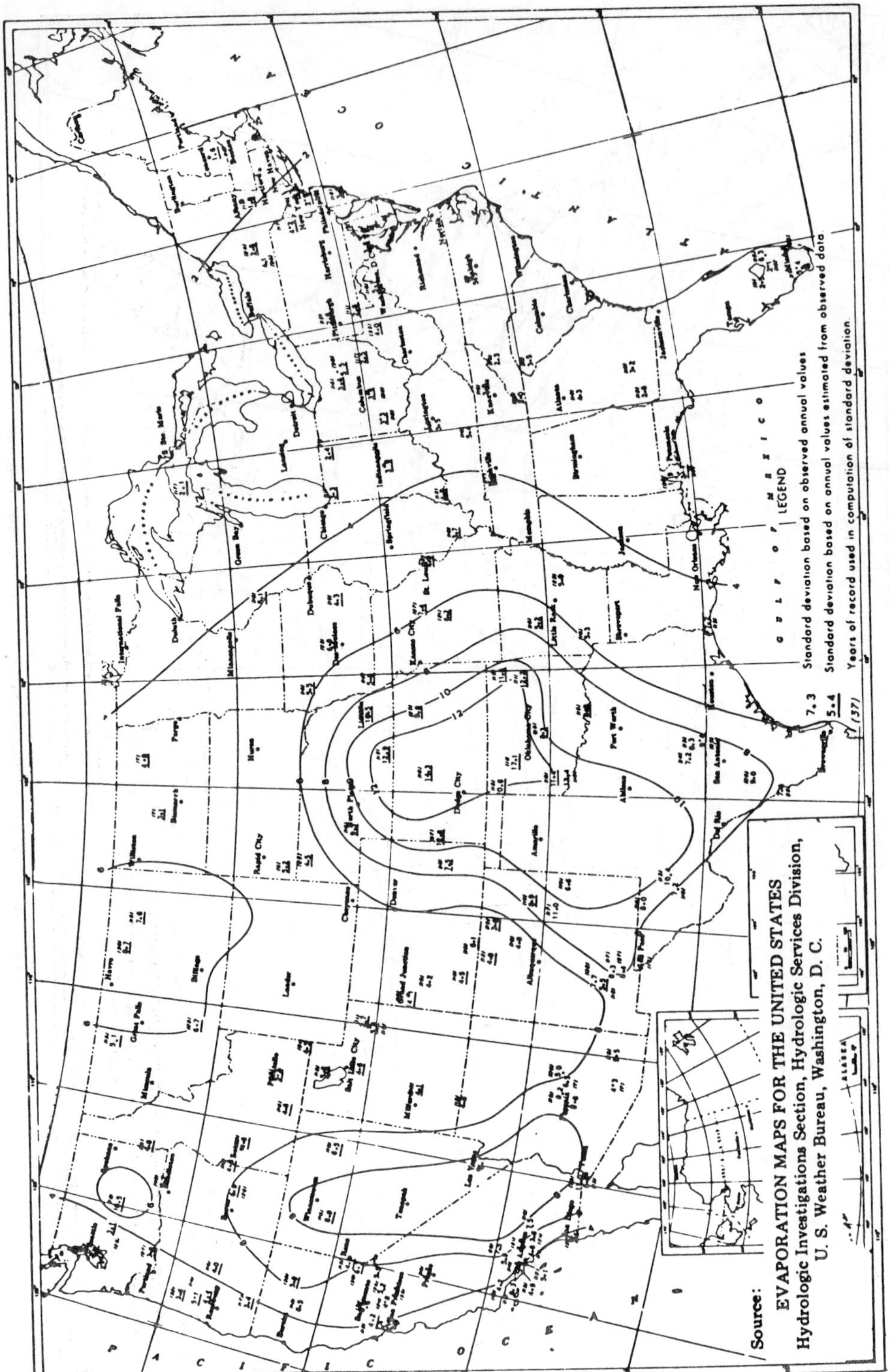

Figure 47. Standard deviation of annual class A pan evaporation (in inches).

APPENDIX C REQUESTS FOR SERVICES

The National Climatic Center may furnish special services for private clients under authority of an Act of Congress which permits the NCC to provide services at the expense of the requester. The amount the requester is charged in all cases is intended solely to defray the expenses incurred by the Government in satisfying his specific requirements to the best of its ability.

Unit costs have been established for reproduction or processing of data. The product can be in various forms, such as copies of microfilm, magnetic tapes, computer output in the form of tabulations, or other types of material. In the case of the programs discussed in this report, the station record must be serially complete and readily available in the NCC tape library. A day or two of missing record during a month is acceptable over the 20-to 25-year record, but care must be exercised to avoid processing stations with excessive missing record. Upon establishing that an adequate data base exists and costs agreed on, authorization for the NCC to proceed with the work may be made by letter or telephone. A copy of the product will be forwarded, usually within 3 to 4 weeks, along with an invoice. No advance payment is necessary, although private users who have a continuing need for climatological services may make advance deposits to cover the cost of their requirements as they arise. This procedure eliminates the need for separate invoices upon completion of each request. Further information about the programs may be obtained by contacting the Statistical Climatology Branch, National Climatic Center, Federal Building, Asheville, North Carolina 28801 (telephone 704-258-2850, ext. 319).

TECHNICAL REPORT DATA
(Please read Instructions on the reverse before completing)

1. REPORT NO.
EPA-600/2-76-250

2.

3. RECIPIENT'S ACCESSION NO.

4. TITLE AND SUBTITLE
USE OF CLIMATIC DATA IN ESTIMATING STORAGE DAYS FOR SOILS TREATMENT SYSTEMS

5. REPORT DATE
November 1976 (Issuing date)

6. PERFORMING ORGANIZATION CODE

7. AUTHOR(S)
Dick M. Whiting

8. PERFORMING ORGANIZATION REPORT NO.

9. PERFORMING ORGANIZATION NAME AND ADDRESS
Robert S. Kerr Environmental Research Laboratory
Post Office Box 1198
Ada, Oklahoma 74820

10. PROGRAM ELEMENT NO.
1BC611

11. CONTRACT/GRANT NO.
EPA-IAG-D5-F694

12. SPONSORING AGENCY NAME AND ADDRESS
Robert S. Kerr Environmental Research Laboratory
Office of Research and Development
U. S. Environmental Protection Agency
Ada, Oklahoma 74820

13. TYPE OF REPORT AND PERIOD COVERED
Final 4/75 to 7/76

14. SPONSORING AGENCY CODE
EPA-ORD

15. SUPPLEMENTARY NOTES

16. ABSTRACT

Computer programs have been developed to estimate storage needs for Soil Treatment Systems from analyses of daily climatological data. One program uses a set of thresholds for temperature, precipitation and/or snow depth to estimate storage needs in colder regions.

A second program is designed for use in high rainfall regions where saturated soil, rather than severe weather, is the limiting condition. Climatological data for a 20- to 25-year period is examined for each case to produce a summary table. This table presents the mean, the standard deviation, the unbiased third moment about the mean, the coefficient of skewness and storage days for recurrence intervals of 5, 10, 25, and 50 years.

17. KEY WORDS AND DOCUMENT ANALYSIS

a. DESCRIPTORS	b. IDENTIFIERS/OPEN ENDED TERMS	c. COSATI Field/Group
Climatology Effluents Cold-weather operations	Land application Estimating storage	4B 2C

18. DISTRIBUTION STATEMENT
RELEASE TO PUBLIC

19. SECURITY CLASS *(This Report)*
UNCLASSIFIED

20. SECURITY CLASS *(This page)*
UNCLASSIFIED

21. NO. OF PAGES
98

22. PRICE

EPA Form 2220-1 (9-73)

☆ U. S. GOVERNMENT PRINTING OFFICE: 1977-757-056/5447 Region No. 5-11

Appendix B

IRRIGATION WATER REQUIREMENTS (Excerpts)

The first 40 pages of Soil Conservation Service Technical Release No. 21 are reproduced here. This material includes a glossary of terms, a discussion of the Blaney-Criddle formula with modifications to compute short-term and seasonal consumptive water use, peak period calculation, effective rainfall, carryover soil moisture, groundwater contribution, and field application efficiency. These factors are combined to yield a net field irrigation requirement. Sample calculations illustrate the application of all these points to design problems.

The remainder of the Technical Release, not reproduced here, includes other factors which must be considered in a detailed irrigation system design, such as salt leaching water requirements, freeze protection, conveyance and storage losses, and operational losses. An appendix contains growth curves similar to Figures 1 and 2 for other crops.

U. S. Department of Agriculture
Soil Conservation Service
Engineering Division

Technical Release No. 21 (Rev. 2)
September, 1970
and (Rev. 1) February, 1967

IRRIGATION WATER REQUIREMENTS

Introduction

It is essential that the water requirements and consumptive use of water be known in irrigation planning for soil conservation and irrigation districts and for individual farms. Conservation of water supplies, as well as of soils, is of first importance in the agricultural economy. In basin-wide investigations of water utilization and in water conservation surveys, consumptive water requirement is one of the most important factors to be considered. There is an urgent need for information on irrigation requirements in connection with farm planning programs for areas where few data are available.

A knowledge of consumptive use is necessary in planning farm irrigation system layouts and improving irrigation practices. Irrigation and consumptive water requirement data are used more and more widely by water superintendents as well as state, federal, and other agencies responsible for the planning, construction, operation and maintenance of multiple-purpose projects and by those responsible for guiding and assisting farmers in the solution of their irrigation problems.

Scope

This release covers the procedures used to estimate irrigation water requirements on a farm or on a project. Irrigation application efficiencies are discussed briefly. More detailed information is presented in applicable chapters of Section 15 of the National Engineering Handbook. Procedures for measuring losses in existing farm distribution and project conveyance systems and for estimating losses in such systems as may be proposed are included. Irrigation water storage requirements may be estimated by use of the procedure contained in Technical Release No. 19.

Definition of Terms

Some of the terms used in this release are defined as follows:

Consumptive Use.
Consumptive use, often called evapo-transpiration, is the amount of water used by the vegetative growth of a given area in transpiration and building of plant tissue and that evaporated from adjacent soil or intercepted precipitation on the plant foliage in any specified time. If the unit of time is small, consumptive use is usually expressed as

acre inches per acre or depth in inches, whereas, if the unit of time is large, such as a growing season or a 12-month period, it is usually expressed as acre feet per acre or depth in feet.

Consumptive Water Requirement.
The amount of water potentially required to meet the evapo-transpiration needs of vegetative areas so that plant production is not limited from lack of water.

Effective Rainfall.
Precipitation falling during the growing period of the crop that is available to meet the consumptive water requirements of crops. It does not include such precipitation as is lost to deep percolation below the root zone nor to surface runoff.

Consumptive Irrigation Requirement.
The depth of irrigation water, exclusive of precipitation, stored soil moisture, or ground water, that is required consumptively for crop production.

Net Irrigation Requirement.
The depth of irrigation water, exclusive of precipitation, stored soil moisture, or ground water, that is required consumptively for crop production and required for other related uses. Such uses may include water required for leaching, frost protection, etc.

Peak Period Consumptive Use.
Peak period consumptive use is the average daily rate of use of a crop occurring during a period between normal irrigations when such rate of use is at a maximum.

Irrigation Efficiency.
The percentage of applied irrigation water that is stored in the soil and available for consumptive use by the crop. When the water is measured at the farm headgate, it is called farm-irrigation efficiency; when measured at the field, it is designated as field-irrigation efficiency; and when measured at the point of diversion, it may be called project-efficiency.

Irrigation Water Requirement.
The net irrigation water requirement divided by the irrigation efficiency.

Field Capacity.
The moisture percentage, on a dry weight basis, of a soil after rapid drainage has taken place following an application of water, provided there is no water table within capillary reach of the root zone. This moisture percentage usually is reached within two to four days after an irrigation, the time interval depending on the physical characteristics of the soil.

Wilting Point.
The wilting point is the moisture percentage, also on a dry weight basis, at which plants can no longer obtain sufficient moisture to satisfy moisture requirements and will wilt permanently unless moisture is added to the soil profile.

Carryover Soil Moisture.
Moisture stored in soils within root zone depths during the winter, at times when the crop is dormant, or before the crop is planted. This moisture is available to help meet the consumptive water needs of the crop.

Influence of Various Factors on Water Use

Many factors operate singly or in combination to influence the amounts of irrigation water consumed by plants. Their effects are not necessarily constant but may differ with locality and fluctuate from time to time. The more important influences are climate, water supply, and plant growth characteristics.

Precipitation.
The amount and rate of precipitation will have an effect on the amount of irrigation water consumptively used during any season. Under certain conditions, precipitation may be a series of frequent, light showers during the hot summer. Such showers may add little or nothing to the soil moisture for use by the plants through transpiration but do decrease the withdrawal from the stored moisture. The precipitation may be largely lost by evaporation directly from the surface of the plant foliage and from the land surface. Some of the precipitation from heavy storms may be lost by surface runoff. Where storms occur within a relatively short period after completion of an irrigation, a high percentage of precipitation is lost due to surface runoff, deep percolation or both. Other storms may be of such intensity and amount that a large percentage of their precipitation will enter the soil and become available for plant transpiration. Such a condition materially reduces the amount of irrigation water needed.

Temperature.
The rate of consumptive use of water by crops in any particular locality is probably affected more by temperature, which for long-time periods is a good measure of solar radiation, than by any other factor. Abnormally low temperatures may retard plant growth and unusually high temperatures may produce dormancy. Consumptive use may vary even in years of equal accumulated temperatures because of deviations from the normal seasonal distribution. Transpiration is influenced not only by temperature but also by the area of leaf surface and the physiologic needs of the plant, both of which are related to stage of maturity.

Growing Season.
The growing season, which is tied rather closely to temperature, has a major effect on the seasonal use of water by plants. It is frequently considered to be the period between killing frosts, but for many annual crops, it is shorter than the frost-free period, as such crops are usually planted after frosts are past and mature before they recur.

For most perennial crops, growth starts as soon as the maximum temperature stays well above the freezing point for an extended period of days, and continues throughout the season despite later freezes.

Sometimes growth persists after the first so-called killing frost in the fall. In the spring, and to less extent in the fall, daily minimum temperatures may fluctuate several degrees above and below 32° F. for several days before remaining generally above or below the freezing point. The hardier crops survive these fluctuations and continue unharmed during a few hours of subfreezing temperature. In fact, many hardy crops, especially grasses, may mature even though growing season temperatures repeatedly drop below freezing. Although the frost-free season may be used as a guide for estimating consumptive use, actual dates of planting and crop maturity are important in determining the consumptive irrigation requirements of the crops.

Latitude and Sunlight.
Because of the earth's movement and axial inclination, the hours of daylight during the summer are much greater in the higher latitudes than at the Equator. Since the sun is the source of all energy used in crop growth and evaporation of water, this longer day may allow plant transpiration to continue for a longer period each day and to produce an effect similar to that of lengthening the growing season.

Other Climatic Factors.
Other climatic factors that have an effect on the amount of irrigation water consumed by plants are as follows:

Humidity.--Evaporation and transpiration are accelerated on days of low humidity and slowed during periods of high humidity. If the average relatively humidity percentage is low during the growing season, a greater use of water by vegetation may be expected.

Wind movement.--Evaporation of water from land and plant surfaces takes place more rapidly when there is moving air than under calm air conditions. Hot, dry winds and other unusual wind conditions during the growing period will affect the amount of water consumptively used. However, there is a limit in the amount of water that can be utilized. As soon as the land surface is dry, evaporation practically stops and transpiration is limited by the ability of the plants to extract and convey the soil moisture through the plants.

Advection.--Crops grown in irrigated areas surrounded by large arid or semi-arid areas can receive additional energy for vaporization of water by advection. A high percentage of net solar radiation received in arid areas is used in heating the atmosphere. As this warm air mass moves over irrigated areas that are generally cooler, energy contained in the air as sensible heat can be used to evaporate water by vertical turbulent transfer. Thus an "oasis" effect is created. This evaporation of water by vertical turbulent transfer may cause a considerable increase in normal consumptive use in arid areas. It is not believed to be of significance in humid areas.

Stage of Plant Growth.
Other factors being equal, the stage of a crop's growth has a very considerable influence on its consumptive-use rate. This is particu-

larly true for annual crops which generally have three rather distinct stages of growth. These are (1) emergence and development of complete vegetative cover during which time the consumptive-use rate increases rapidly from a low value and approaches its maximum; (2) the period of maximum vegetative cover during which time the consumptive-use rate may be near or at its maximum if abundant soil moisture is available and (3) crop maturation where, for most crops, the consumptive-use rate begins to decrease. During the maturation period, the plant becomes the limiting factor in the transpiration rate.

Available Irrigation Water Supply.
All the above-mentioned factors influece the amount of water that potentially can be consumed in a given area. However, there are other factors that also cause important differences in the consumptive-use rates. Naturally, unless water is available from some source (precipitation, natural ground water, or irrigation), there can be no consumptive use. In those areas of the arid and semi-arid West where the major source is irrigation, both the quantity and seasonal distribution of the available supply will affect consumptive use. Where water is plentiful and cheap, there is a tendency for farmers to over-irrigate. If the soil surface is frequently wet and the resulting evaporation is high, the combined evaporation and transpiration or consumptive use may likewise increase. Also, under more optimum soil moisture conditions, yields of crops may be higher than average and more water consumed.

Quality of Water.
Some investigations have shown that, besides the quantity and seasonal distribution of the water supply, the quality of the water also has a minor effect on the consumptive use. Whether or not plants require more or less water, if the supply is highly saline may be debatable.

Soil Fertility.
If a soil is made more fertile through the application of manure or by some other means, the yields may be expected to increase with an accompanying increase in water use. However, this increase is so small that it is seldom considered when estimating consumptive use.

Estimating Consumptive Use

In areas for which few or no measurements of consumptive use are available, it is usually necessary to estimate consumptive use of crops from climatological data. For this purpose the Soil Conservation Service uses the Blaney-Criddle method with some modifications.

Blaney and Criddle found that the amount of water consumptively used by crops during their normal growing season was closely correlated with mean monthly temperatures and daylight hours. They developed coefficients that can be used to transpose the consumptive use data for a given area to other areas for which only climatological data are available. The net amount of irrigation water necessary to satisfy consumptive use is found by subtracting the effective precipitation from the consumptive water requirement during the growing or irrigation season.

As previously indicated, numerous factors must be taken into consideration if the consumptive use of water is to be determined accurately. Of the climatic factors, the effect of temperature and sunshine upon plant growth as measures of solar radiation is without doubt the most important. Temperature and precipitation records are more readily available than most other climatic data. Records of actual sunshine are not generally available, but the effect of sunshine is very important on the rate of plant growth and the amount of water plants will consume.

The effect of sunshine can be introduced by using the length of days during the crop-growing season at various latitudes. As an example, the length of the daytime at the Equator varies little throughout the year, whereas at 50° N. latitude, the length of the day in summer is much longer than in winter. Thus, at equal temperatures, photosynthesis can take place for several hours longer each June day at the north latitude than at the Equator. Crop growth and water consumption vary with the opportunity for photosynthesis.

The Blaney-Criddle procedure has generally given sufficiently accurate results when used for the purpose for which it was originally developed, that is for estimating seasonal consumptive use. However, the design of irrigation systems, distribution systems, and water storage facilities require that estimates of consumptive use be made for short-time periods of from 5 to 30 days. It has been found that the seasonal crop coefficients previously mentioned are not constant for consecutive short periods throughout the growing season of a crop. Thus it became necessary to make two modifications in the original procedure in order to obtain reasonably accurate estimates of short-period consumptive use.

One modification requires the use of climatic coefficients that are directly related to the mean air temperature for each of the consecutive short periods which constitute the growing season. The other requires the use of coefficients which reflect the influence of the crop growth stages on consumptive-use rates. Both of these modifications are explained in more detail in later paragraphs.

The Blaney-Criddle Formula

Disregarding many influencing factors, consumptive use varies with the temperature, length of day, and available moisture regardless of its source (precipitation, irrigation water, or natural ground water.) Multiplying the mean monthly temperature (t) by the possible monthly percentage of daytime hours of the year (p) gives a monthly consumptive-use factor (f). It is assumed that crop consumptive use varies directly with this factor when an ample water supply is available. Expressed mathematically $u = kf$ and $U = \text{sum of } kf = KF$ where,

U = Consumptive use of the crop in inches for the growing season.

K = Empirical consumptive-use crop coefficient for the growing season. This coefficient varies with the different crops being irrigated.

F = Sum of the monthly consumptive-use factors for the growing season (sum of the products of mean monthly temperature and monthly percentage of daylight hours of the year).

u = Monthly consumptive use of the crop in inches.

k = Empirical consumptive-use crop coefficient for a month (also varies by crops).

f = Monthly consumptive-use factor (product of mean monthly temperature and monthly percentage of daylight hours of the year).

$f = \frac{t \times p}{100}$, where

t = Mean monthly air temperature in degrees Fahrenheit.

p = Monthly percentage of daylight hours in the year. Values of (p) for latitudes 0 to 65 degrees north of the Equator are shown in Table 1.

Note: Values of (t), (p), (f), and (k), can also be made to apply to periods of less than a month.

Following are modifications made in the original formula:

$k = k_t \times k_c$, where,

k_t = A climatic coefficient which is related to the mean air temperature (t).

$k_t = .0173t - .314$. Values of k_t for mean air temperatures from 36 to 100 degrees are shown in table 4.

k_c = A coefficient reflecting the growth stage of the crop. Values are obtained from crop growth stage coefficient curves such as those shown in figures 1 and 2.

The consumptive-use factor (F) may be computed for areas for which monthly temperature records are available, if the percentage of hours that is shown in table 1 is used. Then, the total crop consumptive use (U) is obtained by multiplying (F) by the empirical consumptive-use crop coefficient (K). This relationship allows the computation of seasonal consumptive use at any location for those crops for which values of (K) have been experimentally established or can be estimated.

Seasonal Comsumptive-Use Coefficients.
Consumptive-use coefficients (K) have been determined experimentally at

numerous localities for most crops grown in the western states. Consumptive-use values (U) were measured and these data were correlated with temperature and growing season. Crop consumptive-use coefficients (K) were then computed by the formula, K = U/F. The computed coefficients varied somewhat because of the diverse conditions (such as soils, water supply, and methods) under which the studies were conducted. These coefficients were adjusted where necessary after the data were analyzed. The resulting coefficients are believed to be suitable for use under normal conditions.

While only very limited investigations of consumptive use have been made in the eastern or humid-area states, studies made thus far fail to indicate that there should be any great difference between the seasonal consumptive-use coefficients used there and those used in the western states.

Table 2 shows the values of seasonal consumptive-use crop coefficients currently proposed by Blaney and Criddle for most irrigated crops. It will be noted that ranges in the values of these coefficients are shown. These, however, are not all inclusive limits. In some circumstances K values may be either higher or lower than shown.

Monthly or Short-Time Consumptive-Use Coefficients.

Although seasonal coefficients (K) as reported by various investigators show some variation for the same crops, monthly or short-time coefficients (k) show even greater variation. These great variations are influenced by a number of factors which must be given consideration when computing or estimating short-time coefficients. Although these factors are numerous, the most important are temperature and the growth stage of the crop. These factors are discussed in succeeding paragraphs.

Growing Season.

In utilizing the Blaney-Criddle formula for computing seasonal requirements, the potential growing season for the various crops is normally considered to extend from frost to frost or from the last killing frost in the spring to the end of a definite period of time thereafter. For most crops, this is adequate for seasonal use estimates, but a refinement is necessary to more precisely define the growing season when monthly or short-time use estimates are required. In many areas, records are available from which planting, harvesting and growth dates can be determined. These should be used where possible. In other areas, temperature data may be helpful for estimating these dates. Table 3 contains some guides which may be helpful in determining these dates.

Since the spring frost date corresponds very nearly with a mean temperature of 55 degrees, it is obvious that many of the common crops use appreciable amounts of water prior to the last frost in the spring and may continue to use water after the first frost in the fall.

Climatic Coefficient (k_t).

While it is recognized that a number of climatological factors have an

Table 1.--Monthly percentage of daytime hours (p) of the year for latitudes 18° to 65° north of the equator.

Latitude North	Jan.	Feb.	Mar.	Apr.	May	June	July	Aug.	Sept.	Oct.	Nov.	Dec.
65°	3.52	5.13	7.96	9.97	12.72	14.15	13.59	11.18	8.55	6.53	4.08	2.62
64°	3.81	5.27	8.00	9.92	12.50	13.63	13.26	11.08	8.56	6.63	4.32	3.02
63°	4.07	5.39	8.04	9.86	12.29	13.24	12.97	10.97	8.56	6.73	4.52	3.36
62°	4.31	5.49	8.07	9.80	12.11	12.92	12.73	10.87	8.55	6.80	4.70	3.65
61°	4.51	5.58	8.09	9.74	11.94	12.66	12.51	10.77	8.55	6.88	4.86	3.91
60°	4.70	5.67	8.11	9.69	11.78	12.41	12.31	10.68	8.54	6.95	5.02	4.14
59°	4.86	5.76	8.13	9.64	11.64	12.19	12.13	10.60	8.53	7.00	5.17	4.35
58°	5.02	5.84	8.14	9.59	11.50	12.00	11.96	10.52	8.53	7.06	5.30	4.54
57°	5.17	5.91	8.15	9.53	11.38	11.83	11.81	10.44	8.52	7.13	5.42	4.71
56°	5.31	5.98	8.17	9.48	11.26	11.68	11.67	10.36	8.52	7.18	5.52	4.87
55°	5.44	6.04	8.18	9.44	11.15	11.53	11.54	10.29	8.51	7.23	5.63	5.02
54°	5.56	6.10	8.19	9.40	11.04	11.39	11.42	10.22	8.50	7.28	5.74	5.16
53°	5.68	6.16	8.20	9.36	10.94	11.26	11.30	10.16	8.49	7.32	5.83	5.30
52°	5.79	6.22	8.21	9.32	10.85	11.14	11.19	10.10	8.48	7.36	5.92	5.42
51°	5.89	6.27	8.23	9.28	10.76	11.02	11.09	10.05	8.47	7.40	6.00	5.54
50°	5.99	6.32	8.24	9.24	10.68	10.92	10.99	9.99	8.46	7.44	6.08	5.65
49°	6.08	6.36	8.25	9.20	10.60	10.82	10.90	9.94	8.46	7.48	6.16	5.75
48°	6.17	6.41	8.26	9.17	10.52	10.72	10.81	9.89	8.45	7.51	6.24	5.85
47°	6.25	6.45	8.27	9.14	10.45	10.63	10.73	9.84	8.44	7.54	6.31	5.95
46°	6.33	6.50	8.28	9.11	10.38	10.53	10.65	9.79	8.43	7.58	6.37	6.05
45°	6.40	6.54	8.29	9.08	10.31	10.46	10.57	9.75	8.42	7.61	6.43	6.14
44°	6.48	6.57	8.29	9.05	10.25	10.39	10.49	9.71	8.41	7.64	6.50	6.22
43°	6.55	6.61	8.30	9.02	10.19	10.31	10.42	9.66	8.40	7.67	6.56	6.31
42°	6.61	6.65	8.30	8.99	10.13	10.24	10.35	9.62	8.40	7.70	6.62	6.39
41°	6.68	6.68	8.31	8.96	10.07	10.16	10.29	9.59	8.39	7.72	6.68	6.47
40°	6.75	6.72	8.32	8.93	10.01	10.09	10.22	9.55	8.39	7.75	6.73	6.54
39°	6.81	6.75	8.33	8.91	9.95	10.03	10.16	9.51	8.38	7.78	6.78	6.61
38°	6.87	6.79	8.33	8.89	9.90	9.96	10.11	9.47	8.37	7.80	6.83	6.68
37°	6.92	6.82	8.34	8.87	9.85	9.89	10.05	9.44	8.37	7.83	6.88	6.74
36°	6.98	6.85	8.35	8.85	9.80	9.82	9.99	9.41	8.36	7.85	6.93	6.81
35°	7.04	6.88	8.35	8.82	9.76	9.76	9.93	9.37	8.36	7.88	6.98	6.87
34°	7.10	6.91	8.35	8.80	9.71	9.71	9.88	9.34	8.35	7.90	7.02	6.93
33°	7.15	6.94	8.36	8.77	9.67	9.65	9.83	9.31	8.35	7.92	7.06	6.99
32°	7.20	6.97	8.36	8.75	9.62	9.60	9.77	9.28	8.34	7.95	7.11	7.05
31°	7.25	6.99	8.36	8.73	9.58	9.55	9.72	9.24	8.34	7.97	7.16	7.11
30°	7.31	7.02	8.37	8.71	9.54	9.49	9.67	9.21	8.33	7.99	7.20	7.16
29°	7.35	7.05	8.37	8.69	9.50	9.44	9.62	9.19	8.33	8.00	7.24	7.22
28°	7.40	7.07	8.37	8.67	9.46	9.39	9.58	9.17	8.32	8.02	7.28	7.27
27°	7.44	7.10	8.38	8.66	9.41	9.34	9.53	9.14	8.32	8.04	7.32	7.32
26°	7.49	7.12	8.38	8.64	9.37	9.29	9.49	9.11	8.32	8.06	7.36	7.37
25°	7.54	7.14	8.39	8.62	9.33	9.24	9.45	9.08	8.31	8.08	7.40	7.42
24°	7.58	7.16	8.39	8.60	9.30	9.19	9.40	9.06	8.31	8.10	7.44	7.47
23°	7.62	7.19	8.40	8.58	9.26	9.15	9.36	9.04	8.30	8.12	7.47	7.51
22°	7.67	7.21	8.40	8.56	9.22	9.11	9.32	9.01	8.30	8.13	7.51	7.56
21°	7.71	7.24	8.41	8.55	9.18	9.06	9.28	8.98	8.29	8.15	7.55	7.60
20°	7.75	7.26	8.41	8.53	9.15	9.02	9.24	8.95	8.29	8.17	7.58	7.65
19°	7.79	7.28	8.41	8.51	9.12	8.97	9.20	8.93	8.29	8.19	7.61	7.70
18°	7.83	7.31	8.41	8.50	9.08	8.93	9.16	8.90	8.29	8.20	7.65	7.74

Table 1.--Monthly percentage of daytime hours (p) of the year for latitudes 0° to 20° north of the equator

Latitude North	Jan.	Feb.	Mar.	Apr.	May	June	July	Aug.	Sept.	Oct.	Nov.	Dec.
20°	7.75	7.26	8.41	8.53	9.15	9.02	9.24	8.95	8.29	8.17	7.58	7.65
19°	7.79	7.28	8.41	8.51	9.12	8.97	9.20	8.93	8.29	8.19	7.61	7.70
18°	7.83	7.31	8.41	8.50	9.08	8.93	9.16	8.90	8.29	8.20	7.65	7.74
17°	7.87	7.33	8.42	8.48	9.04	8.89	9.12	8.88	8.28	8.22	7.68	7.79
16°	7.91	7.35	8.42	8.47	9.01	8.85	9.08	8.85	8.28	8.23	7.72	7.83
15°	7.94	7.37	8.43	8.45	8.98	8.81	9.04	8.83	8.27	8.25	7.75	7.88
14°	7.98	7.39	8.43	8.43	8.94	8.77	9.00	8.80	8.27	8.27	7.79	7.93
13°	8.02	7.41	8.43	8.42	8.91	8.73	8.96	8.78	8.26	8.29	7.82	7.97
12°	8.06	7.43	8.44	8.40	8.87	8.69	8.92	8.76	8.26	8.31	7.85	8.01
11°	8.10	7.45	8.44	8.39	8.84	8.65	8.88	8.73	8.26	8.33	7.88	8.05
10°	8.14	7.47	8.45	8.37	8.81	8.61	8.85	8.71	8.25	8.34	7.91	8.09
9°	8.18	7.49	8.45	8.35	8.77	8.57	8.81	8.68	8.25	8.36	7.95	8.14
8°	8.21	7.51	8.45	8.34	8.74	8.53	8.78	8.66	8.25	8.37	7.98	8.18
7°	8.25	7.53	8.46	8.32	8.71	8.49	8.74	8.64	8.25	8.38	8.01	8.22
6°	8.28	7.55	8.46	8.31	8.68	8.45	8.71	8.62	8.24	8.40	8.04	8.26
5°	8.32	7.57	8.47	8.29	8.65	8.41	8.67	8.60	8.24	8.41	8.07	8.30
4°	8.36	7.59	8.47	8.28	8.62	8.37	8.64	8.57	8.23	8.43	8.10	8.34
3°	8.40	7.61	8.48	8.26	8.58	8.33	8.60	8.55	8.23	8.45	8.13	8.38
2°	8.43	7.63	8.49	8.25	8.55	8.29	8.57	8.53	8.22	8.46	8.16	8.42
1°	8.47	7.65	8.49	8.23	8.52	8.25	8.53	8.51	8.22	8.48	8.19	8.46
0°	8.50	7.67	8.49	8.22	8.49	8.22	8.50	8.49	8.21	8.49	8.22	8.50

Table 2.--Seasonal consumptive-use crop coefficients (K) for irrigated crops

Crop	Length of Normal Growing Season or Period 1/	Consumptive-use coefficient (K) 2/
Alfalfa	Between frosts	0.80 to 0.90
Bananas	Full year	.80 to 1.00
Beans	3 months	.60 to .70
Cocoa	Full year	.70 to .80
Coffee	Full year	.70 to .80
Corn (Maize)	4 months	.75 to .85
Cotton	7 months	.60 to .70
Dates	Full year	.65 to .80
Flax	7 to 8 months	.70 to .80
Grains, small	3 months	.75 to .85
Grain, sorghums	4 to 5 months	.70 to .80
Oilseeds	3 to 5 months	.65 to .75
Orchard crops:		
Avocado	Full year	.50 to .55
Grapefruit	Full year	.55 to .65
Orange and lemon	Full year	.45 to .55
Walnuts	Between frosts	.60 to .70
Deciduous	Between frosts	.60 to .70
Pasture crops:		
Grass	Between frosts	.75 to .85
Ladino whiteclover	Between frosts	.80 to .85
Potatoes	3 to 5 months	.65 to .75
Rice	3 to 5 months	1.00 to 1.10
Soybeans	140 days	.65 to .70
Sugar beet	6 months	.65 to .75
Sugarcane	Full year	.80 to .90
Tobacco	4 months	.70 to .80
Tomatoes	4 months	.65 to .70
Truck crops, small	2 to 4 months	.60 to .70
Vineyard	5 to 7 months	.50 to .60

1/ Length of season depends largely on variety and time of year when the crop is grown. Annual crops grown during the winter period may take much longer than if grown in the summertime.

2/ The lower values of (K) for use in the Blaney-Criddle formula, $U = KF$, are for the more humid areas, and the higher values are for the more arid climates.

effect on consumptive use by crops, seldom is complete climatological data on relative humidity, wind movement, sunshine hours, pan evaporation, etc., available for a specific site. Thus it is necessary to rely on records of temperature which are widely available.

In 1954 J. T. Phelan attempted to correlate the monthly consumptive-use coefficient (k) with the mean monthly temperature (t). It was noted that a loop effect occurred in the plotted points; the computed values of (k) were higher in the spring than in the fall for the same temperature. The effects of this loop were later corrected by the development of a crop growth stage coefficient (k_c). The relationship between (k) and (t) was adopted for computing values of (k_t), the temperature coefficient. This relationship is expressed as $k_t = .0173_t - .314$. Table 4 gives values of (k_t) for temperatures ranging from 36 to 100 degrees Fahrenheit.

Crop Growth Stage Coefficients (k_c).

As previously stated, another factor which causes consumptive use to vary widely throughout the growing season is the plant itself. Stage of growth is a primary variable that must be recognized since it is obvious that plants in the rapid growth stage will use water at a more rapid rate than will new seedlings. It is also obvious that these variations in consumptive use throughout the growing season will be greater for annual crops than for perennial crops such as alfalfa, permanent pasture grasses and orchards.

In order to recognize these variations in consumptive use, crop growth stage coefficients (k_c) have been introduced into the formula. Values of these coefficients are calculated from research data. When values of (k_c) are plotted against time or stage of growth, curves similar to those shown in figures 1 and 2 will result. Such curves are used to obtain values of (k_c) which, when used with appropriate values of (k_t), will permit a determination of values of monthly or short-time consumptive-use coefficients (k).

It is also recognized that value of (k_c) might, to some extent, be influenced by factors other than the characteristics of the plant itself. For this reason, it is not expected that these curves can be used universally. They should, however, be valid over a considerable area and certainly should be of value in areas where no measured consumptive-use data is available.

With annual crops, such as corn, values of the coefficient (k_c) are best plotted as a function of a percentage of the growing season. Figure 1 shows the suggested values of (k_c) for corn.

With perennial crops, values of the coefficient (k_c) are usually best plotted on a monthly basis. Figure 2 shows the plotting of such values for alfalfa. Crop growth stage coefficient curves for all crops for which data are available are contained in the appendix.

Table 3.--A guide for determining planting dates, maturity dates and lengths of growing seasons as related to mean air temperature

Crop	Earliest moisture-use or planting date as related to mean air temperature	Latest moisture-use or maturing date as related to mean air temperature	Growing Season Days
Perennial Crops			
Alfalfa	50° mean temp.	28° frost	Variable
Grasses, cool	45° mean temp.	45° mean temp.	Variable
Orchards, decid.	50° mean temp.	45° mean temp.	Variable
Grapes	55° mean temp.	50° mean temp.	Variable
Annual Crops			
Beans, dry	60° mean temp.	32° frost	90 - 100
Corn	55° mean temp.	32° frost	140 - Max.
Cotton	62° mean temp.	32° frost	240 - Max.
Grain, spring	45° mean temp.	32° frost	130 - Max.
Potatoes, late	60° mean temp.	32° frost	130 - Max.
Sorghum, grain	60° mean temp.	32° frost	130 - Max.
Sugar beets	28° frost	28° frost	180 - Max.
Wheat, winter			
(Fall season)		45° mean temp.	
(Spring season)	45° mean temp.		

Assumptions in Applying the Formula.
In order to apply results of a consumptive-use-of-water study in one area to other areas, it is usually necessary to make certain assumptions. As previously indicated, if sufficient basic information is available locally, such actual data should be used. But rarely are all needed data known in sufficient detail. Where necessary information is lacking, the following assumptions must be made in applying the consumptive-use formula to transfer data between areas:

1. Seasonal consumptive use (U) of water varies directly with the consumptive-use factor (F).

2. Crop growth and yields are not limited by inadequate water at any time during the growing season.

3. Growing periods for alfalfa, pasture, orchard crops, and "natural" vegetation, although usually extending beyond the frost-free periods, are usually indicated by such periods. Yields of crops dependent only upon vegetative growth vary with the length of the growing period.

Application to Specific Areas.
The application of the Blaney-Criddle formula to specific areas can best be illustrated by examples. Two have been chosen for this purpose. The first is an annual crop, corn, grown in a humid area, Raleigh, North Carolina. The second is a perennial crop, alfalfa, grown in an arid area, Denver, Colorado.

Corn at Raleigh, N. C.--The procedure for estimating the average daily, monthly and seasonal consumptive use by corn at this location is shown in Sample Calculation No. 1. The average length of the growing season for corn grown in the vicinity of Raleigh is 120 days beginning about April 20.

The estimate is made on a monthly basis, the months and fractions thereof being shown in column 1. The midpoint date for each month or fraction is shown in column 2. The accumulated number of days from the planting date, April 20, to the midpoint of each month or period is shown in column 3. The percentage of the 120-day growing season represented by these midpoint dates is shown in column 4. Thus Col. 4 = Col. 3 ÷ 120.

Mean monthly air temperature values, shown in column 5, are taken from Weather Bureau records. The mean temperature is assumed to occur on the 15th day of each month. The mean air temperature for a part of a month can be obtained mathematically or graphically by assuming that the increase or decrease in temperature between the 15th day of any consecutive month is a straight-line relationship. For example, at Raleigh, the mean monthly air temperature for April is 60.6° and that

Table 4.--Values of the climatic coefficient, k_t, 1/ for various mean air temperature, t.

t °F	k_t	t °F	k_t	t °F	k_t
36	.31	61	.74	86	1.17
37	.33	62	.76	87	1.19
38	.34	63	.78	88	1.21
39	.36	64	.79	89	1.23
40	.38	65	.81	90	1.24
41	.40	66	.83	91	1.26
42	.41	67	.85	92	1.28
43	.43	68	.86	93	1.30
44	.45	69	.88	94	1.31
45	.46	70	.90	95	1.33
46	.48	71	.91	96	1.35
47	.50	72	.93	97	1.36
48	.52	73	.95	98	1.38
49	.53	74	.97	99	1.40
50	.55	75	.98	100	1.42
51	.57	76	1.00		
52	.59	77	1.02		
53	.60	78	1.04		
54	.62	79	1.05		
55	.64	80	1.07		
56	.66	81	1.09		
57	.67	82	1.11		
58	.69	83	1.12		
59	.71	84	1.14		
60	.72	85	1.16		

1/ Values of (k_t) are based on the formula, $k_t = .0173\ t - .314$

for mean temperatures less than 36°, use $k_t = .300$.

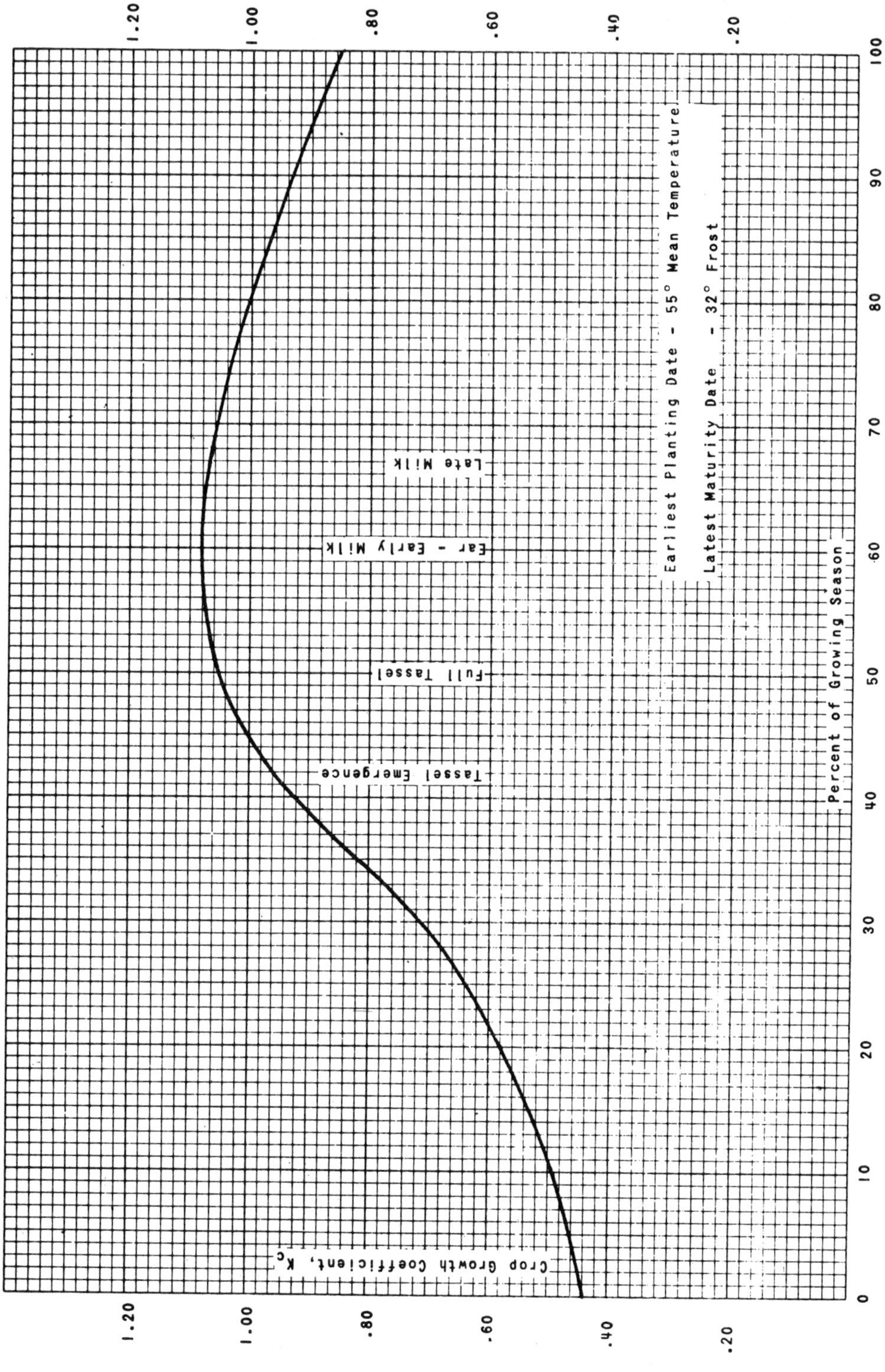

Crop growth stage coefficient curve for corn (grain)

Curve No. 1

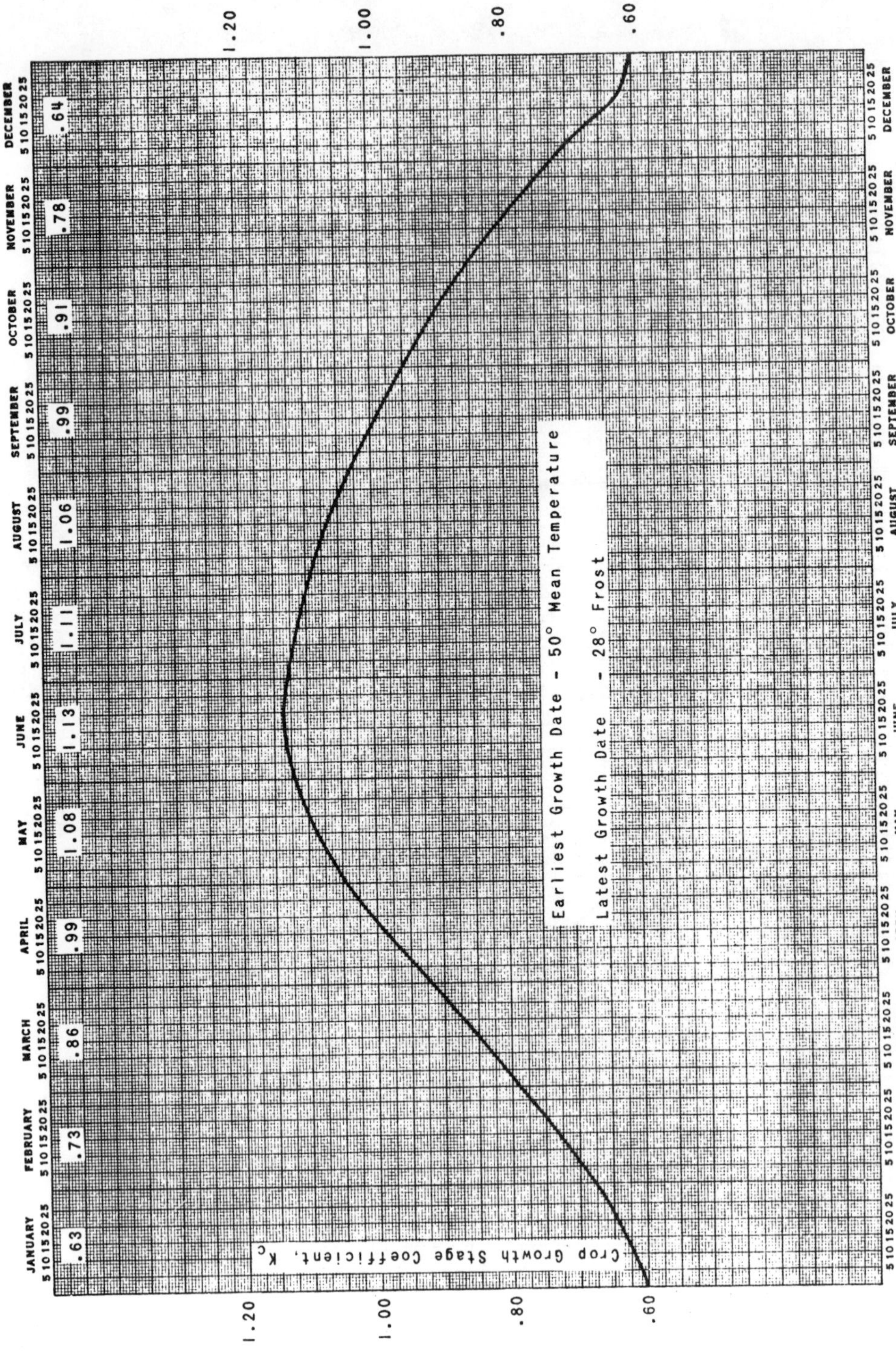

Crop growth stage coefficient curve for alfalfa

Curve No. 2

for May is 69.2°. The mean air temperature for the midpoint date is calculated as follows:

$$60.6° + \frac{10 \text{ days } (69.2° - 60.6°)}{30 \text{ days}} = 63.5°$$

Raleigh is located at Latitude 35° - 47' N. The monthly percentages of daylight hours, shown in column 6, are taken from table 1. For parts of a month the values of these percentages can be obtained in a similar manner as described for mean air temperatures. For example, at Raleigh, the monthly percentage of daylight hours for April is 8.84 and that for May is 9.79. For the period April 20 through April 30, the monthly percentage of daylight hours is calculated as follows:

$$\left[8.84\% + \frac{10 \text{ days } (9.79\% - 8.84\%)}{30 \text{ days}}\right] \frac{10 \text{ days}}{30 \text{ days}} = 3.05\%$$

The values of consumptive use factors (f) shown in column 7 are the product of (t) and (p) divided by 100. Values of the climatic coefficient (k_t), shown in column 8, are taken from table 4. Values of the crop growth stage coefficient (k_c), shown in column 9, are taken from the curve shown in figure 1. The values of the monthly consumptive-use coefficient (k), shown in column 10, are the product of (k_t) and (k_c). Values of monthly consumptive use (u), shown in column 11, are the product of values of (k) and (f). The average daily rates of consumptive use shown in column 12 are the monthly values of (u) (column 11) divided by the number of days in the month.

At this point the computed value of the seasonal consumptive-use coefficient (K) should be determined. This is done by totaling column 11 to obtain the seasonal consumptive-use value, $U = \Sigma u$, and by totaling column 7 to obtain the seasonal consumptive use factor, $F = \Sigma f$. The value of the seasonal consumptive-use coefficient (K), shown at the bottom of column 10, is then obtained by dividing the seasonal value of (U) by the seasonal value of (F). If the resulting value of (K) falls within the range of values for corn shown in table 2, the calculations may be assumed to have provided a reasonable estimate of consumptive use. If the resulting value of (K) falls outside the range shown in table 2, an adjustment of the crop growth stage coefficient curve may be indicated. In such cases, a qualified irrigation engineer or specialist should be consulted.

Alfalfa at Denver, Colo.--The procedure for estimating the average daily, monthly and seasonal consumptive use by alfalfa in this location is shown in Sample Calculation No. 2. The growing season for alfalfa grown near Denver is considered to be that period from the date corresponding to 50° mean temperature in the spring to the date corresponding to 28° frost in the fall. This period is from April 24 to October 25.

The procedure illustrated by Sample Calculation No. 2 is the same as that heretofore described for corn and illustrated by Sample Calculation

No. 1. The values of the crop growth stage coefficient, (k_c), shown in column 8 are taken from the curve for alfalfa shown in figure 2.

Peak Period Consumptive Use

Information on peak period rates of consumptive use is needed to properly design irrigation systems. It is used to determine the minimum capacity requirements of main and lateral canals, pipelines, and other water conveyance or control structures. The peak period rates of water use by crops also influence the administration of streams and reservoirs from which irrigation water supplies are obtained.

In irrigation project design, the peak period of consumptive use is the period during which the weighted average daily rate of consumptive use of the various crops grown in the project area is at a maximum. Different crops may have their peak rates of use at different times. Therefore, some crops may not be using water at their maximum rate during the project peak period. In fact, some of the crops may not even be grown during this period.

Factors Influencing Peak Period Use Rates.
While other factors may have a minor influence on peak period rates of consumptive use, peak period air temperature and net depth of irrigation application have the greatest influence.

Temperature.--An analysis of daily mean air temperature records for any month at any location will show that the mean temperature for the warmest consecutive 5-day period will be greater than that for the warmest 10-day period. Likewise, the mean temperature for the warmest consecutive 10-day period will be greater than that for the warmest 15-day period, and so on. All will be greater than the mean monthly temperature. Since consumptive use, as estimated by the Blaney-Criddle formula, is directly related to air temperature, it is obvious that the shorter the peak period is in days, the greater will be the mean temperature and therefore the greater will be the consumptive-use rate.

Net irrigation application.--The length of the peak period is that number of days in which the normal net irrigation application will last under the peak rate of use for the period. Thus smaller net irrigation applications will last for smaller periods of time and, as shown above, will result in greater peak period-use rates. Conversely, higher net irrigation applications will result in lower peak period-use rates.

Table 5 shows peak-period average daily consumptive-use rates as related to estimated monthly use and net irrigation application. As an illustration of the use of this table, the case of alfalfa irrigated near Denver may be used. From Sample Calculation No. 2, it will be noted that the peak-use month is July and that the average consumptive use for that month is 7.8 inches. From the Colorado irrigation guide, it

Sample Calculation No. 1.--Estimate of average daily, monthly and seasonal consumptive-use
by
Corn at Raleigh, North Carolina
(Harvested for grain)

Lat. 35° 47' North

(1) Month or Period	(2) Midpoint of Period	(3) Accum. Days to Midpoint	(4) Percent of Growing Season	(5) Mean Air Temp. t. ° F.	(6) Daylight Hours p. Percent	(7) Cons. Use Factor f.	(8) Climatic Coeff. k_t	(9) Growth Stage Coeff. k_c	(10) Cons. Use Coeff. k	(11) Monthly Cons. Use u. Inches	(12) Daily Cons. Use u. In./Day
Apr. 20											
	Apr. 25	5	4.2	63.5	3.05	1.94	.79	.46	.36	.70	.070
May											
	May 15	25	20.8	69.2	9.79	6.77	.88	.59	.52	3.52	.114
June											
	June 15	56	46.7	76.9	9.81	7.54	1.02	1.02	1.04	7.84	.261
July											
	July 15	86	71.7	79.4	9.98	7.92	1.06	1.05	1.11	8.79	.284
Aug.											
	Aug. 9	111	92.5	78.3	5.52	4.32	1.04	.91	.95	4.10	.228
Aug. 18											
Season Totals						28.49			.88	24.95	

Sample Calculation No. 2.--Estimate of average, daily, monthly and seasonal consumptive-use by Alfalfa at Denver, Colorado

Lat. 39° 40' North

(1) Month or Period	(2) Midpoint of Period	(3) Days in Period	(4) Mean Air Temp. t. ° F.	(5) Daylight Hours p. Percent	(6) Cons. Use Factor f	(7) Climatic Coeff. k_t	(8) Growth Stage Coeff. k_c	(9) Cons. Use Coeff. k	(10) Monthly Cons. Use u Inches	(11) Daily Cons. Use u In./Day
Apr. 24										
	Apr. 27	6	51.1	1.87	.96	.57	1.03	.59	.57	.095
May										
	May 15	31	56.3	9.99	5.62	.66	1.08	.71	3.99	.129
June										
	June 15	30	66.4	10.07	6.69	.84	1.13	.95	6.36	.212
July										
	July 15	31	72.8	10.20	7.43	.95	1.11	1.05	7.80	.252
Aug.										
	Aug. 15	31	71.3	9.54	6.80	.92	1.06	.98	6.66	.215
Sept.										
	Sept. 15	30	62.7	8.39	5.26	.77	.99	.76	4.00	.133
Oct.										
	Oct. 12	25	53.5	6.31	3.38	.61	.91	.56	1.89	.076
Oct. 25										
Season Totals					36.14			.87	31.27	

is determined that the net irrigation application is 4.2 inches. Thus, by interpolation from the table, the peak-period use rate is found to be 0.28 inch per day.

A suggested procedure for using table 5 to estimate project peak period consumptive-use requirements is outlined below and is illustrated by Sample Calculation No. 3.

1. Determine the net irrigation applications (I) required for the crops in the project area.

2. Determine the monthly consumptive-use rate (u_m) for each crop in the project area for the month of greatest overall water use. (Note that if some crop is using water for only a portion of the month, its rate of use must be computed by dividing the estimated requirement by the fraction of the month when water is used).

3. From table 5, using appropriate values of (I) and (u_m), determine the peak period consumptive-use rate (u_p) for each of the three crops.

4. Using the crop and soil distribution patterns established for the project, compute the weighted peak period consumptive-use rate for the project.

Effective Rainfall

Effective rainfall supplies a portion of the consumptive use by crops. It may be a nearly insignificant portion in arid areas such as the Salt River Valley of Arizona or it may be a major portion in humid areas such as the Atlantic Coastal Plain of the Carolinas. The engineer engaged in estimating irrigation water requirements of a crop is confronted with the problem of determining what portion of total consumptive use will be furnished by effective rainfall and what portion will have to be supplied by irrigation. Since there are no records of effective rainfall available, it is necessary to utilize total rainfall records and estimate the portion of total rainfall that is effective. A procedure for doing this is described in succeeding paragraphs.

Factors Influencing Rainfall Effectiveness.

Total rainfall.--In arid areas where total growing season precipitation is light, the moisture level in the soil profile at the time precipitation occurs is usually such that almost all of it enters the soil profile and becomes available for consumptive use. Losses due to surface runoff or to percolation below the crop root zone are usually negligible. Thus the effectiveness of rainfall in these areas is relatively high.

In humid areas, storms of large magnitude and high intensity occur frequently during the growing season. These storms often produce water in excess of that which can be stored in the soil profile for consumptive

Table 5.--Peak period average daily consumptive use rates (u_p) as related to estimated actual monthly use (u_m)

Net Irrigation Application I (Inches)	Computed Peak Monthly Consumptive Use Rate (u_m) in Inches 1/																
	4.0	4.5	5.0	5.5	6.0	6.5	7.0	7.5	8.0	8.5	9.0	9.5	10.0	10.5	11.0	11.5	12.0
	Peak Period Daily Use Rate (u_p) in Inches per Day																
1.0	.15	.18	.20	.22	.24	.26	.28	.31	.33	.35	.37	.40	.42	.44	.46	.49	.51
1.5	.15	.17	.19	.21	.23	.25	.27	.29	.32	.34	.36	.38	.41	.43	.45	.47	.50
2.0	.15	.16	.18	.20	.23	.25	.27	.29	.31	.33	.35	.37	.39	.41	.44	.46	.48
2.5	.14	.16	.18	.20	.22	.24	.26	.28	.30	.32	.34	.36	.39	.41	.43	.45	.47
3.0	.14	.16	.18	.20	.22	.24	.26	.28	.30	.32	.34	.36	.38	.40	.42	.44	.46
3.5	.14	.16	.18	.19	.21	.23	.25	.27	.29	.31	.33	.35	.37	.39	.41	.44	.46
4.0	.14	.15	.17	.19	.21	.23	.25	.27	.29	.31	.33	.35	.37	.39	.41	.43	.45
4.5	.14	.15	.17	.19	.21	.23	.25	.27	.29	.31	.33	.35	.37	.39	.41	.43	.45
5.0	.13	.15	.17	.19	.21	.23	.25	.26	.28	.30	.32	.34	.36	.38	.40	.42	.44
5.5	.13	.15	.17	.19	.21	.22	.24	.26	.28	.30	.32	.34	.36	.38	.40	.42	.44
6.0	.13	.15	.17	.19	.20	.22	.24	.26	.28	.30	.32	.34	.36	.38	.40	.41	.43

1/ Based on the formula $u_p = 0.034\ u_m^{1.09}\ I^{-.09}$ where
u_p = Average daily peak period consumptive use in inches.
u_m = Average consumptive use for the peak month in inches.
I = Net irrigation application in inches.

Sample Calculation No. 3.--Estimate of project peak consumptive use rates for a project near Boise, Idaho

Item		Unit	Soil Mapping Units 1 M 3	1 S 6	Total
Soil Areas		Acres	930	570	1500
		Percent	62	38	100
Available Moisture Holding Capacity		In./ft.	2.2	1.4	
Crops					
Alfalfa		Acres			975
Small grain		Acres			375
Potatoes		Acres			150
Crop Root Zone Depths					
Alfalfa		Ft.	5	5	
Small grain		Ft.	4	4	
Potatoes		Ft.	3	3	
Net Irrigation Application (I)					
Alfalfa		In.	5.5	3.5	
Small grain		In.	4.4	2.8	
Potatoes		In.	2.6	1.7	
Consumptive Use Rate for Peak Month (u_m)					
Alfalfa		In./mo.	8.5	8.5	
Small grain (1.89" in 17 days)		In./mo.	3.4	3.4	
Potatoes		In./mo.	7.4	7.4	
Peak Period Consumptive Use Rates (u_p) (from table 5)					
Alfalfa		In./day	.30	.31	
Small grain		In./day	.13	.13	
Potatoes		In./day	.28	.29	
Weighted for crop distribution					
Alfalfa	65%	In./day	.195	.202	
Small grain	25%	In./day	.033	.033	
Potatoes	10%	In./day	.028	.029	
All crops	100%	In./day	.256	.264	
Weighted for soils distribution		In./day	.159	.100	.259

use. This excess is lost either to surface runoff or to percolation below the root zone depth. When such storms occur soon after an application of irrigation water has been made, almost all of the rainfall may be lost. Thus in areas of high total growing season rainfall, the effectiveness of rainfall is low by comparison.

For example at Albuquerque, New Mexico, where the average total growing season rainfall is only 8.0 inches, the average rainfall effectiveness is 92 percent. At Baton Rouge, Louisiana, the average total growing season rainfall is 39.4 inches but the average rainfall effectiveness is only 64 percent.

Consumptive-use rate.--Where the consumptive-use rate of a crop is high, available moisture in the soil profile is depleted rapidly, thus providing storage capacity at a relatively rapid rate for receiving rainfall. Should a substantial storm occur, the amount of water required to bring the moisture in the profile back to the field capacity level would be relatively large and the losses due to runoff and/or deep percolation would be relatively small. Conversely, where the consumptive-use rate is low, storage capacity for rainfall is provided at a slower rate. When a storm occurs, there is less capacity in the profile available to receive water and thus the losses will be relatively large. Said briefly, the higher the rate of consumptive use, the greater will be the rainfall effectiveness. Conversely, the lower the rate of consumptive use, the lower will be the effectiveness of rainfall.

Net irrigation application.--As previously stated, the net irrigation application is dependent upon the capacity of the soil profile at root zone depth to store readily available moisture for plant use. When this capacity is low and a storm of considerable magnitude occurs, only a small percentage of the precipitation may be needed to fill the soil profile to field capacity and the resulting rainfall effectiveness will be low. Conversely, if the capacity is high, all or most of the rainfall resulting from such a storm might be stored in the profile before the field capacity level is reached. In this case, the effectiveness of rainfall would be relatively high.

Monthly Effective Rainfall.

Curves and tables have been developed to show the relationship between effective rainfall and the three variable factors discussed previously (see figure 3 and table 6). Either the curves or the table may be used with the same result and both are presented in order to give the user a choice. The curves and the table show the relationship between average monthly effective rainfall (r_e), mean monthly rainfall (r_t), and average monthly consumptive use (u). The values of (r_e) are based on a 3-inch net irrigation application. Factors for converting to other net depths of application are presented. For example, a crop of corn grown on a sandy soil has a net depth of application of 2.0 inches. Average consumptive use for the month of July is 8.79 inches and mean July rainfall is 5.85 inches. From figure 3 or table 6, the average effective rainfall for July is 4.91 x 0.93 = 4.57 inches.

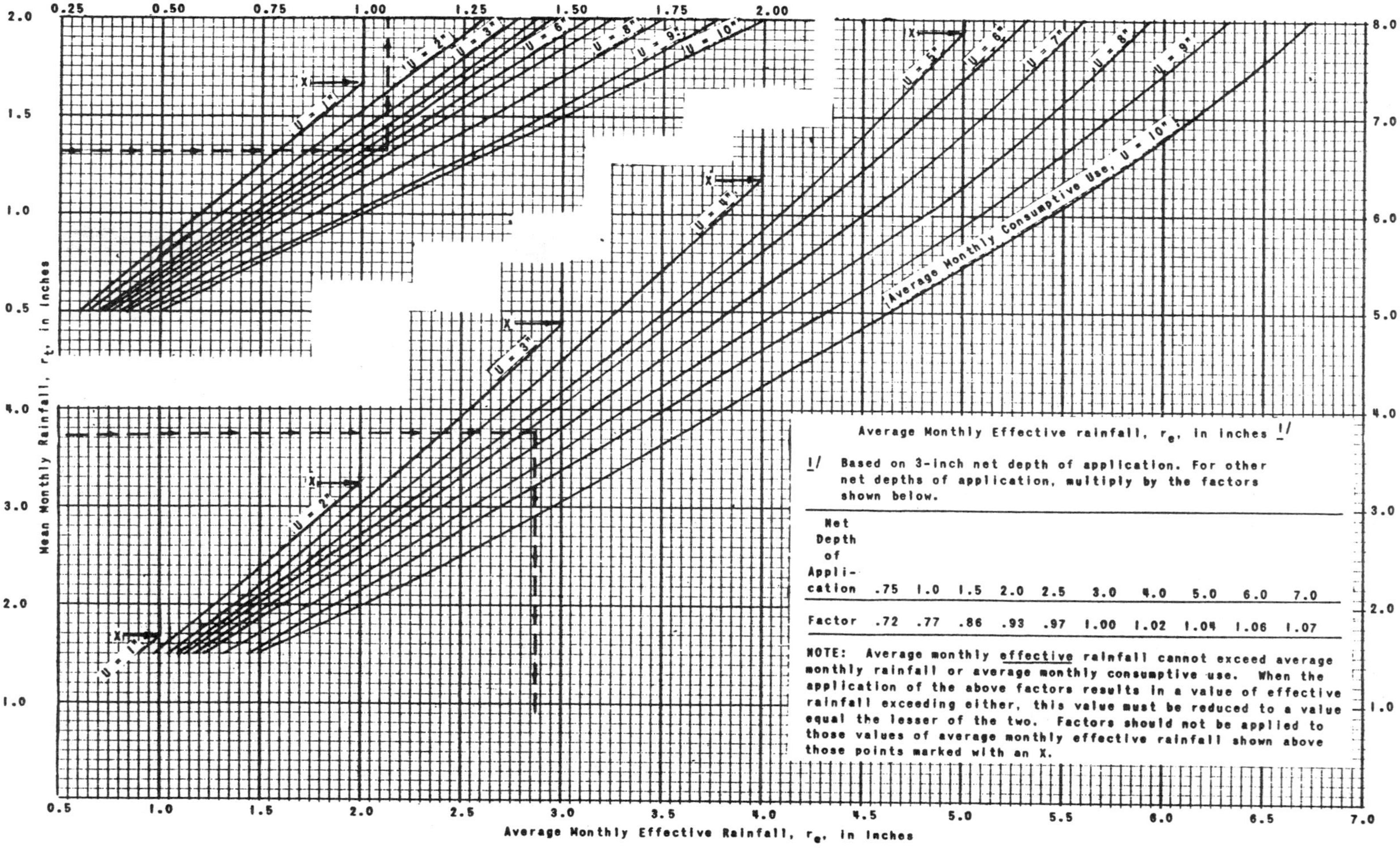

Figure 3 - Average monthly effective rainfall as related to mean monthly rainfall and average monthly consumptive use

Table 6.--Average monthly effective rainfall 1/ as related to mean monthly rainfall and average monthly consumptive-use

Monthly Mean Rainfall r_t Inches	Average Monthly Consumptive-Use, u, in Inches									
	1.0	2.0	3.0	4.0	5.0	6.0	7.0	8.0	9.0	10.0
	Average Monthly Effective Rainfall, r_e, in Inches									
0.5	.30	.32	.35	.36	.37	.40	.42	.45	.47	.50
1.0	.60	.65	.70	.72	.74	.79	.82	.88	.98	1.00
1.5	.90	.98	1.05	1.10	1.13	1.17	1.22	1.32	1.45	1.50
2.0	1.00/1.67	1.29	1.38	1.43	1.47	1.56	1.62	1.75	1.88	2.00
2.5		1.59	1.70	1.78	1.84	1.94	2.02	2.15	2.30	2.50
3.0		1.85	1.99	2.11	2.20	2.30	2.41	2.55	2.70	2.95
3.5		2.00/3.23	2.27	2.41	2.55	2.64	2.79	2.95	3.11	3.38
4.0			2.55	2.71	2.88	2.97	3.15	3.32	3.51	3.80
4.5			2.82	3.00	3.21	3.30	3.49	3.71	3.92	4.22
5.0			3.00/4.87	3.26	3.51	3.62	3.83	4.09	4.32	4.63
5.5				3.55	3.81	3.95	4.17	4.45	4.71	5.04
6.0				3.81	4.09	4.24	4.50	4.80	5.08	5.44
6.5				4.00/6.37	4.35	4.52	4.80	5.12	5.42	5.81
7.0					4.60	4.80	5.10	5.41	5.72	6.15
7.5					4.84	5.06	5.36	5.68	6.03	6.45
8.0	1.00	2.00	3.00	4.00	5.00/7.89	5.31	5.60	5.93	6.32	6.74

1/ Based on 3-inch net depth of application. For other net depths of application, multiply by the factors shown below.

Net Depth of Application	.75	1.0	1.5	2.0	2.5	3.0	4.0	5.0	6.0	7.0
Factor	.72	.77	.86	.93	.97	1.00	1.02	1.04	1.06	1.07

Note: Average monthly effective rainfall cannot exceed average monthly rainfall or average monthly consumptive use. When the application of the above factors results in a value of effective rainfall exceeding either, this value must be reduced to a value equal the lesser of the two. Factors should not be applied to the values of average monthly effective rainfall shown below those encircled. Where mean monthly rainfall (r_t) is less than 0.5 inch it may be assumed to be 100 percent effective.

Seasonal Effective Rainfall.
Average growing season effective rainfall is determined by adding the values of average effective rainfall for the several months and fractions thereof that cover the total growing season of the crop in question. (See Sample Calculations No. 4 and No. 5.)

Caution in the Use of the Curves and the Table.
Figure 3 and table 6 are the result of a comprehensive analysis of 50 years of precipitation records at each of 22 Weather Bureau stations so selected that all climatic conditions throughout the 48 continental states were represented. These studies were made by using the daily soil moisture balance procedure whereby a soil moisture balance is computed for each day by subtracting consumptive use and adding effective rainfall and/or irrigation to the previous day's balance. This procedure necessarily fails to consider two factors which, in some instances, may have a bearing on the effectiveness of rainfall. These factors, soil intake rates and rainfall intensities, are not considered for two reasons: (1) sufficient data are not available; and (2) the complexity involved in their consideration would make such a study impractical. In some areas where soil intake rates are low and rainfall intensities are consistently high, large percentages of rainfall may be lost to surface runoff without the moisture level in the soil profile being raised appreciably. In such areas the values obtained from figure 3 and table 6 may need to be modified.

Frequency Distribution of Effective Rainfall.
It can safely be assumed that, for any given crop at a particular location, monthly and seasonal consumptive use will vary only slightly from year to year provided the crop is planted at about the same time each year. On the other hand, monthly and seasonal effective rainfall can be expected to vary widely from year to year. Since by definition the net irrigation requirement is that portion of total consumptive use not supplied by effective rainfall or other natural sources, it will also vary widely from year to year as effective rainfall varies.

In view of this wide variation in net irrigation requirements from year to year, it is obvious that the development of a dependable water supply cannot be based on average requirements, since this would provide an adequate supply approximately half the time. It is common practice, therefore, to estimate effective rainfall and irrigation water requirements on a probability basis, the percent chance of occurrence used being an economic consideration. For example, it might be economical to provide a water supply that is adequate in nine out of ten years for a high-value vegetable crop or tobacco. For a low-value hay crop or pasture, it may not be economical to provide an adequate supply in more than six out of ten years.

The procedure for determining the frequency distribution of effective rainfall is based on the assumption that, for any fixed period of time or growing season at a given location, other factors being equal, effective rainfall will vary from year to year in direct proportion to the variance in total rainfall. Thus the frequency distribution of total rainfall may be used as a measure of the frequency distribution of effective rainfall. The procedure is as follows:

For the growing season of any given crop at a particular location, Weather Bureau records are used to determine the total rainfall that occurred during the growing season for each year over a period of 25 years or longer. These growing-season rainfall totals are then ranked in order of magnitude and plotted on log-normal probability paper as illustrated by figure 4. A straight line that most nearly fits all of the plotted points is drawn to establish the frequency distribution of growing-season total rainfall. Instructions for plotting the points and drawing the frequency distribution line are contained in the SCS National Engineering Handbook, Section 4, Supplement A, Part 3.18.

The desired percent chance of the developed water supply being equaled or exceeded by the gross irrigation water requirements of the crop is then selected. In the case of corn grown near Raleigh, North Carolina, as illustrated by Sample Calculation No. 4, this selected chance is 20 percent. Thus the developed water supply would be adequate in 8 years out of 10 or 80 percent of the time. The ratio of 80 percent chance growing-season rainfall to average growing-season rainfall is then determined. It will be noted in figure 4 that the 80 percent chance growing-season rainfall is 14.0 inches. From Weather Bureau records, it is determined that the average rainfall for the growing season for corn is 17.87 inches. Thus the aforementioned ratio is 14.0/17.87 or .783. This ratio, when applied to the monthly and seasonal average rainfall values, as shown in column 3 of Sample Calculation No. 4, determines the 80 percent chance monthly and seasonal rainfall values shown in column 9 of the same calculation.

The monthly effective rainfall that can be expected for any frequency of occurrence can be estimated by the use of figure 3 or table 6 when monthly consumptive use and monthly total rainfall for that frequency of occurrence are known. Again using Sample Calculation No. 4 as an example, the monthly consumptive-use values shown in column 2 and the 80 percent chance monthly total rainfall shown in column 9 are used with figure 3 or table 6 to obtain the 80 percent chance monthly effective rainfall shown in column 10.

<u>An Alternate Procedure</u>.

In cases where the degree of accuracy desired does not warrant the time required to plot a growing season rainfall frequency distribution curve for each crop under consideration, an alternate procedure may be used. This procedure involves the application of an average ratio to the average growing season effective rainfall to obtain the growing season effective rainfall for any given percent chance of occurrence. These average ratios vary with the desired percent chance of occurrence and with average annual rainfall values as shown in table 7.

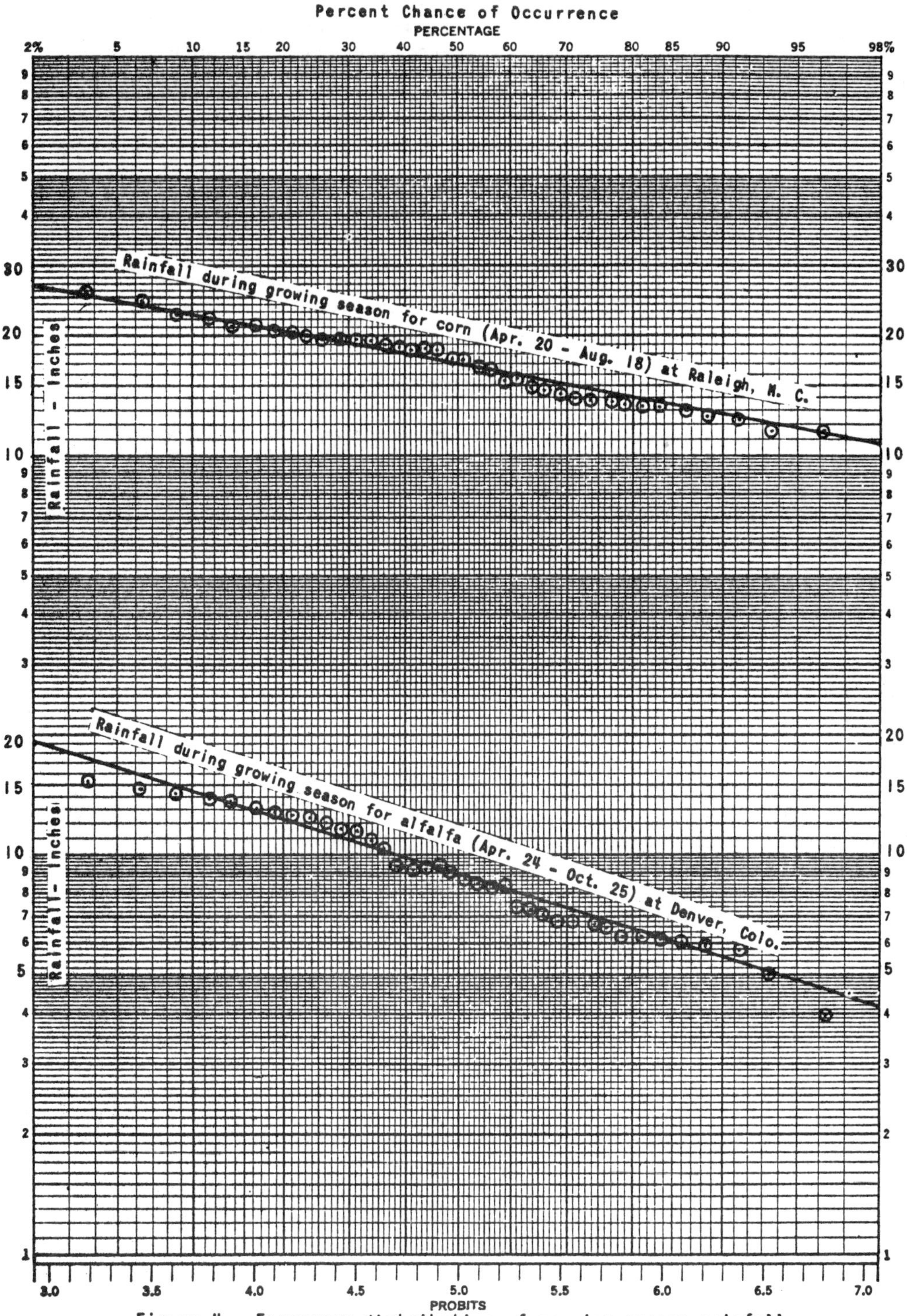

Figure 4 - Frequency distribution of growing season rainfall.

Again using corn at Raleigh, North Carolina, as an example, it is desired to find the growing season effective rainfall that will have an 80 percent chance of occurrence. Average total annual rainfall at Raleigh is 45.92 inches and it has been determined that the average growing season effective rainfall for corn is 12.72 inches (see Sample Calculation No. 4). From table 7 it will be noted that the average ratio applicable to effective rainfall is .842. Thus the growing season effective rainfall that may be expected to occur or be exceeded in eight out of ten years would be .842 x 12.72 or 10.71 inches.

The frequency distribution of effective rainfall for months or other short-time periods may be determined by applying these same ratios shown in table 7.

Irrigation Water Requirements

Consumptive-use data are used in estimating the irrigation water requirements of existing or proposed projects and for crop production on individual farms. The consumptive irrigation water requirement is dependent not only on the total consumptive need, but also on the amount of moisture contributed from such natural sources as effective growing-season rainfall, carryover soil moisture from winter rains and any contribution from ground water. Effective rainfall has been discussed in preceding paragraphs. The other two natural sources are discussed below.

Carryover Soil Moisture.

The contribution of carryover soil moisture resulting from winter rains to the seasonal water requirements is difficult to estimate. In some areas, winter precipitation is sufficient to bring the soil moisture in the root zone depth of the profile up to field capacity. This is particularly true in the humid area states where it is the custom to deduct this readily available moisture (equivalent to the net irrigation application) from seasonal consumptive use when estimating seasonal consumptive irrigation requirements. Where late-season water supplies are short, (usually arid areas) the soil moisture is often well below field capacity and possibly down to the wilting point in the fall.

For crops with a 6-foot root zone, the amount of usable water that could be stored might range from 1 to 2 inches of water per foot depth of soil, or 6 to 12 inches in the 6-foot root zone. This is a major part of the annual requirement of some crops and can be supplied by winter precipitation in some areas in wet years. However, in areas where irrigation water is plentiful, it is not unusual to find the soil moisture content at the end of the season nearly as high as at the beginning. Thus, there is no storage capacity left in the root zone and the contribution from winter precipitation is negligible. Nevertheless, the quantity of moisture carried over in the soil from winter precipitation tends to offset any deficiency in the estimated irrigation water requirements.

Table 7.--Average ratios applicable to effective rainfall

Average Annual Rainfall (Inches)	Percent Chance of Occurrence				
	50	60	70	80	90
3	0.80	0.68	0.56	0.45	0.33
4	.84	.72	.61	.50	.38
5	.87	.76	.65	.54	.42
6	.88	.78	.68	.57	.45
7	.89	.79	.69	.60	.48
8	.90	.81	.71	.62	.51
9	.91	.82	.73	.63	.53
10	.92	.83	.75	.65	.55
12	.93	.85	.78	.69	.58
14	.94	.86	.79	.71	.61
16	.95	.88	.81	.73	.63
18	.95	.89	.82	.74	.65
20	.96	.90	.83	.75	.67
22	.96	.90	.84	.77	.69
24	.97	.91	.84	.78	.70
26	.97	.92	.85	.79	.71
28	.97	.92	.86	.80	.72
30	.97	.93	.87	.81	.73
35	.98	.93	.88	.82	.75
40	.98	.94	.89	.83	.77
45	.98	.94	.90	.84	.78
50	.98	.95	.91	.85	.79
55	.99	.95	.91	.86	.80
60	.99	.95	.91	.87	.81
70	.99	.95	.92	.88	.83
80	.99	.95	.92	.89	.85
90	.99	.96	.93	.90	.86

Example of Use.

It is desired to find the growing season effective rainfall that will occur or be exceeded in 8 out of 10 years at a location where the average total annual rainfall is 30 inches and for a growing season where the average effective rainfall is 12 inches. From the table, the applicable ratio is found to be 0.81. Thus the 80% chance growing season effective rainfall is 0.81 x 12 = 9.72 inches.

Ground Water Contribution.
In areas of high natural ground water, the irrigation requirement may be materially less than if ground water were not available. However, if the high ground water is the result of excess irrigation, the overall demand on the irrigation supply by the crops is not decreased. In such a case, part of the irrigation is obtained by underground methods. As an example, studies in San Fernando Valley in Southern California indicated a consumptive use of water by alfalfa of 37 inches during the irrigation season. In areas of high water table in this valley, only 24 inches of surface irrigation water was required to produce a good yield of alfalfa. The additional 13 inches came from underground water supplies and a small amount of summer precipitation. As with carryover soil moisture, the contribution of ground water to seasonal water requirements is difficult to estimate.

Net Field Irrigation Requirements.
Net field irrigation water requirements for any period of time are estimated by subtracting from potential consumptive and other uses that moisture that is supplied by one or more of the three natural sources previously mentioned.

As previously stated, the effective rainfall studies were made by using the daily soil moisture balance method in which a balance is computed for each day by subtracting consumptive use and adding effective rainfall and/or irrigation to the previous day's balance. In using this method, daily balances are calculated from an assumed soil moisture level at the beginning of the growing season. In these studies, it was assumed that this level was field capacity. In using figure 3 or table 6 to determine net irrigation water requirements, an estimate of anticipated soil moisture conditions at the start of the growing season must be made. The depth of water, if any, required to bring the moisture level in the soil profile up to field capacity must be added to the irrigation water requirements obtained from the use of figure 3 or table 6.

Sample Calculation No. 4 illustrates the procedure for estimating both average net irrigation requirements and those net requirements than can be expected to be equaled or exceeded in two out of ten years using a crop of corn grown in an area near Raleigh, North Carolina. In this case it has been assumed that winter precipitation will bring the soil moisture level up to field capacity and provide 2.0 inches of carryover soil moisture. This 2.0 inches represents the amount of moisture between the 50 percent level and field capacity and is also equivalent to the net depth of application. It has been assumed that each farmer in a project starts irrigating when the soil moisture reaches the 50 percent level and applies enough water to bring the moisture level up to field capacity. Then, at any given time the average soil moisture level over the entire project area will approximate 75 percent. The amount of moisture between the 50 and 75 percent levels (equivalent to one-half the net depth of application or 1.00 inch in this case) has been carried over from month to month and finally consumed at the end of the growing season as shown in the calculation. This leaves the moisture in the soil profile at approximately the 50 percent level at that time. The procedure for estimating gross irrigation requirements is also illustrated.

Sample Calculation No. 5 illustrates the same procedures using a crop of alfalfa grown in an area near Denver, Colorado. In this instance it has been assumed that winter precipitation will provide 2.0 inches of carryover soil moisture and that a net pre-irrigation of 2.2 inches will be needed to bring the soil moisture level up to field capacity. The sum of the carryover moisture and the pre-irrigation, or 4.2 inches, is the net depth of application. As in the previous calculation, one-half the net depth of application or 2.1 inches has been carried through the growing season and used at its end.

Field Application Efficiencies.
Due to unavoidable losses, no field application of irrigation water can ever be 100 percent efficient. Thus more water than is needed to satisfy net irrigation requirements must be applied. A reasonably accurate estimate of field application efficiencies must therefore be made in order to estimate gross field irrigation requirements.

Application losses include evaporation, deep percolation, and surface runoff. The extent of such losses will depend on a number of different factors. The principal ones are discussed in succeeding paragraphs.

Intake characteristics of soils.--In general, considerable loss of water due to deep percolation may be expected when coarse-textured soils with high intake rates are irrigated by surface methods. On the other hand, when fine-textured soils with very low intake rates are irrigated by these methods, considerable losses will occur in the form of excess surface runoff. In either case, field application efficiencies are adversely affected.

Variations in soil intake rates are also a factor in lowering application efficiencies. The intake rate of most soils on which a rotation of crops is grown will vary widely both throughout the growing season and from year to year within the rotation period. Unless considerable flexibility is designed into the irrigation system and the irrigator has the skill required to adjust stream sizes to these changing intake rates, field application efficiences will be materially lowered.

Topography.--It is more difficult to control the flow of water on sloping land than it is on level or near-level land. When relatively steep slopes are irrigated by either of the furrow, corrugation, border or contour ditch methods, excessive surface runoff can be expected where enough water is applied to meet crop requirements. Highest application efficiencies are attained where the land is nearly level and all irregularities are removed by land leveling.

SAMPLE CALCULATION NO. 4
Estimate of monthly and seasonal irrigation requirements
by
Corn at Raleigh, North Carolina

(1) Month	(2) Average Monthly Cons. Use, u. Inches	(3) Mean Monthly Rainfall, r_t Inches	(4) Average Monthly Effective Rainfall r_e, Inches	(5) Average Carryover Soil Moisture Inches	(6) Average Net Irrig. Requirement i_n, Inches	(7) Estimated Field Appl. Efficiency Percent	(8) Average Gross Irrig. Requirement i_g, Inches	(9) 80% Chance Monthly Rainfall, r_t, Inches	(10) 80% Chance Monthly Effective Rainfall r_e, Inches	(11) 80% Chance Carry-over Soil Moisture, Inches	(12) 20% Chance Net Irrig. Require. i_n Inches	(13) 20% Chance Gross Irrig. Require. i_g, Inches
				2.00						2.00		
Apr. 20	0.70	1.20	0.61		None	70	None	0.94	0.54		None	None
				1.91						1.84		
May	3.52	3.62	2.23		0.38	70	0.54	2.83	1.80		0.88	1.25
				1.00						1.00		
June	7.84	4.05	3.08		4.76	70	6.80	3.17	2.47		5.37	7.67
				1.00						1.00		
July	8.79	5.85	4.57		4.22	70	6.03	4.58	3.64		5.15	7.36
				1.00						1.00		
Aug. 18	4.10	3.15	2.23		0.87		1.24	2.47	1.80		1.30	1.86
				0.00						0.00		
Season Totals	24.95	17.87	12.72	2.00	10.23	70	14.61	13.99	10.25	2.00	12.70	18.14

Note: Explanation of column headings appears on the reverse side of this page.

EXPLANATION OF COLUMN HEADINGS - SAMPLE CALCULATION NO. 4

(2) Consumptive-use values shown in this column are taken from Sample Calculation No. 1, column 11.

(3) Mean monthly rainfall values are taken from Weather records.

(4) Values of monthly effective rainfall are obtained by using the values shown in columns 2 and 3 together with table 6 (using a 2.0-inch net depth of application for corn at Raleigh, N.C.). Values in table 6 are for whole months only. To obtain a value for a part of a month, the values shown in columns 2 and 3 must first be converted proportionately to whole month values and table 6 then used to obtain effective rainfall for the entire month. This latter value is then converted back proportionately to obtain the effective rainfall for the actual number of days involved.

(5) Carryover soil moisture must be estimated. In this case it is assumed the winter rains will bring the soil profile up to field capacity, thus the amount of carryover soil moisture will be equal to the net depth of application or 2.0 inches. On an average, one-half of this carryover soil moisture will be consumptively used before irrigation is started and one-half will be carried over for use at the end of the growing season.

(6) The average net irrigation requirement for any month is obtained by subtracting the sum of the values shown in columns 4 and 5 from the value shown in column 2.

(7) Values of obtainable field application efficiencies are taken from the conservation irrigation guide covering the area concerned.

(8) Gross irrigation requirements are obtained by dividing the values shown in column 6 by those shown in column 7.

(9) Values of monthly rainfall for any frequency of occurrence are obtained by first plotting a rainfall frequency distribution curve (see curve for Raleigh, N. C., figure 4) and then obtaining from the curve the value of the growing season rainfall for the desired frequency of occurrence, in this case 8 out of 10 years. This latter value divided by the average growing season rainfall will give a percentage factor which, when applied to the values shown in column 3, will give the values of monthly rainfall shown in column 9 on a frequency basis.

(10) The values of monthly effective rainfall shown in this column are obtained by using the values shown in columns 2 and 9 together with table 6. See explanation of column 4.

(11) See explanation of column 5.

(12) The net irrigation requirements for any month are obtained by subtracting the sum of the values shown in columns 10 and 11 from the value shown in column 2.

(13) Gross irrigation requirements are obtained by dividing the values shown in column 12 by those shown in column 7.

SAMPLE CALCULATION NO. 5
Estimate of monthly and seasonal irrigation requirements
Alfalfa at Denver, Colorado

(1) Month	(2) Average Monthly Cons. Use, u. Inches	(3) Mean Monthly Rainfall, rt Inches	(4) Average Monthly Effective Rainfall re, Inches	(5) Average Carryover Soil Moisture Inches	(6) Average Net Irrig. Requirement in, Inches	(7) Estimated Field Appl. Efficiency Percent	(8) Average Gross Irrig. Requirement ig, Inches	(9) 80% Chance Monthly Rainfall, rt Inches	(10) 80% Chance Monthly Effective Rainfall re, Inches	(11) 80% Chance Carryover Soil Moisture Inches	(12) 20% Chance Net Irrig. Require. in, Inches	(13) 20% Chance Gross Irrig. Require. ig, Inches
				2.00						2.00		
Pre-Irrigation					2.20	75	2.93				2.20	2.93
				4.20						4.20		
Apr. 24	0.57	0.49	0.33		None	75	None	0.34	0.23		None	None
				3.96						3.86		
May	3.99	2.70	1.98		0.15	75	0.20	1.87	1.47		0.76	1.01
				2.10						2.10		
June	6.36	1.44	1.15		5.21	75	6.95	1.00	0.79		5.57	7.43
				2.10						2.10		
July	7.80	1.53	1.32		6.48	75	8.64	1.06	0.91		6.89	9.19
				2.10						2.10		
Aug.	6.66	1.28	1.03		5.63	75	7.50	0.89	0.71		5.95	7.93
				2.10						2.10		
Sept.	4.00	1.13	0.83		2.45	75	3.27	0.78	0.54		2.92	3.89
				1.38						1.56		
Oct. 25	1.89	0.81	0.51		--			0.56	0.33		--	
				0.00						0.00		
Season Totals	31.27	9.38	7.15	2.00	22.12	75	29.49	6.50	4.98	2.00	24.29	32.38

Note: Explanation of column headings appears on the reverse side of this page.

EXPLANATION OF COLUMN HEADINGS - SAMPLE CALCULATION NO. 5

(2) Consumptive-use values shown in this column are taken from Sample Calculation No. 2, Column 10.

(3) Mean monthly rainfall values are taken from Weather Bureau records.

(4) Values of monthly effective rainfall are obtained by using the values shown in columns 2 and 3 together with table 6 (using a 4.2-inch net depth of application for alfalfa at Denver, Colorado). Values in table 6 are for whole months only. To obtain a value for a part of a month, the values shown in columns 2 and 3 must first be converted proportionately to whole month values and table 6 then used to obtain effective rainfall for the entire month. This latter value is then converted back proportionately to obtain the effective rainfall for the actual number of days involved.

(5) In this case it is assumed that there is a 2.0-inch soil-moisture carryover and that a 2.2-inch net irrigation will be required to bring the moisture level in the profile up to field capacity. The sum of these, or 4.2 inches, will equal the net depth of application and is treated as carryover moisture in the calculations. On the average, one-half of this carryover soil moisture will be consumptively used before irrigation is started and one-half will be carried over for use at the end of the growing season.

(6) The average net irrigation requirement for any month is obtained by subtracting the sum of the values shown in columns 4 and 5 from the value shown in column 2.

(7) Values of obtainable field application efficiencies are taken from the conservation irrigation guide covering the area concerned.

(8) Gross irrigation requirements are obtained by dividing the values shown in column 6 by those shown in column 7.

(9) Values of monthly rainfall for any frequency of occurrence are obtained by first plotting a rainfall frequency distribution curve (see curve for Denver, Colorado, figure 4) and then obtaining from the curve the value of the growing season rainfall for the desired frequency of occurrence, in this case 8 out of 10 years. This latter value divided by the average growing season rainfall will give a percentage factor which, when applied to the values shown in column 3, will give the values of monthly rainfall shown in column 9 on a frequency basis.

(10) The values of monthly effective rainfall shown in this column are obtained by using the values shown in columns 2 and 9 together with table 6. See explanation of column 4.

(11) See explanation of column 5.

(12) The net irrigation requirements for any month are obtained by subtracting the sum of the values shown in columns 10 and 11 from the value shown in column 2.

(13) Gross irrigation requirements are obtained by dividing the values shown in column 12 by those shown in column 7.

Where fields are subirrigated, the difficulty in maintaining a water table approximately parallel to the land surface increased rapidly as slopes increase above one-half percent.

Climate.--In arid and semi-arid areas where air temperatures and wind velocities are high, appreciable losses may be expected from the resulting evaporation. These tend to lower application efficiencies of all methods of irrigation except subirrigation. Sprinkler irrigation is particularly affected. High wind velocities so distort the distribution pattern that high application efficiencies are not attainable.

Net depth of irrigation.--The amount of water applied at one irrigation and stored in the soil profile for plant use will affect the application efficiency with some methods of irrigation. In the case of sprinklers, for example, the water retained on the plant foliage and that evaporated from the ground surface while sprinkling is in process will be approximately the same regardless of the depth of application. Thus these losses will be greater percentage-wise for light applications. In the case of graded furrows or corrugations, the amount of water lost to deep percolation will be approximately equal for both light and heavy applications. Generally speaking, then, with these methods, lighter applications will be made at lower efficiencies than will heavier applications.

Irrigation methods.--Relatively high application efficiencies can be attained by most methods of irrigation where the soils, topographic, and climatic conditions are favorable. However, for any given set of conditions, usually a higher application efficiency can be attained with one method than can be attained with another. For example, a close-growing crop on a near-level field where wind velocities exceed 15 miles per hour could be irrigated by the border method with a high application efficiency. Under the same conditions a sprinkler system would have a much lower application efficiency. If the same crop were to be irrigated on a sloping field with relatively calm wind conditions, the sprinkler system would prove to be more efficient. Thus in order to attain a high application efficiency, it is important that the most adaptable method of irrigation be selected.

Adequacy of system design and installation.--In order to attain a high application efficiency, any irrigation system, regardless of method, must be adequately designed and properly installed. The system must include all structures and other devices necessary for controlling the irrigation stream. The extent to which this is accomplished will, in large measure, determine the application efficiency that can be reached.

Skill of the irrigator.--A most important factor influencing field application efficiency is the skill of the irrigator and his interest in using that skill to practice good water management. All of the influential factors mentioned above may be favorable but, unless the irrigator operates the system according to plan, applying water as needed by the crop and at a rate commensurate with the soil intake rate, a high application efficiency will not be attained.

Tailwater recovery systems.--In some instances where the graded furrow and border methods of irrigation are used, relatively large percentages of surface runoff cannot be avoided due to low soil intake rates. In such cases, field application efficiencies are low. Approximately 50 to 65 percent of these losses can be recovered, however, for re-use where tailwater recovery systems are installed. These are systems whereby the runoff from graded furrow and border systems is collected and either pumped back for re-use on the same field or allowed to flow by gravity onto other fields of lower elevation. The use of such a system will materially increase the overall application efficiency on the farm.

Estimating field application efficiencies.--After all of the aforementioned influential factors have been given due consideration, field application efficiencies may best be estimated by referring to that chapter of Section 15 of the SCS National Engineering Handbook covering the specific method of irrigation contemplated or by referring to applicable local irrigation guides.

Gross field irrigation requirements.--Sample Calculations Nos. 4 and 5 also illustrate the procedure involved in estimating gross field irrigation requirements. To determine average gross requirements, the average net requirements shown in column 6 are divided by the estimated field application efficiency shown in column 7. To determine the gross field requirements that can be expected to be equaled or exceeded 20 percent of the time, the 20 percent net requirements shown in column 12 are divided by the same estimated field application efficiency shown in column 7.

Requirements for Related Purposes

In irrigated agriculture, there are occasions where water is needed for purposes other than irrigation but where irrigation systems must be used to apply the water. Where water is used for these additional purposes, their annual requirements must be estimated and added to those for irrigation. Water requirements for the more important of these related purposes are discussed in succeeding paragraphs.

Leaching Requirements

The removal of harmful soluble salts from the crop root zone is essential in irrigated soils if sustained high crop production is to be maintained. Without removal, salts accumulate in direct proportion to the salt content of the irrigation water and the depth of water applied. The concentration of the salts in the soil solution results principally from the extraction of moisture from the soil by the processes of evaporation and transpiration. Such salt concentrations can only be removed by passing enough water through the soil profile to dissolve the harmful soluble salts and transport them, by the downward movement of the water, out of or beyond the crop root zone. This process is known as leaching.

Module II-7

CROP SELECTION AND MANAGEMENT ALTERNATIVES

SUMMARY

This module enumerates the benefits to be derived from cropping at a waste application site and the criteria to be used in selecting a crop for use in a particular situation. Following basic discussions of the requirements of various crops for water, soil-plant-air moisture potentials, crop water tolerance, nutrient removals by various crops, and tolerance to nutrients and salts, practical management techniques are discussed which will extend the usefulness of crop systems in land treatment of wastes. This material is included in the section on "Practical Considerations."

This is a lengthy module which presents much detailed information on certain crops. Some of this data are in the form of tables, sometimes quite involved. It is not necessary or even desirable for the engineer, planner, and designer to try to absorb all this information in one reading. Much of the material is intended to create an awareness in the land application site planner of some of the potential benefits and pitfalls of crop selection and management, the value of careful selection of vegetative cover, and the importance of professional advice on questions in this area. The module covers a body of knowledge which probably is not common to most site planners.

CONTENTS

OBJECTIVES

Upon completion of this module, the reader should be able to:

1. List at least 6 of the 9 criteria for selection of a crop for land treatment of waste.
2. Explain the irrigation water computation given in the example at the end of the section entitled "Evapotranspiration," and use the data given on these pages to duplicate the example for any of the other crops listed.
3. Discuss the major advantages of proper drainage and list at least 8 of the 15 benefits of drainage.
4. List and discuss several of the most widely used crops at land application sites.

INTRODUCTION

The removal of nutrients in harvested crops, higher renovation efficiencies, increased soil stabilization, increased infiltration rate, improved public relations and cost effectiveness are the areas in which cropping may contribute most to the treatment of raw sewage, sewage sludge, and wastewaters. The cost effectiveness of the cropping system will depend upon many things, especially land values, management skills, and cash return from the sale of crops. Crop irrigation has greater land area and management requirements than does a rapid infiltration system. However, sites suited to cropping are generally easier to locate than sites suited to rapid infiltration.

> **The cost effectiveness of a cropping system depends mainly upon land values, management skills, and cash return from harvested crops.*

Farmers and growers try to optimize, within economic constraints, the environment surrounding their crops in order to maximize yields. The primary objectives of a land application project may or may not include maximum crop yield. Maximum economic renovation of wastes is the more direct objective. As stated Module I-1, "Soil as a Treatment Medium," " waste-

*This and other italicized summaries are intended to highlight key ideas, provide a basis for later review or to aid in skimming sections that are relatively familiar. They can be ignored in a complete reading of the text.

water is applied at a rate designed to optimize the renovation capacity of the soil/plant medium and to maximize the utilization of the available nutrients within the wastes." Providing less than optimal conditions will result in less than maximum yields and less than maximal nutrient utilization. However, improving drainage and providing irrigation can potentially improve crop yield, crop quality, and nutrient removal in both wet and dry years. This point should be stressed in selling a land application system to a farming community.

The objective of land application is to maximize cost effective renovation. This can, as a byproduct, improve agricultural yields.

Sopper (1973) enumerated some of the criteria which should be considered in selecting vegetative cover for a land application site. These included:

1. Water requirement and tolerance.
2. Nutrient requirement and tolerance.
3. Nutrient utilization and renovation efficiency.
4. Sensitivity to potentially toxic elements and salts.
5. Insect and disease problems.
6. Season of growth and dormancy requirement.
7. Natural range.
8. Ecosystem stability.
9. Demand or market for the product.

All of these aspects should receive consideration. Many are interrelated.

As discussed in Module I-5, "Vegetative Cover," crop selection is a factor which must be defined before evapotranspiration and percolation (from a given soil, with a given climate) may be estimated with a good degree of accuracy. Before intelligent crop selection can be made, moreover, analyses of the waste to be applied and of the soils involved *must* be made. Analysis of the crops grown *may* also be necessary to protect the food chain or monitor site longevity if an unusually high level of any nutrient or potentially toxic element is applied.

Crop selection depends upon waste and soil characteristics. The amount of evapotranspiration and percolation depend upon crop selection.

Analysis of wastewater and sludge should include the tests which have been developed and are routinely made to assess the quality of irrigation waters including electrical conductivity, sodium adsorption ratio (SAR), HCO_3^-, B, and Cl^-(Table 1) as well as the tests usually used to analyze wastewater. Analysis of soils should be made to check pH, levels of toxic elements, and native concentrations of nitrogen and phosphorus (which can affect renovation efficiency). In addition, soil tests should include CEC and all exchangeable cations, especially K, since it is reported to reduce yields when deficient. In arid areas, soil properties which have been correlated with reduced yields are high levels of salts, boron, and sodium.

The wastewater and sludge should be analyzed for electrical conductivity, SAR, HCO_3^-, B, and Cl^- when it is to be used for irrigation. These have all been associated with reduced yields when applied in excess.

Crops have been broadly classified (Pound, Crites, and Griffes, 1975) as follows: Perrenials (forage or fruit crops), annuals (field crops and vegetables), landscape vegetation (ornamental

Table 1. The Constituents Usually Determined in An Irrigation Water Analysis, Their Abbreviations, and the Units in which They Are Reported.

Determination	Abbreviation	Unit
Electrical conductivity	EC $\times 10^6$ at 25C	micromhos per cm
Soluble-sodium percentage	SSP	percent
Sodium-adsorption-ratio	SAR	
Boron	B	parts per million (ppm)
Dissolved solids	DS	ppm
pH		
Cations		
Calcium	Ca	Milliequivalents/liter meq/liter
Magnesium	Mg	meq/liter
Sodium	Na	meq/liter
Potassium	K	meq/liter
Sum of cations		meq/liter
Anions		
Carbonate	CO_3	meq/liter
Bicarbonate	HCO_3	meq/liter
Sulfate	SO_4	meq/liter
Chloride	Cl	meq/liter
Nitrate	NO_3	meq/liter
Sum of anions		meq/liter

Source: Wilcox and Darum, 1967.

trees, shrubs, and turf grasses), and forests. The latter two "crops" are discussed in a separate module (Module II-8, "Non-crop and Forest Systems"). Aquaculture will not be discussed but has been considered a form of cropping (EPA, 1974). In this module consideration will be directed primarily to highlighting the complexity of tradeoffs which must be made in selecting and managing crops irrigated with wastewater.

IRRIGATED CROPS

Table 2 indicates examples of some of the crops reportedly grown with wastewater irrigation in the U.S. It is evident from Table 2 that for the most part similar crops are irrigated with wastewater as are irrigated with other water. Specific problems however have been reported with wastewater irrigation of some crops as discussed below under, "Crop Tolerances to Nutrients, Salts, and Potentially Toxic Elements." Table 3, taken from Sullivan, *et al.* (1973) describes the crops grown outside Mexico City. It is evident from this table that practically any crop adapted to the prevailing temperature regime could be grown with wastewater irrigation if the wastewater were applied sparingly so as to meet the crop's water requirements without overloading the soil.

Crop Water Requirements

It is well known that management, climate, and plant variety all influence crop yield. Tanner (1974) has recently stressed the importance of water in limiting yield. "(In humid areas) precipitation is sufficient almost always to produce some yield, and yet in most years yield is

Table 2. Examples of Facilities Using Land Application of Wastewater for Crop Production.

	Population	Flow (mgd)	Gal/Cap/Day	Acres	In./Acre/Yr	Crop
Kansas						
Scott City	4,325	0.432	(99.8)	25	(228)	Rice
New Mexico						
Almagordo	25,000	2.5	(100)	260	(130)	Alfalfa, maize, oats, sorghum
Clovis	75,000	3.5	(46)	1,150	(40)	Milo, alfalfa, maize, millet, wheat
Raton	2,300	0.5	(217)	200	(33)	Alfalfa
Roswell	40,000	2-3 (2.5)	(62.5)	770	(43)	Alfalfa, barley, maize, cotton
Santa Fe	45,000	5.3	(118)	740	(95)	Alfalfa, apples
Oklahoma						
Duncan	20,000	2.5	(125)	180	(184)	Wheat, bermuda grass
Texas						
Abilene	100,000	9.0	(90)	2,019	(59)	Cotton, maize, bermuda grass
Dumas	9,770	1.0	(102)	585	(22.7)	Wheat, maize
Kingsville	30,000	3.0	(100)	606	(65.7)	Maize
Midland	62,000	4.3	(69)	1,000	(57)	Bermuda grass, alfalfa, milo, cotton
San Angelo	64,000	5.0	(78)	740	(89.7)	Barley, milo, rye, oats, fescue, alfalfa
Uvalde	9,000	0.9	(100)	150	(79.6)	Maize, oats
Dalhart	5,700	0.64	(112)	240	(35)	Hay
Denver City	4,200	0.15	(35.7)	180	(11)	Cotton, grass
Idalou	1,800			40		Feed, cotton
Morton	3,760			50		Cotton
Seagraves	2,500	0.2	(80)	160	(17)	Cotton

Source: Sullivan, Cohn, and Baxter, 1973.

limited by insufficient water. Much larger yield increases per unit of irrigation are possible in humid regions than in arid regions, which rely almost totally on irrigation." Table 4 presents some yield data from Pennsylvania State University. It is evident that application of wastewater substantially increased the yield of most crops in most years. The differences between the 1 and 2 in./wk treatments were not as marked in this particular study. Pound and Crites (1973) state, "A loading of about 4 in./wk would seem to be the upper limit of a true irrigation system." This is the design rate for seasonal wastewater application to the sandy soils of Muskegon, Michigan.

It is generally agreed that water is limiting, in most of the world, for crop production. Even in humid areas yield is limited by insufficient water in most years.

Table 3. Summary of Agriculture Production 1971-1972.
03 Irrigation District—Tula Hidalgo, Mexico.

Crop	Hectares	Metric Tons
Alfalfa	12,396.40	1,181,376.920
Garlic	94.50	258.456
Peas (large)	12.89	20.820
Green Oats	2,998.75	54,426.714
Squash (small)	674.33	7,282.764
Barley grain	1,865.43	3,645.410
Barley (forage)		4,059.874
Onion	23.79	168.909
Parsley seed	3.53	4.589
Cabbage	27.95	501.843
Peas	1.00	7.900
Green hot peppers	768.80	8,231.350
Flowers	10.41	
Navy beans	1,259.02	1,563.870
American string beans	58.30	151.580
Spinach	0.82	9.020
Fruit trees	25.08	213.180
Sunflower	37.19	230.578
Lima beans	95.84	1,990.070
American tomato	1,554.65	49,437.870
Lettuce	74.47	1,457.764
Corn (kernels)	17,053.60	70,260.525
Corn (forage)		65,179.023
Corn (sweet)	101.20	7,084.000
Forage turnips	112.37	1,011.330
Melon	1.00	7.100
Cucumber	34.74	166.752
Meadow grass	12.80	2,080.000
Tomato	216.90	2,051.706
Wheat grain	7,293.79	13,865.494
Watermellon	0.40	3.960
	46,809.95	1,476,749.371
	Acres	U.S. Tons
	115,620.65	1,624,424.30

Note: 1 metric ton = 1.1 (U.S.) tons.
1 (U.S.) ton = 0.907 metric tons.
1 hectare = 2.47 acres.

Note: The crop hectares listed are more than the hectares of land available since a second crop in some instances has been produced on the same land.

Source: Sullivan, Cohn, and Baxter, 1973.

In traditional irrigation planning the gross volume of water required at the diversion point is made up of a number of components.

1. Crop needs (function of potential evapotranspiration and crop).
2. Soil needs (a function of the salinity problem and tillage requirements).

Table 4. Crop Yields at Various Levels of Wastewater Application in Penn State University Studies.

Crop	Unit	0 in./wk		1 in./wk		2 in./wk	
1963							
Wheat	bu/acre	48		45		54	
Corn	bu/acre	73		103		105	
Alfalfa	tons/acre	2.18		3.73		5.12	
Red clover	tons/acre	2.48		4.90		4.59	
1964							
Red clover	tons/acre	1.76		5.30		5.12	
Corn	bu/acre	81		121		116	
Corn stover	tons/acre	3.58		7.29		8.48	
Oats	bu/acre	82		124		97	
1965							
Alfalfa	tons/acre	2.27		4.67		5.42	
Corn	bu/acre	63		114		111	
Corn silage	tons/acre	3.11		3.93		4.32	
Oats	bu/acre	45		80		73	
Reed canary grass	tons/acre	–		–		6.13	
1966							
Alfalfa	tons/acre	1.95		3.86		4.38	
Corn	bu/acre	18[a]	33[b]	115[a]	98[b]	140[a]	115[b]
Corn silage	tons/acre	2.75[a]	2.47[b]	9.02[a]	4.45[b]	7.53[a]	5.68[b]
Reed canary grass	tons/acre	–	–	–	–	4.32[a]	–
1967							
Corn	bu/acre	93[c]	87[d]	96[c]	79[d]	116[c]	80[d]
Corn silage	tons/acre	4.67[c]	–	4.47[c]	–	4.42[c]	–
Reed canary grass	tons/acre	–	–	–	–	7.03[c]	–

[a] 19-in. row.
[b] 38-in. row.
[c] 20-in row.
[d] 40-in row.

Source: Pound and Crites, 1973.

3. Transmission and distribution losses (runoff and deep seepage from fields, deep seepage and evaporation from distribution channels, and administrative losses).
4. Storage losses (evaporation and seepage from reservoir).

Of all of these elements, crop needs has received the greatest amount of study and may be estimated with a high degree of accuracy.

Crops differ in their rooting habits, and thus soil drainage characteristics can influence their response to irrigation. Figure 1 is reproduced from SCS (1964) and illustrates the rooting habits of many common crops. Spinach has a shallow rooting habit, sugar beet an intermediate, and alfalfa a deep rooting habit. Shallow rooted plants thrive with frequent irrigations, though they will not necessarily stand waterlogging. In mixed stands the requirements of the most shallow rooted species (usually clovers) dictate the required frequency of irrigations.

Deep rooted plants take up a similar amount of water as shallow rooted species when moisture is adequate. Deep rooted plants should not be hurt by frequent irrigations if drainage is adequate. Whenever irrigation is practiced, care must be taken to wet the soil profile uniformly at least through the effective rooting depth. Keeping the subsoil too wet or too dry will restrict

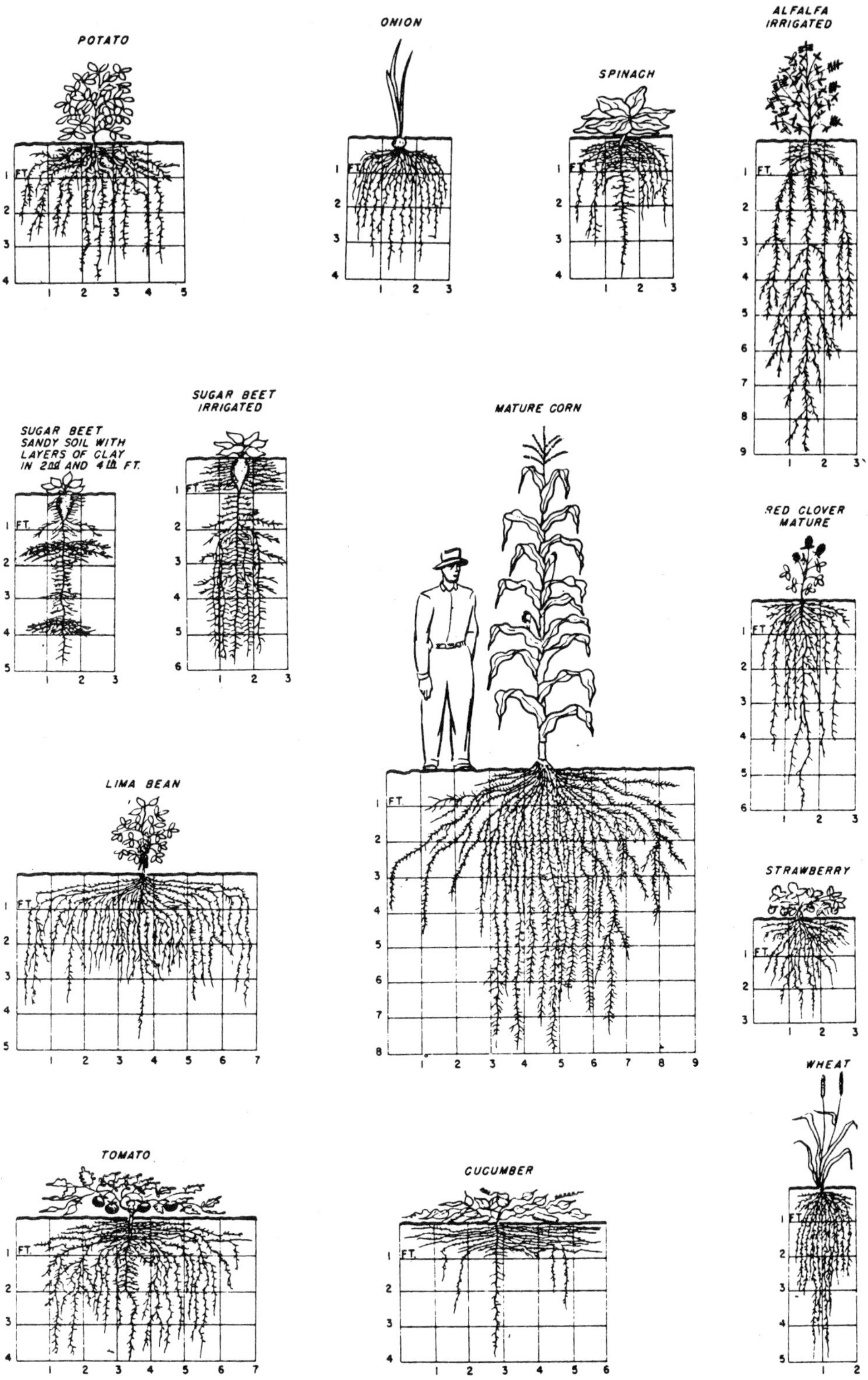

Figure 1. Root systems of field and vegetable crops in deep irrigated soils of central United States. Source: SCS (1964).

root depth and increase the degree of drought damage which might occur if for some reason water supplies are interrupted. The effect of wet subsoil may not be so great on annual crops which may be considered to be shallow rooted until they approach flowering.

Crops differ in rooting habits. Deep rooted plants take up moisture from a greater depth. They can be harmed, however, by water-logged subsoil.

Tanner (1974) has shown that yield has been well correlated with transpiration. Since in the Pennsylvania State study (Table 4) yields between the 1 and 2 in./wk. treatments were similar, the second inch of the 2-inch treatment applied in addition to natural precipitation was probably lost as percolate rather than as increased evapotranspiration. Figure 2 illustrates "production functions," plots of crop yield versus water application, which are useful to engineers determining irrigation water requirements. Such functions are empirical and must be generated in the area to which they will be applied. (The State Extension Service or agricultural college can provide these curves for locally irrigated crops.) From this type of data, culturally and economically optimum amounts of water can be derived. The shape of this curve gives a good indication of the qualitative effects which may be expected from increasing water applications. Maximum yield is associated with maximum transpiration. Maximum transpiration can be approached by providing irrigation that makes up the deficit between precipitation and potential evapotranspiration. Increasing water application after this increases the volume of percolating water. As discussed below, reduced yield with increasing water application could be due to poor aeration adversely affecting root growth and stand (Figure 2). It could also be due to a reduced supply of available nitrogen as a result of leaching losses or the inhibition of minerali-

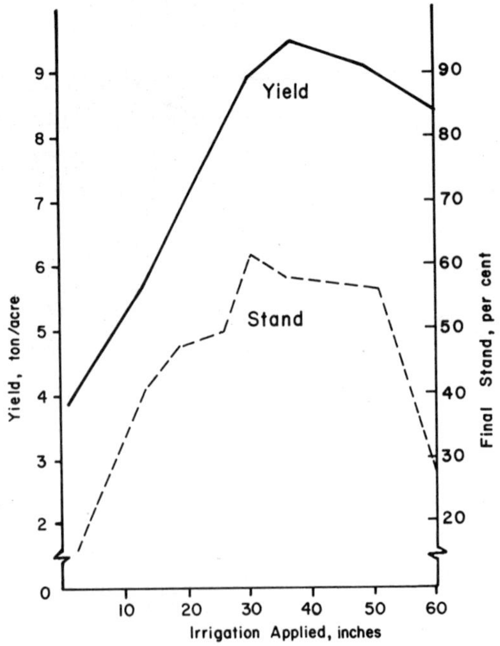

PRODUCTION FUNCTION FOR ALFALFA

Figure 2. Six-year average yield, and final stand, of alfalfa at Davis, California, in response to different amounts of annual irrigation. (Perfect stand contained 1.5 plants/ft^2) (Taken from Beckett and Robertson, 1917). The growing season is 242 days.

zation of soil organic nitrogen. The latter response may not be seen if the wastewater contains considerable nitrogen.

Production functions can be generated to indicate optimum irrigation requirements. Water applied above the optimum would be expressed as percolate through the soil.

Evapotranspiration

Evaporation from wet surface and transpiration from well irrigated crops is limited by the energy available to evaporate water. Energy budget studies have shown:

$$R_N + l - E - H - S - PS = 0$$

when R_N = net radiation, l = latent heat of vaporization, E = evapotranspiration, H = sensible heat stored in air, S = sensible heat stored in soil, and PS = photosynthesis. Evapotranspiration from a dense sod is essentially equal to evaporation from a wet surface and is limited by climate, i.e., available energy. The two major sources of energy driving evapotranspi:ation are net radiation and convection of warm air with less than 100% relative humidity. Conduction of heat from the soil may be an additional source of energy, especially if water is applied to a hot, dry soil. Figure 3 illustrates the nature of the water potential gradient which exists be-

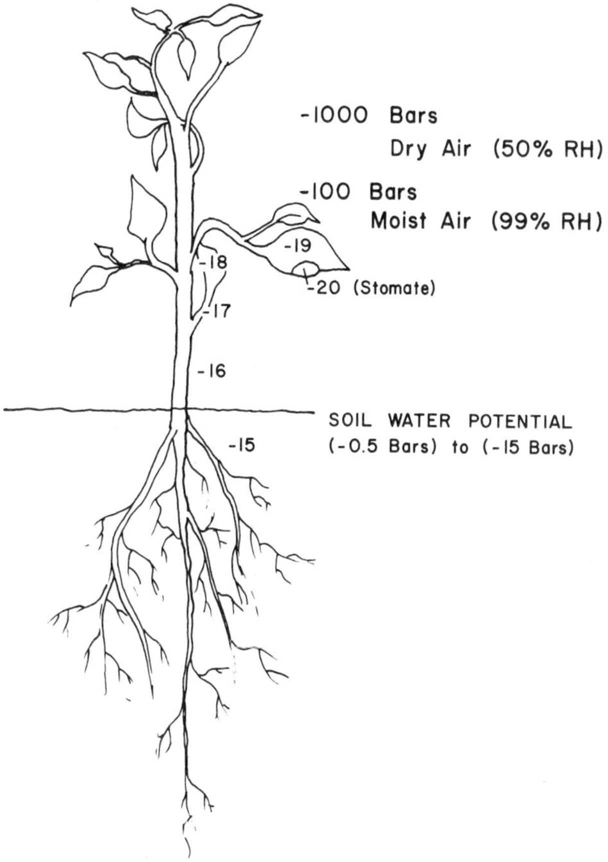

Figure 3. Typical soil-plant-air moisture tensions.

tween soils-plant-air. It will be noted that as soil moisture decreases, the potential gradient is lessened to a certain extent (10 to 20 bars) but this does not greatly affect the overall potential. Evapotranspiration is decreased due to decreased capillary flow as the soil surface dries. The degree to which a plant may satisy a high transpiration demand depends primarily upon the size and density of its root system. Anything which restricts the depth or vigor of root penetration may restrict transpiration.

Transpiration rate depends upon the size and density of the root system. Anything which restricts the depth or vigor of the roots restricts transpiration.

Net radiation is the largest source of energy available for evaporating water. This is true even of advective situations. Net radiation is defined as:

$$R_N = (1 - a)R_G + R_{NT}$$

Table 5. Principal Federal Sources of Data Useful in Planning and Operation of Irrigation Developments Served from Surface Sources.

Type	Agency
Humidity	WB,* FS, TVA, IBWC, ARS
Pan evaporation	WB,* BR, IBWC, TVA, CE, ARS, FS
Precipitation	WB,* TVA, IBWC, SCS, ARS, FS, BR, CE
Chemical quality of water	GS,* PHS, BR, ARS, TVA, IBWC
Suspended load	GS,* BR, TVA, CE, IBWC, ARS, PHS
Reservoir sedimentation	SCS, BS, BR, TVA, CE, ARS, FS
Snow	SCS,* WB,* CE, FS, BR, TVA, ARS, MPS
Solar radiation	WB,* ARS, BR, FS
Streamflow	GS,* ARS,* IBWC,* BR, CE, FS, TVA
Temperature air	WB,* ARS, IBWC, TVA, FS
Wind	WB,* TVA, ARS, CE, FS, BR
Soils	SCS,* FS, BLM, BIA, ARS
Topography	GS*
Geology	GS*
Land use	BC,* ERS, FS, SCS, BLM, BIA, SRS

*Asterisk indicates agencies that regularly publish data. Other agencies publish data intermittently in research or other reports.

Key to agencies:

Department of Agriculture
- ARS–Agricultural Research Service
- ASCS–Agricultural Stabilization and Conservation Service
- ERS–Economic Research Service
- FS–Forest Service
- SCS–Soil Conservation Service
- SRS–Statistical Reporting Service

Department of the Interior
- BIA–Bureau of Indian Affairs
- BLM–Bureau of Land Management
- BR–Bureau of Reclamation
- GS–Geological Survey
- NPS–National Park Service

Department of Commerce
- BC–Bureau of Census
- WB–Weather Bureau

Department of Health, Education and Welfare
- PHS–Public Health Service

Department of the Army
- CE–Corps of Engineers

Other Federal Organizations
- IBWC–International Boundary and Water Commission, United States and Mexico
- TVA–Tennessee Valley Authority

Source: Wilcox and Duram, 1967.

where R_N = net radiation, R_G = solar (global) radiation, R_{NT} = net thermal radiation, and a = albedo or reflectance of solar radiation (a function of crop and soil cover). Net radiation may be calculated on the basis of information available in tables and measurements of temperature and net solar radiation (Tanner, 1974). Table 5 indicates sources of available information. Solar radiation measurements have been made at only a few weather stations, however, they may be made conveniently and inexpensively in a fashion similar to that of Kerr, *et al.* (1967).

Relative humidity, wind velocity and turbulence influence the amount of advected energy. Heat exchange will be influenced by the aerodynamic relations of the crop canopy. Open row crops can increase surface area, turbulence, and mixing and hence evaporation.

Determination of Evapotranspiration and Irrigation Requirements Using Climatic Data

ET may be determined accurately with micrometeorological (energy budget) methods, lysimeters, and soil water measurements. Because of the expense of equipment and/or labor involved with these methods, simple climatological estimates of ET have been developed. All purely climatological methods estimate the potential evapotranspiration, ET_p, that would be produced by the prevailing climate if available moisture were not limiting, which is usually the case within a land application site. The climatological estimate may be modified by various schemes to account for the influence of declining soil water on evaporation and transpiration. The empirical climatological methods can be divided into groups according to the measurements which they require: temperature correlation, radiation methods, humidity methods, and pans and anemometers.

ET may be determined with micrometeorological methods, lysimeters and soil water measurements. It may be estimated with climatological data to give potential evapotranspiration (ET_p)

Temperature Correlations. Their main value has been in estimating ET in areas where the only climatic record is temperature.

1. Blaney-Criddle (1950) assumes a linear relationship between ET_p and temperature.

$$ET_p = CL_dT$$

   C is a monthly "consumptive use" crop and location coefficient, L_d is a day length factor, and T is mean temperature. This formula was derived in the irrigated west and has been used for many years. Tables are available. Blaney-Criddle is not a good estimator of short term water demand but is an empirical estimator for the area for which it was developed. A complete discussion of the Blaney-Criddle formula, including modifications for short-term use, is found in Appendix B of Module II-6, "Climate and Wastewater Storage."

   Temperature correlation methods for estimating ET_p are easy to use but lack accuracy for short term ET. These can be used when only temperature records are available.

2. Thornthwaite (1948) assumes non-linear relation between ET_p and temperature.

$$ET_p = 1.6\,(10\,T/I)^a$$

   T is mean temperature, I is a month heat index (available in Tables) and "a" is a function of I. This formula was derived in the humid east. Like other mean temperature methods it is not satisfactory for ET modeling on a short-term basis. The Thornthwaite method underestimated ET by about 50% during the growing season at Davis, California.

Other methods are more complex, but yield more accurate short and long term ET_p information. These include the methods developed by Penman, Priestley and Taylor, and Tanner. A local college of agriculture can provide the information necessary to use these formulas. In land application systems, accurate estimates of evapotranspiration along with known amounts of precipitation and applied wastewater would allow good estimates of the volume of water percolating through the root zone.

Figure 4 demonstrates several relationships which exist between soil moisture tension and available water depletion. Using Figure 4 together with the data contained in Table 6 allows a reasonable approximation of available water held by a given soil type at any given soil moisture tension. Table 7 illustrates the soil moisture tensions at which water should be applied to produce maximum yields (satisfy potential evapotranspiration). The range of values given for each crop recognizes the effects of high and low evaporative demand and expected weather conditions. Table 8 indicates typical values of ET observed with various crops and depths of root penetration.

The amount of data on available moisture for different soil types, soil moisture tension, the moisture tension for each crop at which water should be applied and daily ET can be used to approximate the amount of wastewater that crops need for optimum production.

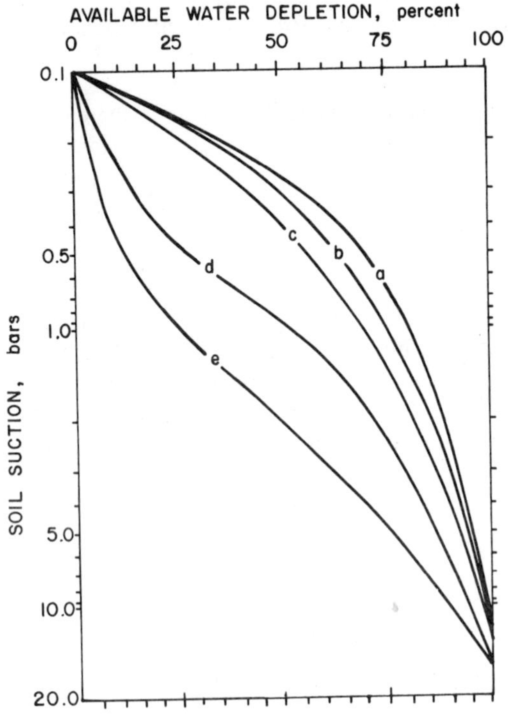

AVAILABLE SOIL MOISTURE STORAGE*

	Range	Average
	---------in./ft------	
a. Loamy sand	0.50-1.00	0.75
b. Fine sandy loam	0.75-1.25	1.00
c. Sandy loam	1.25-1.75	1.50
d. Loam	1.50-2.30	2.00
e. Clay	1.60-2.50	2.30

* Storage between field capacity (1/10 to 1/3 ATM) and wilting point (15 ATM).

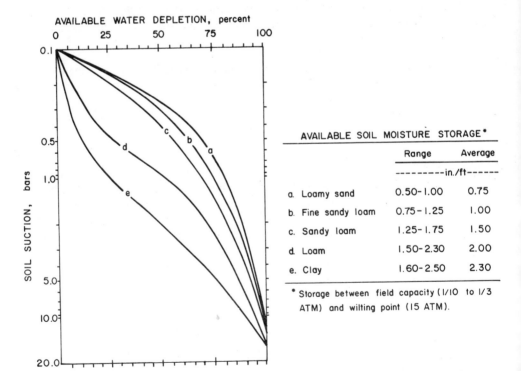

Figure 4. Water retention curves for several soils plotted in terms of percent available water removed. Source: Haise and Hagan, 1967.

Table 6. Typical Range of Available Soil Moisture by Soil Textural Class.

	Available Soil Moisture Storage[a]		
	Range	in./ft	Average
Very coarse textured sands and fine fine sands	0.50–1.00		0.75
Coarse textured loamy sands and loamy fine sands	0.75–1.25		1.00
Moderately coarse textured sandy loams and fine sandy loams	1.25–1.75		1.50
Medium textured very fine sandy loams, loams, and silt loams	1.50–2.30		2.00
Moderately fine textured sandy clay loams and silty clay loams	1.75–2.50		2.20
Fine textured sandy clays, silty clays and clays	1.60–2.50		2.30

[a]Storage between field capacity (1/10 to 1/3 ATM) and wilting point (15 ATM).

Source: Metcalf and Eddy, Inc., 1977.

The utility of this information will be evident from an example: consider potatoes grown on a sandy loam soil. Potatoes are shallow rooted and require about 0.2 inch of water per day (Table 8). They should be irrigated when soil moisture tensions are about 0.4 bar (Table 7). A sandy loam will yield 50% of its available moisture capacity before soil moisture tensions reach 0.4 (Figure 4). A moderately coarse textured sandy loam stores on average 1.5 in./ft (Table 6). If the roots explore a depth of 18 in., containing 2.25 inches of "available moisture," they would use up 50% of this moisture in less than 6 days. For maximum yields these potatoes would require irrigation or precipitation more often than once a week no matter how much water was applied. They would require water applications providing 1.4 in./wk available in the root zone. If potatoes were grown on a loam soil, where 0.4 bar is reached at 25% water depletion and about 3 inches of "available moisture" is stored in a depth of 18 inches, 25% of 3 inches is 0.75 inch and this would be used in only 4 days. Additions of 0.8 inch of precipitation or irrigation water would be needed every 3 to 4 days. In such a situation, care would have

Table 7. Soil Water Suction at Which Water Should Be Applied for Maximum Yields of Various Crops Grown in Deep, Well Drained Soil Fertilized and Managed for Maximum Production.

Crop	Soil Suction, bars*
Vegetative Crops	
Alfalfa	1.50
Beans (snap, lima)	0.75– 2.00
Cabbage	0.60– 0.70
Canning peas	0.30– 0.50
Celery	0.20– 0.30
Grass	0.30– 1.00
Lettuce	0.40– 0.60
Tobacco	0.30– 0.80[a]
Sugar cane	0.25– 0.30
Sweet corn	0.50– 1.00

Table 7. *(Continued)*

Crop	Soil Suction, bars*
Root Crops	
Onions, early	0.45– 0.55
Onions, bulbing	0.55– 0.65
Sugar beets	0.40– 0.60
Potatoes	0.30– 0.50
Carrots	0.55– 0.65
Broccoli, early	0.45– 0.55
Broccoli, postbud	0.60– 0.70
Cauliflower	0.60– 0.70
Fruit Crops	
Lemons	0.40
Oranges	0.20– 1.00
Deciduous fruit	0.50– 0.80
Avocadoes	0.50
Grapes, early	0.40– 0.50
Grapes, mature	1.00
Strawberries	0.20– 0.30
Cantaloupe	0.35– 0.40
Tomatoes	0.80– 1.50[a]
Bananas	0.30– 1.50[b]
Grain Crops	
Corn, vegetative	0.50
Corn, ripening	8.00–12.00
Small grains, vegetative	0.40– 0.50
Small grains, ripening	8.00–12.00
Seed Crops	
Alfalfa, prebloom	2.00
Alfalfa, bloom	4.00– 8.00
Alfalfa, ripening	8.00–15.00
Seed carrots, 60-cm depth	4.00– 6.00[c]
Onions, 7-cm depth	4.00– 6.00[c]
Seed onions, 15-cm depth	1.50
Lettuce, productive	3.00[c]
Coffee requires short periods of low potential to break bud dormancy, followed by high water potential	

*Where two values for soil water suction are given, the lower suction value is used when the evaporative demand is high and the higher value when it is low; intermediate values are used when the atmosphere demand for evapotranspiration is intermediate. (these values are subject to revision as additional experimental data become available.)

[a] Based on converting 50% available water to water potential (soil suction) equivalents using curves for appropriate soil textures.

[b] Based on converting 70% available water to water potential (soil suction) equivalents using curves for clay soils.

[c] Resistance values were converted to water potential from calibration of similar plaster resistance units.

Source: Haise and Hagan (1967).

Table 8. Design Moisture-Extraction Depth and Peak-Period Consumption-Use Rate for Various Crops Grown on Deep, Medium-Textured, Moderately Permeable Soils.[a]

Crop	Wisconsin (State)		Indiana (State)		Piedmont Plateau[b]		Virginia (Coastal Plain)		New York (State)	
	Depth In.	Use Rate In./day	Depth In.	Use Rate In./day	Depth In.	Use Rate In./day	Depth In.	Use Rate In./day	Depth In.	Use Rate In./day
Corn	24	0.30	24	0.30	24	0.22	24	0.18	24	0.20
Alfalfa	36	.30	36	.30	36	.25	36	.22	30	.20
Pasture	24	.20	30	.30	24	.25	20	.22		
Grain	18	.25			24	.16				
Sugar beets	18	.25								
Cotton					24	.21				
Potatoes	18	.20	12	.25	24	.18	18	.18	18	.18
Deciduous orchards	36	.30			36	.25	36	.22	36	.20
Grapes			24	.25	30	.20				
Soybeans	18	.25	24	.30	24	.18				
Shallow truck	12	.20	9	.20	12	.14			12	.18
Medium truck	18	.20	12	.20	18	.14	18	.16	18	.18
Deep truck	24	.20	18	.20	24	.18			24	.18
Tomatoes	18	.20	18	.20	24	.21	24	.18	24	.18
Tobacco			24	.25	18	.18	18	.17		

[a] From current irrigation guides.
[b] Parts of Georgia, Alabama, North Carolina, and South Carolina.

Source: SCS, 1964.

to be taken to ensure even distributions of water and avoid waterlogging of low spots in the field.

Drainage Requirements and Crop Water Tolerance

For many years agriculturalists have recognized that excess water in the root zone resulted in reduced crop growth and more difficult and less timely field operations. The benefits to be derived from the removal of excess water were summarized by Van't Woudt and Hagan (1957) (Table 9). Drainage has usually allowed greater early crop growth, an advantage for wastewater application in which the object is to apply the most wastewater consistent with renovative capacity. Thus drainage, by removing water, can increase a field's irrigation requirement.

> *Drainage can allow earlier, more rigorous crop growth. This improves renovative capacity and increases the amount of wastewater that can be applied.*

When equipment is driven over a wet soil, the soil can be compacted and macropore space is reduced. A compacted layer will aggrevate the problems of wet fields by:

1. Restricting water percolation.
2. Restricting root penetration.
3. Restricting gaseous diffusion.
4. Reducing the effective volume of soil available for renovation.

Following a rain or irrigation a soil may appear dry in the top 2 or 3 inches and yet be too wet for successful tillage. The weather is always uncertain, and allowing the soil to dry sufficiently for field operations to be completed will demand cessation of water applications for variable periods of time. Cropping systems and cultural practices which minimize field traffic should minimize compaction problems. Sod crops generally have greater efficacy in maintain-

Table 9. Benefits to Be Derived from the Removal of Excessive Soil Moisture.

Item #	Description of Benefit
1	Improved aeration
2	Improved nitrogen economy in the soil
3	Increased benefit from topdressing, particularly phosphorus
4	Increased activity of earth worms
5	Reduced weed infestation
6	Improved workability of the soil
7	Increased depth of rooting
8	An earlier warming up of the surface soil, leading to a prolonged growing season
9	Improved soil structure
10	Increased storage capacity of the soil and increased infiltration rate of the soil, and therefore reduced erosion
11	Prevention of damage to roots by frost heaving
12	Increased availability of moisture during a dry spell because of deeper root penetration leading to increased drought resistance of plants on drained soils
13	Prevention of alkali accumulation
14	Decreased danger of plant diseases
15	Improved public and animal health

Source: Van't Woudt and Hagan, 1957.

ing soil structure than have other crops, if care is taken not to harvest forage when the soil is too wet.

Soil compaction—resulting in lower percolation—can be protected against by crop selection and cultural practices.

In a land application system a significant fraction of the water applied may be intended for percolation. The amount of such water may be quite variable and could be limited by the ability of a given soil to drain the root zone within one or two days, or by the nitrogen or phosphorus concentrations of the percolate. These limitations will be compared in the subsequent section, "Crop Tolerances to Nutrients, Salts, and Toxic Elements."

As discussed in Module II-3, "Organic Matter," aeration of the soil depends primarily on diffusion within the air-filled pore space. Diffusion within the liquid phase is very slow. The diffusion coefficient of oxygen in water is about 2.56×10^{-5} cm^2/sec and about 1.89×10^{-1} $cm^2/sec.$ in air. Diffusion through saturated soils is even slower than that through water (Letey, *et al.*, 1967). When a soil becomes flooded, respiration of roots and microorganisms within the soil can quickly exhaust the available oxygen. High temperatures accelerate the depletion. If the soil can drain efficiently and fresh air refills the pore space, brief flooding should not reduce crop growth. If flooding persists, O_2 concentrations will decline and CO_2 increase. This may directly influence crops by reducing root growth,.water absorption and nutrient absorption. Following irrigation, entrapped air within the flooded root zone may maintain aerobic conditions for a day or two. The saturated zone during the period from about two days after irrigation until the next irrigation is, therefore, critical for plant production (Van den Berg, (1973). This emphasizes the importance of soil porosity and drainage characteristics.

Oxygen is rapidly depleted from a flooded soil; CO_2 increases. This can reduce root growth, water, and nutrient adsorption.

According to Letey, *et al.*, (1967) the roots of many species will not grow in an environment where the oxygen diffusion rate is less than about 0.20 $\mu g/cm^2/min$ (about 12 lb/acre/24 hr). In a study conducted in the San Joaquin Valley of California, oxygen diffusion rates were measured at various times and depths after irrigation and at various distances from a wetted furrow (Figure 5).

In this experiment, conducted on a Panoche clay loam, the ODR was adequate for good root growth to a depth of less than 30 cm on the 4th day after irrigation. In the same experiment, the ODR under sprinkler irrigation was slightly higher than furrow irrigation, but not significantly higher. In other experiments the rate of recovery of ODR has been shown to be dependent on crop rotation and cultural practices. The natural porosity characteristics of a soil are probably the most important factor influencing soil aeration.

In addition to the direct effects attributable to low O_2 and high CO_2 levels under anaerobic conditions, toxic organic compounds may be formed and phosphorus and toxic elements may be mobilized. The effect of water-logging on crops has been shown to be affected by crop, variety, stage of growth, temperature, light intensity, and fertility (Grable, 1966). It is therefore difficult to make meaningful generalizations, as can be seen by studying Figure 6 which indicates response of several crops to flooding. Complete information on the response of each crop to these variables is not recorded. Some practical experience is available. Reviews of available information have been published by Russell (1952), Van't Woudt and Hagan (1957), Letey, *et al.* (1967), Van den Berg (1973), and Wesseling (1974).

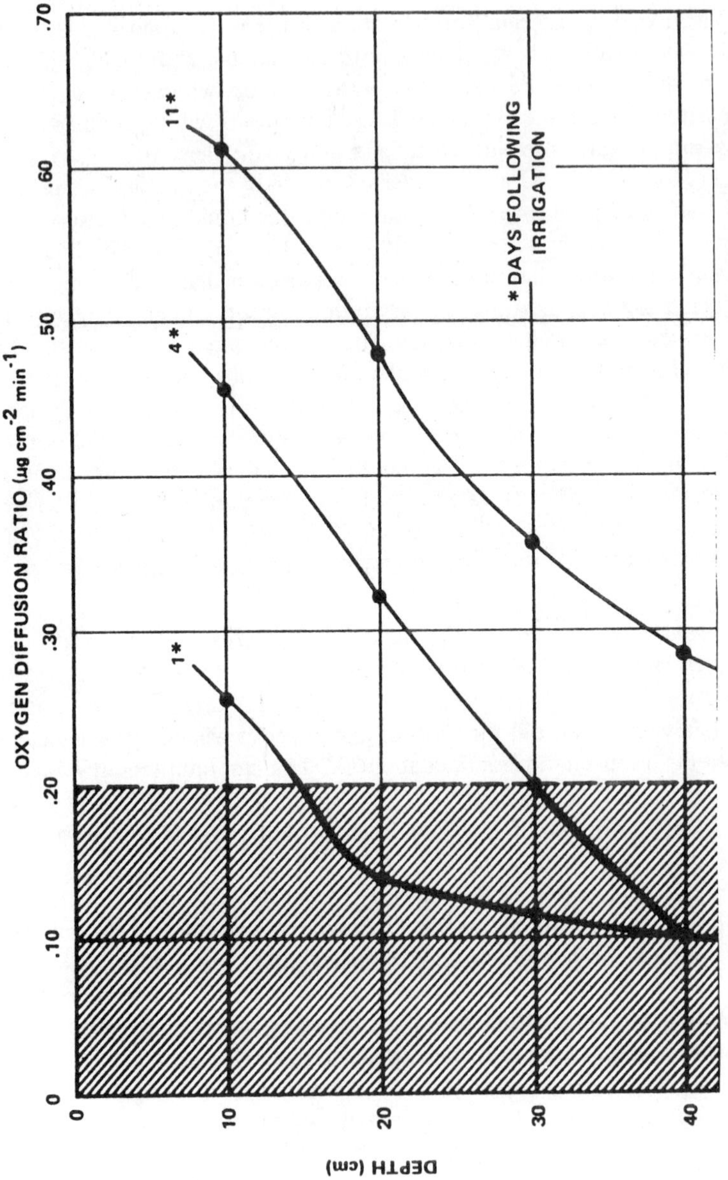

Figure 5. Oxygen diffusion rate (ODR) as a function of soil depth on various days following irrigation in a cotton field.

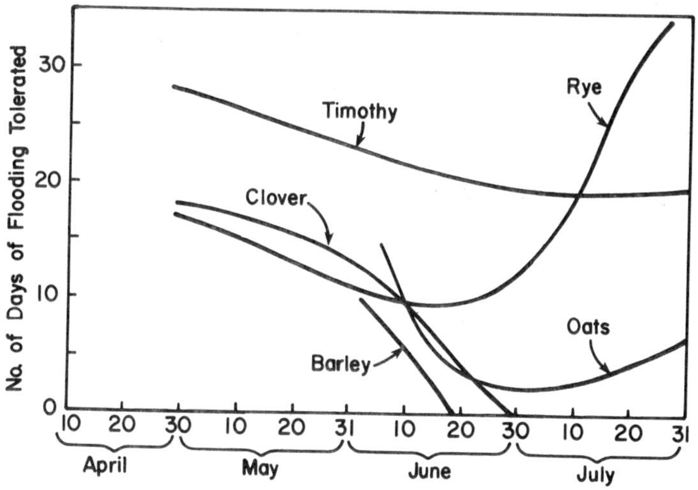

Figure 6. The estimated number of days flooding is tolerated by various crop plants at different times during the growing season under Finnish conditions, without the plants being destroyed. 1—timothy; 2—rye; 3—clover; 4—barley; 5—oats. Source: Van't Woudt and Hagan, 1957.

Anaerobic conditions caused by water logging can lead to the formation of phytotoxic compounds and the mobilization of phosphorus and potentially toxic elements. The effect of water-logging on crops is affected by crop, variety, stage of growth, temperature, light intensity, and fertility.

Root Systems and Soil Water Conditions

It has been observed from these and other experimental results that small grains are most tolerant of flooding at early stages of growth and when approaching maturity. Most perennial grasses can tolerate some flooding when dormant, although frost heaving of tap rooted crops may be a serious problem under these conditions in cold areas. The tolerance of most crops declines as temperatures rise and growth commences. Most crops are especially sensitive as they approach flowering. Some grasses appear to be the most tolerant of all crops. Shallow rooted clovers are among the least tolerant to flooded soils.

Crops, in varying degrees, develop roots adapted to saturated conditions. The most adapted to these conditions is rice, which is specialized for flooded fields. Some species have roots which will penetrate through a well aerated profile and extend below the saturated capillary fringe. These plants include alfalfa, sweet clover, soybeans, rye grass, wheat, and some fruit trees on special rootstock. These plants, however, do not necessarily demonstrate tolerance to flooding as does paddy rice. Other crops, including sugar cane, barley, corn, oats and tomatoes, respond to poor aeration by developing roots which are similar to those of rice. The tolerance of these crops to flooding is variable.

Some plant species develop root systems adapted to saturated conditions. However, they may not be adapted to flooded conditions.

In some crops such as corn and sorghum the most susceptible stage to damage by flooding is the seedling stage. This may be a temperature effect. In later stages of growth, corn can tolerate some flooding and sorghum is even more tolerant (Robins, 1967). Cotton and grain sorghum are unusual in that they may be flooded for several days during hot weather without being destroyed. Sugar cane and rice are the most adapted crops if waterlogged conditions are likely to prevail. These crops are commonly grown at land application sites in arid areas.

Among the grasses, Dallas grass, Reed canary grass, tall fescue and timothy appear to be the best adapted to moist conditions, much more so than ryegrass and orchard-grass (Van't Woudt and Hagan, 1957). Narrow leaf trefoil is the most tolerant legume grown under irrigation in California. The adaptation of different varieties within a species can be quite variable.

The effect of wet soils on crops differs according to the species and the stage of growth of particular species.

Disease and Insect Breeding Conditions Are Enhanced in Wet Soils

Under irrigation, favorable moisture relations increase the potential for disease and insect problems and these should be considered in the crop selection process. Menzies (1967) has reviewed the disease problem and Klostermeyer (1967) that of insects. Long low rates of water application or supplemental irrigation in already humid climates may maintain an environment for spore germination, infection and sporulation for secondary infections. Other types of irrigation present more of a problem in this regard than furrow irrigation because they wet the foliage and spread spores. In arid areas, however, experience has shown that disease problems did not develop with the anticipated degree of severity. Insect problems have been serious in various places, and are one of the reasons alfalfa is often left unirrigated during 2 to 3 summer months in Arizona (Van't Woudt and Hagan, 1957). For information on controlling insects consult a state college of agriculture, USDA office or agricultural extension personnel.

A final problem under irrigation is "scalding" damage. As reviewed by Van't Woudt and Hagan (1957), "scalding" (also known as "white spot disease") is the other major reason why alfalfa in Arizona is left unirrigated. "Scald" is associated with conditions of excessive moisture when high temperatures prevail. Symptoms develop suddenly, 12 to 24 hours after irrigation or heavy rain. Chlorophyll is destroyed first between leaf veins and along leaf margins. Entire leaves and entire plants may be affected with a resultant reduction in stand. The mechanisms involved in producing scald characteristics are not well understood, but ponding aggravates the situation. The following crops tend to be subject to "scalding": sweet clover, ladino clover, strawberry clover, birdsfoot trefoil, potatoes, barley, beans and lettuce. Alsike clover, Reed canary grass and Dallas grass were reported to be relatively resistant to scalding.

Irrigation can cause plant disease and insect problems. This problem is less severe with ridge and furrow irrigation. "Scalding" damage is possible where high temperatures prevail, although some crops are resistant to scald.

NUTRIENT REMOVAL AND RENOVATION EFFICIENCY

One of the primary objectives of land application is the removal of soluble nitrogen and phosphorus. When applied to land to fertilize vegetation, if the vegetation is not removed, then nutrients will recycle within the biomass, accumulate, and in spite of increased denitrification, eventually "leak" to the environment in drainage waters. The removal of macro- and micronutrients allows higher nutrient application rates without environmental degradation, and can be regarded as prolonging the effective life of a site as a waste treatment medium. If nutrients are applied only to meet crop requirements, then the site should remain effective indefinitely.

Weeds or natural vegetation could be allowed to grow in lieu of a crop of a land treatment site and would take up nutrients which would be removed if the weeds were themselves cut and removed. There is some doubt about the amount of nutrients which may be assimilated by weeds, but weeds are fast growing and vigorous competitors for water and nutrients. To the extent that weeds are quick to mature, they might not make advantageous use of the full growing season. Perhaps the biggest drawback to the use of weeds for nutrient uptake is the lack of a ready market for the harvested product. However, as pointed out by Wolcott and Cook (1976), certain weeds may nevertheless be useful in certain situations. In forage crops, quackgrass (*Agropyron repens L*) is often already established in many fields and forms on rhizome-laced sod which may help hold alfalfa from frost heaving and heavy equipment from rutting the field. Reed canary grass, though slow to establish from seed, also forms a tough rhizome sod.

Weeds might be advantageous for nutrient uptake and ease of establishment. There may not be a ready market for the harvested crop.

Tables 10 and 11 represent some of the values reported by various workers for yields and nutrient uptake of agronomic crops. In general it may be observed that crops take up only about $\frac{1}{8}$ to $\frac{1}{10}$ as much phosphorus as nitrogen. Potassium uptake by forage crops may be close to that of nitrogen but much of this is returned as crop residues if only edible seeds are removed. Forage crops also remove greater amounts of minor elements and trace elements from a soil than do edible seeds.

Crops vary in the amount of nitrogen and phosphorus taken up. These figures range from 72 to 450 lbs N/acre depending upon the part of the plant harvested, the type of crop and the yield.

Based on the information contained in Tables 10 and 11, comparisons of land area requirements, nutrient loading, and hydrologic loading can be made. Table 12 considers a design flow of 1 mgd, typically produced by 10,000 people. This design flow is with typical municipal sewage.

In humid areas, 12 in./acre/yr in addition to natural precipitation should, when applied as needed, meet ET requirements of most crops. Twelve inches/acre/year will also contribute substantially to the fertilizer requirements of most crops, although supplemental potassium

Table 10. Harvested Removal of Nutrients for Selected Crops and Yield Goals.

	Crop Yields and Nutrients Harvested (Lb /Acre)						
Nutrient	Corn Grain	Corn Silage	Wheat Grain	Soybeans	Alfalfa-Brome	Reed Canary Grass	Hardwood Forest (Annual Uptake, Lb/Acre)
Yield	150 bu	25 tons	60 bu	35 bu	5 tons	5.5 tons	
Nitrogen	125	165	72	120	220	408	84
Phosphorus	22	30	13	12	30	56	8
Potassium	28	150	14	36	166	247	26
Calcium	3	45	2	5	90	44	22
Magnesium	10	30	4	6	37	40	5

Source: Wolcott and Cook (1976).

Table 11. Annual Nitrogen, Phosphorus, and Potassium Utilization by Selected Crops (Lb/Acre).[a]

Crop	Yield	Nitrogen	Phosphorus	Potassium
Corn	150 bu	185	35	178
	180 bu	240	44	199
Corn silage	32 tons	200	35	203
Soybeans	50 bu	257[b]	21	100
	60 bu	336[b]	29	120
Grain sorghum	8,000 lb	250	40	166
Wheat	60 bu	125	22	91
	80 bu	186	24	134
Oats	100 bu	150	24	125
Barley	100 bu	150	24	125
Alfalfa	8 tons	450[b]	35	398
Orchard grass	6 tons	300	44	311
Brome grass	5 tons	166	29	211
Tall fescue	3.5 tons	135	29	154
Bluegrass	3 tons	200	24	149

[a] Values reported above are from reports by the Potash Institute of America and for the total above-ground portion of the plants. Where only grain is removed from the field, a significant proportion of the nutrients is left in the residues. However, since most of these nutrients are temporarily tied up in the residues, they are not readily available for crop use. Therefore, for the purpose of estimating nutrient requirements for any particular crop year, complete crop removal can be assumed.

[b] Legumes can get most of their nitrogen from the air, so additional nitrogen sources are not normally needed. There is evidence that alfalfa utilizes the nitrogen applied in wastewater when it is available.

Source: Sommers and Nelson, (1976).

would be required if forage crops were to be harvested regularly and supplemental nitrogen would be needed to supply the peak uptake demands of a corn crop. The utilization efficiency of the nutrients applied should be good, particularly if the wastes are applied during the hottest part of the growing season, and if applied in a fashion which does not injure the crop. Larger amounts, up to 0.3 inch of water per day, are needed in arid areas for optimal crop

Table 12. Hydraulic and Nutrient Loading at Various Application Rates.[a]

Flow = 1 MGD
Population = 10,000
Total yearly flow ≅ 1000 acre-foot

in./wk	Land Area Required	acre-in./acre-yr	Lb-N/acre-yr	Lb-P/acre-yr	Lb-K/acre-yr	Lb-Zn/acre-yr
0.4	1000 (acres)	12	116	49	43	3.8
0.8	500	24	232	98	86	7.6
1.2	330	36	348	147	129	11.4
1.6	250	48	464	196	172	15.2
2 in./wk	200	60	580	245	215	19.0

Wastewater characteristics: N-43 mg/l, P-18 mg/l, K-16 mg/l, Zn-1.5 mg/l which can be considered medium strength untreated sewage.

[a] Applied over 30 weeks in addition to natural precipitation.

growth. Larger amounts can be applied in humid areas as well, as long as percolation through the desired rooting depth is accomplished within about two days so that soil aeration is not restricted. When 24 inches or more of waste-water is applied during the growing season, phosphorus and zinc will begin to accumulate and may eventually limit the ability of the site to renovate wastewaters and to produce crops under general farming. At the rates cited in Table 12, it is highly unlikely that a site receiving domestic wastes would be limited for phosphorus removal until well past the site design period. During this period the total amount of phosphorus in the plow layer may be doubled and redoubled. Phosphorus will also be increased at lower soil depths. Evidence suggests, however, that the ultimate reversion products of phosphorus are very stable (Taylor and Kunishi, 1974). Module II-2, "Phosphorus Considerations," discusses this aspect of land treatment in detail.

Potentially Toxic Element Uptake Can be Beneficial in Lower Concentrations

Nitrogen uptake by a crop is a passive process and may be regarded as a function of transpiration and the nitrogen concentration of the water transpired. The incorporations of nitrate into amino acids, and the uptake of cations from the soil solution, on the other hand, are active aerobic processes dependent upon available photosynthate and oxygen. Plants demonstrate varying degrees of selectivity in the active uptake and translocations of cations to above ground plant parts and especially to reproductive organs. These observations have practical implications in that high nitrate forages can be toxic to livestock and seeds, fruits and edible roots tend to exclude toxic elements.

> *Plants actively and selectively take up potentially toxic elements. These elements are also selectively transported to the various plant parts. This results in different plants and the parts within the plants having varying concentrations of potentially toxic elements.*

The uptake and harvest of potentially toxic elements may seem undesirable from the standpoint of protecting the food chain. Uptake and removal in non-toxic (i.e., normal) concentrations, however, might be thought to prolong site longevity. Taking Zn as an example of a potentially toxic element, however, it is evident that the amounts applied in waste (Table 12) are large compared to the less than 0.5 pound of Zn per acre removed in most harvested crops, and large compared to the amounts added as fertilizers to alleviate deficiencies. Allaway (1975) has indicated that fertilizers supplying as little as 10 pounds of Zn per acre can produce optimum crop growth although heavier applications are needed to increase food and feed crops as sources of dietary Zn. A considerable margin of safety exists between the rates used to increase levels of dietary Zn and those which result in plant toxicity and yield reduction. Typical estimates of "safe" amounts would allow application of domestic wastewater for a number of years (Table 13). Heavily contaminated urban/industrial wastewaters, on the other hand, may exceed these limits in a short time.

Turf grass sod and ornamental nursery production has an apparent advantage on sludge and wastewater disposal sites in that a good deal of soil is harvested and removed with each crop. While this tends to deplete soils at unamended sites, on a land application site the removal of nitrogen, phosphorus and toxic elements in these specialty crops at rates much higher than in agronomic crops would be a decided advantage in extending site longevity. Spreading these elements over landscaped areas would tend to remove the contained toxic elements from food

Table 13. Total Amount of Sludge Metals Allowed on Agricultural Land.

	Soil Cation Exchange Capacity (meq/100 g)[a]		
	0–5	5–15	>15
Metal	Maximum Amount of Metal (Lb/Acre)		
Pb	500	1000	2000
Zn	250	500	1000
Cu	125	250	500
Ni	50	100	200
Cd	5	10	20

[a]Determined by the pH 7 ammonium acetate procedure.

Source: Sommers and Nelson (1976).

chains. The net result would be similar to that achieved if sludge were dried or compacted and sold as a soil amendment to landscapers and golf courses.

Irrigating sod and ornamental nursery stock for transplanting can extend site life and allow higher rates of waste application.

Aside from specialty crops in which soil is harvested, and while influenced by many factors including plant species and soil ionic balances, nutrient uptake into crops remains a function of crop growth. Farmers generally grow the crops best adapted to the length of season and amount of available moisture in their fields. Crop selection for maximum growth and nutrient uptake at land application sites, where available moisture should be abundant, will be limited primarily by yearly temperature fluctuations and solar energy.

Crop selection is limited by temperature fluctuation and solar energy at a land application site.

Perennial crops have an advantage in that they can begin growth and develop a crop canopy as early in the year as temperatures permit. They do not have to wait for preparatory tillage operations which can be delayed by wet weather. The early growth of perennials is also rapid because they often enter the growing season with considerable food reserves.

Perennial crops begin growth early and do not have to wait for preparatory tillage.

Figure 7 and Table 14 indicate how climatic regions influence the distribution of grass species grown in the United States. (The same type of information concerning legumes can be found in Kipps, 1970.)

Tropical grasses have a greater capacity to stand heat and utilize intense radiation than do temperate species. Under these conditions tropical grasses have been established at land application sites and utilize maximal amounts of water and nutrients. Because they are sensitive to cold, however, they grow only during warm seasons and cannot even be successfully maintained further north than the Carolinas.

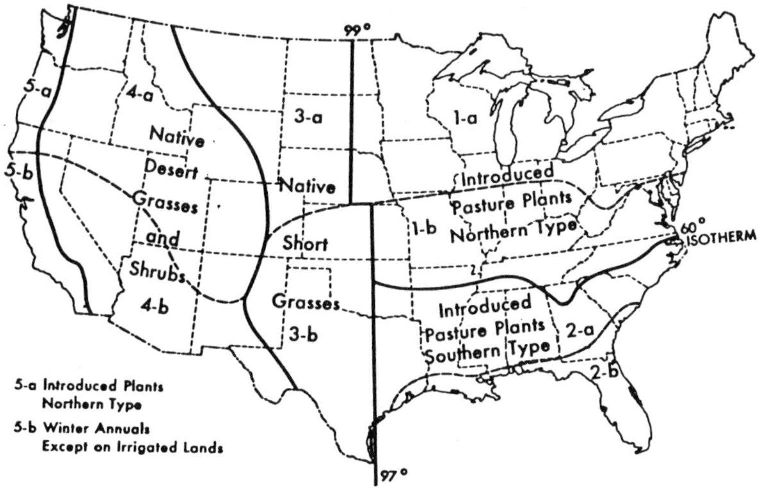

Figure 7. Map showing the five natural pasture regions, determined largely by climate, of the United States, and the types of pasture plants that provide most of the pasturage in each. Each region is divided into sections *a* and *b*, mainly because of temperature relations, except on the Pacific Coast. Here section 5*a* is the more or less humid Northern area, and 5*b* the rather arid Southern area. *(U.S. Dept. Agr. Misc. Pub. 194. 1946.)*

Techniques Exist for Extending Active Growing Seasons

In the south, the growth of cool season crops may allow plant wastewater renovation throughout the year; in the north periods exist when no crops will grow actively. Nevertheless, the season of active growth can be extended through crop selection (Figure 8). Where the length of growing season permits, double cropping of short season varieties of annual crops will generally rival the nutrient uptake of perennial crops, and invariably surpasses that of long season annuals. Fall seeded small grains grow later in the fall and earlier in the spring than pasture and meadow crops. Rye is the most cold and water tolerant of the small grains. Fall seeded small grains are

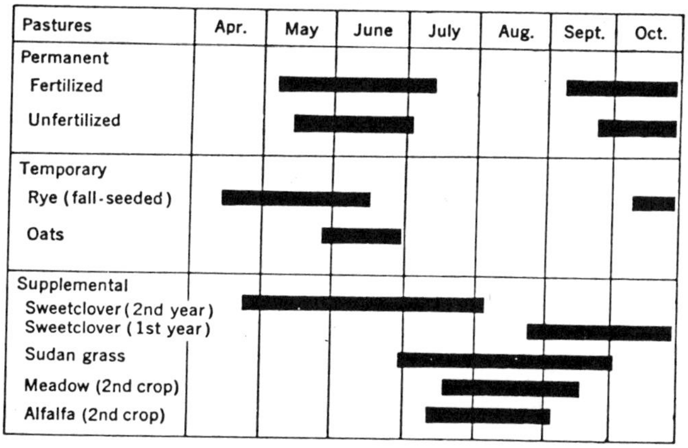

Figure 8. Combined use of permanent, temporary, and supplemental pastures designed to provide adequate grazing throughout the season. (USDA, 1939)

Table 14. Information Regarding Grasses for Permanent Pastures (USDA, 1946).

Name	Climatic Adaptation[a]	Degree of Palatability	Season for Grazing	Time and Rate of Seeding per Acre (Lb)	Soil Adaptation[b]	Remarks
Bahiagrass (*Paspalum notatum*)	Section 2-b	High	Early spring to late fall	Early spring, 10–15	Sandy loam to sand	Seed expensive and of low germination
Bermudagrass (*Cynodon dactylon*)	Region 2 and sections 3-b and 4-b	Medium	Late spring to early fall	Early spring, 5–8	Loams, clays, and silts	Propagated to a large extent vegetatively
Bromegrass or smooth brome (*Bromus inermis*)	Western part of section 1-a and sections 3-a and 4-a	High	Very early spring to late fall	Early spring or early fall, 15–20	Practically any type	Excellent grass for use with alfalfa
Canada bluegrass (*Poa compressa*)	Region 1-a and sections 3-a and 4-a	High	Early spring to late fall	Early spring or fall, 15–20	Almost any type	Succeeds on poor soils
Carpetgrass (*Axonopus compressus*)	Region 2	Medium	Spring to fall	Early spring, 8–12	Moist sands or sandy loam	Makes a very tight turf
Centipedegrass (*Eremochloa ophiuroides*)	Southern half of region 2	Medium	Spring to fall	Early spring; use sod or stolons; no seed available	Almost any type	Makes a close turf and is very aggressive, when once established, crowding out weeds, legumes, and other grasses
Crested wheatgrass (*Agropyron cristatum*)	Sections 3-a and 4-a	High	Very early spring to late fall	Early spring, 12–15	Almost any type	Drought resistant, easy to get a stand
Dallisgrass (*Paspalum dilatatum*)	Region 2 and sections 3-b, 4-b, and 5-b where irrigated	High	Early spring to late fall	Early spring or fall, 8–12	Any fairly productive soil	Seed expensive and often of low germination; difficult to get a stand
Johnsongrass (*Sorghum halepense*)	Region 2 and section 3-b; also sections 4-b and 5-b where irrigated	High	Spring to fall	Early spring, 20–25	Loams and clays	Productiveness decreases rapidly when grazed; very difficult to eradicate
Kentucky bluegrass (*Poa pratensis*)	Region 1 and section 5-a; also sections 3-a and 4-a where moisture is plentiful	High	Spring to late fall	Early fall, 15–20	Sandy loams to clays of high productivity	The leading pasture grass on good soils in the North
Meadow fescue (*Festuca elatior*)	Region 1 and section 5-a; also sections 3-a and 4-a where moisture is plentiful	High	Early spring to late fall	Early fall, 20–25	Loams to heavy clays	Valuable in section 5-a; of limited value elsewhere, disappearing rather quickly except on heavy, moist clays

Meadow foxtail (*Alopecurus pratensis*)	Sections 1-a and 5-a; also 4-a at high altitudes	High	Early spring to late fall	Early fall, 20–25	Moist sandy loams to clay	Very useful in pasture mixtures on wet soils, especially in 5-a
Orchardgrass (*Dactylis glomerata*)	Region 1; also sections 3-a, 4-a, and 5-a where moisture is plentiful	Medium to high	Early spring to fall	Early fall or early spring, 20–25	Any soil type except sand, if not too wet	Inclined to grow in bunches unless seeded thickly
Paragrass (*Panicum purpurascens*)	Section 2-b	High	Spring to fall	Early spring, no seed available	Wet soils	Propagated by planting pieces of stem or sod
Perennial ryegrass (*Lolium perenne*)	Southern half of region 1 and in section 5-a	High	Early spring to late fall; winter grazing in section 2-a to limited extent	Very early fall (or spring in the North), 20–25	Sandy loams to clays of medium to good fertility	Used but little in pasture except in section 5-a and in New England States
Redtop (*Agrostis alba*)	Regions 1, 2, and section 5-a; also under irrigation and in mountain meadows, sections 3-a and 4-a	Medium	Early spring to late fall	Early fall best, early spring fair, 10–12	Grows on majority of soil types; prefers moist soils	Of most value on poorly drained soils too wet for other grasses
Reed canarygrass (*Phalaris arundinacea*)	Sections 1-a and 5-a; also sections 3-a and 4-a where moisture is plentiful	Medium	Spring to fall	Very early spring, 8–12	Loams to heavy clays	Very good for wet lands, will endure submergence
Rescuegrass (*Bromus catharticus*)	Region 2, and where moisture is sufficient in sections 3-b and 4-b	Medium	Fall to spring (winter pasture)	Early fall, 20–25	Sandy loam to clay loam	An annual used in some localities for winter and spring grazing
Rhodesgrass (*Chloris gayana*)	Section 2-b; also southern parts of sections 3-b and 4-b	Medium	Spring to fall	Spring, 10–12	Loams to clays	Found most useful in the dry sections of southern Texas where other grasses fail
Slender wheatgrass (*Agropyron pauciflorum*)	Northern parts of sections 3-a and 4-a	High	Early spring to late fall	Early spring, 15	Practically any type, except sand	Better for hay than pasture; inclined to be stemmy
Tall oatgrass (*Arrhenatherum elatius*)	Sections 1-b and 5-a	Medium	Early spring to late fall	Early fall, 20–25	Practically any type, except sand	Better in hay mixtures than for pasture; stemmy; used most for pasture in section 5-a
Timothy (*Phleum pratense*)	Region 1 and section 5-a; also sections 3-a and 4-a where moisture is sufficient	High	Early spring to late fall	Early fall (or early spring), 12–15	Practically any type, except sand	Comes quickly and furnishes much pasturage at first but is not permanent

[a] The region and section numbers refer to those in Figure7.
[b] Specifications of soil types in this and the following tables are necessarily very general because of space limitations.

Table 15. Fate of N Applied to Corn-Rye Rotation. (Army Corps of Engineers, 1972)

Harvest Removal	Pounds of N
Corn grain	195
Rye forage	85
Corn stover (partially removed with rye forage)	20
Volatilization and Denitrification[a]	150
Tolerable Leaching Losses	50
Total	500

[a]Thought to occur mainly during the first 30 days of corn growing season.

usually seeded after the harvest of an intertilled crop such as beans or early corn. They may be successfully seeded after a sod if the sod is plowed after the first cutting "No tillage" or "minimum tillage" techniques may extend the options available to the farm manager on a land application site. The Army Corps of Engineers (1972) selected a minimum tillage corn-rye cropping system for illustrative purposes in their attempt to integrate crop production with wastewater disposal in northern Illinois. In this scheme corn was grown each year, rye was areally broadcast 1 to 2 weeks before corn harvest. Rye forage was harvested in mid November and was killed with a contact herbicide prior to no-till corn planting in the spring. The Corps determined that 500 lb N/acre/yr could be applied without degrading groundwater quality (Table 15).

The growth of cool season crops in the south may allow plant wastewater renovation throughout the year. Double cropping of short season annuals may rival the nutrient uptake of perennials in temperate areas. No tillage techniques also provide for uninterrupted crop rotations.

In general, it is important that the land application site designer consult the local agricultural extension office, state college, or department of agriculture office for crop varieties and species suited to the area and the wastewater application situation.

CROP TOLERANCES TO NUTRIENTS AND SALTS

When additions of elements in excess quantities are made, the natural functionings of plants may be disrupted. Such effects may be apparent from the reduced growth, delayed maturity, lodging, or reduced quality of crops. There are many mechanisms which can cause these effects. Reduced growth may be due to high total salt concentrations or to high concentrations of a specific ion. Reduced quality may vary all the way from a lack of proper color in fruit to actual toxicities in animals consuming the plants.

Excess nutrient problems differ from problems due to high concentration of total salts or trace elements, because large amounts of these nutrients are needed for normal crop growth. N, P, and K are added routinely by farmers. When irrigating with wastewater, however, the farmer loses a degree of control over the timing and the amount of fertilizer applied.

Phosphorus

Excessive phosphorus may be toxic to plants by inducing deficiencies of Cu, Zn, Fe or Mn. Hinesly, *et al.* (1974) added excessive amounts of liquid digested sludge to several crops, and observed phosphorus toxicity in soybeans during one season. In this experiment, much more sludge was added than would be recommended on the basis of N, P, or potentially toxic element loadings. It is unlikely that phosphorus toxicity would result from moderate applications of wastewaters or sludges.

It is unlikely that phosphorus toxicity would result from moderate applications of wastewaters or sludges.

Potassium

Potassium is generally not toxic to plants or plant consumers although in combination with high nitrate levels it has been associated with reduced magnesium uptake in forages and subsequent deficiencies in lactating cows and sheep (grass tetany). Potassium is generally low in sludges and present in moderate amounts in wastewaters and effluents. Potassium should not cause toxicity problems although deficiency problems may arise.

Nitrogen

Most of the problems associated with excess and untimely fertilizer application center around nitrogen. Crop logging, or analyzing plant samples to determine nutritional status, has resulted in correlations between high nitrogen levels and a variety of problems. Abundant nitrogen and water results in lush growth. This can result in lodging of small grains and buckwheat (inability of the soft stems to support the oversized, heavy seeds), excessive foliage growth of potatoes and cotton at the expense of tubers and fruits, decreased sugar content of sugar beets, and increased fruit rot and decreased vitamin C, due to shading, in melons, squash and tomatoes.

Excess nitrogen can be damaging to several types of crops. These include small grains, tubers, cotton, sugar beets, melons, squash and tomatoes.

Excess nitrates in forages is a special problem. Nitrate accumulation in plant tissues varies with plant type, climatic conditions, and fertilization. Certain weeds (i.e., pigweed) accumulate nitrate up to 1-1.5% of their dry weight. Under conditions with slow growth (drought, cold, or extended periods of cool cloudy weather) corn, sorghum, and succulent annual grasses may accumulate abnormally high concentrations of nitrate. Applications of high-nitrate-containing wastewaters a short time prior to harvest may increase nitrate levels in crops. High nitrate forages are a problem in two ways: (1) Nitrogen oxide gases that can be deadly to humans and farm animals in enclosed areas can be formed immediately after ensiling high nitrate forages, and (2) nitrate may be converted to nitrite in alkaline digestive tracts (i.e., within ruminants and human infants) which may react with hemoglobin and reduce its capacity to transport oxygen, resulting in methemoglobinemia or cyanosis. At nitrate levels higher than 0.1% of animal body weight there is a 50% fatality rate (see Module II-1, "Nitrogen Considerations").

High nitrates in forage crops can produce nitrous oxide upon being ensiled or cause cyanosis in ruminants.

Other miscellaneous problems have been reported. Tall fescue, one of the most palatable grasses well adapted to moist conditions, has been observed to cause fatty tumors on the intestines of cattle ("fat necrosis") when fertilized with heavy rates of manure (Hart, 1974).

The milling and baking qualities of wheat grain produced with wastewater have been found to be lower than the qualities of grain grown with well water and commercial inorganic fertilizers (Day, 1973).

At Bakersfield, California, management factors which have been employed to counteract the effect of excessive and untimely applications of nitrogen on cotton include: (1) Thin cotton more than usual, (2) Stress the plants between irrigation periods to stimulate fruitive growth; and (3) Blend effluent with well water to reduce the total N concentration (Crites, 1974).

Modifications to counteract the effect of excessive and untimely applications of nitrogen have included thinning the crops, stressing them between irrigation periods and diluting the wastewater.

Salts

Besides excess amounts of macro-nutrients, the most general and most extensively studied reason for reduced crop growth is referred to as "salts." "Total salts" means essentially the same thing as "total dissolved solids," a commonly measured characteristic of wastewaters and effluents. Studies have shown that from 247-421 mg/l of total dissolved solids (average 300 mg/l) is added to water by one municipal use (Baier and Fryer, 1973). When a solid dissolves in water it generally breaks apart into positively and negatively charged parts or ions. The mobility of an ion in solution depends on its charge, its mass, its size and shape, as well as on temperature and other factors. Because they carry charges and are mobile, ions in solution will conduct electricity and are often measured as electrical conductivity (EC) in water and wastewater.

Salt problems have been investigated in relation to experience gained in arid areas under irrigation, in greenhouse culture, and in the reclamation of land from the sea. In spite of the variation in ionic balances which might be presumed to exist between these different situations, salts, as measured by electrical conductivity, have been consistently correlated with yield reductions.

Excess salts have been correlated with yield reductions, mainly in arid areas.

Regardless of the mechanism, accumulation of salts in arid soils has been a recognized problem for some time, and a body of practical knowledge has been accumulated, "Salts" are generally measured as the conductivity of a soil water extract. Electrical conductance is expressed in mhos which are reciprocal ohms. Electrical conductivity has the dimensions mhos/cm. In practice millimhos turn out to be a convenient size unit and results are usually expressed as mmhos/cm or EC $\times 10^3$.

As with tolerance to waterlogging, crops vary in their tolerance to excess salts in the soil solution (Table 16), and tolerances vary with stage of growth. In Table 16 it should be noted that reed canary grass, while tolerant to flooding, drought and scald, and while efficient in taking up large amounts of nutrients, is only moderately tolerant of salts. Barley, which will not tolerate waterlogging, is very tolerant of salts. Day (1974), however, noted that barley was found to be more sensitive to the constituents in wastewater than was wheat or oats.

Crops differ in their tolerance to excess salts. Tolerance also can depend on the stage of growth of the crops.

Table 16. Relative Tolerances of Field and Forage Crops to Salt or Electrical Conductivity.[a]

Tolerant (partial listing)	Semitolerant (partial listing)	Sensitive (complete listing)
Barley	Alfalfa	Alsike clover
Bermuda grass	Corn (field)	Burnet
Birdsfoot trefoil	Flax	Field beans
Canada wildrye	Oats	Ladino clover
Cotton	Orchard grass	Meadow foxtail
Rape	Reed canary grass	Red clover
Rescue grass	Rice	White dutch clover
Rhodes grass	Rye	
Sugar beet	Sorghum (grain)	
	Sudan grass	
	Tall fescue	

[a]Tolerant–10,000–18,000 umhos/cm;
Semitolerant–4,000–12,000 umhos/cm;
Sensitive–2,000–4,000 umhos/cm.

Source: Pound and Crites, 1973.

Criteria for Salt Content of Irrigation Water Have Been Developed

Unless diluted by rain, salinity of the soil solution will at least equal that in irrigation waters. Due to evaporation, soil solution concentrations in arid areas generally exceed those of applied waters. Based on experience in a variety of climatic and drainage situations, criteria have been developed which should be directly applicable to the evaluation of effluents and wastewaters as sources of irrigation waters. Table 17 summarizes those of the U.S. Salinity Laboratory. Based on U.S. criteria, an effluent containing 500-1,500 ppm TDS would be classified as "high-salinity water" not suitable for use in arid areas on soils with restricted drainage. In light of experience with use of very low quality irrigation water, new criteria have been proposed by workers in Algeria (Table 18). These criteria, based on soil type as well as plant tolerance, assume ideal irrigation conditions (permeable soil and good drainage) and so may have applicability on intensively managed land application sites. It is clear from Table 18 that the excellent drainage characteristics of coarse soils greatly outweigh the relative dilution advantage of clay soils in avoiding toxic salt concentrations. When a significant fraction of the applied wastewater moves through the soil as percolation, salt buildup may not be a problem even in arid areas unless the wastewater TDS is exceptionally high and sensitive crops are grown.

Soil solution concentrations of salts are generally higher than the concentration in applied wastewater. With good drainage and management, crops can be irrigated with high salinity wastewater.

In addition to the effect of total salts on crop growth, high concentrations of certain elements have been recognized to be particularly harmful. Sodium is perhaps the most critical ion to be considered in waters containing appreciable amounts of TDS. Na is critical because it affects plant growth directly, and indirectly through its undesirable effect on soil structure and permeability. These effects are summarized in Table 19, wherein sodium concentration is ex-

Table 17. U.S. Salinity Laboratory's Grouping of irrigation water.

	Classification of Water	Electrical Conductivity in µmhos/cm at 25°C (EC)	Salt Concentration in g/l (approximate)
C1	*Low salinity water* can be used for irrigation with most crops on most soils, with little likelihood that a salinity problem will develop. Some leaching is required, but this occurs under normal irrigation practices, except in soils of extremely low permeability.	$0 < EC \leqslant 250$	<0.2
C2	*Medium-salinity water* can be used if a moderate amount of leaching occurs. Plants with moderate salt tolerance can be grown in most instances without special practices for salinity control.	$250 < EC \leqslant 750$	0.2–0.5
C3	*High salinity water* cannot be used on soils with restricted drainage. Even with adequate drainage, special management for salinity control may be required and plants with good salt tolerance should be selected.	$750 < EC \leqslant 2250$	0.5–1.5
C4	*Very high salinity water* is not suitable for irrigation under ordinary conditions but may be used occasionally under very special circumstances. The soils must be permeable, drainage must be adequate, irrigation water must be applied in excess to provide considerable leaching, and very salt tolerant crops should be selected.	$2250 < EC \leqslant 5000$	1.5–3

Source: FAO (1973).

pressed in terms of exchangeable-sodium percentage (ESP). This is a measure of the percentage of active cation exchange sites in the soil which are occupied by Na^+ ions. For instance, when 15% of the exchange complex is occupied by sodium, the exchangeable sodium percentage equals 15.

Excessive sodium affects both plant growth and soil structure. Plants most tolerant to sodium can be grown when the ESP is over 60.

Boron has been identified as a problem in some irrigated areas, and is one of the most troublesome elements discussed in Module II-4, "Potentially Toxic Elements." Information on the crop tolerance to boron is reproduced in Table 20. In soils where B is already high, careful analysis of wastewaters should be made, as concentrations are often high enough to cause problems. Boron concentrations may increase in the future if borates are increasingly substituted for

Table 18. Upper Limits of Electrical Conductivity, in Micromhos, of Irrigation Water for Three Groups of Plants and Five Soil Groups.

	Plant Tolerance Groups					
			III (High)			
Soil Textures	I (Low)	II (Moderate)	Dutt	Horticultural	Forage	Field Crops
Sandy	2500	6500	15,000–20,000	8000	12,000	10,000
Loamy sand	1600	4000	6000–10,000	4500	7000	6000
Loamy	1000	3000	8000	3500	5000	4500
Loamy clay	800	2000	6000	2400	–	3500
Clay	400	1000	3000	1800	1800	1600

Group I = Electrical conductivity of soils saturation extract < 4000 umhos/cm.
Group II = Electrical conductivity of soils saturation extract 4000 to 10,000.
Group III = Electrical conductivity of soils saturation extract > 10,000.

Source: FAO (1973).

Table 19. Tolerance of Various Crops to Exchangeable-Sodium Percentage (ESP) under Nonsaline Conditions.

Tolerance to ESP and Range at which Affected	Crop	Growth Response under Field Conditions
Extremely sensitive (ESP = 2–10)	Deciduous fruits Nuts Citrus Avocado	Sodium toxicity symptoms even at low ESP values
Sensitive (ESP = 10–20)	Beans	Stunted growth at low ESP values even though the physical condition of the soil may be good
Moderately tolerant (ESP = 20–40)	Clover Oats Tall fescue Rice Dallisgrass	Stunted growth due to both nutritional factors and adverse soil conditions
Tolerant (ESP = 40–60)	Wheat Cotton Alfalfa Barley Tomatoes Beets	Stunted growth usually due to adverse physical condition of soil
Most tolerant (ESP = more than 60)	Crested and Fairway wheatgrass Tall wheatgrass Rhodes grass	Stunted growth usually due to adverse physical condition of soil

Source: Bernstein (1974).

Table 20. Crop Tolerance Limits for Boron in Saturation Extracts of Soil. For Each Group, Tolerant, Semitolerant, and Sensitive, the Range of Tolerable Boron Concentration is Indicated. Tolerance Decreases in Descending Order in Each Column.

Tolerant	Semitolerant	Sensitive
Sugar beet	Potato	Pecan
Alfalfa	Cotton	Walnut
Turnip	Barley	Plum
	Wheat	Pear
	Corn	Apple
	Milo	Grape
	Oat	Cherry
		Peach
		Orange
2.0 ppm of boron	1.0 ppm of boron	0.3 ppm of boron

Source: Bernstein (1974).

phosphates in detergents (Dowdy, *et al.*, 1976). Borates, like nitrates, move readily through soils with percolating waters.

Excessive boron is phytotoxic. Borates also leach readily.

Chloride is generally not recognized as a problem in field crop production, and Eaton (1971) states that anions are generally harmless. Its effects interact with those of total salts and Na. When chlorinated effluents are applied to cropland the largest effect may be an increased lime requirement (Chaney, 1973).

PRACTICAL CONSIDERATIONS

Thus far in this module, the requirements and tolerance of a variety of crops have been discussed with reference to the conditions which are likely to prevail at a land application site: high hydraulic loading with water high in nutrients, salts and potentially toxic trace elements. In this section a variety of miscellaneous observations are made, which may be of some benefit in anticipating returns and avoiding unexpected problems.

Costs

The cost effectiveness of a cropping system depends upon many things, not the least of which is the manager's ability to market his crops. Tables 21 and 22 are an indicator of the potential value of a cropping operation.

In Tables 21 and 22 figures were taken appropriate to the most productive soils receiving recommended management. This results in optimum yields (Table 22) and when these are combined with favorable prices, considerable profit margin may theoretically be realized. If unfavorable price relationships exist at harvest time then the situation may be dramatically changed. If wheat brought only $1.50/bu then a yield of over 60 bu/acre would be required just to break even. The costs in Table 21 and 22 do not include any charges for land costs or taxes or interest on investment. The values assigned to haylage and corn silage by Knoblauch

Table 21. New York State Production Costs, 1977.

		Corn		Alfalfa[d]		Wheat
Tillage[b]						
Plow @ $10/acre		$25.00		$6.25		$25.00
Disc. @ $6/acre		(10.00)		(2.50)		(10.00)
Harrow @ $3/acre		(3.00)		(1.00)		(3.00)
Planting[b] @ $5/acre		5.00				
Drill[b] @ $5/acre				1.25		5.00
Apply fertilizer, lime or pesticides[b] @ $3/acre		3.00		1.50		3.00
Cultivate[b] @ $5/acre		5.00				
	Subtotal	($38.00)		($9.00)		($33.00)
Seed[a]	(.3 bu)	13.80	(12 lb)	4.25	(2.3 bu.)	14.50
Fertilizer[a]						
N @ $.18/lb	(130 N)	23.40			(60 N)	10.80
P_2O_5 @ $.19/lb	(60 P_2O_5)	11.40	(75 P_2O_5)	14.25	(60 P_2O_5)	11.40
K_2O @ $.10/lb	(60 K_2O)	6.00	(100 K_2O)	10.00	(40 K_2O)	4.00
Lime, cover crop, other[a]		7.00		3.00		3.00
Herbicide		6.85		2.00		
Pesticides		12.30[c]				
	Subtotal	($80.75)		($33.50)		($43.70)
Harvest						
	combine corn	20.00[b]	mow, bale haul and store hay	100.00[e]	combine small grain	15.00[b]
	dry corn	21.60[a]				
	sell corn	6.80[a]			sell wheat	5.50[a]
	Subtotal	($48.40)				($20.50)
	Total dry shell corn	$167.15	*Total hay*	$142.50	*Total wheat*	$97.20
	Field chop corn	(30.00)[b]	Chop haylage	(30.00)[b]		
	Total haylage	$148.75	Total haylage	$72.50		

[a] From Knoblauch and Milligan, 1977.
[b] From Knoblauch and Snyder, 1977.
[c] Third year and thereafter of cont. corn-spray for rootworm @ $12.30/acre.
[d] Tillage and seeding costs are reported as $\frac{1}{4}$ of the actual cost because the stand is assumed to be productive for 4 years without renewal.
[e] based on yield of 5 ton/acre at a cost of $.50/50 lb bale (from Ref. 2).

and Milligan (1977) represent their value to a farmer feeding dairy cattle. These products are difficult to transport long distances and cannot be readily stored once removed from the silo. If animals are fed silage, pasture, or other crops as part of the land application system, most of the nutrients contained in the crop would have to be recycled back to the land in the form of animal manures. If harvested as hay, forage crops may more easily be sold off the farm. For such sale, however, quality is important and the type of hay which is in greatest demand (probably alfalfa) may reasonably influence crop selection. Making hay, however, has inherent risks as it is not possible to control the weather. In fact, on land application sites where soils may never become hot and dry, the curing of hay would be relatively slow. The quality of the hay

Table 22. N.Y.S. Production Costs and Returns, 1977.

Crop	Cost[a]	Return[b]	Margin/Acre
Dry shell corn	$167.15/acre	120 bu/acre X $2.50/bu = $300/acre	$132.85
Corn ensilage	$148.75/acre	20 tons/acre X $17.00/ton = $340/acre	$191.25
Hay	$142.50	5 tons/acre X $70.00/ton = $350/acre	$207.20
Haylage	$ 72.50	15.6 tons X $21.15/ton = $330/acre	$257.20
Wheat	$ 97.20	60 bu X $3.00/bu = $180/acre	$ 82.80

[a] From Table 21.
[b] Based on estimates as indicated.

harvested might not always be the best, but it could still be profitable in areas where demand is strong.

When well managed and with a favorable market price, crops can yield a favorable profit margin. Low prices can result in a loss.

Crop Management

In evaluating crop relations for a land application site, all of the subjects previously addressed in this module should receive consideration. When applied to the best agricultural soils, wastes will be of value for their N, P and water content. As shown in Table 21, savings in fertilizer may be substantial. Practices such as irrigation and drainage represent additional costs to the farmer but increase the farmer's control over the environment surrounding his crops. In some locations, improved drainage may confer benefits comparable to that of the supplied fertilizer. With good drainage, additional crops become adapted to a site, and all crops may be planted and harvested earlier. In some locations cropping is limited or impossible without irrigation waters. In humid areas on deep soils a benefit from irrigation would be realized in all but the ideal crop growing years. Also the availability of irrigation waters would act as insurance protecting a farmer's investment in seed and tillage and this could do much toward stabilizing the agriculture in an area. Available water would be especially beneficial when starting small seeded forage species. Seeding a forage mixture in the fall following wheat harvest is a risky practice if adequate moisture is not assured.

Wastewater and sludge have considerable fertilizer value. Improved drainage and irrigation capability can allow a wider variety of crops to be planted, the ground can be worked earlier and harvest can take place earlier.

In selecting a crop the question of the actual physical management of the wastes is very important. Foliar applications of sludge to standing hay may do more to contaminate the crop than any amount of waste additions made directly to the soil. Chaney, *et al.* (1976) reported that a considerable fraction of the elements applied when sludge was spread onto a freshly cut tall fescue meadow were not washed off the plants by subsequent rains although their concentrations were diluted by subsequent growth. Applications of sludge on actively growing forage crops, while being desirable in terms of N and P renovation, may result in contaminated forage if the sludge is allowed to dry and form a crust on plant leaves. This could greatly reduce the value of harvested hay or haylage. Analysis of applied wastes, field observations, and analysis of plant samples are necessary for intelligent management. Where one waste might result in crops with elevated levels of cadmium, another waste might not even correct boron deficiencies.

Foliar applications of sludge may contaminate forage. The residue may not be washed off by rainwater.

Crop Establishment

Once a crop has been selected it must still be managed correctly. Knowledge of how to do this is shared among farmers throughout the country and agriculturalists at many agencies and land grant institutions. While reed canary grass may be chosen because of its high nutrient uptake and water tolerance, it should also be recognized that reed canary grass loses palatability quickly as it approaches maturity and must be cut three times per season for maximum production. Reed canary grass is also slow to establish. In one overland flow experiment at Pennsylvania State University, where reed canary was shown with Italian ryegrass, the quickly germinating ryegrass completely dominated the first year's forage growth (Butler, *et al.*, 1974). Reed canary grass is often sown in other grasses only in a few widely separated rows with the intent that it will spread out and dominate over time.

If an annual crop is selected then in most cases a conventional seed bed will be prepared. In most soils, planting must be delayed until the soils are dry enough and warm enough to avoid compaction and ensure rapid germination. Because sludges may inhibit germination, they may have to be spread some time ahead of tillage operations. To allow for the drying and warming of the soil, wastewater application may have to be withheld until weather conditions allow planting and germination.

Agricultural factors may necessitate cessation of irrigation for a period of time.

An important aspect of the establishment of any crop is weed control. Weeds are controlled by a combination of tillage and herbicide application. Any successful tillage operation requires that the soil be dry enough at a 6-inch depth to avoid compaction and thus, like making hay, will require suspension of wastewater applications for variable periods, depending on weather. It would seem, under these circumstances, that herbicide would be an easier way to control weeds. Herbicide activity, however, may be significantly reduced by leaching a soil with large volumes of wastewater. Degradation by a large population of soil microorganisms maintained in moist, fertile soils can also reduce herbicide activity. Since the activity of the various herbicides may be differentially affected by the moist conditions, it will be necessary to consult an expert in this area.

Herbicides may be affected by conditions at a land treatment site.

CROP SELECTION—CONCLUSION

This module is intended to serve as a nontechnical introduction to many of the complex factors which have been recognized by crop scientists. It is not intended to provide complete instruction in farm management. It has attempted to supplement information contained in other modules. In any applied situation, an expert will no doubt be employed to manage farming operations, including crop selection, at a land application site. The information in this module is intended to provide a basis for communication between agricultural specialists and planners. It is hoped that it will provide some feeling for the kind of knowledge that is available, and the kind of problems that may be anticipated and avoided at preliminary design stages of a land treatment system involving crop production.

BIBLIOGRAPHY

Allaway, W. H. 1975. The effects of soils and fertilizers on human and animal nutrition. Agriculture Information Bulletin No. 378. ARS and SCS, USDA, Washington, D.C. 52 p.

Army Corps of Engineers. 1972. Chicago-South end of Lake Michigan area wastewater management study. Chicago, Ill.

Baier, D. C. and W. D. Fryer. 1973. Undesirable plots responses with sewage irrigation. Journal of the Irrigation and Drainage Division, Proceedings of the American Society of Civil Engineers, Vol. 99, No. IR2. pp. 133–141.

Bernstein, L. 1974. Crop growth and salinity. pp. 39–54, *In* Drainage for agriculture. *Agronomy*, No. 17. ASA. Madison, Wis. 700 p.

Butler, R. M., J. V. Husted, and J. N. Walter. 1974. Grass filtration for final treatment of wastewater. pp. 256–272. *In* Wastewater use in the production of food and fiber. Proceedings. EPA-660/2-74-041. 568 p.

Chaney, R. L. 1973. Crop and food chain effects of toxic elements in sludges and effluents. pp. 129–142. *In* Recycling municipal sludges and effluents on land. National Association of State Universities and Land Grant Colleges, Washington, D.C. 244 p.

Chaney, R. L., S. B. Hornick, and P. W. Simon. 1976. Heavy metal relationships during land utilization of sewage sludge in the northeast. pp. 283–316. *In* Land as a waste management alternative. R. C. Loehr, editor. Proceedings of the 1976 Cornell Agric. Waste Man. Conf. Ann Arbor Science Publishers, Inc. Ann Arbor, Mich. 811 p.

Crites, R. W. 1974. Irrigation with wastewater at Bakersfield, Calif. pp. 229–239. *In* Wastewater use in the production of food and fiber. Proceedings. EPA 660/2-74-041. 568 p.

Day, A. D. 1973. Recycling urban effluents on land using annual crops. pp. 155–160. *In* Recycling municipal sludges and effluents on land. Natl. Assn. of State Universities and Land Grant Colleges. Washington, D.C. 244 p.

Dowdy, R. H., R. E. Larson, and E. Epstein. 1976. Sewage sludge and effluent use in agriculture. pp. 138–153. *In* Land application of waste materials, Soil Conservation Society of America, Ankony, Iowa. 313 p.

Eaton, F. M., W. R. Olmstead, and O. C. Taylor. 1971. Salt injury to plants with special reference to cations versus anions and ion activities. *Plant and Soil* 35, pp. 533–547.

EPA. 1974. Water use in the production of food and fiber–proceedings of the conference held at Oklahoma City, Oklahoma. March 5–7, 1974. Environmental Protection Technology Series. EPA-660/2-74-041. 568 p.

FAO. 1973. Irrigation, drainage and salinity, an international resource book. Hutchinson & Co., Ltd. London. 510 p.

Grable, A. R. 1966. Soil aeration and plant growth. *Advan. Agron.*, 18:57–106.

Haise, H. R. and R. M. Hagan. 1967. Soil, plant, and evaporative measurements as criteria for scheduling irrigation. pp. 557–604. *In* Irrigation of agricultural lands. *Agronomy*, No. 11. ASA, Madison, Wis. 1180 p.

Hart, R. H. 1974. Crop selection and management. pp. 178–200. *In* Factors involved in land application of agricultural and municipal wastes. (DRAFT) ARS. USDA. Beltsville, Md. 200 p.

Hinesly, T. D., O. C. Braids, R. I. Dick, R. L. Janes, and J-A. E. Molina. 1974. Agricultural benefits and environmental changes resulting from the use of digested sludge on field crops. University of Illinois Urbana. 375 p.

Keller, W. and C. W. Carlson. 1967. Forage crops. *In* Irrigation of Agricultural Lands. *Agronomy*, No. 11, ASA, Madison, Wis. 1180 p.

Kerr, J. P., G. W. Turtell, and C. B. Tanner, 1967. An integrating pyranometer for climatological observations and meso-scale networks. *J. Appl. Meteoral.* 6:688–694.

Kipps, M. S. 1970. Production of field crops. McGraw-Hill Book Co., New York. 790 p.

Klostermeyer, E. C. 1967. Insect problems of irrigated lands, pp. 1065–1069. *In* Irrigation of agricultural lands. *Agronomy*, No. 11, ASA, Madison, Wis. 1180 p.

Knoblauch, W. A. and R. A. Milligan. 1977. An economic analysis of N.Y. Dairy Farm Enterprises. Agric. Econ. Research 77-1, Cornell University, Ithaca, N.Y. 66 p.

Knoblauch, W. A. and D. P. Snyder. 1977. Custom and rental rates for farm machinery in N.Y.S., Agric. Econ. Extension 77-10. Cornell University, Ithaca, N.Y.

Letey, J. Jr., L. H. Stalzey. and W. D. Kemper. 1967. Soil aeration. pp. 941–949. *In* Irrigation of agricultural lands. *Agronomy*, No. 11. ASA, Madison, Wis. 1180 p.

Menzies, J. D. 1967. Plant diseases related to irrigation, pp. 1058–1064. *In* Irrigation of agricultural lands. *Agronomy*, No. 11, ASA, Madison, Wis. 1180 p.

Metcalf and Eddy, Inc. 1977. Process design manual for land treatment of municipal wastewater. EPA, Army Corps of Engineers, and USDA. EPA-625/1-77-008.

Pound, C. E. and R. W. Crites. 1973. Wastewater treatment and reuse by land application. Vol. II. Environmental Protection Technology Series. EPA-660/2-73-006b. 249 p.

Pound, C. E., R. W. Crites, and D. A. Griffes. 1975. Evaluation of land application systems. Technical Bull. EPA-430/9-75-001. 182 p.

Richards, L. A., ed. 1954. Diagnosis and improvement of saline and alkali soils. Agric. Handbook No. 60. USDA. Washington, D.C. 160 p.

Robins, J. S., J. T. Musick, D. C. Finfrock, and H. F. Rhoades. 1967. Grain and field crops. *In* Irrigation of agricultural lands. *Agronomy*, No. 11, ASA, Madison, Wis. 1180 p.

Russell, M. B. 1952. Soil aeration and plant growth. pp. 253–301. *In* B. T. Shaw, ed. Soil physical conditions and plant growth. Academic Press, Inc. New York.

SCS National Engineering Handbook. 1964. Section 15. Irrigation. U.S. Govt. Printing Office. Washington, D.C.

Sommers, L. E. and D. W. Nelson. 1976. Analyses and their interpretation for sludge application to agricultural land. pp. 3.1–3.7. *In* Application of sludges and wastewaters on agricultural land: A planning and educational guide. B. D. Kneyek and R. H. Miller, eds. Research Bull. 1090. Ohio Research and Development Center, Wooster, Ohio.

Sopper, W. E. 1973. Crop selection and management alternatives-perennials. pp. 143–154. *In* Recycling municipal sludges and effluents on land. Natl. Assn. of State Universities and Land Grant Colleges, Washington, D.C. 244 p.

Sullivan, R. H., M. M. Cohn, and S. S. Baxter. 1973. Survey of facilities using land application of wastewater. Office of Water Program Operations. EPA-430/9-73-006. Washington, D.C. 377 p.

Tanner, C. B. 1974. Water, crop yield, and potato production. Chapter 5, p. 63. *In* Report to environmental data service, NOAA. Relation of climate to leaching of solutes and pollutants through soil systems.

Tanner, C. B. 1974. Water, crop yield, and potato products. Chapter 5. *In* C. B. Tanner and W. R. Gardner, Relation of climate to leaching of solutes and pollutants through soils. Report submitted to the National Oceanic and Atmospheric Administration, Washington, D.C.

Taylor, A. W. and H. M. Kunishi. 1974. Soil adsorption of phosphates from wastewater. pp. 66–96. *In* Factors involved in land application of agricultural and municipal wastes. Agricultural Research Service, National Program Staff, Soil, Water, and Air Sciences, Beltsville, Md., USDA, 200 p. (DRAFT).

Van den Berg, C. 1973. Crop growth under influence waterlogging. pp. 260–267. *In* irrigation, drainage, and salinity, an international source book. FAO/UNESCO. Hutchinson & Co., Ltd. London. 510 p.

Van't Woudt, B. D. and R. M. Hagan. 1957. Crop responses at excessively high soil moisture levels. pp. 514–579. *In* Drainage of agricultural lands. James N. Luthin, ed. *Agronomy*, No. 7. ASA., Madison, Wis. 620 p.

Wessling, J. 1974. Crop growth and wet soils. pp. 7–37. *In* Drainage for agriculture. J. van Schilfgaarde, ed. *Agronomy*, No. 17, ASA, Madison, Wis. 700 p.

Wilcox, L. V. and W. H. Durum. 1967. Quality of irrigation water. pp. 104–124. *In* Irrigation of agricultural lands. R. M. Hagan, H. R. Haise, and F. W. Edminster, eds. *Agronomy*, No. 11. ASA., Madison, Wis. 1180 p.

Wolcott, A. R. and R. L. Cook. 1976. Crop and system management for wastewater application to agricultural land. pp. 7.1–7.12. *In* Application of sludges and wastewaters on agricultural land. A planning and educational guide. D. Kreyek and R. H. Miller, eds. Research Bulletin 1090, Ohio Agricultural Research and Development Center, Wooster, Ohio.

Module II-8

NON-CROP AND FOREST SYSTEMS

SUMMARY

Although agricultural land is most often utilized for land application of wastes, an alternative site may be necessary because agricultural land is unavailable or too expensive. Forest land may be the only choice in some regions because of its proximity, its abundance, and its lower price tag. The needs of the community may determine the choice of a land treatment site. Citizens may desire more open space, such as green belts, parks, or golf courses. Especially in arid regions, there may be pressure to irrigate these areas with reclaimed wastewater. Some cities have used reclaimed wastewater of high quality for swimming. Another alternative is to manage the land treatment site as a wildlife habitat, thereby increasing public acceptance.

Still another community demand may be to use wastewater and sludge to reclaim surface or strip mined areas. Reclamation usually results in the area being restored to natural vegetation conditions, although it may be possible in some cases to establish agriculturally productive land.

This module discusses the characteristics of alternate sites and management schemes and attempts to evaluate the efficiency of each alternative in terms of waste treatment. Following the breakdown of alternatives listed above, this module is divided into three distinct sections: Forest Lands, Park and Recreational Applications, and Land Reclamation and Revegetation.

CONTENTS

OBJECTIVES

After completing this module the reader should be able to:

1. Describe the characteristics of forest land which favor and disfavor it as a land treatment site.
2. List the different options for managing forest land used for a waste application site.
3. Describe how wastewater can be used to create and enhance recreation sites.
4. Describe the chemical and physical characteristics of mine spoil material and how it can be amended with the addition of sludge and/or wastewater.
5. Describe one study which experimented with adding sludge and/or wastewater to former mining areas, specifying the results of such additions.

INTRODUCTION

EPA (1975) broadly classifies crops as follows: Perennials (forage and fruit crops), Annuals (field crops and vegetables), Landscape Vegetation (including parks and golf courses), and

Forestland. The latter two "crops," along with reclamation of mine spoils, are discussed in this module.

A major limitation to the cropping alternatives discussed in Module II-7, "Crop Selection," is that in most places the land most suitable for large-scale field operations is already in active agriculture. The price of this land has increased steadily in recent years, to the point where it may no longer be possible to pay for land with the crops produced. Furthermore, as land pressures intensify, farmers may strongly resist any transfer of large land tracts from private to public ownership.

Most farmers could benefit from improved drainage and an assured adequate supply of moisture. These benefits, however, would be least on the best agricultural lands. Benefits are potentially greatest on sites where a few severe limitations can be overcome at an otherwise manageable location. Such was the situation in Muskegon County, Michigan, where barren, excessively well drained sandy soils were available nearby at low cost. In many situations, such sites may not be available.

In some areas, the possibility of reclaiming coal mine spoil and acid wastes may preclude any thought of alternate management schemes. In mountainous areas forests may be the only areas available for land application. Forest lands are usually much less expensive than croplands, and as discussed below, may be managed effectively if not overloaded. In urban/suburban areas, parks and golf courses may be the only open areas available for wastewater irrigation. In such situations, the creation of additional green belts in outlying areas may be desirable for more than one reason.

EPA has recently noted (Bastian and Whittington, 1977) that land application of sludge remains a controversial issue among researchers primarily because of possible effects upon increasing cadmium in the human food chain. The use of sludges in non-food chain crop production, such as the options outlined in this module, as well as the sod and nursery production options mentioned in Module II-7, "Crop Selection and Management Alternatives," allow the recovery of the nutrients and soil building value of municipal wastes while minimizing risks. Being able to avoid controversy among experts on this issue may help to settle the fears of the public over waste pathogens and toxic substances.

FOREST LANDS

Forest lands prove useful as land application sites. Besides being relatively inexpensive, they are abundant, occupying more than 33% of the total land area in the U.S.

Forest Soils Have High Infiltration/Percolation

Undisturbed forest soils are superior in minimizing surface runoff if the underlying soil is not saturated. Infiltration rates often exceed 6 in/hr. This is largely due to high macroporosity of the upper mineral soil and its protection by vegetation and surface litter against the impact of water drops. The litter layer also furnishes food for organisms which permeate the soil and maintain favorable conditions for water entry. In fact, if the forest floor remains in a relatively undisturbed state, infiltration rates continue high even when the overhead canopy is removed. Disruption of the surface, however, by trampling and compaction from human activity, livestock, or machinery decreases infiltration markedly and may lead to surface runoff.

The initial phases of wastewater application at a food processing facility in New Jersey demonstrated how great the differences between forest and cropland may be: On a cultivated field, 2 inches of wastewater per day produced soil saturation and considerable surface runoff;

in a hardwood forest some 400 feet away on the same soil, over 150 inches were applied during a 10-day period without noticeable saturation or surface runoff (Mather, 1953). This and many less dramatic examples establish the high acceptance rate for water by forest soils. But they also emphasize a related consideration, that is, the ultimate destiny of the applied water.

**Undisturbed forest soils have a high rate of infiltration and percolation.*

Because of high infiltration capabilities, forest lands are particularly suited to year-around application of wastewater. Due to the insulating properties of the forest floor, soil frost may not even form. Even where soil freezing does occur, it is often of the "honeycomb" type, which maintains a high porosity (Trimble, *et al.*, 1958). Waste renovation, however, requires more than just infiltration. Suspended and dissolved waste constituents must also be considered.

Forest Soil Limited by Its Drainage Potential

Despite excellent infiltration capacities, forest lands may be limited by the ultimate fate of applied water and nutrients. These constraints may require lower loading rates on forest lands than on croplands. Unlike agricultural lands, it is unlikely that extensive underdrainage will be undertaken due to difficulties which would be encountered with excavation work in trees, undergrowth, and roots. It is, therefore, very important that a site be naturally well drained or that water applications be limited to little more than potential evapotranspiration (ET_p) demands.

Unamended forest land may be poorly drained compared to agricultural land.

Due to this perennial growth habit and well developed root system and canopy structure, a forest community should have ET_p demand greater than most crops. As with crop irrigation, ponding of water on the soil surface must be avoided if growth, nutrient uptake, and transpiration are to be maximized and nuisances avoided. If water tables are maintained too close to the surface, trees will develop a shallow rooting habit and be subject to toppling by the wind. For this reason Sopper and Kardos (1973) recommended that non-irrigated buffer zones of 50-100 feet be left on the windward side of any irrigated forest area to act as a windbreak.

Tree Species Differ in Tolerance to Water

Tables 1 and 2 present some information on the tolerance of a few tree species to high water tables and flooding. Note that, as with forage species, the shallow rooted species often found in moist soils are not necessarily the most tolerant to flooding.

The tolerance of different forest species to salts and toxic elements is not well documented but variation may be expected. Red Pine appears to be sensitive to boron (Sutherland *et al.*, 1974). A Pennsylvania study indicated that among the seedlings of 8 coniferous tree species, European and Japanese Larch and White Pine responded best to irrigation with sewage effluent (Table 3).

Forests are commonly located on steep rocky hillsides where soils are often shallow and imperfectly drained. Water applied on a hillside may move laterally within the soil. This may

*This and other italicized summaries are intended to highlight key ideas, provide a basis for later review or to aid in skimming sections that are relatively familiar. They can be ignored in a complete reading of the text.

Table 1. Correlation of the Ground Water Table with the Composition and Growth of Forest Vegetation in Wisconsin.

Depth to Ground Water, ft	Composition of Forest Stand	Estimated Yield at 100 Years (cu ft)
	A. *Podsol region, silty clay loams derived from granitic drift*	
1-1.5	Balsam fir, white spruce, some black ash and red maple. Understory of mountain ash, tag alder, willows and dogwood	2,200
2.0-3.0	Hard maple, rock elm, red maple, some basswood, yellow birch, balsam fir and white spruce. Hard maple and basswood inferior.	3,500
4.0-5.0	Hard maple, basswood, some white pine. Leatherwood and numerous other shrubs. Vigorous growth of sprouts.	4,800
	B. *Prairie–Forest Region. Silt loams derived from calcareous drift*	
0.7-1.2	Lowland Meadow	–
2.0-3.0	Bur oak, black oak, some red oak, aspen and box elder. Abundant walnut in understory.	1,900
4.0-5.0	White oak; red oak, some black oak, walnut, hickory, white ash	3,200

Source: Van't Woudt and Hagan (1957).

Table 2. Tolerance of Various Trees and Shrubs to Flooded Soil in the Upper Mississippi Basin.

Species	Years Survived	Remarks
Sand bar willow	2	Mostly dead in last year
River birch	2	Survived well 1st year
Cottonwood	2	Survived well 1st year
Silver maple	3	Mostly dead in 2nd year
Elm	3	Mostly dead in 2nd year
Hackberry	3	Fair growth in 2nd year
Red oak	3	Scarce in bottoms
Bur oak	3	Mostly dead in 2nd year
Swamp white oak	3	Fair growth 2nd year
Pin oak	3	Mostly on higher ground
Alder	3	Hardy to second year
Green ash	4	Hardy 2nd year, fair in 3rd
Black willow	4	Hardy to 3rd, all died 4th
Deciduous holly		Hardy to 4th year
Swamp privet		Hardy to 4th year
Buttonbush		Hardy after 4 years
Red osier dogwood		Hardy after 7 years

Source: Van't Woudt and Hagan (1957).

Table 3. Total Height Growth of Surviving Tree Seedlings During the Period 1965 to 1970.

	Irrigated Plots		Control Plots	
Species	Survival (%)	Height (ft)	Survival (%)	Height (ft)
European larch	23	8.7	0	–
Japanese larch	17	8.4	0	–
White pine	70	6.3	17	2.6
Red pine	40	5.4	20	2.0
White spruce	30	4.3	3	2.8
Pitch pine	3	3.8	3	1.4
Austrian pine	13	3.6	13	2.1
Norway spruce	47	3.6	7	1.6

From: Sopper and Kardos, Recycling treated municipal wastewater and sludge through forest and cropland, Pennsylvania State University Press, University Park, PA., 1973, p. 287.

increase soil contact and improve treatment efficiencies. This water, however, may emerge at localized spots downslope, and drainage must be provided from these locations or a manageable pond or marsh created.

Forest Soil Limited in Nitrogen Uptake

Aside from the ability of a site to drain itself naturally, the fate of added nutrients, particularly nitrogen, is a major constraint in the use of forest land for waste application, regardless of season. To understand this constraint requires some familiarity with normal nutrient relations in unamended forest systems.

Nitrogen loading is often the limiting factor in using forest lands for waste treatment sites.

Natural forest communities have adapted to relatively infertile soil conditions through efficient nutrient cycling mechanisms. Recent work (Switzer and Nelson, 1972; Bormann, Likens and Melillo, 1977) has shown considerable internal recycling of nutrients. Prior to leaf fall, considerable nitrogen is translocated from the leaves to woody tissue. The annual uptake of nutrients compares with certain agronomic crops but only a small percentage of this is ever removed from the area. Nutrient rich leaves, twigs, bark, and roots are returned to the soil rather than being hauled away in harvests.

These materials remain on or in the soil until decomposition again renders the nutrients available for plant uptake. Small natural losses from the system are essentially balanced by small natural inputs. At any given time, the majority of nutrients are tied up in the forest floor or in the soil organic matter. Figure 1 depicts the distribution and annual cycling of nitrogen in an oak-hickory forest in Tennessee.

A further consideration is the changing pattern of nutrient retention with time. Forests do not have uniform nutrient requirements throughout their lifespans. As shown in Figure 2, annual nutrient retention in the plant tissue increases up to some maximum during the early life of a new forest stand. Beyond this point, annual retention decreases so that at great age, no further net accumulation of nutrients occurs from one year to the next. The actual point of maximum nutrient accumulation varies among species. For example, loblolly pine, a common

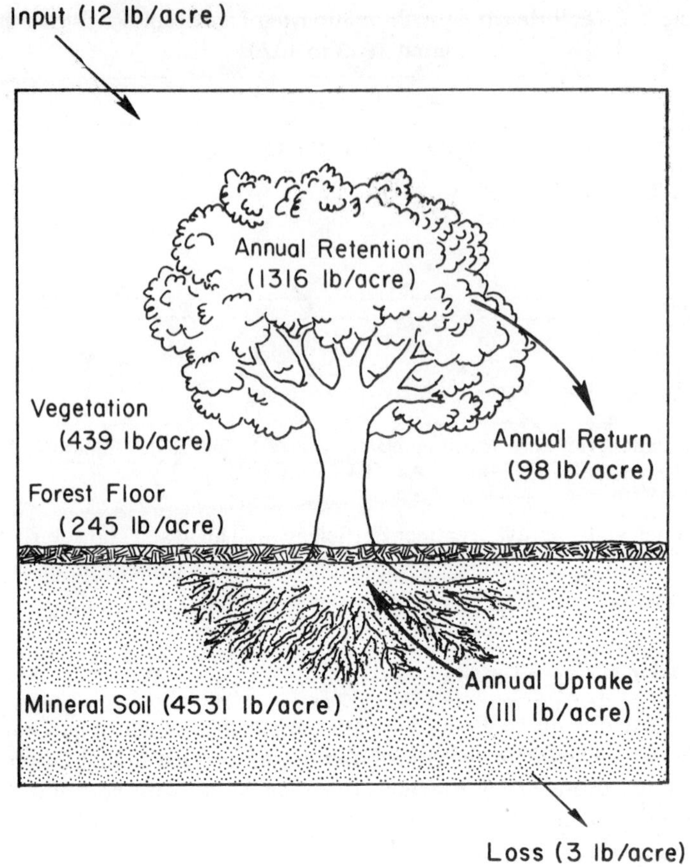

Figure 1. Within natural forest systems, most nitrogen is tied up in the mineral soil, with lesser amounts in the forest floor and the vegetation. Annual cycling of nitrogen means that only a small portion of the annual uptake is actually retained in the plant tissue. In this study, measured atmospheric input exceeds leaching loss. Additional gains from N fixation and losses through denitrification are likely but the amounts are unknown (after Henderson and Harris, 1975).

species in the Southeast, achieves maximum accumulation for all macronutrients (N, P, Ca, Mg, K, S) during the first 30 years, with much of this being concentrated in the initial 10 years of growth (Switzer, *et al.*, 1968; Switzer and Nelson, 1972). By age 45, nutrition in a well nourished loblolly pine stand is largely a matter of cycling, with little or no net accumulation. Hardwood species such as birch, maple, and oak have somewhat longer periods of nutrient accumulation, although the overall trends remain as shown in Figure 2.

Young stands of trees take up nutrients to a greater degree than mature stands.

Harvest of wood crops removes accumulated nutrients from the forest site, much as with an agronomic crop. The amounts removed, however, are generally low when averaged over the time required for accumulation. Table 4 compares harvest removals and annual accumulations of two short rotation forest crops with corn.

Although harvesting wood removes accumulated nutrients, the long periods between tree harvests makes nutrient removal small when compared with corn.

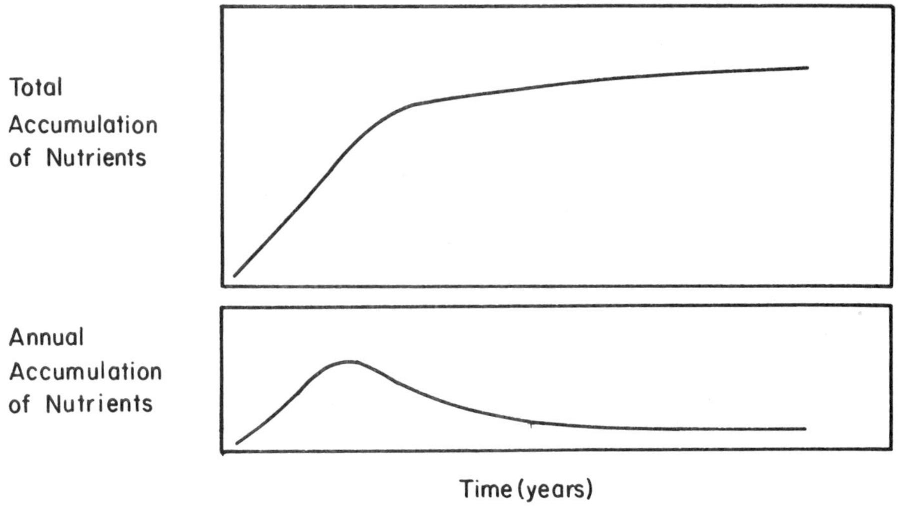

Figure 2. The annual accumulation of nutrients in forest stands varies with age. Higher rates of annual accumulation occur in younger stands; in old-growth stands, mortality offsets growth and net accumulation approaches zero. From Tamm, Holmen, Popovic, and Wiklander, leaching of plant nutrients from soils as a consequence of forestry operations, AMBIO, 3 (6) 214 (1974).

Forests Can Be Altered to Enhance Waste Treatment

Waste renovation on forest land might be promoted by replacing native trees with species better suited to specific application schemes. For example, hardwoods in general accumulate greater amounts of nitrogen and other elements than conifers. Young stands retain a greater proportion of the annual uptake than do older trees.

European investigators are studying irrigation of poplar plantations for the dual purposes of wood production and wastewater renovation. This option could become attractive within a few years if current experiments on whole tree removal and ultra-short rotations prove economically feasible. Fast growing hardwood species such as sycamore, hybrid poplars, and cottonwood may be harvested by some version of a silage chopper, and then regenerated by coppicing, that

Table 4. Comparative Nutrient Removal by Two Forest Species and Corn.

	Nutrient Removal through Harvest lb/acre		Average Annual Nutrient Removal from Site lb/acre-yr	
	N	P	N	P
Corn				
Annual harvest (16 years)	1600	288	100	18
Annual harvest (30 years)	3000	540	100	18
Loblolly Pine, age 16 years				
Whole tree (above ground)	227	27	14.2	1.7
Debarked pulpwood[a]	66	12	4.1	.8
Aspen, age 30 years				
Whole tree (above ground)	170	25	5.7	.8

[a]Includes only stemwood to 3-in. top with bark removed.
Source: Jorgensen, *et al.*, 1975; Boyle, 1975.

is by rapid sprouting from established root systems (Boyle, 1975; Herrick and Brown, 1967). This management technique maximizes production of pulp fiber. Limited data indicate that such short rotations can provide nutrient uptake comparable to corn (Urie, 1975). Despite these intensive management options, the data in Table 4 indicate a basic limitation in the capability of traditional forest crops to remove added nutrients. This emphasizes the importance of the soil itself in retaining waste constituents, and as a site for other renovation mechanisms such as denitrification.

On-going studies on forest land indicate that tree species may be planted and managed so as to increase nutrient uptake, making it comparable to the uptake of corn.

Forest Soils Take Up Considerable Phosphorus

Phosphorus is tightly bound as iron and aluminum compounds in most acid forest soils. Soil analyses from a mixed conifer forest in California indicate that up to 1,400 lb P/acre could be held in the surface 6 inches and still maintain a soluble phosphorus concentration of less than 0.20 ppm (Powers, *et al.*, 1975). Total phosphorus retention, of course, is strongly related to soil depth, with deeper soils capable of binding more total phosphorus, all other factors being equal. Resting periods between applications increase P retention, as explained in Module II-2, "Phosphorus Considerations." An exception occurs in certain highly weathered sands where low iron and aluminum levels preclude appreciable phosphorus retention (Ballard and Fiskell, 1973).

Many forest soils are capable of handling high loadings of phosphorus.

Forest Soils Remove Nitrogen Through Nitrification-Denitrification

Adsorption of ammonium ions on soil colloids and fixation by soil organic matter can remove large amounts of nitrogen from applied effluent. The soil's ability to remove ammonium depends upon the amount of competing (primarily divalent) cations contained in the soil and in the applied wastewater. Lance (1972) has reviewed a method of estimating this, and using his method, if a typical effluent containing 25 ppm NH_4^+–N, 25 ppm Ca^{+2} and 20 ppm Mg^{+2} were applied, ammonium would be adsorbed up to about 10% of the CEC. A potentially much greater amount of ammonia nitrogen may be fixed by soil organic matter, especially under alkaline conditions.

The fate of adsorbed ammonia -N is to be nitrified by soil microorganisms. Following nitrification, percolating waters may carry a pulse of high nitrate water which, if not managed, may reach ground waters. The rate of nitrification is dependent upon temperature, moisture, pH and other factors which are not fully understood. Nitrification has generally been found to be very slow during the winter, and often maximum nitrate concentrations, resulting from the mineralization of added manure, are not observed in soils until summer. This, however, may be influenced by the type of plant community. Woodwell, *et al.* (1974) at Brookhaven, New York observed nitrification (and nitrate leaching) to continue during the winter under a late successional hardwood forest, but not under red pine, abandoned field, or timothy ground covers.

Nitrification in forest soils depends on temperature, moisture, and perhaps the type of plant community.

Whether denitrification is significant in waste amended forest soils has not yet been determined. Studies of this point are currently in progress (Nutter, *et al.*, 1975; Schultz, *et al.*, 1975). The process of denitrification requires the presence of both nitrate and decomposable organic substances under anaerobic conditions. Nitrate is seldom abundant in natural forest soils, although specific conditions governing nitrate formation are not fully understood. Acid conditions apparently retard nitrification but commonly do not prevent it. The primary control in nitrate production may be simply the supply of ammonium ions. In unamended forest soils, the small amounts of available ammonium contributed by precipitation inputs or mineralization of organic nitrogen are quickly utilized by plant roots and microorganisms. Populations of nitrifying bacteria are, therefore, often small. An influx of ammonium in wastewater additions, however, would likely result in population changes and lead to increased nitrification. If so, the fresh carbon provided by annual litter fall and root turnover, together with periodic anaerobic conditions caused by heavy wastewater applications, makes denitrification look promising as a nitrogen renovation mechanism.

On-going studies indicate that denitrification may be a major nitrogen renovation mechanism in wastewater amended forest soils.

Careful scheduling of wetting and drying may be used to create periodic cycles of aerobic and anaerobic conditions and so promote nitrification-denitrification reactions in the soil. The same technique should also increase phosphorus retention by allowing for reversion reactions between applications. A study is now in progress in Georgia to develop design guidelines for waste application on steep forested slopes. It is hoped that periodic saturation of soil at the base of the slope, caused by subsurface lateral flow, will result in denitrification (Nutter, *et al.*, 1975).

Studies Show Low Rate of Mature Forest Nitrogen Removal Compared to Herbaceous Vegetation

Work with the "Living Filter" land application system (Hook and Kardos, 1976) has demonstrated that a growth of sparce white spruce and dense annual herbage ("Old Field Site") on a clay loam soil treated with 2-inch effluent/week from April through November had a much greater nitrogen removal efficiency, presumably due to denitrification or organic matter accumulation, than did a mature hardwood community ("New Gameland Site") on a sandy loam soil treated with 2 in./wk year around. Nitrate concentrations under the hardwood community were consistently greater than 10 ppm suggesting again that nitrification continued year around. When a previously untreated hardwood plot adjacent to the "new gameland" site was irrigated identically with that site, using an oxidized effluent, steady state conditions with essentially zero renovation were established within 18 months.

A red pine plantation on the same soil and receiving the same treatment as the "old field" site had an average nitrate nitrogen concentration of 24.2 ppm at the 120 cm depth in the year preceding clear cutting. Following clear cutting, when volunteer herbaceous vegetation became established, nitrate nitrogen concentrations at the 120 cm depth dropped to an average of 8.3 ppm the first year and 3.8 ppm the second year. Thus, regardless of the mechanism, herbaceous vegetation appears to be effective, at least in the short run, in preventing nitrate leaching. Table 5 indicates the changes in predominant herbaceous species which occurred at the "old field" site as a result of 19 years of irrigation.

Herbaceous vegetation at the forest site seems to prevent nitrate leaching to groundwater.

Table 5. Predominant Herbaceous Vegetation Species on the Irrigated and Control Plots of the White Spruce Area in 1972.

	Irrigated Plot		Control Plot	
Species	Percent Cover (%)	Average Height (ft)	Percent Cover (%)	Average Height (ft)
Goldenrod (Solidago juncea)	<1	2.8	5	1.8
Aster (Aster spp.)	0	–	5	1.1
Dewberry (Rubus flagellaris)	0	–	40	0.8
Strawberry (Fragaria vesca)	<5	0.9	10	0.5
Poverty grass (Danthonia spicata)	0	–	20	0.3
Everlasting (Antennaria spicata)	0	–	5	0.1
Goldenrod (S. rugosa, S. graminifolia, S. juncea)	<5	5.3	0	–
Milkweed (Asclepias rubra)	<5	5.1	0	–
Indian hemp (Apocynum cannabinum)	<5	3.3	0	–
Night shade (Solanum dulcamara)	10	2.3	0	–
Clearweed (Pilea pumila)	75	1.5	0	–

From: Sopper and Kardos, Recycling treated municipal wastewater and sludge through forest and cropland, Pennsylvania State University Press, University Park, PA., 1973. p. 286.

If it is determined that increased litter accumulation is the major mechanism responsible for nitrogen removal in "old field" type situations, then some form of litter removal could be adopted to prevent eventual nutrient recycling from the nitrogen rich litter and resultant leaching to groundwater.

In certain types of forests, periodic "controlled" burns might be used to remove a portion of the surface organic debris during this period of accumulation. In the process, much of the nitrogen would be volatilized. The techniques are available because prescribed fire is already used in Southeastern pine forests for fuel reduction, game management and site preparation, and in the West for fuel reduction (to reduce the danger of wildfire).

Nitrogen can be removed from forest land by periodic controlled burning in which N is volatilized.

Applications of wastewater contaminated with cadmium or mercury should perhaps not be made to acid forest soils. Munro, *et al.* (1977) concluded that "solubility limits and leaching of heavy metals from forest floor litter is a significant process in the movement of these metals (Pb, Cd, Zn, and Cu) through a forested watershed." Solubility of these metals increases as pH goes down.

Success of Forest Sites Varies

Like other plant communities, forests are a product of both genetic and environmental factors. The potentials for growth, development, and survival are controlled by inherent genetic properties, whereas environmental conditions determine the degree to which these potentials are realized. Environmental changes, such as additions of water and nutrients, can alter the performance of individuals and, eventually, the overall species composition of the site. An important point is that some alteration in vegetation is inevitable, and the more accurately the change can be predicted, the more likely that specific management objectives can be achieved. Few waste application projects allow analysis of long term effects and thus predictions are largely speculative. Results from three application systems, however, illustrate some of the effects and management difficulties to be considered:

Very few forest waste application sites have been studied and monitored to determine long-term effects.

1. Seabrook Farms, a vegetable processing facility at Seabrook, New Jersey, generates up to 12 million gallons of wastewater per day. This wastewater has been applied to a 70 to 84 acre woodland area since 1950. The effect of such applications is particularly striking due to the small area involved. Essentially, an oak-hickory forest type has been transformed into a near-swamp community. The original trees and associated undergrowth were either destroyed by the high pressure nozzle jets or were unable to withstand the increased moisture regime. These have died and been replaced with dense stands of water-tolerant grasses (up to 6 feet tall), vines, shrubs, and herbs (Stevens 1972). The changes in soil and vegetation characteristics after the first seven years of operation with from 400–600 inches of wastewater applied each season are shown in Figure 3.

As the graphs indicate, changes in both soil and vegetation have been major. Despite the dieback of much of the native cover, however, the hydrologic capacity of the forest floor has

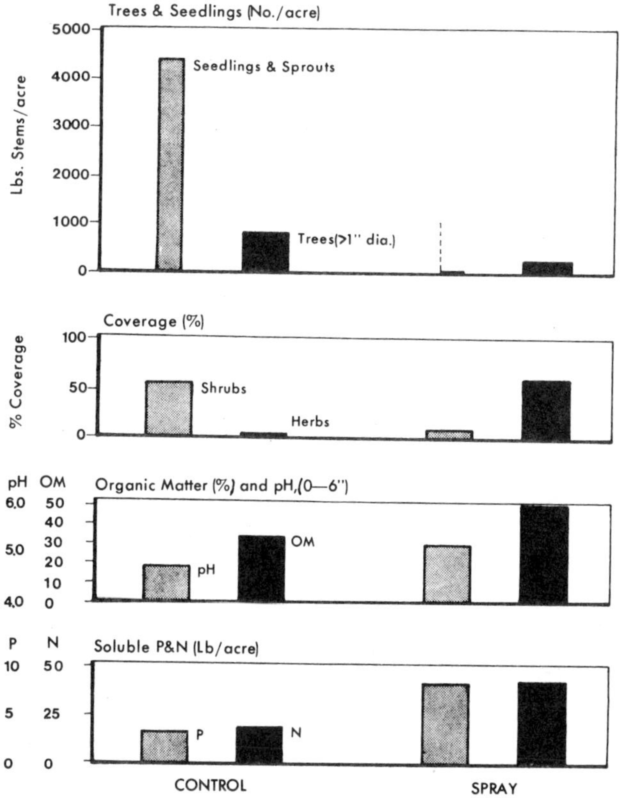

Figure 3. Changes in soil and vegetative characteristics at Seabrook Farms as a result of spray irrigation.

been maintained. Another survey (Stevens, 1972) identifies impacts on nearby woodlands outside the immediate application area. Here accumulations of water in shallow ponds and resulting dieback of trees indicate that seepage at lower depths is a continuing effect. As of 1977, the Seabrook Farms woodland irrigation system has been discontinued. Changes in plant operation have resulted in greatly reduced flows, which are now treated in a lagoon made up of part of the old system.

Although the site was changed from an oak-hickory forest to a swamp, Seabrook Farms obtained a high degree of wastewater purification and a high rate of infiltration/percolation for more than 20 years.

2. In 1953, the Gerber Products Co. of Fremont, Michigan, planted four species of trees as part of a cannery wastewater application study. Seedlings of boxelder (*Acer negundo*) and cottonwood (*Populus deltoides*) and unrooted cuttings of black willow (*Salix nigra*) and balsam poplar (*Populus balsamifera*) were used. Wastewater was applied at a rate of 2 in./wk for 24 to 32 weeks (Approximately April-October). Table 6 gives the results after three years.

Surviving trees of all species grew much better in the irrigated soil as compared to the control. Winter-killing appears to have been a major factor in the low survival of cottonwood trees. The abundant soil moisture available late in the growing season together with added nitrogen probably delayed normal hardening of the plant tissues in the fall season (Rudolph and Dils, 1955). But another cause of over-winter kill was mouse damage. Summer irrigation promoted a heavy growth of grasses around the trees which, combined with snow, provided excellent rodent harborage. Mice killed many trees by complete removal of bark from the lower stem (Fisk, 1964). Subsequent year-round spraying on these sites eventually eliminated all trees through ice accumulation and breakage (Urie, 1971).

It appears that in both the Seabrook and Gerber systems the tree cover disappeared mainly because of management practices, such as poor matching of the water distribution systems to the structural characteristics of the trees.

In the Seabrook and Gerber forest systems the tree cover was destroyed partly because water distribution systems were not adjusted to structural characteristics of the trees.

3. The most intensive study to date is that begun at Pennsylvania State University in 1963. Treated municipal sewage effluent has been spray-irrigated in forested areas on two different soils for over 10 years. The sites include two mixed-hardwood stands (mainly oak), a red pine plan-

Table 6. Survival and Growth of Forest Species Under Irrigation with Cannery Wastewater.

	Survival (%)		Ave. Height Growth (ft)	
Tree Species	Control	Irrigated	Control	Irrigated
Boxelder	5.6	11.5	1.5	3.4
Cottonwood	7.0	0.1	1.0	6.0
Black Willow	15.5	58.3	1.5	9.0
Balsam Poplar	–	0.4	–	5.0

Source: Rudolf (1957).

Table 7. Renovation Efficiencies[a] for Treated Municipal Sewage Effluent Applied to Corn and Hardwood Forest, Pennsylvania State University.

Nutrient	Corn Silage (%)	Hardwood Forest (%)
N	145	39
P	143	19

[a]Percent of element applied which is removed by vegetation itself. Application rate 2 in./wk. Additional phosphorus fixed by soil.

Source: Sopper, 1975.

tation, and a young plantation of white spruce in an old field that initially supported a variety of native grasses also. Application rates ranged from 1-4 in./week over the growing season and in some cases extended the year around. Many detailed results are available in publications (Parizek, *et al.*, 1967; Sopper and Kardos, 1972; Murphy, *et al.*, 1973; Wood, *et al.*, 1973; Sopper and Kardos, 1973; Sopper, 1975; Richenderfer, *et al.*, 1975; Kardos and Hook, 1976).

Wastewater renovation by the forest vegetation itself is generally acceptable, but not nearly as efficient as by agronomic crops (Table 7). Nevertheless, tree growth was generally stimulated by irrigation (Table 8). The major exception was red pine irrigated with 2 in./wk. This species is subject to root damage on wet soils and after five years of irrigation, a wind storm following wet snow felled every tree on the plot.

Additional results indicate that effluent irrigation somewhat enhanced wood fiber quality of red pine and red oak as raw material for pulp and paper (Murphy, *et al.*, 1973). In contrast, irrigation with 1 to 2 in./wk of wastewater had a detrimental effect on the mechanical properties of red pine crown wood (Murphy and Brisbin, 1970). Effluent irrigation has also reduced the number of tree seedlings and increased mortality of older saplings in hardwood stands at Penn State. After one year of irrigation, seedling numbers were 16- and 14 thousand per acre on the

Table 8. With One Exception, Wastewater Irrigation Materially Increased Tree Growth in the Pennsylvania State University Studies.

	Ave. Annual Diameter Growth (mm)		Ave. Annual Height Growth (cm)	
Species and Rate	Control	Irrigated	Control	Irrigated
White Spruce				
2 in./wk.	4.5	10.0	13	46
Red Pine				
1 in./wk.[a]	1.5	4.3	42	58
2 in./wk.[a]	1.8	1.5	52	49
Mixed Hardwoods				
1 in./wk.[a]	4.1	4.6		
2 in./wk.[b]	3.3	5.6		
4 in./wk.[c]	3.8	5.3		

[a]Growing season only.
[b]Year around.
[c]Various schedules.

Source: Sopper, 1975.

control and irrigated hardwood plots, respectively. Ten years later the comparable numbers were 14- and 2 thousand, respectively. During the same period, mortality of saplings and large trees was about 20% higher in irrigated areas. The decrease in seedling numbers may have resulted from less light at the forest floor because of stimulated overstory growth. Increased mortality of older trees was largely due to ice damage during periods of winter irrigation. Ice damage must be expected if application systems operate during winter months, although it can be minimized with proper design of sprinkler systems (Myers, 1973).

Wastewater irrigation increased tree growth in most species in a Penn State study.

Tree Seedling and Cutting Irrigation Shows Promise

Recent studies conducted at the University of Washington (Breuer, Cole and Schiess, 1977) show good results in terms of nitrogen leaching to groundwater, and other wastewater renovation goals, for Douglas fir seedling and poplar cutting plots compared to a predominantly reed canary grass plot. These three test areas, along with a bare soil plot, were established in 1974 on land which had previously been clear-cut of Douglas fir.

Application of secondary treatment plant effluent was made to all plots at 5 cm (2 in.) per week year-round during 1975 and 1976, and percolate was sampled at various depths via lysimeters. Mean annual applied NH_4^+-N concentration was 16.5 mg/l and total N concentration was about 18.5 mg/l. All of the vegetated plots produced groundwater quality well within the 10 mg/l standard for nitrate concentration, at 6 foot depths.

The Douglas fir plot also included a lush growth of native grasses and herbs, which had to be cut several times during 1975 and 1976, and may have contributed to the overall nutrient removal efficiency of this plot. The poplar plot also included native grasses which were also well established during 1975. However, much improved growth of the tree cuttings in 1976 resulted in greatly reduced growth of the grass, and nitrogen uptake efficiency as shown by the percolate studies was dramatically improved.

This work may have positive implications for the technique of wastewater application to fast-growing tree species, then harvesting in relatively short rotations. Longer term data from this and similar studies would be very useful, as well as more specific studies to determine plant uptakes versus soil mechanisms, particularly for nitrogen.

Options exist in managing forest systems

Over-all objectives will play a major role in the way forest land is employed in the design and operation of waste application systems. Three basic options are:

1. Maximize waste application and accept the resulting site changes in vegetation. Essentially this approach amounts to letting the natural vegetation adapt to a new regime of high soil moisture, high fertility, and mechanical damage.
2. Maximize renovation by adapting waste application to existing soil and vegetation. Increased growth yields may constitute an added benefit.
3. Manage so as to maximize quantity and/or quality of timber, pulpwood, or other product. In this approach, highly intensive forest management techniques are required.

Obviously, intermediate alternatives exist but these three options delineate the general range of choices.

PARK AND RECREATIONAL APPLICATIONS

Treatment Sites Can Be Used for Recreation

Recreation includes dozens of activities in which people commonly engage for relaxation. There are water sports—fishing, boating, swimming, diving, etc., and land sports—tennis, cycling, walking, golf, and so on. Water sports are commonly split into two categories pertinent to the present discussion: contact and non-contact sports. For example, a non-contact sport is boating; contact sports are swimming and diving. The water quality required for non-contact sports is lower than for contact sports. The water quality standards for recreational uses of water normally are published by each state, and should be referred to by those planning to incorporate water-based recreation in the land application system.

Water quality standards for recreational use of reclaimed wastewater are set by each state.

Parks, Golf Courses Use Effluent for Irrigation

Land-based recreational sites can utilize wastewaters for irrigation and fertilization. Golf courses, university campuses, city parks, and so on have successfully been developed and maintained using effluent waters and sludges. Providing irrigation for golf courses and landscape areas such as campuses and highway median strips is a costly practice in the arid areas of the country. The use of treated wastewater for such irrigation has been practiced for several years in many of these states. Wastewater for landscape irrigation is normally a secondary-treated and disinfected effluent. Wastewater irrigation of these areas serves a dual purpose: it saves the municipality the cost of irrigating with potable water, and provides additional treatment for the wastewater before discharge into surface or ground waters.

Using reclaimed wastewater for irrigating golf courses, campuses, and other landscaped areas is especially attractive in arid areas.

Perhaps the most well-known park developed using effluent is the Golden Gate Park in San Francisco. The rich soil of the park was developed on sand dunes using the street sweepings of decades ago, when horses provided the main energy for transportation. The park was originally irrigated by an outfall sewer traversing the park, but with development of the area came complaints of objectionable odors, and the practice was discontinued. The purchase of potable water was too expensive. An activated sludge sewage treatment plant was completed in 1932 and supplied 1 mgd treated effluent for irrigation of the 1,013 acre park.

Several golf courses have been established as well as maintained using effluent from secondary treatment plants. Table 9 lists some of the communities which irrigate golf courses with municipal effluent and the date when their land application projects were begun (EPA, 1973). Many of the designs of recent projects have taken advantage of earlier experience and their own test plots. Experience at the Ventura, California golf course (Sullivan, 1970) resulted in the following recommendations:

1. Hardy species indigenous to the area should be planted whenever possible for landscaping, golf course hazards, etc. Keeping the ground moist for the first 2–3 weeks after seeding

maintains low soluble salt concentrations and insures quicker ground cover and thicker stands.

2. Adequate disinfection is necessary to protect the public health. California considers adequate disinfection to be achieved when the number of coliforms does not exceed a MPN of 23/100 ml. Careful monitoring of daily flow and chlorine residual should be practiced.
3. Pipelines should be designed without cross connections between potable and effluent supplies. Valves and sprinkler heads should be tagged and colored so as to warn the public. Special protection should be provided for drinking fountains on the golf course.
4. A wastewater application rate of 60 inches/acre/year was appropriate at Ventura.

Following a pilot study, the decision to irrigate the golf course with effluent was made. The program was later expanded to include water hazards on the golf course and a recreational lake. A public information program run concurrently with the pilot study was effective and there was no problem of acceptance by the public by the time the project was completed.

Suhr (1971) has reviewed case studies of treatment works which discharge effluent to potable water supplies. Experience has shown that it is possible for land treatment to meet water quality standards for swimming and other contact sports. If not overloaded, land treatment can even result in effluent of drinking water quality. In fact most communities rely upon such natural purification to maintain the quality of their groundwater supplies. This does not mean, however, that it is easy to renovate water to the point that it may be used for a recreational lake.

Wastewater may be renovated by land treatment to a quality high enough for swimming.

Table 9. Location of Golf Courses Irrigated with Municipal Effluent (and Date Land Application Was Started).

Arizona	
City of Prescott	(1960–62)
City of Lake Havasu	(1971)
California	
City of Ontario, San Bernardino County	(1915)
Santee County Water District, San Diego County	(1959)
City of San Clemente	(1957–68)
City of San Bernardino	(1962)
City of Laguna Hills	(1964)
City of Livermore	(1965)
City of Ventura	(1965)
Colorado	
City of Colorado Springs	(1953)
New Jersey	
City of Cranbury	(1967)
New Mexico	
City of Santa Fe	(1962)
Texas	
City of Abilene	(1920)
City of La Mesa	(1960)

Source: EPA, 1973.

Each state establishes water quality standards which include designation of certain waters within the state for specific purposes. The classification of the water would then determine the quality of effluent which may be discharged into the various waterways.

Designers of land application systems who plan to utilize the wastewater for recreational activities which involve the water itself, be they contact or non-contact activities, should be aware of local and state regulations, and be prepared to meet the standards for water quality at the recreational impoundment. It is beyond the scope of this program to provide the participants with water quality standards for all the states; these standards can be requested from state health departments or pollution control agencies.

Each state sets standards for purity of recreational waters.

California has promulgated specific water quality standards for reclaimed wastewater for various uses in the reference *Wastewater Reclamation Criteria* (1975). Article 6 refers specifically to recreational impoundments. For non-contact sports such as fishing, boating, etc., the reclaimed wastewater shall be "adequately disinfected (and) filtered." It "shall be considered adequately disinfected if at some point in the treatment process the median Most Probable Number (MPN) of coliform organisms does not exceed 2.2/100 ml. The median value shall be determined from the bacteriological results of the last 7 days for which analyses have been completed." For contact sports (swimming, wading) the wastewater must be adequately disinfected and oxidized. Disinfection will be deemed adequate on the same basis as for non-contact water. For a landscape impoundment, that is, one which is used only for aesthetic enjoyment and with which the public has no contact, the wastewater must be adequately disinfected and oxidized. MPN of coliform organisms in non-contact waters should not exceed 23/100 ml (Foster, 1969).

Several land application systems are being managed so as to produce recreational water areas. Not surprisingly, it is the arid western and southwestern regions of our nation where techniques to reclaim and reuse wastewater have been most fully explored. The water hazards at the Ventura golf course and the lakes at Lancaster, California, north of Los Angeles are examples of non-contact recreational areas.

Lancaster Recreational Lakes Illustrate Nutrient Problems

Dryden and Stern (1968) provided an extensive review of the waste quality criteria which were adopted by them in the planning of the recreational lakes at Lancaster, California. A waste flow of 3 MGD received primary treatment followed by biological stabilization in oxidation ponds. The most severe limitation encountered in attempts to renovate the oxidation pond water to a point where it was suitable for use in the lakes was the removal of phosphorus to a concentration of 0.5 mg/l or less. At dissolved phosphorus concentrations higher than 0.5 mg/l algae growth was so fast that fish could not survive. During the day the algae used up all CO_2 dissolved in the water resulting in a pH rise to between 8.5 and 9.5 which may have been toxic to fish as a result of ammonia toxicity. During the nights, respiration of large algae and bacterial populations reduced dissolved O_2 of the water to as low as 1 mg/l, which is too low for fish. At 0.5 mg/l algae growth continued but at a reduced rate and few extreme conditions were encountered. Still, in order to assure population stability it was considered necessary that some method of supplementing the oxygen content of recreation lake water should be provided for emergency use.

Land Treatment Sites Can Be Wildlife Habitats

Increased interest in the conservation of our natural resources has led to a greater awareness of how our lifestyle affects the other life forms which share our environment. With the expansion of our population, we are in fact crowding out many plants and animals, and many are even being driven to extinction. Happily, the land application alternative to waste management in most cases does not reduce wildlife habitat; indeed, it may even serve to provide new habitat in areas that have been devoid of wildlife for decades. Little work has been done on managing land application sites for wildlife habitat. However, work at the Pennsylvania State University has shown that chlorinated effluent sprayed on wildlife areas has little or no effect on the wildlife occupying these areas. Some results of these studies will be reviewed in this module.

A study shows application of chlorinated effluent in wildlife areas has little or no effect on wildlife.

Many land application system operators have noted an influx of wildlife, particularly waterfowl (duck, geese, grebes, and so on), when water is allowed to accumulate in ponds or marshy areas. Leicester, New York, is the site of a food processing plant where wastewater is sprayed onto forested areas, and tends to pond in low areas and ditches. These wet spots harbor several broods of black ducks and mallards.

It must be remembered that standing water in wastewater irrigation areas may lead to increased mosquito populations and other nuisances. This can be of overriding concern, and must be carefully evaluated against the practice of allowing water to stand stagnant in these areas.

The members of a community may be interested in creating or expanding wildlife habitat for several reasons. An important ecological concept is that diversity of populations is crucial to ecosystem stability. A greater number of different species of plants and animals (diversity) will lead to a more stable natural community than a community with only a few different plant and animal species. In agricultural areas, natural scrub or woodland interspersed with large fields may encourage small birds and mammals to remain in an area. This may have a moderating influence on insect pest problems.

Incorporating a wildlife habitat into the land application scheme may make land treatment more acceptable to the community.

Hedgerows have been used, traditionally, as fences to separate small fields. Increasingly, hedgerows have been removed from agricultural fields because they may harbor insect pests and provide alternate hosts for diseases, because they increase the number of turns required in working fields, and because they accumulate snow in northern winters which can delay field operations in the spring. Hedgerows may make a contribution to land application sites in that they often make a site more attractive to visitors, because they help to control water and wind erosion and because they are effective in blocking aerosol drift from spray irrigation. Areas which are too steep or too close to a stream for field operation should also not receive wastewater irrigation but may yet provide a wildlife habitat (Figure 4). With proper management, an effluent-irrigated area could thus provide a very important and useful adjunct to the community which it serves.

Management of land application systems for wildlife habitat must of course take into account the design criteria developed for protection of the public health and the environment. In most cases, wildlife attraction will be a secondary benefit of a forested site or a site with marshy areas.

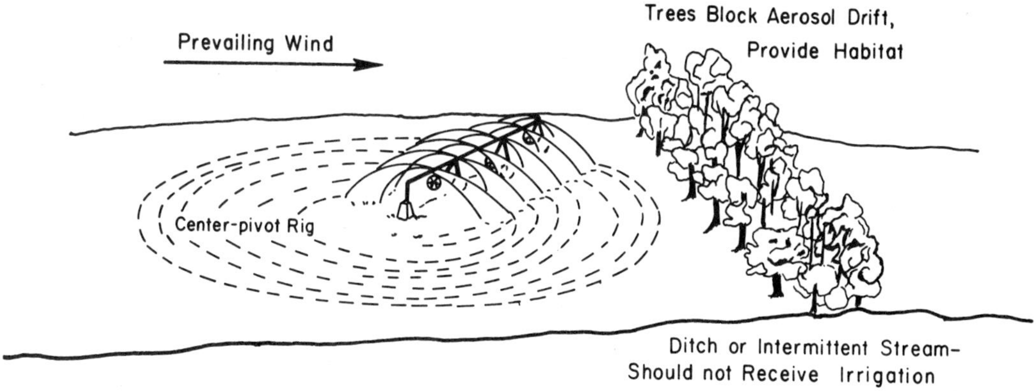

Figure 4. Hedgerow vegetation may be planted along spray fields to aid in aerosol control and provide habitat for small mammals and birds.

An example of a marsh system which is successfully polishing secondary effluent while providing a diversified habitat for many wetland species is the Mt. View Sanitary District operation at Martinez, California (Demgen and Nute, 1977; Demgen and Blubaugh, 1977). This site has attracted several kinds of waterfowl, as well as a diversified population of aquatic animals and plants. Groups such as the Audubon Club have shown interest in this system, and visits by groups or individuals to observe wildlife are common. The Mt. View site is bounded by industry and a highway, which makes the presence of a wildlife area more unusual and welcome. If a wastewater treatment marsh were not in such a constrained site, and the wishes of the community were considered, hunters could be included among those using the area. A full discussion of community interaction in land treatment planning is included in Module I-8, "Societal Constraints."

Marshes enhanced by treated wastewater have been shown to attract wildlife and people who are interested in the animals.

As mentioned previously, the only extensive research reported as of 1977 on effects of land application systems on wildlife has been conducted by Penn State. For detailed information, the reader is referred to the work of Wood, Glantz and Kradel (1975).

The study included several investigations of responses of small mammals and deer to the conditions prevailing at a land application site. Studies included uptake and the concentration of heavy metals, incidence of disease, changes in populations, changes in feeding habits and other activities. The Abstract, Conclusions and Recommendations of the study group appear in Appendix A.

The study at Penn State showed that, in general, wildlife was not adversely affected by spray irrigation of wastewater. The only significant difference was an increase in the body weight of young rabbits in the fall. It must be remembered, however, that other geographical areas will represent different situations, and that changes occur gradually. In the early years of irrigation at Seabrook, New Jersey, it was also concluded that the increased herbage growth provided a better habitat, more shelter and food, for small mammals (Mather, 1953).

Just as a soil scientist should be engaged to appraise the soil at a site, so should a wildlife biologist be consulted to evaluate the stability of a desirable wildlife habitat. Such professional help will allow the system designer to maximize the benefit to be gained from such management.

Questions to be kept in mind for such consultation will include which species should be managed, what vegetation will attract these species while remaining amenable to the wastewater irrigation situation, and so on. To be sure, creating a lake in the desert will have a greater effect on neighboring wildlife than would creation of a marsh in the northeast, where marshes are relatively abundant.

> *The land treatment operator should consult a wildlife biologist to evaluate the site as a possible wildlife habitat.*

Recommendations Exist for Recreation Areas Using Wastewater

The recommendations listed below should be kept in mind when designing a land application site which will be managed to derive a secondary benefit of recreation or wildlife habitat. They are non-specific in nature. Specific guidelines for individual sites must be generated with the site in mind. As discussed earlier, public health regulations in which each state sets water quality for water sports must be consulted. For wildlife habitat expansion, a wildlife biologist should be retained. He or she can provide valuable information as to how to manage a site for wildlife in general, or for particular species of interest.

Recommendations for Land Activities. Recreation areas which utilize effluent irrigation must have controlled access. The public should not be allowed into the area when spraying is being conducted. The site must be laid out so that when certain areas are being sprayed, these areas will not be available for use. Adequate warning signs should be erected to keep the public away from areas during irrigation. The site must have sufficient time to rest between applications of wastewater and entrance by the public.

- Plant species should be chosen both for their suitability to the recreation area and for their tolerance to effluent irrigation. The high nutrient content of the effluent will encourage luxuriant growth. This must be kept in mind when determining mowing schedules and other daily operating procedures.
- Loading rates must not exceed the infiltration capacity of the soil. The ultimate fate of applied water should be determined.

Recommendations for Water Activities.

- Individual state public health regulations must be obeyed for contact and non-contact water sports.
- If a body of standing wastewater is incorporated into the recreation area, such as a water hazard at a golf course, warning signs should be erected to warn against drinking of the water by the public.
- Constant monitoring and record-keeping are necessary where wastewater is reclaimed and used for swimming. The facility should be able to close down immediately on detecting a health hazard in its water.

Recommendations for Wildlife Habitat.

- Any standing water, such as marshes and ponds, should be posted against drinking and body contact to protect the public health.
- Any effluent containing pesticides or other chemicals which are known or suspected of being harmful to wildlife must not be disposed of on wildlife areas.

- Wildlife populations should be monitored. Dramatic increases in populations should be counteracted by harvesting (hunting or trapping). Disease outbreaks must also be monitored, as wildlife disease organisms may not be detected in normal public health monitoring.
- If public use of the wildlife areas is encouraged, the recommendations listed above for land activities must also be considered.

LAND RECLAMATION AND REVEGETATION

Wastewater Can Be Used for Reclaiming Mined Areas

Thousands of acres in this country have been disturbed to get at the valuable coal and mineral deposits which lie beneath the surface. In the past, little thought has been given to reclaiming the barren wastelands which are the result of these mining operations. However, all states now mandate—and environmental concern reiterates—that newly mined areas must be restored to an acceptable condition by those responsible for mining them. Surface mining of coal, gravel, sand, iron ore, phosphates, and kaoline clay is the most common method of retrieving these materials from the earth. Underground mining of coal also results in surface degradation. The low grade coal that is not economical to process, and the rock associated with the coal, is all brought to the surface. This spoil is heaped into mountains, where it mars the landscape, and causes run-off problems after rains.

Surface and underground mining in the U.S. have degraded large areas of land; federal law is increasingly requiring that this land be reclaimed.

While today most states have laws requiring some degree of restoration of strip mined land, a U.S. Department of the Interior report (1967) indicated that 66% of the 3.2 million acres which had been disturbed was considered unreclaimed. This represents an area approximately equal in size to the state of Delaware. It was considered that 20% of these existing spoils would be difficult to vegetate with traditional methods due to their physical or chemical characteristics.

Reclamation is an expensive and difficult process. A promising method of reclamation utilizes sludge and wastewater to restore precious nutrients to the land so that vegetation can be supported. Other waste materials which are being evaluated as amendments to mine spoil include composted garbage and fly ash (Bennett, *et al.*, 1976). Utilization of these materials in conjunction with each other may prove to be cost effective treatment although environmental questions have not been fully answered.

The addition of sludge and wastewater to mine spoil and mined areas can restore nutrients to the land so vegetation can grow.

Surface Mine and Spoil Material Lack Organic Matter

Land that has been surface mined, or stripmined, has had many or all of the soil layers removed, intermixed with each other, and dumped in piles. As we know from Module I-1, "Soil as a Treatment Medium," the soil is a delicate system which must be kept intact in order for it to properly carry on its processes. Spoil material may be considered to be soil parent material rather than soil. Nearly all essential minerals are found to some extent in this material. Organic matter, however, is virtually nonexistent. Table 10, taken from Czapowskyj (1973), lists the major nutrient contents of some coal mine spoils.

Table 10. Main Nutrient Contents of Some Coal-Mine Spoils.

Spoil	pH	N (%)	P (ppm)	K	Ca (mg/100 gm)	Mg
		Kentucky–Bituminous				
Shale–black	2.2	–	–	<0.1	1.0	2.3
Shale–gray	5.0	–	–	0.8	5.5	4.8
Shale–gray	6.2	–	–	0.6	8.6	4.2
		Indiana–Bituminous				
Sandstone	4.0	<0.1	16.6	<0.1	1.8	0.6
Shale–gray	4.7	<0.2	6.6	<0.1	3.5	1.5
Glacial till	7.3	<0.1	15.6	0.1	17.5	3.5
		Pennsylvania–Anthracite				
Breaker refuse	2.9	<0.1	10.7	0.4	3.2	0.3
Sandstone	6.3	0.1	41.9	0.2	2.9	0.3
Glacial till	7.1	<0.1	64.2	0.2	2.4	0.4

Source: Czapowskyj, 1973.

In terms of texture, two thirds of surface spoil material is finer than sandy loam, one third is coarser. On coarse piles, in spite of steep slopes, surface runoff is essentially never observed. Subsurface leachates may nevertheless cause serious water pollution problems, and surface creep may result in the continued exposure of fresh spoil material (Harrison, 1974). Vegetation establishment on coarse material is particularly hampered by a lack of water. Even the finer spoil materials tend to have a low available water holding capacity. The finer material has practically no aggregate stability and tends to form crusts which deter vegetative establishment (Bennett *et al.*, 1976). The finer material is subject to surface runoff, and even light suspended solids loads can be quite damaging. Harrison (1974) explains, "Coal and coal shales have a density of less than 2.0 g/cm^3, and often approach 1.0 g/cm^3. Therefore, the maximum particle size that will stay in suspension under given conditions is much larger than for other materials . . . a small amount of suspended coal dust in a stream or lake has a much greater effect on water color and the transmission of light than an equal weight of other mineral matter."

Because mine spoil is devoid of organic matter, has low water holding capacity and little aggregate stability, it cannot support vegetation.

In mountainous regions, spoils are often dumped in a fashion similar to that depicted in Figure 5. Numerous measurements of dump piles have established that 37° is the average angle of repose of spoil (Harrison, 1974). Major slides of this material can occur if material underlying the toe of a dump pile fails. Seepage waters may increase the likelihood of such failures [Figure 5 (b)]. If no provision is made for drainage of abandoned pits [Figure 5(a)], accumulated water will raise the water table within the spoil and contribute to failure as a result of the increased weight of the spoil mass, reduction of the effective intergranular pressure due to the buoyancy effect, and seepage pressure (Harrison, 1974).

Spoil banks often present a chemical environment hostile to life. While some spoil banks may be readily revegetated, others remain unvegetated for many years. The pH of these "orphan

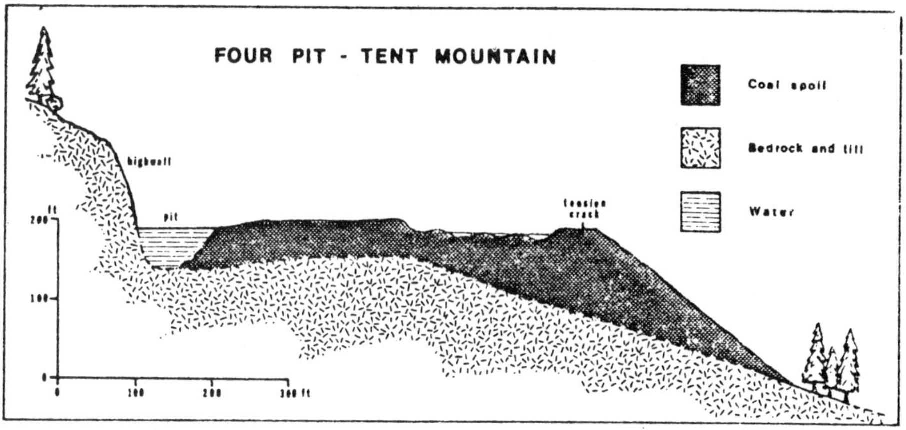

(a)

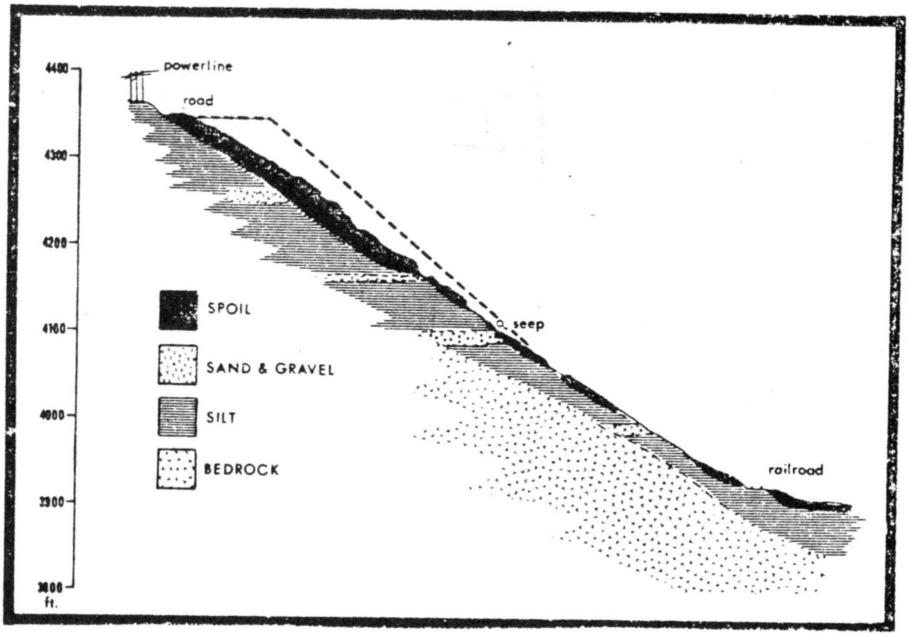

(b)

Figure 5. Cross sections through waste dumps: (a) cross section from headwall to natural hillside. Four Pit, Tent Mountain, Alberta; (b) cross section from dump to railway, Erickson Pit, Michael, B. C. *Source:* Harrison, 1974.

lands" is usually very low. This acidity is an important factor in mobilizing toxic elements, particularly Al, Fe, and Mn, thus preventing revegetation. Experience in Alberta, Canada indicates that most cultivated crops, and especially legumes such as alfalfa, encounter toxicity problems with levels of aluminum greater than 2.0 ppm and levels of manganese greater than 15.0 ppm (Taylor and Gill, 1974).

Spoil banks often present a chemical environment hostile to life.

Iron pyrite brought to the surface as part of the mine spoil is responsible for much of the acidity at very acid sites. Exposed to the atmosphere, iron pyrite is readily oxidized to ferrous sulfate and sulfuric acid (Hill, 1971) according to the following reaction:

$$2FeS_2 + 2H_2O + 7O_2 \longrightarrow 2FeSO_4 + 2H_2SO_4$$

Because of the instability of steep spoil piles, new pyrites are continually brought to surface, thus maintaining the acidity of some spoil banks.

Reclamation Efforts Aimed at Revegetation

The most important decision which must be made prior to beginning a reclamation project is the final use of the site once it has been reclaimed. The great majority of reclaimed mine areas in humid regions of this country have been planted to forest. Some of these forests are managed for fiber production, while others are allowed to develop in natural succession. Other management schemes for reclaimed areas have included airstrips, housing developments, parklands, industrial parks, and so forth. Many of these uses could be compatible with continued wastewater application and site monitoring. In almost all types of potential land use considered, establishment of vegetation is essential. This is most difficult on very acid and very steep spoils in very dry areas.

Before reclamation is begun, the final use of the site must be determined.

Before reclamation of spoil bank material is attempted, the bank may need to be leveled, graded and perhaps contoured. Whereas the angle of respose of spoil material averages 37°, the "biological angle of repose" appears to be between 25°-30° (Harrison, 1974). On steeper slopes the rate of growth of a plant's root system is not rapid enough to allow it to anchor itself against the slow downhill movement of the soil cover (surface creep). Shrubs tend to dam material on their uphill side, while material below moves away, exposing roots. Trees are bent and fall over before reaching sufficient size to resist this movement.

While spoil banks naturally lie at about a 37° angle, a 25°-30° slope is much better for establishing vegetation.

Even on sites graded to 25° it may not be easy to establish vegetation. Aids to vegetative establishment which have been effectively used include mulches, nurse crops, lime, fertilizer, seed bed preparation, seed mixture, timeliness and method of sowing, and seeding rate. The following observations have been made. Mulches and nurse crops aid establishment by reducing surface creep, increasing moisture at the soil surface, reducing temperature and protecting seedlings from wind driven abrasive material (Bennet, *et al.*, 1976). Lime helps to neutralize the acidity of the site and reduces the availability of toxic elements. Rock phosphate has been used effectively because it contains 40% CaO equivalents as well as phosphorus. Rock phosphate is normally reacted with sulfuric acid to form super-phosphate fertilizer. The same neutralizing reaction occurs when it is added to very acid mine spoils (Bennett, *et al.*, 1976). Nitrogen and phosphorus are the main elements deficient on coal mine spoils. Potassium may also be needed on kaoline spoils. The supply of micronutrient Mn, Zn, Ca, and S range from toxic levels in extremely acid spoils to adequate in moderately acid spoils (Czapowskyj, 1973).

Many additives and management techniques can be employed to promote spoil bank revegetation.

Excessively smooth surface grading of fine textured spoils can lead to sheet erosion and poor seedling establishment. A preferred procedure is to leave furrows running across the slope to catch and hold seed and fertilizer (Bennett, *et al.*, 1976). Hydro-seeding, using water as a carrier of seed and fertilizer, will result in a more uniform distribution of seed where cyclone seeding is hampered by high winds or steep slopes (Bennett *et al.*, 1976). Most studies have concluded that only species which are likely to dominate should be seeded, that these should be sown at rates sufficient to provide good cover, and that the best planting time is just before the wettest season of the year (based on probability) (Schuman *et al.*, 1976).

The selection of adapted species has received considerable investigation. Recent reviews of some of this information were presented by Schuman, *et al.* (1976) for the western U.S. and by Bennett, *et al.* (1976) for the eastern regions. The latter review contains information on some little-known legumes which may prove useful.

In recent years, some mining operations have attempted to remove the surface soil layers carefully, pushing them into separate piles, so that they can be replaced on top of the spoil pile once one has been removed from the site. This is a costly but effective method of assuring that revegetating plant roots may contact organic matter. As discussed in Module I-1, "Soil as a Treatment Medium," organic matter is beneficial in a number of ways: it holds water, forms complexes with toxic elements, and it provides a steady balanced supply of nitrogen, phosphorus and potassium.

Sludge and Wastewater Alter Spoil Banks

Municipal wastes tend to have a pH of about 6.8-7.2. In addition to their value as a source of water, they are high in the dissolved minerals, nutrients and organic matter that plants need, and that are lacking in the spoil bank material. Characteristics of these wastes are more fully discussed in Module I-3, "Waste Characteristics."

Municipal wastes have structural and chemical characteristics that spoil bank material lacks.

Case Studies on Reclamation Field Work

Penn State Studies. In 1968, a study was undertaken at Pennsylvania State University to determine if municipal sewage sludge and wastewater could be used to aid in the revegetation of spoil bank material (Sopper, Dickerson, Hunt, and Kardos, 1970). It was hypothesized that the slightly alkaline sludge would help to neutralize the highly acid spoil material, and that the nutrient-rich wastewater would leach acids and toxic elements to below plant root depth, as well as provide organic colloids, which would detoxify the soluble iron, aluminum and manganese. Table 11 lists the average composition of effluents and sludges used in some experiments at Penn State.

Initial experiments consisted of establishing ten test plots, each containing about 25 tons of highly acid (pH 2.0–3.0) spoil material from the lower Kittanniny bituminous coal seam in Clearfield County, Pennsylvania. This material was chosen because several attempts at revegetating it had failed over the past 23 years. The plots were boxed in plywood lysimeters each 32 feet long, 4 feet wide, and 4 feet deep, open on the bottom, and located in an open field. Six inches of sand were placed in the bottom of the boxes to facilitate monitoring of the percolate.

In the spring of 1969, liquid sludge and wastewater were applied to the boxes. The wastewater was applied at the rate of 0.5 in./hr by a rotating sprinkler. Liquid sludge was applied from a tank truck. Seven species of trees were randomly planted in the boxes, including Japanese

Table 11. Average Chemical Composition of Sewage Effluent and Liquid Digested Sludge (from Sopper and Kardos, 1972).

	Sewage Effluent (mg/l)		Digested Sludge (mg/l)	
Constituent	1969	1970	1969	1970
Nitrate–N	6.1	11.8	–	–
Kjeldahl N	34.4	6.2	932	1,430.0
NH_4^+	11.9	2.9	–	–
Phosphorus	5.24[a]	5.55[a]	111.0[b]	118.0[b]
Potassium	13.7	22.4	79.6	48.5
Calcium	34.9	29.5	136.5	58.8
Magnesium	16.0	15.1	33.8	19.0
Sodium	35.2	31.7	33.8	30.3
Manganese	0.04	0.10	1.30	0.98
Boron	0.36	0.20	0.40	0.59
Chloride	43.2	37.3	–	–
Sulfate	–	38.6	–	–
Aluminum	0.9	1.3	57.5	30.5
Iron	0.5	0.6	14.8	8.5
Zinc	0.11	0.22	1.8	1.7
Copper	0.17	0.1	1.4	1.2
pH	7.2	6.9	7.6	7.5
Dry solids (%)	–	–	3.2	3.8

[a]Soluble orthophosphate.
[b]Total phosphorus.

larch, white spruce, Norway spruce, white pine, European alder, hybrid poplar, and black locust. In addition, two grass species, K-31 tall fescue and orchard grass, and two legumes, crown vetch and birdsfoot trefoil, were planted.

Several application rates were used to determine the best application regimen. The control boxes were not irrigated at all after planting. Other treatments included (1) 1 inch of effluent plus one inch of sludge per week, (2) 2 inches of effluent per week, (3) 2 inches of sludge per week, and (4) 2 inches of effluent plus 2 inches of sludge per week. Treatment continued for 24 weeks.

Vegetative growth in several plots was impressive. The best treatments for germination and growth included application of both sludge and effluent. The wastewater irrigation washed the sludge solids from the plants, which prevented the foliage damage experienced by the plants receiving only sludge application. Black locust was the most successful tree in terms of survival percentage and height growth. The only treatment which affected a significant change in the pH of percolate collected at a depth of 3.5 feet was the 2 inches effluent and sludge per week treatment. Under this regimen, the pH went from 2.2-2.8 to 4.1 (Sopper, 1970). This treatment had the best height growth for tree seedlings and the greatest dry matter production of grasses and legumes. This treatment also resulted in the highest values for phosphorus and nitrate-nitrogen, and lower for manganese, iron and aluminum in the percolate. Reductions in K, Ca, Mg, Zn, B, and Cu concentrations in the percolate were also noted in the irrigated boxes as compared to the control boxes. According to Sopper and Kardos (1972) "irrigation with effluent and sludge leached and diluted the native salts, and solubilities of the Mn, Fe, Al, Cu,

Table 12. Dry Matter Production and Height Growth of Grasses and Legumes for the Two Treatments. No Grass or Legume Germination Occurred on the Controls.

	1-Inch Combination Treatment		2-Inch Combination Treatment	
Species	DMP (lb/acre)	Height (ft)	DMP (lb/acre)	Height (ft)
Grass				
Weeping lovegrass	6127	1.45	11067	1.35
Blackwell switchgrass	2702	1.60	4689	2.80
Deertongue	2212	0.50	4624	0.65
Reed canarygrass	636	0.75	3215	1.25
Redtop	259	0.45	1845	0.55
Garrison creeping foxtail	0	–	1284	0.50
Climax timothy	83	0.55	1112	1.60
Saratoga smoothbrome	52	0.40	929	0.65
Legume				
Ladino clover	214	0.50	3078	0.70
Iroquois alfalfa	405	0.65	3060	1.10
Sericea lespedeza	501	0.75	2887	1.05
Pennscott red clover	519	1.10	2717	1.15
Bristly black locust	1193	1.15	2431	1.95
Sweet clover	0	–	1424	1.15
Chemung crownvetch	51	0.30	907	0.50
Lathro flatpea	123	0.40	761	0.55

Source: Edgerton, B. R., W. E. Sopper, and L. T. Kardos, 1975.

and Zn were suppressed by the dual action of alkalinity of the effluent and sludge and humic precipitation by the organic colloids of the sludge."

One study showed application of both sludge and wastewater to spoil material enhanced seed germination and vegetative growth.

The results of a subsequent study initiated in 1971 (Edgerton, *et al.*, 1975) utilizing similar applications to 8 grass and 8 legume species are presented in Table 12. The small-scale lysimeter studies at Penn State are ongoing, but they have already demonstrated that with proper site preparation and application of sludge and wastewater, very acid strip mine spoil bank material may be revegetated. Revegetation is the crucial first step in reclaiming land, for without vegetation, there can be no stabilization of the piles of spoil material. Eventually, this reclaimed land can be put to a useful purpose, whether as a forest, recreation area, housing, or other kind of development. As was discussed earlier in this module, the decision as to how a site will be managed will lie in the hands of the community in which the site is located.

Scranton Tests Show Good Results with Sludge Application

Scranton, Pennsylvania. Once the study group at Penn State found that strip mine spoil material could be revegetated in a laboratory situation, they decided to start a small scale revegetation project. The Scranton, Pennsylvania area was the scene of much mining activity in the past. Deep-mining for coal has left mountains of spoil material throughout the area and runoff from them turns the Lackawanna River bright yellow.

The city of Scranton and its neighboring boroughs are built around and on piles of waste from the old mining operations. An economical method of reclaiming these sites and putting them to beneficial use would certainly be welcome.

In 1974, three plots were laid out at the Cedar Avenue Mine Bank by the Penn State workers. The first received 15 inches of effluent on a 60 foot by 60 foot area over a 2-month period. Sludge was applied to subplots at the rate of 0, 20, 40, and 80 tons per acre. A second plot was treated with the same amount of sludge but fresh city water. A third plot received only sludge and natural precipitation. A fourth control plot received no sludge and no irrigation of any sort. Nine inch tree seedlings were planted along with grass and legume seeds.

Although there was no treatment during the second season, all the planted vegetation did well on the treated plots. The greatest success has been on the plots which received only dried sludge. The sludge was applied and mixed into the spoil material by hand. No odor or other deleterious effect was noted in the area. This field study of reclaiming spoil banks is hoped to be the first of many attempts to revegetate such waste areas and make them productive, or at least pleasant to look at.

Scranton study showed dried sludge application to be most effective in promoting growth of tree seedlings on strip mine spoil material.

Palzo Study Shows Need for High Rates of Sludge Application

The Palzo Tract, Illinois. A tract in the Shawnee National Forest in Illinois was established in 1970 as a demonstration site to evaluate sludge application on strip-mined land (Lejcher and Kunkle, 1973). Anaerobically digested sludge from the Metropolitan Sanitary District of Greater Chicago (MSDGC) was used. Within the 77 hectare site, known as the Palzo Tract, 4 plots 500 square meters each were delineated. Water quality and plant response were measured continually.

The plots were treated with 304, 178, 78, and 0 dry metric tons of sludge per hectare. The next spring, the plots were seeded with K-31 tall fescue and weeping lovegrass at rates of 22 and 8 kg/hectare.

Germination occurred within 30 days on all but the control plot. At the end of the first growing season the most heavily treated plot showed 90% vegetative cover. Plots with lower treatment rates had less than 50% cover. Similarly, only the plot with the highest application rate was able to support the same amount of cover the second year as it did the first.

Table 13. Effect of Sludge Treatment on pH of Spoil Within the Plots at the Palzo Tract, Illinois.

Treatment Rate (dry metric tons/hectare)	Plot Average Soil Surface pH
304	6.2
178	5.2
78	4.7
0	2.3

From: Lejcher, R. and S. R. Kunkle. 1973. Restoration of acid spoil banks with treated sewage sludge. pp. 184–186. *In* W. E. Sopper and L. T. Kardos, eds. Recycling treated municipal wastewater and sludge through forest and cropland. Pennsylvania State University Press, University Park, PA. p. 190.

Table 14. Average Concentration of Certain Parameters in Subsurface Runoff from the Sludge Treated Plots at Palzo Before and After Treatment ("After" Values Based on 1970–1972 Averages) (Lejcher and Kunkle, 1973).

	Treatment (metric tons/hectare)						
	304		178		78		0
	Before	After	Before	After	Before	After	Average
Element	-milligrams per liter-						
Al	1,240	402	395	138	440	346	548
Cd	1.14	1.92	0.31	1.18	0.70	1.13	0.66
Cr	3.5	4.8	1.3	1.6	2.3	3.4	2.1
Cu	11.6	7.5	3.8	3.3	4.0	3.6	4.5
Fe	3,700	1,260	1,280	320	1,000	822	1,620
Mn	70	30	51	17	71	42	36
Pb	0.33	0.23	0.16	0.18	0.42	0.18	0.22
SO_4	11,000	7,740	8,400	3,730	7,100	6,770	7,980
Zn	24.4	36.4	8.1	24.8	14.1	26.0	13.3
Acidity	22,940	8,900	7,310	3,320	7,040	5,900	9,770

From: Lejcher, R. and S. R. Kunkle. 1973. Restoration of acid spoil banks with treated sewage sludge. pp. 184–186. *In* W. E. Sopper and L. T. Kardos, eds. Recycling treated municipal wastewater and sludge through forest and cropland. Pennsylvania State University Press, University Park, PA. p. 187.

The Palzo Tract studies indicated that sludge application on strip mine spoil must be at a level high enough to neutralize the acidity of the spoil to allow establishment of vegetation. Table 13 presents data on pH changes on sludge-amended spoil material. Table 14 reports the concentrations of selected elements in percolating waters.

Palzo Tract studies show that sludge application on strip mine spoil must be high enough to neutralize the acidity of the spoil.

Conclusions from Field Studies Suggest Possibilities for Research Work

It is evident from these results as well as from those obtained at Penn State that despite contributions of large amounts of toxic and trace elements, waste additions can actually decrease the amount of some of these elements in the soil solution. All elements, however, do not behave in the same way. In the Penn State study, waste application decreased concentrations of Al, Fe, Mn, and Ca. In Illinois, however, Zn as well as Cd and Cr were increased. These results suggested that a serious problem of heavy metal leaching may be encountered if the alkalinity of wastewaters alone is relied upon to neutralize acid mine spoils. Note that Cd and Cr levels measured in the percolate from both treated and controlled plots exceed by a considerable amount the levels specified in the National Interim Primary Drinking Water Regulations (EPA, 1976).

Treatment of mine spoils with the residues from coal-fired power plants (fly ash, bottom ash, and flue gas desulfurization (FGD) sludge) may be another example of additions of potentially toxic elements which may result in reduced overall availability of those elements. Power plant wastes are generally very alkaline and often have good water retention characterisitcs. Concern has been expressed over high Se and Mo concentrations in forage crops grown on mine spoils heavily amended (100–600 tons/acre) with fly ash. Boron toxicity to plants is also a likely problem. As discussed in other modules, however, boron will leach fairly quickly from a site, and Mo and Se toxicities are associated primarily with crops grown on alkaline soils. Moderate fly ash additions to acid mine spoils might result in a much more neutral, much less toxic substrate for plant growth than either substance taken alone. Total additions of toxic elements would be much less than if wastewaters were used to give a comparable degree of neutralization. Wastewater additions to a neutralized spoil-ash mixture should result in a medium with improved water retention characteristics, and the additions of N, P, K, and organic matter should favor plant growth. Bern (1976) has argued that in the near future "economic forces will dictate a combined coal refuse/power plant residuals operation." Such operations would be facilitated by locating power generating plants close to coal deposits, so called "mine-mouth power stations." Applications of wastewaters to such carefully monitored sites may well prove a cost effective treatment in active mining areas.

Wastewater additions to power plant residues would probably improve water retention, add nutrients and build organic matter.

Strip Mining and Reclamation

Agricultural Use of Reclaimed Strip Mine Land. In Illinois, anaerobically digested liquid sludge has been applied to strip mine land. The land was subsequently planted to corn. Attempts at growing corn on unamended strip mine land have had poor results. Average yields for this situation was 13 bu/acre, and the corn was of poor quality. A study was conducted to determine the physical-chemical characteristics of corn grown on amended strip mine land, as well as the heavy metal content of the crop (Garcia, *et al.*, 1974).

Commercial yellow seed corn was used for the study. One plot of strip-mine land was treated with sludge equivalent to a loading of 25 tons of solids per acre. An adjacent plot received no sludge addition. Of particular interest to the workers was the question of whether heavy metals (Zn, Mn, Cu, Pb, Cr, Cd, and Hg) were translocated in the plant, especially to the edible parts.

A physical examination of the corn from the two plots showed a dramatic difference in success. The corn from the untreated plot looked immature, and reflected stress conditions. Kernel size was small to intermediate, and about 20% of the kernels were diseased. Corn kernels from the treated plots showed only minor variation in size, and there were no diseased kernels in the sample. The corn appeared similar to commercial grade corn. The weight of the corn grown on the sludge-amended plots was about four times greater than the weight of corn grown on the untreated plot. The weight of the cob, and the number of kernels per cob, was about three times greater for corn grown on the treated plot.

Chemical analyses similarly revealed greater success for corn grown on the sludge-amended sites. Protein, fat, and fiber values were lower for corn from the untreated sites. The increased protein content showed that the sludge amendment definitely improved the nutritional quality of the corn.

The heavy metal analysis did not reveal any significant difference in heavy metal content between corn from treated and untreated plots in this short term study. The only exception to this was zinc, which was higher in the untreated corn. The concentration of zinc in the cobs and husks from untreated plots was about three times higher than that from treated plots. The workers hypothesized that this high concentration of zinc was due to the organically deficient strip mine spoil on which the corn was grown.

While increased levels of toxic elements were not found in corn in this study, the concentration of cadmium in the spoil material was increased from less than 2 ppm to over 70 (Garcia, *et al.*, 1974). Other studies (see Module II-7, "Crop Selection") have shown that corn is a poor test crop because it particularly excludes Cd and that such elevated levels of soil cadmium may represent a hazard. It would seem that there is no reason to expect mine spoil to accept larger quantities of toxic elements than would prime agricultural soils, while still remaining productive. Indeed, to the extent that they may become acid if not properly managed, they may be deemed less appropriate receptors.

Mine spoil cannot be expected to accept larger quantities of toxic elements from sludge than prime agricultural soils; in fact, they will more likely accept less if not properly managed.

In certain situations, strip mined soils may be reclaimed to the point where they may be used to produce healthful crops. Managed in this fashion, they would be subject to the same constraints discussed in Module II-7, "Crop Selection." No agricultural soil will remain productive if it is too heavily contaminated with industrial wastes.

BIBLIOGRAPHY

Akin, E. W., W. H. Benton, and W. F. Hill. 1971. Enteric viruses in ground and surface water: a review of their occurrence and survival. *In* Virus and water quality: occurrence and control. 13th Water Quality Conference, 59, Univ. of Illinois. 222 p.

Ballard, R. and J. G. A. Fiskell. 1973. Phosphorus retention in coastal plain forest soils. *Soil. Sci. Soc. Amer. Proc.*, **38**:363–366.

Bastian, R. K. and W. A. Whittington. 1977. Current EPA guidance on land application of municipal sewage sludges. pp. 13–14. *In* Disposal of residues on land. Proceedings of the National Conf. on Disposal of Residues on Land, Sept. 13–15, 1976, St. Louis, MO. Information Transfer, Inc. Rockville, Md, 216 p.

Bennett, O. L., W. H. Armiger, and J. N. Jones, Jr. 1976. Revegetation and use of eastern surface mine spoils. pp. 195–215. *In* Land application of waste materials, Soil Conservation Society of America, Ankeny, Iowa. 313 p.

Bern, J. 1976. Residues from power generation: process, recycling and disposal. pp. 226–248. *In* Land application of waste materials. Soil Conservation of America, Ankeny, Iowa. 313 p.

Bormann, F. H., G. E. Likens, and J. M. Melillo. 1977. Nitrogen budget for an aggrading northern hardwood forest ecosystem. *Science*, **196**, pp. 981–983.

Boyle, J. R. 1975. Nutrients in relation to intensive culture of forest crops. *Iowa State J. Res.*, **49**:297–303.

Bryan, F. L. 1974. Diseases transmitted by foods contaminated by wastewater. Presented at the Conference on the Use of Wastewater in the Production of Food and Fiber. Oklahoma City, Okla.

Breuer, D. W., D. W. Cole, and P. Schiess. 1977. Nitrogen transformation and leaching associated with wastewater irrigation in Douglas-fir, poplar, grass, and unvegetated systems. (Unpublished paper). College of Forest Resources, Univ. of Washington, Seattle. 33 p.

Czapowskyj, M. M. 1973. Establishing forest on surface-mined land as related to fertility and fertilization. Forest Fertilization Symposium Proc., USDA Forest Service General Technical Report NE-93, pp. 132–139.

Demgen, F. C. and J. Warren Nute, Inc. 1977. Marsh enhancement program conceptual plan. Rept. to Bd. of Directors, Mt. View Sanitary Dist., Contra Costa Co., Calif. San Rafael, Calif. 41 p. + appendices.

Demgen, F. C. and B. J. Blubaugh. 1977. Mt. View Sanitary Dist. marsh enhancement pilot program. Progress Rept. No. 3. Mt. View San. Dist., Martinez, Calif. 50 p.

Dryden, F. D. and G. Stern. 1968. Renovated wastewater creates recreational lake. *Environ. Sci. Technol.*, **2**(4):268–278.

Edgerton, B. R., W. E. Sopper, and L. T. Kardos. 1975. Revegetating bituminous strip-mine spoils with municipal wastewater. *Compost Science*, **16**(4):20–25.

EPA. 1976. National interim primary drinking water regulations. EPA-570/9-76-003. 159 p.

EPA. 1975. Technical bulletin. Evaluation of land application systems. 430/9-75-001.

EPA. 1974. Land application of sewage effluents and sludges: selected abstracts (abs #015M and 020M, 284I). 660/2-74-042.

EPA. 1973. Survey of facilities using land application of wastewaters. 430/9-73-006.

Fisk, W. W. 1964. Food processing and waste disposal. *Water and Sewage Works*, **111**: 417–420.

Foster, H. B., Jr. and W. F. Jopling. 1969. Rationale of standards for use of reclaimed water. *J. San. Eng. Div., Proc. A.S.C.E.*, SA (3):503–514.

Garcia, W. J., C. W. Belssin, G. E. Inglett, and R. O. Carlson. 1974. Physical-chemical characteristics and heavy metal content of corn grown on sludge-amended strip-mine spoil. *J. Agr. Food Chem.*, **22**(5):810–815.

Harrison, J. E. 1974. Geologic aspects of mountain coal mine waste disposal. *In* Land for waste management, proceedings of the international conference at Ottawa, Canada, Oct. 1973, the Agricultural Institute of Canada, Ottawa. 388 p.

Henderson, G. S. and W. F. Harris. 1975. An ecosystem approach to characterization of the nitrogen cycle in a deciduous forest watershed. pp. 189–193. *In* Bernier, B. and C. H. Winget, ed. Forest soils and forest land management. Laval Univ. Press, Quebec. 676 p.

Herrick, A. M. and C. L. Brown. 1967. A new concept in cellulose production: silage sycamore. *Agr. Sci. Rev.* Fourth Quarter, pp. 8–13.

Hill, R. D. 1971. Restoration of the terrestrial environment. *The ASB Bull.*, **16**(3):107–116.

Hook, J. E. and L. T. Kardos. 1977. Nitrate relationships in the Penn State living filter system. pp. 181–198. *In* R. C. Loehr, ed. Land as a waste management alternative. Proceedings of the 1976 Cornell Agricultural Waste Management Conference, Ann Arbor Science Pub., Inc. Ann Arbor, Mich. 811 p.

Jorgensen, J. R., C. G. Wells, and L. J. Metz. 1975. The nutrient cycle: key to continuous forest production. *J. Forest*, **53**:400–403.

Kardos, L. T. and J. E. Hook. 1976. Phosphorus balance in sewage effluent treated soils. *J. Environ. Qual.*, **5**:87–90.

Lance, J. C. 1972. Nitrogen removal by soil mechanisms. *J. Water Pollut. Contr. Fed.*, **44**:1352–1361.

Lejcher, R. and S. R. Kunkel. 1973. Restoration of acid spoil banks with treated sewage sludge. pp. 184–186. *In* W. E. Sopper and L. T. Kardos, eds. Recycling treated municipal wastewater and sludge through forest and cropland. Penn State Univ. Press, University Park, Pa. 479 p.

Little, S., H. W. Lull, and I. Remson. 1959. Changes in woodland vegetation and soils after spraying large amounts of wastewater, *Forest Sci.*, **5**:18–27.

Mather, J. R. 1953. The disposal of industrial effluent by woods irrigation. *Trans. Amer. Geophys. Union.*, **34**:227–239.

Merrell, J. C., Jr. and P. C. Ward. 1968. Virus control at the Santee, Calif. project. *Amer. Water Works Assn. J.*, **60**:145–153.

Munro, J. K., Jr., R. S. Luxmore, C. L. Begovich, K. R. Dixon, A. P. Watson, M. R. Patterson, and D. R. Jackson. 1977. Transport model to predict the movement of Pb, Cd, Zn, Cu, and S through a forested shed. pp. 45–58. *In* Disposal of residues on land. Information Transfer, Inc. Rockville, Md. 216 p.

Murphey, W. K., R. L. Brisbin, W. J. Young, and B. E. Cutter. 1973. Anatomical and physical properties of red oak and red pine irrigated with municipal wastewater. pp. 295–310. *In* Sopper, W. E., and L. T. Kardos, eds. Recycling treated municipal wastewater and sludge through forest and cropland. Penn State Univ. Press, University Park, Pa. 479 p.

Murphey, W. K. and R. L. Brisbin. 1970. Influence of sewage plant effluent irrigation on crown wood and stem wood of red pine. Pa. Agr. Exp. Sta., University Park, Pa. Bull. 772. 29 p.

Myers, E. A. 1973. Sprinkler irrigation systems: design and operation criteria. p. 324–333. *In* Sopper, W. E. and L. T. Kardos, eds. Recycling treated municipal wastewater and sludge through forest and cropland. Penn State Univ. Press, University Park, Pa. 479 p.

Nute, J. Warren, Inc. 1975. Mt. View Sanitary Dist. marsh enhancement pilot program. Progress Rept. No. 1. San Rafael, Calif. 9 p.

Nute, J. Warren, Inc. 1977. Mt. View Sanitary Dist. marsh enhancement pilot program. Progress Rept. No. 2. San Rafael, Calif. 20 p.

Nutter, W. L. and R. C. Schultz. 1975. Spray irrigation of sewage effluent on a steep forest slope, I. Nitrate renovation. Agron. Abst. Amer. Soc. Agron. Madison, WI.

Parizek, R. R., L. T. Kardos, W. E. Sopper, E. A. Myers, D. E. Davis, M. A. Farrell, and J. B. Nesbitt. 1967. Wastewater renovation and conservation. Penn State Univ. Studies No. 23. University Park, PA. 71 p.

Peterson, M., *et al.* 1973. A guide to planning and designing effluent irrigation disposal systems in Missouri. U. of Mo. Extension Division. NP 337 3/73/1250. 90 p.

Plass, William T. 1974. Factors affecting the establishment of direct-seeded pine on surface-mine spoils. USDA Forest Services Res. Paper NE-290.

Powers, R. F., K. Isik, and P. J. Zinke. 1975. Adding phosphorus to forest soils: storage capacity and possible risks. *Bull. Environ. Contam. Toxicol.*, **14**:257–264.

Richenderfer, J. L., W. E. Sopper, and L. T. Kardos. 1975. Spray-irrigation of treated municipal sewage effluent and its effect on chemical properties of forest soils. Gen. Tech. Rep. NE-17. USDA, Forest Service, NE Forest Exp. Sta. Upper Darby, Pa. 24 p.

Rudolph, V. J. 1957. Further observation on irrigating trees with cannery waste water. *Mich. Agr. Exp. Sta. Quarterly Bull.*, **39**:416–423.

Rudolph, V. J. and R. E. Dils. 1955. Irrigating trees with cannery waste water. *Mich. Agr. Exp. Sta. Quarterly Bull.*, **37**:407–411.

Schultz, R. C., W. L. Nutter, and G. H. Brister. 1975. Spray irrigation of sewage effluent on a steep forest slope, II. Vegetation, forest floor, and soil variations. Agron. Abst. Amer. Soc. Agron. Madison, WI.

Schuman, G. E., W. A. Berg, and J. F. Power. 1976. Management of mine wastes in the western United States. pp. 180–194. *In* Land application of waste materials. Soil Conservation Society of America. Ankeny, Ia. 313 p.

Sopper, W. E. 1975. Wastewater recycling on forest lands. pp. 227–243. *In* Bernier, B. and C. H. Winget, eds. Forest soils and forest land management. Laval Univ. Press, Quebec. 675 p.

Sopper, W. E. and L. T. Kardos. 1973. Vegetation responses to irrigation with treated municipal wastewater. pp. 271–294. *In* Sopper, W. E. and L. T. Kardos, eds. Recycling treated municipal wastewater and sludge through forest and cropland. Penn State Univ. Press, University Park, Pa. 479 p.

Sopper, W. E. and L. T. Kardos. 1972. Effects of municipal wastewater disposal on the forest ecosystem. *J. Forestry*, **70**:540–545.

Sopper, W. E. and L. T. Kardos. 1972. Municipal wastewater aids revegetation of strip-mined spoil banks. *J. Forestry*, **70**(10).

Sopper, W. E., 1970. Revegetation of strip-mine spoil banks through irrigation with municipal sewage effluent and sludge. *Compost Science*, Nov.–Dec. pp. 6–11.

Stevens, L. A. 1974. Clean water. E. P. Dutton, N.Y. 289 p.

Stevens, R. M. 1972. Green land-clean streams. Center for the Study of Federalism, Temple Univ. Philadelphia, Pa. 330 p.

Suhr, L. G. 1971. Some notes on reuse. *Journal AWWA*, **63**, pp. 630–634.

Sullivan, D. L. 1970. Wastewater for golf course irrigation. *Water and Sewage Works*, **117**(5):153–159.

Sutherland, J. C., J. H. Cooley, D. G. Neary, and D. H. Urie. 1974. Irrigation of trees and crops with sewage stabilization pond effluent in southern Michigan. pp. 295–313. *In* Wastewater use in the production of food and fiber, EPA-660/2-74-041. Proceedings of the conference held in Oklahoma City, Okla. March 5–7.

Switzer, G. L., L. E. Nelson, and W. H. Smith. 1972. Nutrient accumulation and cycling in loblolly pine (*Pinus taeda* L.) plantation ecosystems: the first twenty years. *Soil Sci. Soc. Amer. Proc.*, **36**:143–147.

Switzer, G. L., L. E. Nelson, and W. H. Smith. 1968. The mineral cycle in forest stands. pp. 1–9. *In* Forest fertilization. TVA, Muscle Shoals, AL.

Tamm, C. O., H. Holmen, P. Popovic, and G. Wiklander. 1974. Leaching of plant nutrients from soils as a consequence of forestry operations. *Ambio.*, **3**:211–221.

Taylor, K. G. and D. Gill. 1974. Environmental alteration and natural revegetation at a mine site in the Northwest Territories, Canada. pp. 16–25. *In* Land for waste management, proceedings of the international conference at Ottawa, Canada-Oct. 1973. The Agricultural Institute of Canada, Ottawa. 388 p.

Trimble, G. R., Jr., R. S. Sartz, and R. S. Pierce. 1958. How types of soil frost affect infiltration. *J. Soil Water Conserv.*, **13**:81–82.

Urie, D. H. 1971. Opportunities and plans for sewage renovation on forest and wildlands in Michigan. *Mich. Acad.*, **4**:115–124.

Urie, D. H. 1975. Nutrient and water control in intensive silviculture on sewage renovation areas. *Iowa State J. Res.*, **49**:313–317.

U.S. Department of the Interior. 1967. Surface mining and our environment. U.S. Dept. of the Interior. 1967-0-278-800. 124 p.

Van't Woudt, B. and R. M. Hagan. 1957. Crop responses at excessively high soil moisture levels. *In* Drainage of agricultural lands. American Society of Agronomy, Madison, Wis. 620 p.

Webster, B. 1975. Bullrushes being used in artificial marshes to filter water. N.Y. Times, 9 March.

Wood, G. W., P. J. Glantz, D. C. Kradel, and H. Rothenbacher. 1975. Faunal response to spray irrigation of chlorinated sewage effluent. Institute for research on land and water resources. Publication No. 87, Penn State Univ. Press, University Park, Pa. 89 p.

Wood, G. W., D. W. Simpson, and R. L. Dressler. 1973. Effects of spray irrigation of forests with chlorinated sewage effluents on deer and rabbits. *In* Sopper, W. E. and L. T. Kardos, eds. Recycling treated municipal wastewater and sludge through forest and cropland. Penn State Univ. Press, University Park, Pa. pp. 311–323.

Woodwell, G. M., J. Ballard, J. Clinton, M. Small, and E. V. Peron. 1974. An experiment in the eutrofication of terrestrial ecosystems with sewage. Evidence of nitrification in a late successional forest. pp. 215–228 *In* Sutherland, *et al.* (cited above).

Younger, V. B., W. D. Kesner, A. R. Berg, and L. R. Green. 1973. Ecology and physiological implications of greenbelt irrigation with reclaimed wastewater. pp. 396–407. *In* W. E. Sopper and L. T. Kardos, eds. Recycling treated municipal wastewater and sludge through forest and cropland. Penn State Univ. Press, University Park, Pa. 479 p.

APPENDIX A

EFFECTS OF SPRAY IRRIGATION OF TREATED WASTEWATER ON WILDLIFE (WOOD, ET AL., 1975)

ABSTRACT

The only significant difference found in a 3-year comparison of cottontail rabbit (*Sylvilagus floridanus*) populations confined to non-irrigated and sewage effluent irrigated aspen-white pine-shrub habitat was in the significantly heavier body weights in the fall of juveniles on irrigated sites.

Comparison of spring and fall populations of mice (*Peromyscus* spp.) in non-irrigated, effluent irrigated, and sludge-injected effluent irrigated mixed-oak forest habitat revealed no population density differences between irrigated and non-irrigated sites during spring. Fall 1973 populations were significantly different with higher densities on irrigated sites, and these densities were greater on sludge-injected effluent sites than effluent only sites.

Analyses of liver, kidney and bone tissues for cadmium, chromium, copper, nickel, lead and zinc were done on cottontail rabbits from effluent irrigated and non-irrigated sites. Copper was the only element found to be significantly higher ($P > 0.05$) in concentration in rabbits from irrigated sites and this only in kidney tissue. Unexplainably rabbits from irrigated sites showed significantly ($P > 0.05$) lower concentrations of cadmium in liver; cadmium, nickel, and lead in kidney; and copper in bone tissue.

Whole-body analyses of dry ashed white-footed mice carcasses for chromium, nickel, lead and zinc were also made. Ten animals were taken from each of the following: effluent irrigated site, sludge-injected effluent site, and a non-irrigated control. Chromium levels were higher and nickel levels were significantly lower ($P > 0.05$) on the effluent irrigated site. Otherwise there were no statistically significant differences.

Studies of the effects of effluent irrigation on production, nutritive quality, and use of deer forages in mixed-oak forest stands showed no significant differences in site feeding capacity between irrigated and non-irrigated sites in late spring. Summer production of palatable forage on the non-irrigated site far exceeded that of the irrigated site. Paradoxically, use of irrigated sites by semi-free ranging white-tailed deer (*Odocoileus virginianus*) exceeded use of the non-irrigated sites in both spring and summer.

Studies of songbirds revealed that their use of irrigated areas declined during the time of irrigation. In aspen-white pine-shrub habitat, activity on irrigated sites was higher than non-irrigated sites on non-irrigation days. This was due primarily to the increased amounts of brush caused by greater winter ice damage to sapling trees on irrigated sites which created a more favorable habitat for certain species of birds. Activity on non-irrigated mixed-oak forest stands was greater than that on similar irrigated stands during the early summer. The reverse was true, in late summer when surface soil macro-organisms that served as a food source were apparently more available on the irrigated sites.

Preliminary investigations of mosquito populations in the irrigation area and at similar remotely located control areas suggested that population densities were greater in the general area of irrigation. There were no demonstrable differences between irrigated sites and sites close to them that were not irrigated. *Adedes vexans* and *Aedes trivittatus* were by far the dominant species present, except at 1 site located about 8 km from the irrigation facility, where species of the genus *Culex* were dominant.

A very limited investigation of the possible influence of wastewater irrigation on disease incidence in several species of wild mammals revealed no evidence of leptospirosis in specimens taken from irrigated or non-irrigated sites. Coccidiosis was common in populations of cottontail rabbits confined to non-irrigated and irrigated sites, and its occurrence was greater among juveniles on irrigated sites. Sereological tests of blood from deer, opossum, cottontail rabbits, humans, and domestic rabbits serving as sentinel animals for California encephalitis, St. Louis encephalitis, and Powasson, revealed only antibodies for California encephalitis in some individuals of all species, but there was no evidence of a difference in incidence of disease that might be caused by wastewater irrigation.

Microbiological examination of the wastewater for total coliform, fecal coliform and fecal streptococci showed that 100 percent of the sludge-injected effluent samples had counts exceeding 1000 per 100 ml for each of these forms. Only 26 percent of the effluent samples had counts of this magnitude. Studies of fecal samples from rabbits, deer and songbirds revealed 16 *E. coli* O-groups in animals from irrigated areas. These O-groups were not found in animals from non-irrigated areas. Eleven of the above mentioned O-groups were found in the wastewater itself. There were 6 *Klebsiella sp.* isolates found in the wastewater and 46 found in the wild birds and mammals. There was no conclusive evidence of increased frequency of this organism due to irrigation, however. *Salmonella sp.* was not found in wastewater samples, although it was isolated from 1 songbird fecal sample.

CONCLUSIONS

The conclusions drawn from these studies of limited intensity should not be considered definitive for 3 reasons. First, the time span of this research is short relative to that during which a wastewater system may create environmental changes in a particular ecosystem. Second, the communities of plants and animals that were available for study were limited to those on sites already being irrigated, which was mostly a mixed-oak forest stand of large pole to small sawtimber, and an early successional community of an aspen-white pine-shrub complex. Third, due to the limited size of study sites and the time span of the study, relatively small numbers of animals have been observed, which seriously limits the researchers' ability to extrapolate the findings to large populations over large areas. Within these limits the findings suggest:

1. Three years of irrigation for aspen-white pine-shrub communities with chlorinated sewage effluent did not result in any net losses in cottontail rabbit reproduction.
2. The changes in the cottontail rabbit habitat in the aspen-white pine-shrub communities appeared to be the result of 3 factors:
   a. General increase of concentrations of crude protein, phosphorus, potassium and magnesium occurred, but concentrations of calcium dropped in both summer and winter forages.
   b. The extensive ice damage to saplings and shrubs caused by winter irrigation increased the amounts of available forage, both winter and summer.
   c. The ice formations on shrubs and trees surrounding irrigation sprinklers increased the amount of winter cover.
3. Irrigation of cottontail rabbit habitat apparently had little influence on body weights of adult animals in the fall. It did appear to result in increased weight in juvenile animals, however.
4. The amounts of deer forage available on irrigated and non-irrigated mixed-oak forest sites in late spring were not meaningfully different. Summer irrigation, however, tended

to enhance greatly the production of unpalatable herbaceous species and effectively lowered the feeding capacity considerably below that of the non-irrigated site. Yet, the preference factors for 5 out of 6 palatable species which grew on both sites were higher on the irrigated site.

5. Irrigation may raise the concentrations of crude protein, phosphorus, potassium, and magnesium; lower the concentration of calcium; and have no effect on crude fiber or *in vitro* dry matter digestibility of deer forage.
6. Chlorinated effluent does not deter the use of irrigated sites by deer. In fact, deer may use these sites more often than non-irrigated sites, possibly due to the greater amounts of vegetative cover on the former.
7. Irrigation apparently has no effect on spring populations of mice *Peromyscus.* Fall populations, however, may be higher on irrigated than non-irrigated sites due to increased density of herbaceous vegetation associated with effluent irrigated and sludge-injected effluent sites.
8. Songbird use for an area appears to be primarily related to the type of habitat that is being irrigated and the time of irrigation. In sapling tree stands that are heavily damaged by ice and consequently produce more brush, one may expect greater songbird use by species such as song sparrows, yellowthroats and gray catbirds. The effect of irrigation on mixed-oak tree stands for songbird use may depend on 2 things. First, winter irrigation may result in the loss of understory saplings and low tree branches due to ice damage. They would otherwise be used for perching and nesting sites during the summer. When food is equally abundant on irrigated and non-irrigated sites, those with perching facilities will be preferred. Second, when seasonal changes in moisture stress result in greater surface soil macro-organism activity, greater songbird use may be expected of the irrigated sites. Songbirds will avoid irrigated areas during the time of irrigation.
9. Irrigation of forested areas will probably enhance the production of mosquitoes, particularly those of the genus *Aedes.* The interaction of the increased mosquito populations and increased songbird activity may provide an important interface for arbovirus transmission.
10. There is no conclusive evidence to indicate an increased incidence of diseases in wild animals as a result of irrigating an ecosystem with municipal wastewater. In systems of neutral to alkaline soil conditions, the changed moisture regime could be favorable for the survival of leptospires, but, so far, if this is happening it apparently has not manifested itself epidemiologically. California encephalitis in the mosquito and wild mammal populations can be expected in wastewater irrigated ecosystems in Pennsylvania. There is no evidence, however, to show that its frequency of occurrence is any greater or less than in similar non-irrigated ecosystems.
11. Bacterial agents (*E. coli*, possibly *Klebsiella* sp. and others) will survive sewage treatment and are introduced into irrigated ecosystems. This consequently affects the bacterial populations of ponds in the area. However, more conclusive data are needed to verify transmission of these micro-organisms from wastewater to wild animals, and their beneficial or detrimental effect to the health of the host.
12. If concentrations of cadmium, chromium, copper, nickel, lead and zinc are increasing in the bodies of small herbivorous mammals feeding on irrigated sites, the changes are either quite small or not detectable by the sampling techniques. This does not mean that concern for this aspect of the study should be minimized, but suggests that some long term monitoring should be used, should these contaminants become saturated in irrigated soils.

RECOMMENDATIONS

The results are not definitive as to what the response of wild animals will be to the practice of irrigating municipal wastewater on forest and farm land. They do point out areas that need continuing research.

First, the effects of wastewater irrigation on the occurrence of biting insects is in great need of intensive investigation. Changes in population size and species composition of these insects may be reflected in the potential for disease transmission in the irrigated areas. In addition, insect populations will influence the comfort of both people and animals who use the area.

Second, changes in the occurrence of potentially pathogenic organisms that might result from wastewater irrigation should have further research and should be part of a constant system of monitoring. The greatest limitation to conclusions from short-term, single-attempt studies is that no basis for comparison of present and previous figures is available. This work has provided some information, but no data existed with which it could be compared. A plan should be devised to provide for systematic sampling in the irrigated area and in a control area, which would not only serve as a public health measure but also provide long-term research data.

Approximately 10 years of research has convinced many people that sewage effluent can be disposed of safely by irrigation. Relative to the thoroughness with which effluent spray irrigation has been researched, little evaluation of sludge-injected effluent has been done. Because of its high bacterial counts and heavy metal concentrations any recommendation to a municipality of this procedure for sludge disposal should be carefully considered in perspective with the other alternatives. In addition it should then only be recommended as an experimental procedure with contingency plans to alleviate adverse environmental effects should they arise.

Third, The Pennsylvania State University Wastewater Renovation Facility is a prototype that is serving as a model. An important area of information that has not come from this facility, but which could be developed, would provide guidelines for choosing wastewater renovation sites, preparing environmental impact statements for the facilities, and environmental parameter monitoring. The lack of data on such aspects as faunal response is partly to blame, but most of the problem is due to the difficulty of assembling an interdisciplinary group of scientists to study the problems and report their findings.

Module II-9

WASTE APPLICATION SYSTEMS

SUMMARY

Land application systems are discussed with reference to the options available for applying wastewater and sludge to the site. Spray systems, surface flow methods, and sludge application schemes are all included with discussions of the advantages and disadvantages of each option within these categories.

A distinction is made between the choice of treatment method and the application system. The selection of an application system is a function of the treatment alternative. Selection of alternative is made on the basis of fundamental soil, plant, waste, and managment factors pointing to overland flow, rapid infiltration, or irrigation as most appropriate for the given situation. Once this decision is made, several options will be available for application of the wastewater or sludge.

Application systems described in this module include seven types of sprinkler irrigation techniques, gated pipe, three ditch systems, and the three most common field surface irrigation methods. Sludge spraying, spreading, and subsurface injection are outlined.

One important aspect of application system design is uniformity of wastewater contact with the treatment medium. This topic is addressed from a theoretical standpoint and related to the practical situation. A brief discussion of the special problems encountered in overland flow and rapid infiltration systems is also included.

CONTENTS

OBJECTIVES

Upon completion of this module, the reader should be able to:

1. List the application systems alternatives which are applicable to each of the following:

   Spray irrigation
   Surface irrigation
   Overland flow
   Rapid infiltration

2. Discuss the advantages and disadvantages of the seven types of spray application covered in the module.
3. Discuss the crop limitations of the surface irrigation methods, and identify the spray systems which impose limitations on the crops that may be grown.
4. List the application methods appropriate for sludge and discuss their advantages and disadvantages.
5. Define "coefficient of uniformity" and discuss the importance of uniformity of application to waste application sites.
6. Name the best sources of engineering advice on waste application systems.

INTRODUCTION

An integral part of land treatment is the physical transport and application techniques utilized in dispersing wastewaters and sludges on the land. The system selected must satisfy many demands imposed upon it by the waste constituents, the application site characteristics, economic considerations, type of land treatment systems, and the law. Because each system design is highly site specific, many variations will emerge across the country.

Techniques used for water distribution in irrigated agriculture have been successfully adapted to land application systems for wastewater renovation. This is good practice in that stocks of pipes, nozzles, valves, gates, etc., are readily available for the initial installation, and parts and maintenance assistance can be obtained easily from the irrigation equipment supplier. However, use of irrigation equipment "off the shelf" without any adaptation for wastes application has led to systems which have been badly designed and in some cases resulted in complete failures (Norum, 1976). Many irrigation equipment manufacturers have developed expertise in land treatment, and it is best to obtain advice from their applications engineers.

Selection of waste application systems should not be confused with the selection of the *method* of land treatment (i.e., spray or surface irrigation, overland flow, rapid or moderate rate infiltration, etc.). Selection of treatment method is a more basic decision which then leads to the selection of application system alternatives. The factors contributing to treatment method evaluation are discussed in several other modules in this program, including Modules I-1, "Soil as a Treatment Medium"; I-6, "Site Evaluation"; I-3, "Waste Characteristics"; II-4, "Potentially Toxic Elements"; II-5, "Pathogens"; II-1, "Nitrogen Considerations"; and II-2, "Phosphorus Considerations"; and will not be discussed here in any detail.

SITE CHARACTERISTICS INFLUENCE SELECTION, DESIGN OF SYSTEM

Site characteristics which greatly influence the selection and design of waste application systems are topography, soil, and the cover crop desired. It is difficult to discuss individual parameters without drawing upon other factors which influence their behavior. The availability of certain soils, topography, and depths to bedrock will be factors affecting the type of treatment system designed. Slope conditions may dictate the type of crop that is needed for the soil encountered. This in turn may dictate the application system needed to attain the required effluent quality. Site characteristics are discussed at length in Module I-6, "Site Evaluation."

**The topography, soil, and vegetative cover of the site greatly influence the selection and design of the waste application system.*

Topography Limits Method of Wastewater Application

The topography of the treatment site limits the method of wastewater application. Systems depending upon rapid infiltration of the wastewater must be relatively level (0 to 1% slope) whereas systems utilizing slow to moderate infiltration rates may vary from 0 to 10+% slope. Large rotating sprinkler irrigation rigs have the mechanical power and flexibility to be utilized on slopes of 15 to 20% (Norum, 1976), although erosive flows have resulted with these big rigs upon steep slopes. Overland flow systems have been successful on slopes ranging from 2 to 8%, and the site may be shaped to meet the needs of this treatment method. With methods other than shaped overland flow, the soils and slopes are generally accepted as they occur, however, and the length of run and volume of water applied are varied as needed to prevent erosive flows and to attain the required effluent quality. Surface irrigation systems are designed to accom-

*This and other italicized summaries are intended to highlight key ideas, provide a basis for later review or to aid in skimming sections that are relatively familiar. They can be ignored in a complete reading of the text.

Table 1. Treatment Alternative vs Site Characteristics.

Treatment Alternative	Soil Permeability	Topography (%)	Cover Crop
Rapid Infiltration	moderate to rapid	0–1	generally none
Overland flow	slow	2–8	pasture grasses
Irrigation			
Sprinkler	slow to rapid	0–15	any
Surface	slow to moderate	0–8	any

modate existing soils, slope, and cover crop conditions. The rate at which water is applied is dependent upon these factors as well as the type of irrigation system utilized.

Soil and slope of the site usually are not altered to fit a certain treatment method; rates of water application are adjusted to fit the site characteristics.

Soil Structure and Permeability Must be Able to Withstand Loadings

Soil conditions at the treatment site play an important role in selecting and managing a wastewater application system. Soil permeability is the major consideration in the infiltration and percolation capacity of the soil profile. Rapidly, moderately, and slowly permeable soils, along with optimum land slopes, relate to the land treatment alternatives as shown in Table 1. The ability of the soil structure to maintain its integrity under hydraulic stresses as well as vertical loadings from irrigation rigs and crop management equipment must be investigated prior to application system selection. This is especially true where wastewater renovation is secondary to crop production schemes. The soil and slope available to the system has a great deal to do with the method of application, especially where crops are surface irrigated and nonerosive flows are to be maintained.

The ability of the soil structure to maintain its integrity as well as permeability under saturated conditions limits the application system selected.

Type of Cover Crop Bears Directly on Choice of Application System

The cover crop planted on the treatment site may be selected with wastewater renovation foremost in mind, or the wastewater renovation system may have to be adapted to crop management schemes suited to the available land. In either case, the cover crop used will have a direct bearing on the application system selected. Surface irrigation systems are generally soil, slope, and crop specific. Flood irrigation is generally used with close grown crops such as pasture and grain crops whereas furrow or corrugation application systems are more adapted to row crops such as corn, sugar beets, or sorghum. Corrugation irrigation is a method which entails the creation of shallow, closely spaced furrows through the use of a special tillage implement. This technique is discussed in a later section of this module.

The cover crop at the site has direct bearing on which application system to select: surface irrigation is suited to close-growing crops, furrow and corrugation systems for row crops and sprinkler irrigation can be adapted to all vegetative covers.

Sprinkler irrigation systems are adaptable to all vegetative cover situations. However, portable sprinkler systems are dependent upon soil and topographical limitations as well as the vegetative cover height. In order to control nonpoint sources of pollution such as sediment and nutrient runoff, cover crops must be evaluated in relation to the irrigation volume, site topography, and soil conditions. Crop and other vegetative cover considerations are examined in depth in Modules II-7, "Crop Selection and Management Alternatives," and II-8 "Non-crop and Forest Systems."

To control runoff, cover crops must be considered in relation to irrigation volume, site topography and soil conditions.

APPLICATION TECHNIQUES

Sprinkler Irrigation

Sprinkler irrigation is being used in many parts of the country to apply wastewaters and sludges to the land. Wastes containing less than 10% solids are readily dispersed by most equipment and slurries up to 15% can be handled by some systems. These solids must be small and uniformly distributed throughout the wastewater in order to achieve uniform application and to avoid system clogging. Gross solids must be removed to protect pumps and spray nozzles from clogging and undue wear. Equipment manufacturers have developed sophisticated systems over many years of sprinkler irrigation design experience. Although systems are available which can meet the requirements of wastewater application, care must be taken that the sprinklers can handle the solids loading or other special needs of a waste treatment system.

Certain sprinkler irrigation systems can handle sludges with up to 15% solids.

Sprinkler irrigation systems in common use today include a variety of design schemes that have survived years of rigorous field trials. The systems discussed in this module are:

1. Solid set, both buried and above ground
2. Center pivot
3. Side-roll, wheel move with or without trail tubes
4. Continous travel
5. Towline laterals
6. Stationary gun
7. Traveling gun

Sprinkler systems commonly used today have a variety of design schemes which have survived years of field use.

The portability of these systems within the treatment site depends upon the irrigation scheduling and management scheme utilized. All the systems listed above, except for the buried solid set system, are designed to be relatively portable. Even the large center pivot systems can be towed from one water source to another by a farm tractor. Because of the intensity of use expected in wastewater renovation systems, center pivot rigs will more than likely remain in one location and be thought of as permanent. Main lines for all of these systems will probably be permanently buried. This provides protection from freezing weather and surface loadings from large farm machinery, as well as other hazards.

Except for a completely underground solid set system, some degree of portability is built into all sprinkler irrigation systems.

Spray application systems are appropriate for crop irrigation and, in some cases, overland flow systems. In general, they allow disposal on sandy soils where high infiltration would make surface application very difficult, and spray systems are often the most economical for irregularly shaped or rolling terrain. Initial installation costs are generally highest for sprinkler systems, but labor requirements can be less than surface irrigation, and automation is possible.

Spray application is appropriate for crop irrigation, some overland flow systems, sites with sandy soil and sites with irregularly shaped or rolling terrain.

Solid set irrigation systems can be permanently buried systems or laid on the ground surface and moved by hand between irrigated fields, in which case the equipment can be collected and stored during the non-application season. The term "solid set" refers to the fact that the system has no moving components (except nozzles) during a spray period, *not* that it is totally immobile. Main lines and laterals are laid out in a grid pattern to cover the irrigated area. Overlap between sprinklers varies between 30 to 60% of the wetted diameter. This overlap is needed for uniform water distribution.

Solid set irrigation systems may be automated and can be portable if laterals are laid above ground.

Equipment manufacturers have developed an array of automatic valves which may be regulated from a main control point. In this manner the entire system could be scheduled on a time clock and programmed to run continuously. Continuous surveillance of the system would still be required to make adjustments and maintain the equipment.

Systems can be programmed to run continuously via an array of automatic valves. This eliminates much hand labor but not the need for constant surveillance.

Figure 1 is taken from a design example in an equipment manufacturer's handbook. This layout represents a permanent solid set system in an orchard. The sprinklers are spaced to give most efficient coverage, and the lateral pipelines are laid diagonally to the main. This configuration is most economical, minimizing overall length versus size of laterals required, and cutting the number of valves and connections to the main line to approximately half of those in a system employing laterals along each row. If a moveable solid set system were used for this application, the rectangular grid layout would probably be favored due to the ease of moving laterals along the straight rows. The extra money spent on valves and fittings in this case would be made up by only having enough laterals and nozzles for partial coverage of the site and moving them between sets.

The principal advantages of a permanent solid set system are the possibility of automation and reduced labor demands. Disadvantages are high cost, generally too high to be economically feasible for all but very small systems which can use small diameter pipe, and the necessity of protecting risers and nozzles from equipment and animals.

A permanent solid set system can be automated, thus reducing the amount of labor needed. Movable solid sets have low initial cost but high hand labor requirement.

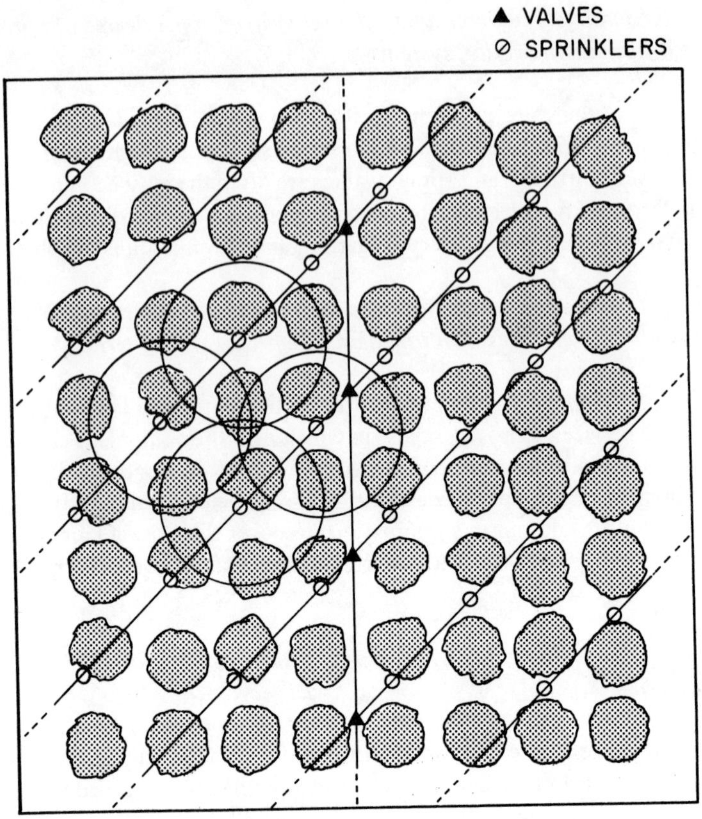

Figure 1. Solid set sprinkler irrigation (Fry and Gray, 1971. Courtesy Rain Bird Mfg. Corp.).

Hand move solid set systems generally use 20 to 40 foot sections of lightweight aluminum pipe joined by couplers with sprinklers built in. Advantages and disadvantages of the hand move solid set systems are:

Advantages

Low initial cost
Used systems available
Few mechanical parts
Low power requirements (50 psi at the sprinklers)
Flexible in coverage of area and shape of site

Disadvantages

High labor requirement
Sections disassembled and moved by hand
Small sprinkler outlets

Center pivot distribution systems, shown in Figure 2, cover up to 640 acres (a one square mile "full section") with a half-mile long boom with end units spraying the corners of the field. They are composed of a complex system of towers, pipes, guy-wires, and motorized wheels. Circular distribution patterns result from center pivot units, although telescoping end units have

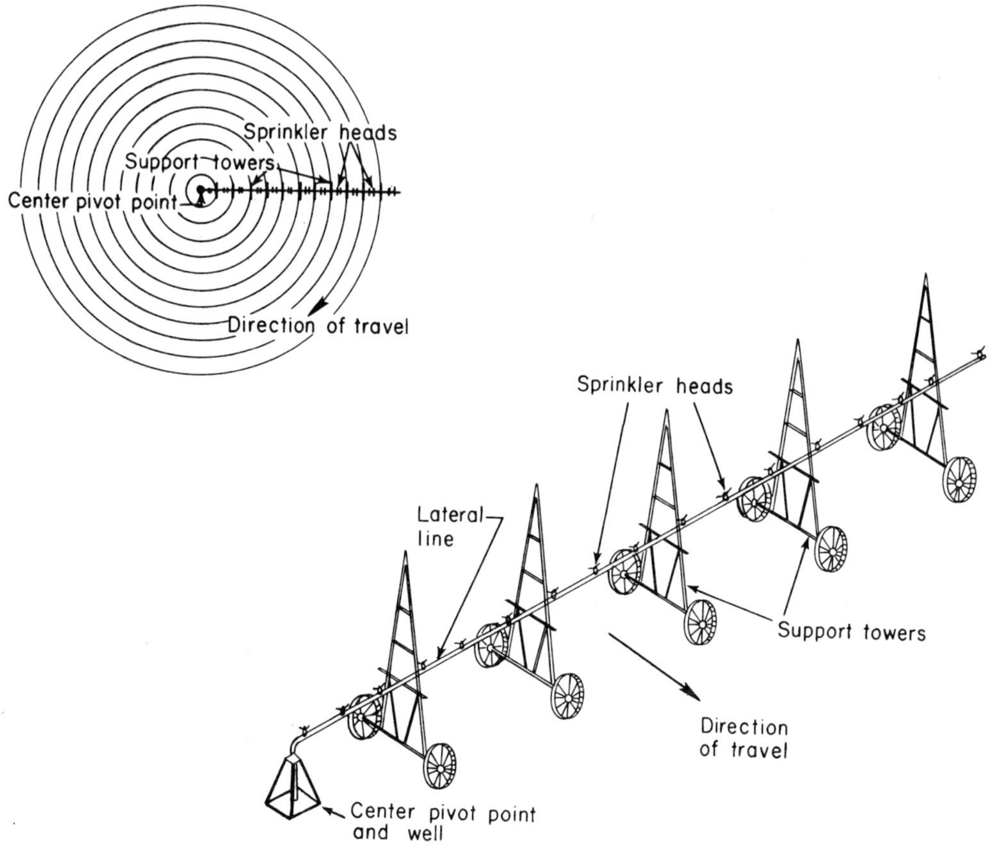

Figure 2. Center pivot system (Fry and Gray, 1971. Courtesy Rain Bird Mfg Corp.).

been developed to cover the triangular corner pieces also. These systems are generally thought of as permanent installations, but quick-disconnects at the "water source" end provide for easy detachment. The wheels may be turned parallel to the main distribution line thereby allowing the system to be towed from one center pivot point to another. These systems distribute large volumes of water which cause localized saturation of the soil. Because the weight of the entire system rests upon the various tower wheels, the wheels are susceptible to becoming mired down in mud. Decreases in application rates and various design changes of the wheels have kept these problems to a minimum.

The Muskegon County Wastewater Management System No. 1 utilizes 54 center pivot irrigation rigs to irrigate approximately 5,400 acres of farmland with municipal and industrial wastewater. The irrigation spray rig costs were approximately \$382/acre whereas the total irrigation spray system costs were approximately \$1,285/acre, including distribution piping, pumps, and other equipment and preparation. Application rates average 2.5 in./wk with maximum rates of 4 in./wk. The system layout is shown in Figure 3.

Center pivot systems can cover up to 640 acres per unit, distributing large volumes of water which cause localized saturation of the soil. Muskegon County Wastewater Management System No. 1 uses 54 of these rigs to irrigate 5400 acres. 100 acres/unit is average for center pivot rigs used in irrigated agriculture.

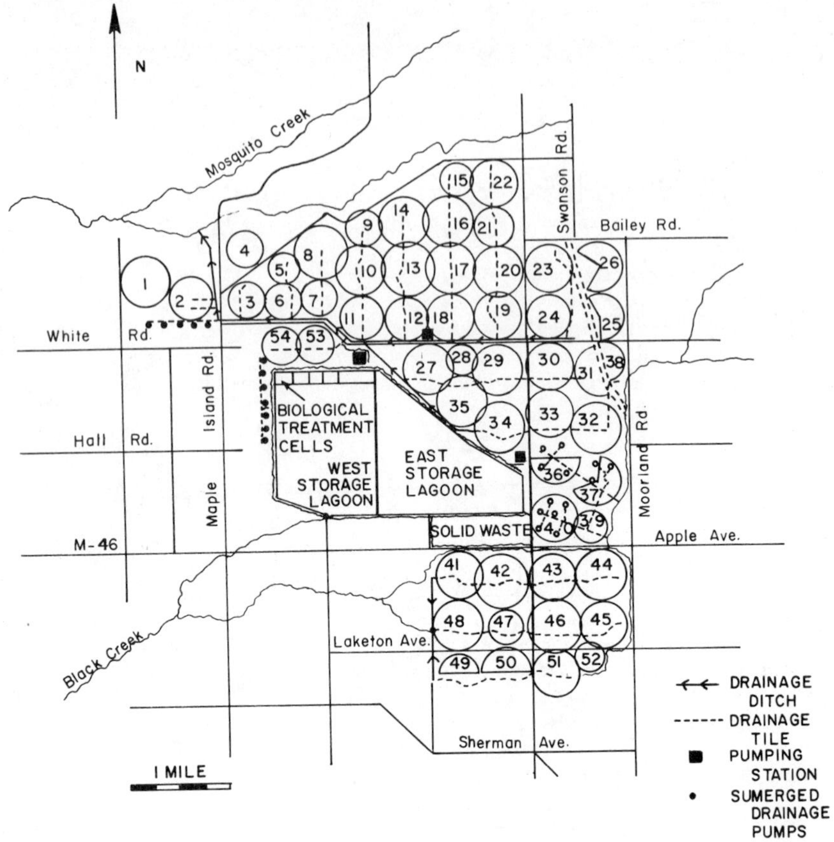

Figure 3. Muskegon county wastewater management system No. 1 (Demirjian, Y. A., 1975).

Figure 4 illustrates the three most common patterns of sprinkler nozzle spacing and sizing in use with center pivot rigs. Advantages and disadvantages of the three patterns are noted. The fixed spacing with variable sprinkler size (Figure 4b) generally would be the least amenable to wastewater application due to the higher operating pressures required, more soil compaction, and less uniform precipitation rate. The first consideration results in higher operating costs for pumping, and the latter two affect the treatment efficiency of the site. Soil compaction will reduce infiltration capacity and can affect the water application rate which may be used. Uniformity in spray application rate is necessary in order not to exceed the design application rate in any area of the field. This topic of application uniformity will be covered in some detail in a later section of this module.

Center pivot systems have at least three options for nozzle design and spacing available from the manufacturer. The option selected must meet the treatment requirements of the site.

The spray flood jet pattern (Figure 4c) uses downward-facing nozzles and results in lower power requirements. Plugging problems are listed if "gravel" is pumped, but wastewater solids should not pose a problem. This type of sprinkler pattern is used at Muskegon. An additional advantage of downward spraying is that aerosol production is minimized with lower operating

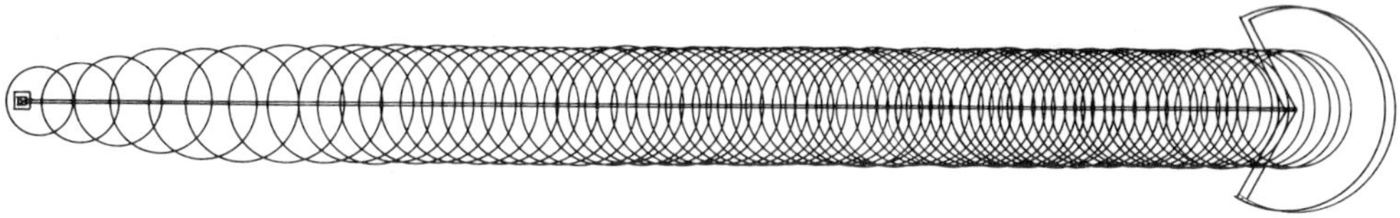

4(a). VARIABLE SPRINKLER SPACING PATTERN (Top View)

1. Uses approximately the same size sprinklers along the entire length of the system with the space between sprinklers decreasing as you move out from the pivot.
2. Requires lower operating pressure than the fixed spacing.
3. Less Soil Compaction than with the fixed spacing due to smaller droplet size.
4. More sprinklers used for a given length system than the fixed spacing. This results in lower G.P.M. per sprinkler which provides a uniform precipitation rate along the length of the system.
5. The width of the water pattern is approximately the same throughout, this results in a slightly higher precipitation rate than the fixed spacing.

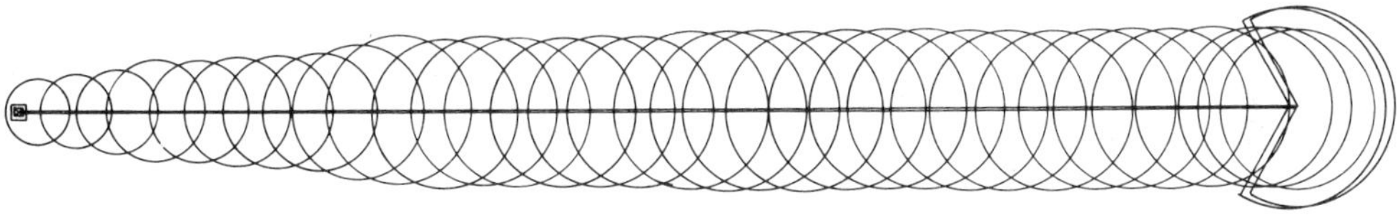

4(b). FIXED SPRINKLER SPACING PATTERN (Top View)

1. Uses sprinklers of varying sizes spaced evenly along the system with the size of the sprinklers increasing as you move out from the pivot.
2. Requires higher operating pressure than the variable spacing.
3. Higher soil compaction than with the variable spacing due to larger droplet size.
4. Less sprinklers used for a given length than with the variable spacing. This results in higher G.P.M. per sprinkler which provides a less uniform precipitation rate along the length of the system.
5. The water pattern is wider at the end of the system getting narrower as you move in toward the pivot. This gives a slightly lower precipitation rate than with the variable spacing.

4(c). SPRAY FLOOD JET PATTERN (Top View)

1. Uses flood jet spray nozzles of different sizes variable spaced along the length of the system. (Nozzles increase in size and are spaced closer as you move out from the pivot.)
2. Causes the least amount of soil compaction because of the extremely small droplet size.
3. Requires lower operating pressure, (20 to 25 P.S.I. lower than comparative fixed or variable spacing), which results in lower H.P. requirements and operating costs.
4. The water pattern is narrower and results in a higher precipitation rate than the fixed or variable spacing, therefore soil intake capabilities will have to be considered.
5. If a well pumps large gravel there may be some problems with the nozzles plugging.

Figure 4. Sprinkler patterns in use with center pivot irrigation systems (Lindsay, 1977).

pressure, and those which are formed are not airborne to be transported by the wind as readily as aerosols from higher pressure systems.

The advantages and disadvantages of center pivot systems can be summarized as follows:

Advantages

Almost complete automation
Can use various types of sprinkler patterns
Can be cost effective for larger systems
Even coverage of field area

Disadvantages

High cost per individual rig
Fixed area coverage
Many mechanical parts
Small sprinkler outlets

Side-roll, wheel move systems with and without trail tubes are shown in Figure 5. Because wheel diameters are available in sizes up to 8 feet, the mature cover crop should not exceed 4 feet in height. The system is stationary while spraying takes place, then is moved to the next set. An engine mounted in the middle of the pipe axle propels the system from one irrigation set to another.

Side-roll, wheel move systems eliminate hand labor in moving and provide good coverage for growing crops.

Advantages

Eliminates hand labor in moving
Good coverage for close-growing crops

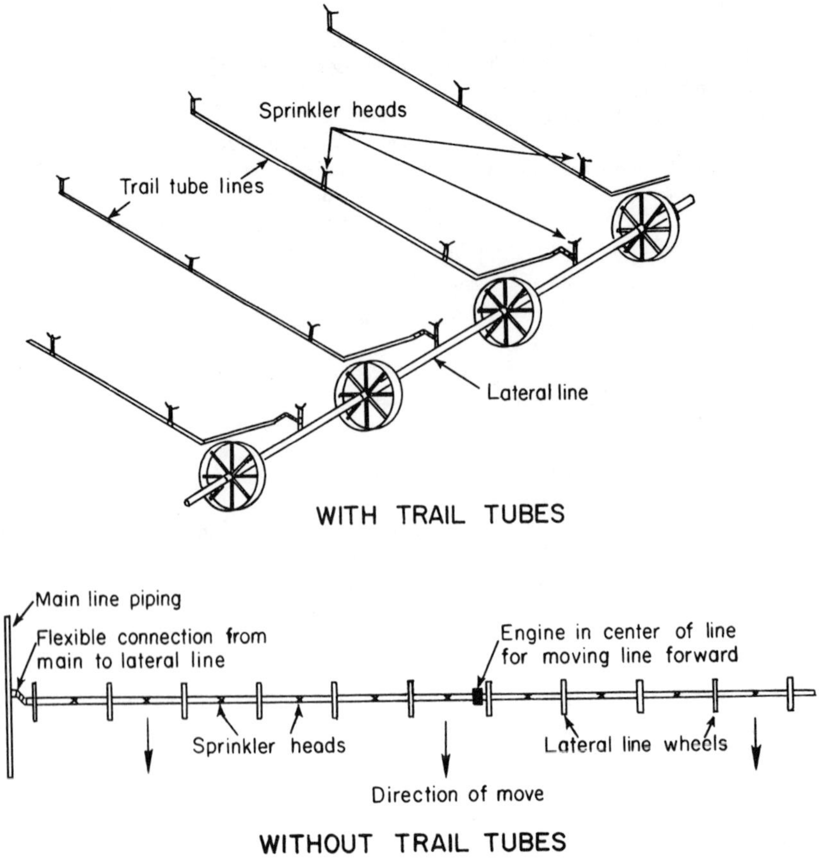

Figure 5. Side-roll, wheel move distribution system (with and without trail tubes) (Fry and Gray, 1971. Courtesy Rain Bird Mfg. Corp.).

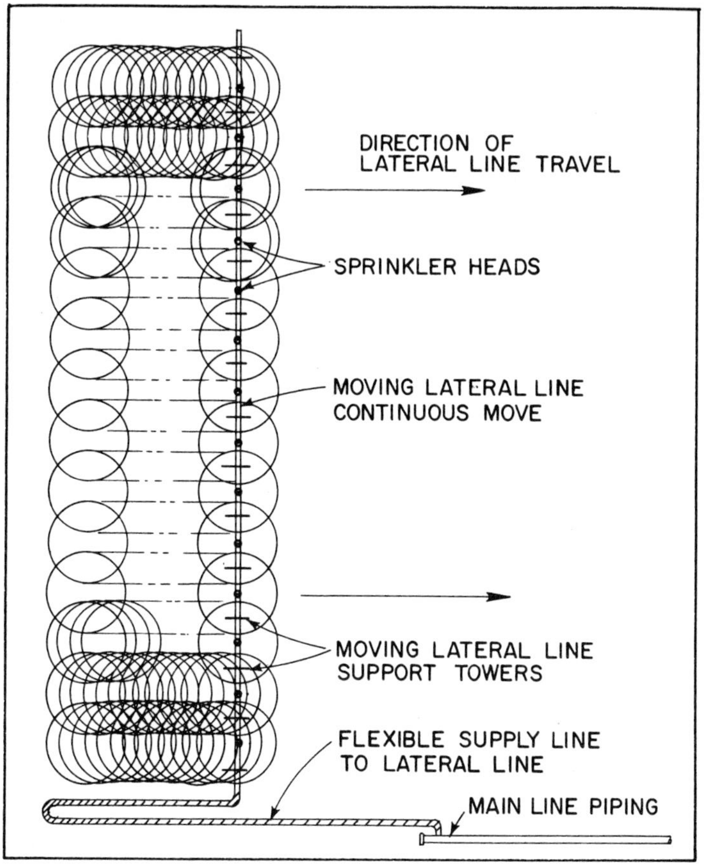

Figure 6. Continuous travel system (Fry and Gray, 1971. Courtesy Rain Bird Mfg. Corp.).

Disadvantages

Rectangular field shape only
Can have alignment problems
Restricted height of crop
Can have mechanical problems

Continuous travel distribution systems, Figure 6, utilize a flexible supply line which is dragged along behind the continuously moving lateral line. This flexible line may allow only one irrigation set per day. Advantages and disadvantages are the same as those for the side-roll; wheel move system.

Continuous travel distribution systems have the same advantages and disadvantages as side-roll, wheel move systems.

Towline laterals, Figure 7, are towed from one irrigation set across the main delivery line to another set, following a scheduled pattern. One suggested pattern is shown in Figure 7(b). Figure 7(c) illustrates a system with a buried main line and hydrant connections for the moveable laterals. The towline lateral system has stronger coupling between the pipe sections, but is otherwise much the same as a hand move solid set system.

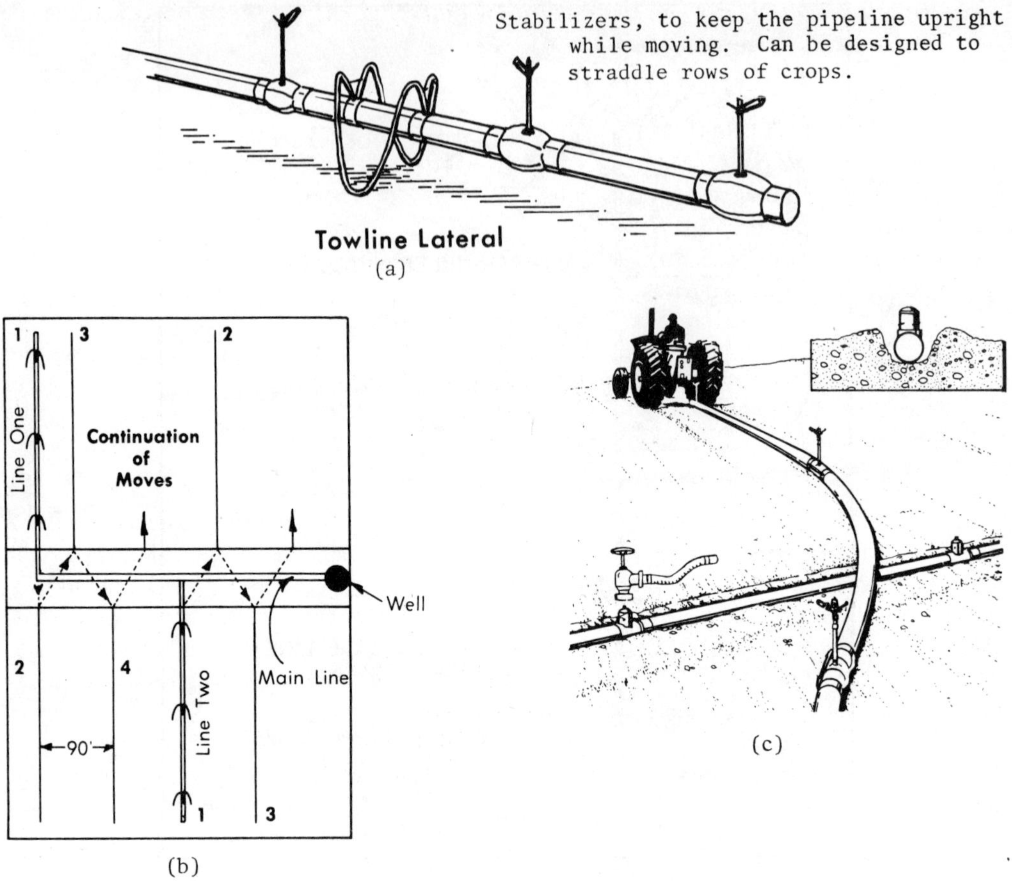

Figure 7. Towline lateral system and details of movement (Lindsay, 1977).

The towline lateral system is a movable solid set layout which minimizes hand labor.

Advantages

Low initial cost, comparable to hand move
Lower labor requirement
Few mechanical parts
Low power requirement (50 psi at the sprinkler)

Disadvantages

Not flexible in land area coverage because of difficulty in adjusting lateral length
Small sprinkler outlets
Tractor required
Driving lanes required for tractor

The *stationary big gun* system consists of a single large-bore rotating sprinkler head which is connected to the distribution piping and moved between sets by hand. Typical application rates are comparable to hand move solid set systems. The big gun is capable of spraying slurries with its large diameter nozzle.

The stationary big gun can spray slurries with its large diameter nozzle and is currently used for sludge application.

Advantages

Low initial investment
Few mechanical parts
Few plugging problems with large nozzle
Flexible in land area
Pipe diameter can be slightly less than with small sprinkler
Capable of spraying sludge slurries

Disadvantages

Higher power required for large outlet (80 psi at the sprinkler)
High labor requirement
Distribution not uniform in windy conditions
High aerosol transport

The *traveling gun* distribution technique, Figure 8, utilizes the same big gun nozzle mounted on a cart. These systems have a flexible supply line which is dragged along behind the cart. This hose is a major cost item in the system. The cart may contain a running gear and be towed by a pull cable or it may have hydraulic motors to propel the wheels. Various configurations are available from several manufacturers. The traveling big gun has been extensively used for wastewater and sludge application. One common size traveling gun will irrigate 10 acres per set, compared to 2 acres per set for the same size stationary gun.

The traveling gun can irrigate up to 5 times the acreage per set than a stationary gun of the same size.

Advantages

Very low labor requirement
Few plugging problems
Flexible in land area and topography
Capable of spraying sludge slurries

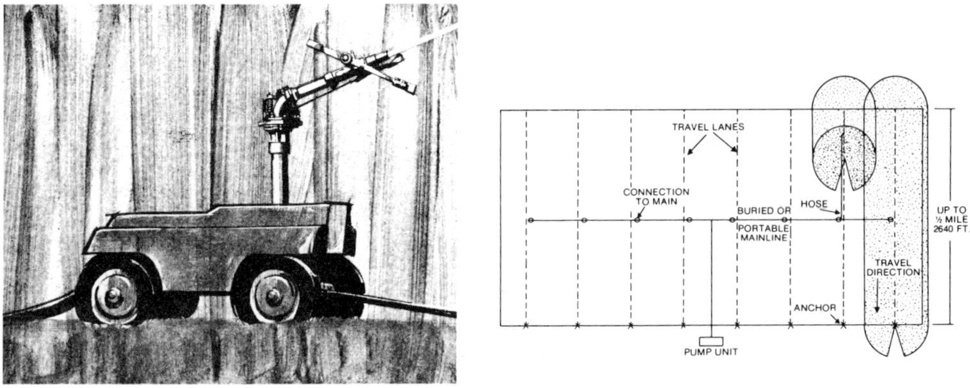

Figure 8. Nelson big gun and typical layout for a traveling sprinkler system (Nelson, 1977).

Disadvantages

High initial cost

High power requirement

Many mechanical parts–can have auxilliary gas engines on the carts

Table 2 presents a comparison of the most widely used spray irrigation systems, in terms of characteristics important to waste application.

WASTEWATER APPLICATION SHOULD BE UNIFORM FOR EFFECTIVE TREATMENT

In the design of spray irrigation systems each unit area of soil is expected to accept a designated volume of wastewater. If, however, the wastewater volume is not applied evenly to the renovation site, treatment is circumvented.

> *Wastewater must be applied uniformly over the soil so that the soil can receive the design application rate without over- or underloading of large areas.*

A means of quantifying the uniformity of application of sprinkler systems is the Uniformity Coefficient (or Christiansen's U.C., after the developer). It is expressed as a percentage and defined by the equation:

$$C_u = 100\left(1.0 - \frac{\Sigma x}{mn}\right)$$

where

Σx = Sum of deviation of individual observations from the mean

m = Mean or average values

n = Number of observations

In order to determine C_u in a field situation, containers are spaced in grid fashion over the wetted area to catch the irrigated water. An ideal application situation would result in the pattern of Figure 9, which indicates the over-all uniformity achieved by the cumulative pattern of several individual sprinklers arranged in a row. Manufacturers often provide suggested sprinkler

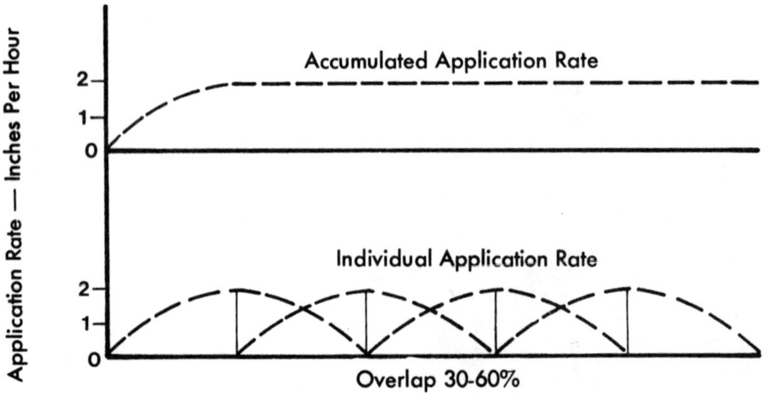

Figure 9. Idealized distribution and accumulation of irrigated water from four sprinkler nozzles.

Table 2. Characteristics of Major Types of Spray Systems (Adapted from Norum, 1976).

Type of System	Maximum Field Slope (%)	Ave. Water App. Rates (in./hr) Min	Max	Field Shape	Field Surface Conditions	Max. Crop Height (ft)	Labor Required (hr/acre irr.)	Size of Single System (acres)	Approximate Cost ($/acre[a])	Wind Drift Hazard	Solids Handling Capability	Comments
Portable solid set[b]	no limit	0.05	2.0	any shape	no limit	no limit	0.20–0.50	no limit	400–900	average	poor	Well suited for field crop and forest disposal sites
Buried solid set	no limit	0.05	2.0	any shape	no limit	no limit	0.05–0.10	no limit	400–1000	average	poor	Well suited for small systems spraying field crops where completely automatic operation is required
Side wheel roll	5–10	0.10	2.0	rectangular	reasonably smooth	3–4	0.10–0.30	20–80	100–300	average	poor	Best suited for periodic spraying of industrial wastes (such as cannery wastes) on field crops
Traveling gun	no limit	0.20	1.0	rectangular	roadway for hose and sprinkler unit	no limit	0.10–0.30	40–100	120–250	high	excellent	Well suited for spraying sludge on field crops as demonstrated by Metrol. San, Dist. of Greater Chicago, Fulton County, IL
Center pivot	5–15	0.005	0.090	circular	clear of high obstructions, path for towers	8–10	0.05–0.15	40–160	180–350	low	average	Well suited for large-scale spraying of wastewater on field crops as demonstrated at Muskegon, Mich.

[a]Does not include cost of water supply, pump, power unit and mainline. Best source of equipment and cost information is irrigation systems vendors.
[b]Towline lateral system is same except that field shape is not as flexible, labor requirements are substantially less.

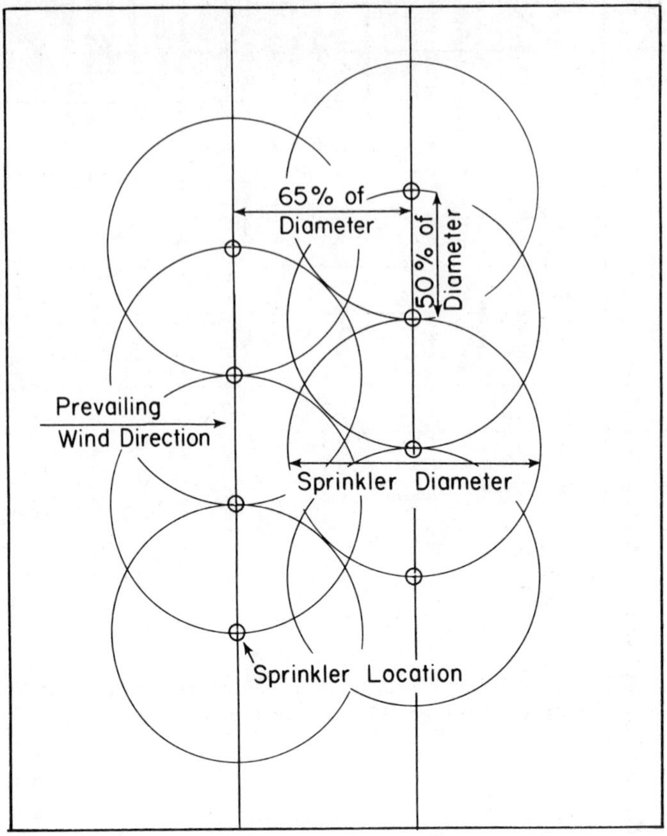

Figure 10. Solid set staggered spacing of sprinkler irrigation system (Nelson, 1977).

patterns to achieve as uniform an application as possible for given conditions. Figure 10 is an example of this, taking wind direction into account. The coefficient of uniformity (C_u) for sprinkler irrigation systems may be as high as 90%. However, values in the mid-70s are more realistic. Higher C_u values may be attained where the need justifies the additional expense.

The factors which influence the uniformity of application are:

1. The sprinkler selected
2. The grid spacing utilized and degree of overlap
3. The nozzle size and operating pressure
4. The wind speed

Sprinkler manufacturers publish data associated with their equipment which aids the designer in determining many design parameters, including C_u. The nozzle size, pressure rating, gallons per minute flow capacity, and the wetted diameter are standard specifications in any sprinkler irrigation design. Typical distribution curves for a given nozzle at a given pressure are also available. These curves are used in conjunction with the desired grid spacing to attain a uniform application.

Manufacturer's information on sprinkler equipment includes nozzle size, pressure rating, flow capacity in gallons per minute, and the wetted diameter.

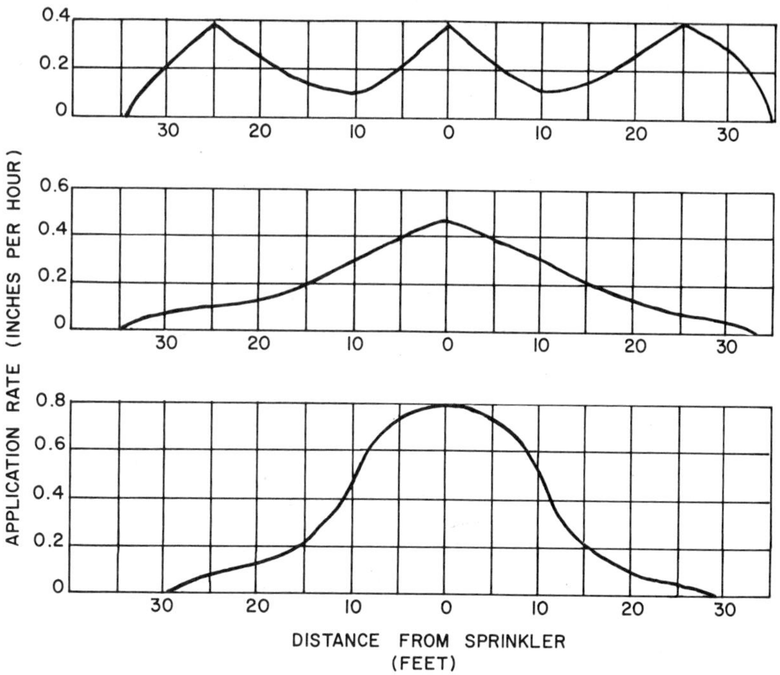

Figure 11. Patterns of water application by a rotating-type sprinkler operating at different pressures: (Top), pressure too low; (Middle), pressure satisfactory; (Bottom), pressure too high. (Quackenbush, 1957).

If operating pressures are not maintained as designed, different distribution patterns result as shown in Figure 11. Very large quantities of water could collect over individual areas, possibly exceeding the infiltration or percolation capacity of that area and causing saturation and ponding.

If operating pressures of sprinklers are not maintained as designed, very large quantities of water could collect over certain areas and exceed the infiltration or percolation capacity of that area, causing saturation and ponding.

The distribution curves resulting from trial runs of various sprinkler models were arrived at through the use of water. If a thicker slurry is to be applied, then field experiments to determine the distribution pattern are required to maintain a uniform application of waste.

An idea of the significance of uniformity in waste application by spray irrigation comes from the following discussion taken directly from Norum (1976).

C_u values commonly used in irrigation design range from 70 to 95%. The lower values are appropriate for lower valued crops in areas of low cost water. The higher values are used on high-valued crops irrigated with high cost water. It has been suggested that a value of 80% or better is appropriate for center pivot systems.

The coefficient of uniformity for center pivot sprinkler systems should be 80% or higher.

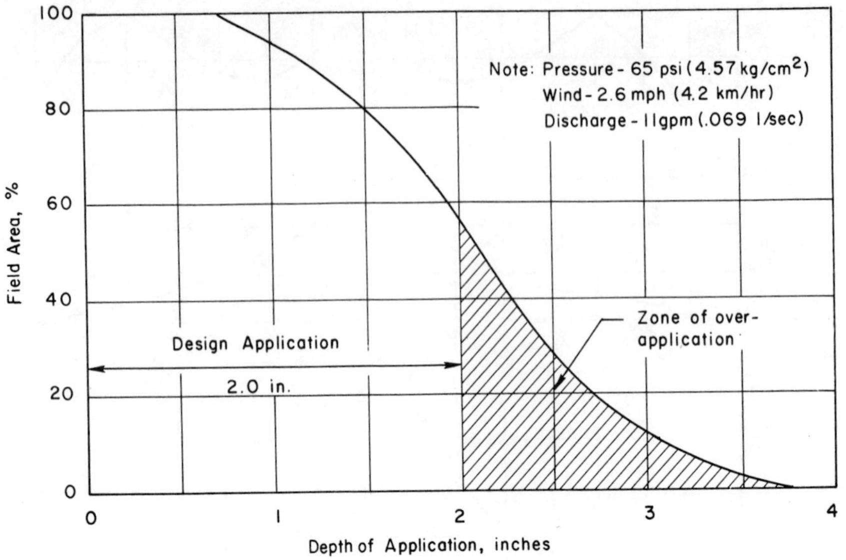

Figure 12. Application uniformity data for a 60 ft × 80 ft solid set system (Norum, 1976).

A modified method of characterizing uniformity has been developed by Norum. The significance of the procedure can best be understood by referring to Figure 12, which shows application data for a commonly used solid set system.

If the design application is 2.0 in. (5.08 cm), it can be seen from Figure 12, that about 50% of the disposal site will receive an over-application ranging up to 3.75 in. (9.52 cm). If the renovating capability of the soil profile has been accurately determined to be 2.0 in. (5.08 cm), then the quality of renovation has been seriously jeopardized in the over-applied areas. This is especially significant if the quality of renovation is to approach drinking water standards. If the design application is based on a nutrient constraint of 200 pounds of nitrogen per acre (224 kg/ha) per year, for example, then an actual range of from about 75 to 375 lb/acre (84 to 420 kg/ha) is possible.

For a given system design many factors affect the actual composite uniformity probable on a disposal site: soil lateral permeability, presence of restricted permeability layers in the soil profile, movement of surface water during application, wind velocity and direction, and operating options. The type of uniformity of application curve to be expected is unique to the system design and site conditions and can only be determined by test. The results of actual tests on equipment commonly used for solid set disposal systems are shown in Figure 13. The abcissa is the actual application divided by the mean application following a method outlined by Norum. The data in Figure 13 help to determine the spacing that should be used. From the curves, the deterioration in uniformity of application is apparent when the same sprinkler is used at increased spacings.

Factors which affect the actual composite uniformity that is probable at a site include soil lateral permeability, presence of restricted layers in the soil, movement of surface water during application, wind velocity and direction, and operating patterns.

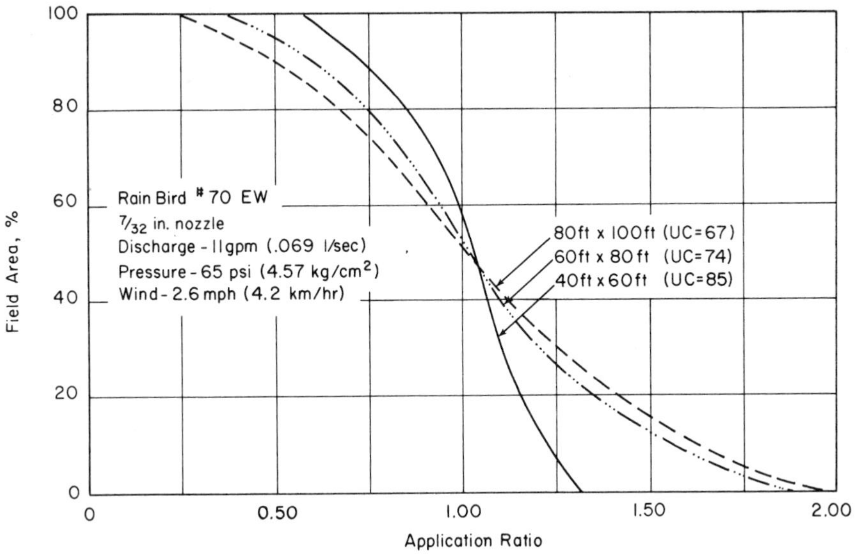

Figure 13. Uniformity of application curves for a typical solid set system (Norum, 1976).

For the spraying of wastewaters, applications to the disposal site should be limited to the renovating capability of the soil. Following a method suggested by Norum, and using the data in Figure 13, the design applications for the indicated spacings to comply with a 2.0 in. (5.08 cm) per week constraint over 90% of the area are:

Application, in. = 2.0 ÷ application ratio at 10% field area

40 ft × 60 ft	2.0/1.23 = 1.63 in.
60 ft × 80 ft	2.0/1.52 = 1.32 in.
80 ft × 100 ft	2.0/1.67 = 1.20 in.

These data can be used in framing a disposal area versus system cost tradeoff. The 40 ft × 60 ft (12.2 m × 18.3 m) spacing achieves good uniformity, and uses a minimum of land, but has a high equipment capital cost. The 80 ft × 100 ft (24.4 m × 30.5 m) spacing has poor uniformity, fails to completely charge the root zone and requires more land, but has a lower equipment capital cost. Ultimately the decision on system uniformity must be made on the basis of this kind of analysis. It varies significantly from irrigation procedures and cannot be solved by rules of thumb or generalizations. The consequences of inadequate design are potentially more serious than with irrigation system design.

SURFACE APPLICATION INVOLVES THE CHOICE OF SEVERAL TECHNIQUES

In addition to spray irrigation, several forms of surface irrigation have been used successfully in land treatment of wastewater. Surface application systems also are appropriate for some overland flow treatment schemes and for rapid infiltration basins.

Wastewater may be applied to the treatment site by gated pipe and by ditches using spile tubes, siphon tubes, or gates for flow distribution. All these systems are gravity fed and require hand labor to operate. None have good solids handling characteristics due to deposition.

Surface application of wastewater via gated pipe, open ditches, siphon tubes, or gates for flow distribution are fed by gravity and operated by hand; none of these methods handles large concentrations of solids well.

Commercial gated pipe (Figure 14) is generally 6 to 12 inch diameter aluminum or plastic pipe with openings every 30 to 80 inches. The openings have hand adjustable gates to regulate flow from each gate. In this way, pressure loss along the pipe can be compensated by larger openings at the far end which will maintain equal volume flow along the length of the pipe. The gated pipe is portable, thus more flexible than ditches for application to the site. For small systems, operators have made "home-made" gated pipe from small diameter polyethylene tubing but cutting holes which are covered with slit segments of the same pipe. Advantages of gated pipe are its portability, low labor requirement compared to other surface application systems, and good management capability of the wastewater. Costs are higher than the open ditch systems.

Gated pipes are portable, capable of managing the wastewater and able to work with low labor input; they are more expensive than open ditch systems.

Gated pipe distribution has been used in a successful row crop irrigation system in California which uses air-actuated valves to open and close the gates on an automatic cycle (Norum, *et al.*, 1976). This system retains the advantages of low pressure operation and good water distribution of other gated pipe systems, while eliminating hand labor of wastewater distribution in the field.

Gated pipe is being used with automatic gate systems to eliminate most hand labor.

Figure 14. Gated pipe incorporates hand-adjustable openings which can be matched to furrow spacing for row crop irrigation. They provide gravity flow operation and portability (adapted from Lindsay, 1977).

Spile tubes are stoppered straight pipes placed through the berm of irrigation ditches. Irrigation siphon tubes are sections of bent tubing which are filled with water, then inverted across the side of the ditch. Both these systems are very low cost, but may require considerable maintenance and labor for normal operation. Ditches feeding these dispersion devices usually are fitted with gates to control flow into the various segments. Ditches also can be fitted with turnout gates which take the place of spile tubes and siphons.

Spile tubes and irrigation siphon tubes are cheap but require a lot a maintenance and labor to work properly.

Any of the distribution methods listed above can be used for border irrigation, furrow irrigation, and corrugations.

SURFACE IRRIGATION UTILIZES GRAVITY FLOW FOR APPLICATION

Border (or border strip) irrigation requires low berms in parallel rows running in the direction of maximum slope of the site. A field is contained between each pair of berms. This method is well suited to close-growing crops such as forages and pasture grasses. The Werribee Farm land treatment system in Melbourne, Australia, has used this method of application since 1897. Their current practice is to irrigate to a depth of 4 inches once every 18 to 20 days. The fields at Werribee are mechanically subsoiled and graded. The effectiveness of this method depends to some degree on obtaining an even sheet of water entering the field area. This is not always possible, and some minor problems have been reported. For example, at Werribee, which applies raw wastewater, solids deposition occurs near the upper end of the fields. Border irrigation fields may be constructed with a grader or similar piece of earthmoving equipment.

Border irrigation is well suited to close growing crops such as pasture grasses; its effectiveness depends on obtaining an even sheet of water entering the field.

Furrow irrigation provides a uniform means of wastewater application for row crops. Contour plowing can enable furrow irrigation on land slopes up to 10%; otherwise a slope of 1% is the usual advisable maximum. Furrow spacing can be matched with gated pipe outlets for direct application from the pipeline.

Furrow irrigation is suited for row crops.

Corrugations are shallow v-notches spaced about 30–40 inches apart and created with a tool bar implement towed by a tractor. Corrugations are well suited to close-growing crops on steep, irregular land.

Corrugations are well suited to close-growing crops on steep, irregular land.

In place of the uniformity coefficient concept in sprinkler systems, surface-applied waters are designed to attain a uniform wetting front within the soil profile. This is accomplished by balancing the infiltration rate of the soil with the irrigation stream size, the area to be irrigated, the amount of water to be applied, and the slope of the land.

In surface-applied waters the goal is to obtain a uniform wetting front throughout the soil profile.

Application efficiencies attained through surface irrigation reflect the uniformity of the wetting front within the soil profile. In systems with good planning, proper construction, and careful operation, an application efficiency of 70% can be expected. However, this requires applying the irrigation water twice per irrigation period which involves twice the labor of a single set. Application efficiencies on the order of 40 to 60% are probably more realistic. Tail-water runoff would have to be collected and returned to the treatment medium if the desired quality were not wet.

Application efficiencies of only 40 to 60% can be expected from most surface irrigation systems.

OVERLAND FLOW SYSTEMS ARE INSTALLED ON GENTLE SLOPES

Of special importance to the overland flow treatment system is the consideration of erosion in design. The slopes involved in this treatment alternative would not normally produce erosion, but large rainfalls coupled with wastewater application could have this result. The erosion hazard is minimized by forming small berms at intervals on the treatment site surface.

Erosion must be kept to a minimum if the overland flow system is to work effectively

The flow of wastewater over gently sloping (2 to 8%) terraces which are covered with grass characterizes the overland flow system. Wastewater distribution generally has been through the use of sprinkler irrigation systems located at the top of the grassed slope. These slopes are broken at distinct intervals by a series of cross terraces which collect the renovated wastewater and deliver it to a central collection and discharge point, as shown in Figure 15.

Design objectives: There are two main objectives in the design of application systems to serve overland flow treatment sites.

1. Prevent erosive flows along the sloping treatment medium and in the graded collection terraces during rainfall/runoff events.
2. Provide a hydraulic loading rate which will result in the desired effluent quality.

Two goals in overland flow application systems are to prevent erosive flows and to maintain the hydraulic loading rate which will result in the desired effluent quality.

Erosive flows such as sheet and channel erosion have been studied for many years. The U.S.D.A. Soil Conservation Service has done extensive work over the last 40 years in the area of sheet and rill erosion. The Corps of Engineers, Bureau of Reclamation, Department of Transportation, Forest and Park Services, etc., have studied slope stabilization and channel flow. The object of these studies has been to provide non-erosive flows of water over a soil surface.

To prevent crosive flows, the Universal Soil Loss Equation has been used to determine the spacing between terraces. This spacing is based upon an allowable soil loss which in turn is based upon both the physical characteristics of the soil profile in question and the economic conditions resulting from soil erosion. A terrace spacing of 100 to 300 feet is generally found in tilled soils. Agricultural Handbook No. 282, "Predicting Rainfall-Erosion Losses From Crop-

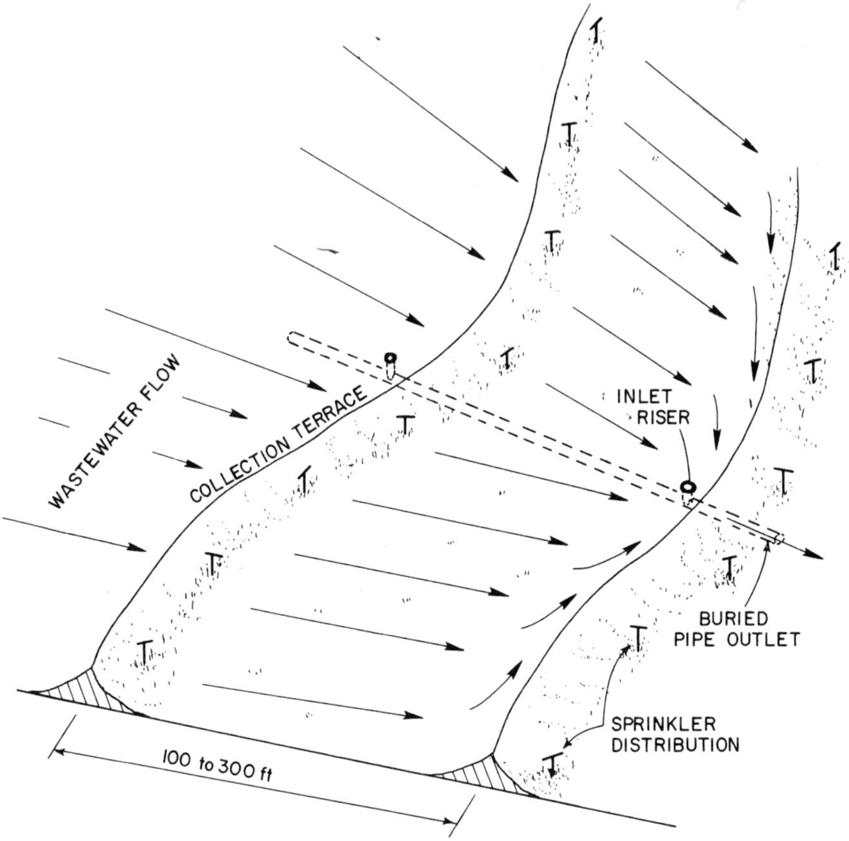

Figure 15. Overland flow wastewater distribution system.

land East of the Rocky Mountains", fully develops the Universal Soil Loss Equation. More site specific technical information may be obtained from the Soil Conservation Service office which covers the geographical area in question. Site specific parameters have not, as yet, been developed for all areas of the United States. The 100 to 300 foot terrace spacing has also been found to be most effective in terms of wastewater renovation efficiency in overland flow systems.

To prevent erosive flows, spaces between terraces should be according to the Universal Soil Loss Equation.

Spaces of 100 to 300 feet between terraces are effective for renovating wastewater in overland flow systems.

RAPID INFILTRATION BEDS GENERALLY USE SURFACE APPLICATION

In rapid infiltration systems, wastewater may be applied by spray, but surface distribution and application systems are more usual. Rapid infiltration beds can be constructed in much the same manner as border strip irrigation fields, with one berm serving as a wall between two beds.

Vegetative cover can also be used in rapid infiltration beds to enhance the infiltration rate of the soil. Other rapid infiltration systems use graded aggregate over the soil surface.

SLUDGE APPLICATION REQUIRES SPECIAL EQUIPMENT

Sludge slurries may be sprayed, spread through outlets mounted on a portable tank, or placed under the soil surface by special plow attachments.

Gun sprinklers, traveling or stationary, can handle sewage sludge or liquid animal manure slurries of around 15% solids (Peterson, 1975). Specifications for this application include a nozzle diameter of at least $\frac{3}{4}$ inch and minimum sprinkler pressure of 70 to 80 psi. Power requirements for pumping to supply such a system would be about 30 hp. Some models of traveling gun sprinklers use turbine and cylinder drive units to move along the cable, and these may restrict the diameter of solid which may pass the unit to as little as $\frac{3}{8}$ inch. No such problem is encountered with travelers using internal combustion engines or drive units mounted on the cable winch at the end of the run.

Gun sprinklers can handle sludge or slurries with about 15% solids.

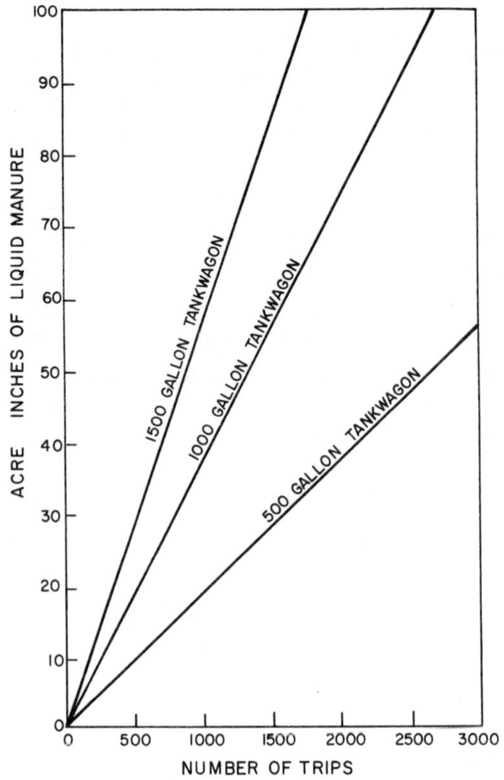

Figure 16. Economic analysis of sludge application systems may be related to the number of trips necessary to apply a certain volume with various sizes of tank wagons (Peterson, 1975).

If possible, it may be advisable to pump wastewater or water from another source (other than a potable supply) through the gun sprinkler following each sludge application. This will wash solids off foliage (some problems have been noted with crop burning due to high chemical concentrations in slurries), clean out the pipe, and clean the sprinkler and the external surfaces nearby. If wastewater is not available, a farm pond is an excellent source if auxilliary pumping equipment can be used.

Water from another source should be pumped through the gun sprinkler after each sludge application if possible to wash off foliage and to clean the pipe and sprinkler.

Tankwagons for manure or sludge slurries are normally 1,000 to 2,000 gallons capacity, with some being over 5,000 gallons. Tanks designed to be pulled through tilled fields use oversize tires. Wagons can be economical for hauling wastes to distant fields, which would be costly to use with irrigation systems. Curves of number of trips necessary to dispose of given volumes of waste have been developed (Figure 16) and aid in economic assessment of this alternative.

Tankwagons can have outlets, usually consisting of a manifold with deflectors on the openings, for direct surface application. Odors and high concentrations of pollutants in runoff waters can ensue from surface spreading. This has led to the development of subsurface plow spreaders. Shallow injection with good mixing of sludge and soil produces rapid drying and maintains aerobic conditions. Injection depths of no more than 4 to 6 inches should be used. Tractors pulling these chisel plow systems may also be fitted with a sludge tank, which must be of a limited size, thus maximizing sludge transport problems. Systems for sludge injection which use the same flexible hose used with traveling big gun sprinklers are commercially available. The hose connects a large mobile tank with the tractor plow attachment. In this way trips between sludge source and field with the tankwagons are minimized, and operating costs reduced.

Shallow subsurface plow spreaders can inject sludge slurries and manure into the soil, producing rapid drying and maintaining aerobic conditions.

BIBLIOGRAPHY

Agricultural Research Service. 1965. Predicting rainfall-erosion losses from croplands east of the Rocky Mountains. USDA Agricultural Handbook 282, U.S. Gov't Printing Office, Washington, D.C.

Christiansen, J. E. 1941. The uniformity of application of water by sprinkler systems. *Agr. Engr. Journal*, **22**, 89–92.

Demirjian, Y. A. 1975. Conference on the Muskegon County, Michigan wastewater management system, Sept. 17–18, 1975. Performance and economics of the system.

Fry, A. W. and A. S. Gray, eds. 1971. Sprinkler irrigation handbook, 10th edition. Rain Bird Mfg. Corp. Glendora, Calif.

Lindsay Mfg. Co. 1977. Equipment brochures. Lindsay, Nebr.

Nelson Irrigation Corp. 1977. Equipment brochures. Walla Walla, Wash.

Narum, Q. A. 1976. An approach to effluent disposal on land. Presented at Nat'l. Council for Air and Stream Improvement, Inc. West Coast Regional Conference, Portland, Ore.

Norum, E. M. 1976. Design and operation of spray irrigation facilities. pp. 251–288. *In* Sanks, R. L. and T. Asano, eds. Land treatment and disposal of wastewater. Ann Arbor Science, Ann Arbor, Mich.

Peterson, M. 1975. Current technology–sludge. p. 1B13/10. *In* E. M. Norum, ed. Wastewater resource manual. Sprinkler Irrigation Association. Silver Spring, Md.

Pound, C. E. and R. W. Crites. 1973. Wastewater treatment and reuse by land application. EPA-660/2-73-006b, p. 74. Prepared for Office or Research and Development, U.S. EPA, U.S. Gov't. Printing Office, Washington, D.C.

Pound, C. E., R. W. Crites, and D. S. Griffes. 1975. Cost of wastewater treatment by land application. EPA-430/9-75-003. Prepared for U.S. EPA Office of Water Programs, Washington, D.C.

Quackenbush, T. H., G. M. Renfro, K. H. Beauchamp, L. F. Lawhon, and G. W. Eley. 1957. Conservation irrigation in humid areas. p. 29. Agricultural Handbook No. 107. USDA Soil Conservation Service. U.S. Gov't. Printing Office, Washington, D.C.

Soil Conservation Service. 1967. National engineering handbook. Section 15, Irrigation. U.S. Dept. of Agriculture, SCS. Washington, D.C.

Module II-10

DRAINAGE FOR LAND APPLICATION SITES

SUMMARY

Drainage for land treatment sites must be evaluated with respect to the purpose the system is to serve. Off-site drainage controls the flow of storm runoff onto the site or groundwater incursion into the soil within the site. On-site drainage is employed for a variety of reasons. These two areas of drainage control, on-site and off-site, must be designed as a system to obtain the desired effluent quality from the land treatment system.

Surface drainage techniques often used for the control of water include diversions, random surface drains, waterways, and ditches. Subsurface drainage is generally accomplished by means of drain piping using plastic, clay, and concrete. These techniques will generally make up the drainage system.

The objectives of drainage systems as they relate to land treatment sites are discussed, including some flexible operating patterns which can be incorporated through the use of on-site drainage.

Factors affecting drainage which must be obtained during reconnaissance surveys fall into four major categories: topography, geology, man-made obstructions, and soils. This type of information is generally available through various goverment agencies as well as private and public utilities and corporations. Many information sources for drainage overlap those used for other aspects of site evaluation. The degree of accuracy required of the information is dependent upon the stage of drainage evaluation or design.

Costs incurred for drainage control are dependent upon the four factors mentioned above—topography, geology, man-made obstructions, and soils—as well as the type of drainage system selected and the equipment used to install it. An example of the drainage that can be used with land application systems is the 5,000 acre layout at the Muskegon Wastewater Management System. Drainage was provided at a cost of $4.3 million dollars ($860/acre) which reflects capital expenditures only (Demirjian, Y. A., 1975).

CONTENTS

OBJECTIVES

Upon completion of this module, the reader should be able to:

1. List and discuss the two major objectives of off-site drainage.
2. Discuss several objectives of on-site drainage systems as they relate to land treatment.
3. Explain the use of drainage in the three operating options involving water table manipulation discussed in this module.
4. List and discuss the four main areas of site characteristics (besides volume of flow to be handled) in which information is required to design a drainage system.

INTRODUCTION

Generally, large areas of land are required for land treatment sites. These areas usually contain low spots, springs, and seeps which will require earthmoving or artificial drainage. Runoff of storm flow onto the treatment site from areas of higher elevation may have to be diverted, usually in channels, to protect the site from excessive hydraulic loads as well as erosive flows. Subsurface drainage may be required to lower the water table, extract the renovated water from the groundwater aquifer, or manipulate the water table for increased wastewater renovation. Therefore, both surface and subsurface drainage systems are required to perform various functions at land application sites. At a given site, it may be necessary to use a combination of the two to achieve the drainage objectives.

A drainage plan of the area under investigation must be made in order that the total drainage requirements may be identified and designed as part of the land application system. This plan should include surface and subsurface drains as well as the indentification of an adequate outlet. An operational scheme must also be included where the water table is manipulated to achieve a higher degree of renovation. Maintenance of the drainage system must also be considered once the land application site is in operation.

Including a drainage system in the site planning can make an otherwise unsuitable parcel of land become a viable alternative. An example of such a site would be one in which all other factors are favorable except a high seasonal water table. The cost of drainage improvements would then be included in the economic analysis versus, for example, transportation costs for a more distant site not requiring extensive drainage.

The purpose of this module is to present drainage concepts in relation to their application in land treatment systems, not to provide all the background required to design drainage systems. Reference is made to detailed manuals or texts where necessary.

A discussion of the benefits of drainage to crop production on agricultural land is found in the section on "Drainage Requirements and Crop Water Tolerance" in Module II-7, "Crop Selection and Management Alternatives." These reasons for providing artificial drainage improvements are listed in tabular form, along with the reaction of various crops to excess water in the root zone.

OFF-SITE DRAINAGE

Collection, transport, and disposal of runoff water which may enter the land treatment site from outside sources must be planned as a system. The objectives of this system are twofold:

1. Protection of the land treatment area from erosive flows.
2. Preclusion of additional hydraulic load through the treatment medium.

**The purpose of off-site drainage is to protect the system from erosion and excessive hydraulic load.*

Diversion Ditches for Interception of Surface Runoff

Often, open ditches serve as diversions to protect lower lands from overland flow. These are generally dug across a slope, or at the toe of a slope with the borrow material diked on the lower side for efficient use of material (Figure 1). The bottom of the diversion terrace could extend quite deep to intercept a seep along a hard pan layer. Diversion terraces are generally graded and grass covered so that the collected water may be delivered at non-erosive flows to a central discharge point. This discharge point is vital to the smooth operation of the whole surface drainage system, and adequate capacity must be provided in the outlet to prevent any backup and overflow of the channel at design flow.

Diversion terraces intercept runoff and deliver it to outlets at non-erosive flows. Outlets must be large enough to prevent backup.

The level of protection for which the diversions are designed depends upon what is to be protected. Roadways and equipment may require safeguarding against a 50-year, 24-hour rainfall runoff event, whereas a grassed slope may require protection from the 25-year, 24-hour rainfall runoff event. In any case, the level of protection and the amount of runoff expected must be identified.

The level of site protection desired and the amount of runoff expected must be identified.

*This and other italicized summaries are intended to highlight key ideas, provide a basis for later review or to aid in skimming sections that are relatively familiar. They can be ignored in a complete reading of the text.

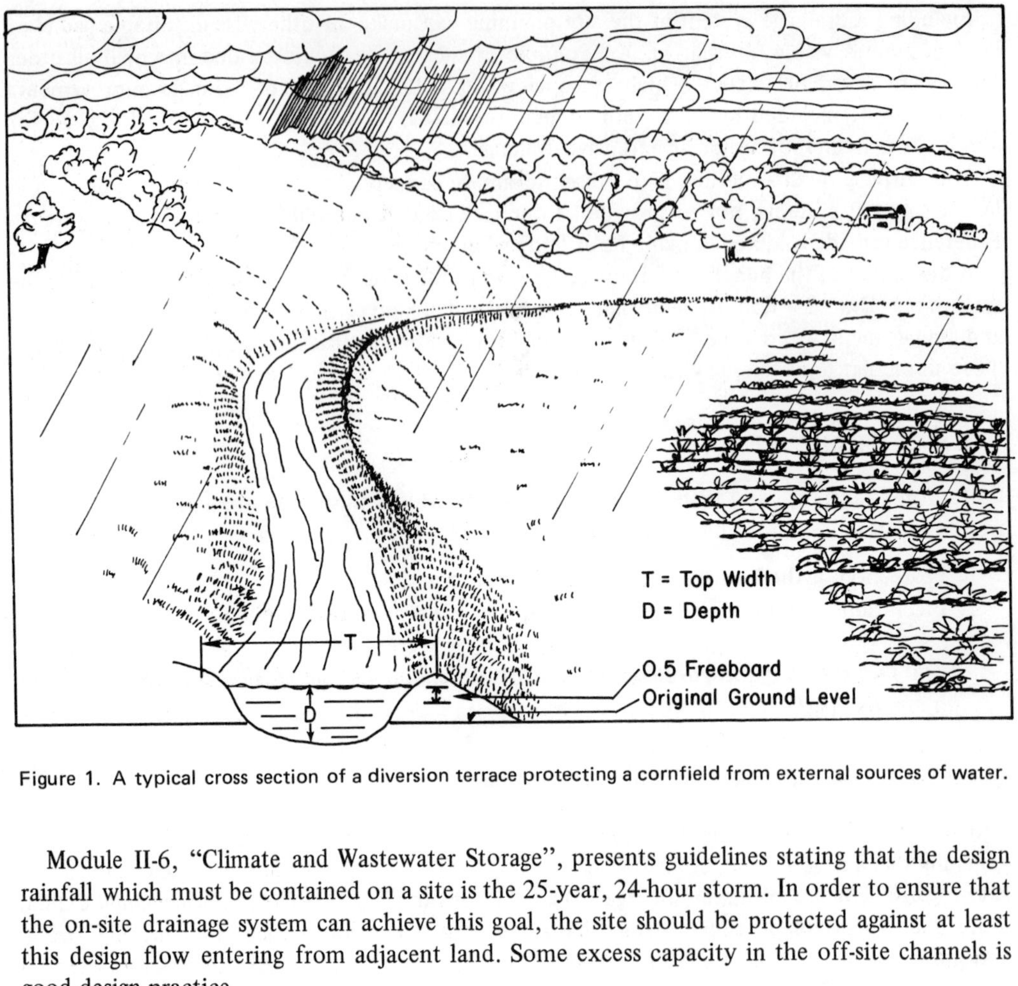

Figure 1. A typical cross section of a diversion terrace protecting a cornfield from external sources of water.

Module II-6, "Climate and Wastewater Storage", presents guidelines stating that the design rainfall which must be contained on a site is the 25-year, 24-hour storm. In order to ensure that the on-site drainage system can achieve this goal, the site should be protected against at least this design flow entering from adjacent land. Some excess capacity in the off-site channels is good design practice.

The off-site drainage system ought to protect against at least as severe a storm as the on-site system is designed to handle.

Runoff from Storms of Varying Duration Should be Evaluated

In addition to the 25-year, 24-hour rainfall, peak flows can occur from intense storms of much shorter duration. Depending on the conformation of the land, such a stormflow may necessitate a larger channel than is required for the longer, less intense rainfall. Design flows for storm runoff of this type are usually obtained from the rational formula, $Q_p = KiA_d$ (Linsley and Franzini, 1964), where Q_p is the peak runoff in acre-inches per hour, K is a runoff coefficient which relates the runoff from a given surface to an impervious standard, i is the intensity of rainfall (inches per hour) which can be expected at a given location for a duration equal to the "time of concentration" (t_c) of the watershed. t_c is derived from the watershed characteristics. A_d is the watershed area in acres. Since 1 acre-in. per hr equals 1.008 cu ft/sec, Q_p is generally assumed to be expressed in cfs.

Peak flows can exceed the design storm expectation if there is a brief, intense storm.

This equation is a simple one, but it does require some discretion in its use. Experience with the rational formula and runoff calculation in general is very helpful, because assumptions must be made about some of the factors in the formula.

Time of concentration is taken as the time required for runoff to reach its peak value when applied to a watershed at some constant rate. For an impervious surface, this peak runoff would equal the rainfall rate. For most watersheds, t_c would consist of two components–overland sheet flow and channel flow. Formulas are used for the overland flow component which take into account slope, length, and retardance coefficient, a measure of flow resistance offered by several types of ground cover. Channel flow t_c is computed by length of longest channel divided by average velocity at bankfull stage.

Intensity of rainfall is taken from rainfall data for a given area using the duration equal to the computed t_c. Curves are usually available from city engineering or highway department offices. The coefficient of runoff is taken from a table of values for various ground surfaces, and there is usually some room for assumptions in choosing this factor. Other formulas are also used in stormflow computation, and a full discussion of runoff rate and volume calculations is found in Schwab, *et al*. (1966).

Intensity of rainfall for particular durations are available from city engineering or highway department offices.

Should the runoff calculated for a short, intense rainfall be great enough to exceed the capacity of offsite channels designed on the 25-year, 24-hour basis, excess capacity should be designed to contain the higher flow, if it is at all economically sound to do so. Thus, localized flooding or excess surface water on the land application site can be avoided.

If a brief intense storm is calculated to give more runoff than the design storm, the off-site channel capacity may have to be greater.

Once the flow is determined, surface channels should be designed according to standard engineering practice, allowing for proper side slopes, bottom widths, freeboards, etc., as shown in numerous texts. One excellent ditch design reference, available free on request from the Soil Conservation Service, is SCS (1971) Chapter 5.

Interceptor Drains to Exclude Excessive Groundwater

Often seeps occur along the toe of a sloping land surface. These seeps are generally the result of the water table surfacing due to rock or hard pan layers constraining flow. Interceptor drains (Figure 2) are an alternative to collect and transport this water away from the land treatment site, most often to a surface discharge channel or natural stream.

Interceptor drains lower the water table.

Groundwater flow enters an interceptor drain line either through perforations in the pipe or through the joint between sections of the pipe. The term "tile drain," which is still widely used to refer to drainage pipe, refers to the fact that for many years almost all such pipe was made of ceramic tile. Other materials now commonly used include concrete, bituminous coated fiber, plastics, asbestos cement, aluminum alloy, and steel. Various factors enter into the selection of the proper pipe for a particular application, including climatic conditions, soil chemical characteristics, depth requirements, and cost, including installation. Specific information on drainage

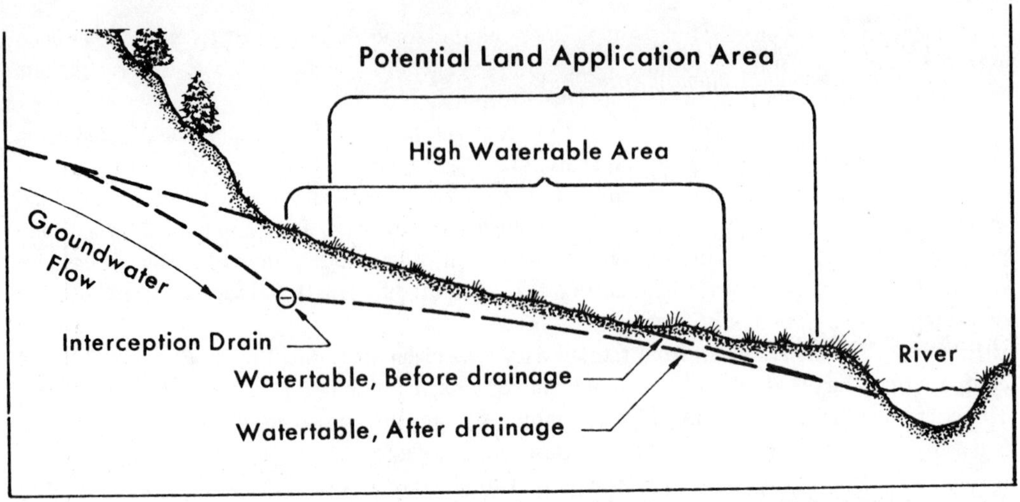

Figure 2. Interceptor drain collecting groundwater flow thus lowering the water table.

pipe is available from many sources, including SCS (1971) Chapter 4. This reference, and the qualified drainage engineers on the staff of the various Soil Conservation Service local offices, can also provide a wealth of information on pipe laying techniques, the use of filter envelopes and other mechanisms of improving drainage system performance, and other information which is useful and tailored to the local situation. This type of data is outside the scope of this module.

Table 1. Drainage Pipe Materials.

Material	Advantages	Disadvantages
Ceramic tile	• Resists action of acids and sulfates • Smooth walls ease flow in smaller sizes	• May absorb excess water • May be damaged by freezing and thawing
Concrete	• Convenient length for use as culverts • Can be poured in place	• May be damaged by acids and sulfates • Heavy
Bituminized fiber	• Light weight • Corrosion resistant	• May soften in sun • Needs smooth bed
Plastics	• Ease of handling • Light weight • Roll type can be installed by machine	• Must be carefully bedded
Asbestos cement	• Not subject to chemical damage • Broad range of sizes	• Relatively costly • Careful handling important
Aluminum alloy and steel	Good for: road crossings • unstable soil • outlets for other pipes • surface installations	• Expensive

There is wide choice in interceptor drain pipe material. Factors such as soil chemistry, depth requirements, climatic conditions, and cost influence the choice.

Advice on specific drainage questions should be sought from drainage experts. Written references give general information for system design.

Table 1 briefly summarizes some advantages and disadvantages of commonly used drainage pipe materials.

ON-SITE DRAINAGE

Objectives

On-site drainage may be accomplished through the use of surface or subsurface drains. These, singly or in combination, may be used to drain isolated springs, low spots, or seeps; or they may be used in parallel fashion in a total site drainage scheme. The objectives of a system may include the following:

1. Elimination of wet areas caused by springs or seeps.
2. Elimination of septic pools and anaerobic conditions to prevent odors.
3. Elimination of standing surface water to decrease potential runoff of pollutants.
4. Lowering of the water table to make a greater depth of the soil treatment medium accessible.
5. Water table manipulation for increased wastewater renovation, e.g., nitrogen removal.
6. Drainage outlet control of retention time through the site.
7. Switching of application basins through manipulation of the underdrain system.

Objectives 5-7 are examples of operational flexibility which can be built into certain systems for special applications, and will be discussed in subsequent sections. The first four objectives conform to standard agronomic practice, which aims to maximize crop production by eliminating surface ponding and waterlogging of the soil. In order to establish a good stand of most crops, alleviation of waterlogged soil conditions in the early growth stage is essential. This enables plants to develop healthy roots so they can continue to transfer sufficient nutrients to the upper plant parts for full development when the water table naturally drops in the summer. Therefore, drainage systems designed to lower the water table may contribute an agronomic benefit to the land as well as making land application feasible on a marginal site. Thus, in cases where the waste treatment authority contracts with private farmers to operate application sites on their farms, the decision to install such drainage for waste treatment reasons may also be greeted with favor by the farmer. He would derive benefit for future use of the land by improving his marginal fields.

A well designed drainage system will increase the acceptability of a site for waste treatment and crop production. This latter aspect could be a key point in convincing landowners to lease their land for a site.

Drainage systems to achieve objectives 4-7 are subsurface systems. Elimination of wet areas and surface ponding, however, can be handled by totally surface systems, subsurface piped systems with risers and ground level inlets, or systems incorporating both surface ditches and subsurface piping.

Both surface and/or subsurface drainage methods can be used for eliminating wet areas and ponding.

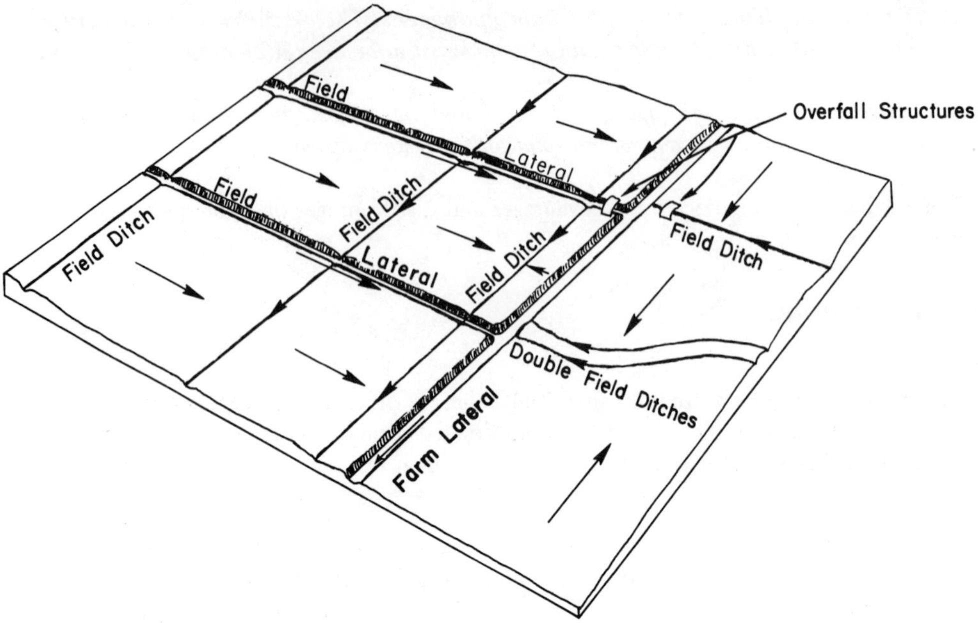

Figure 3. Typical layout-individual farm surface drainage system. (Where the ground surface is undulating, ditches and drains will meander) (SCS, 1971).

Surface drainage schemes can be laid out either on a grid (or similar configuration) as in Figure 3, or as a random system (Figure 4).

Complete Coverage Surface Drainage

The farm drainage system illustrated in Figure 3 shows the types of surface drainage ditches in common use on agricultural land. In the case of land application sites, the rows could act as distribution channels in surface irrigation. Application could also be by a spray system. The field ditches act as collection channels and the field laterals and farm lateral act as disposal channels. The laterals can also lead to holding basins for recycling through the system. Collection channels must be graded smoothly with the field slope so that they may accept non-erosive flow of water over their sides. Disposal channels are built for optimum flow characteristics and should receive flow only at junctions with field drains. This sort of system would be typical of farming operations in wet climates. It would also be suitable for overland flow wastewater treatment systems which would use close growing crops rather than the row crops indicated in Figure 3.

Drainage ditches that are laid out in a grid pattern can also be used for wastewater application and/or collection.

Disposal channels are built for optimum flow and should receive flow only at junctions with field drains.

Random Field Drains for Isolated Springs, Low Spots, or Seeps

Random field drains are used to drain isolated wet spots within the land treatment area, to prevent odors, to decrease the pollutant runoff potential, and to maximize the use of the land

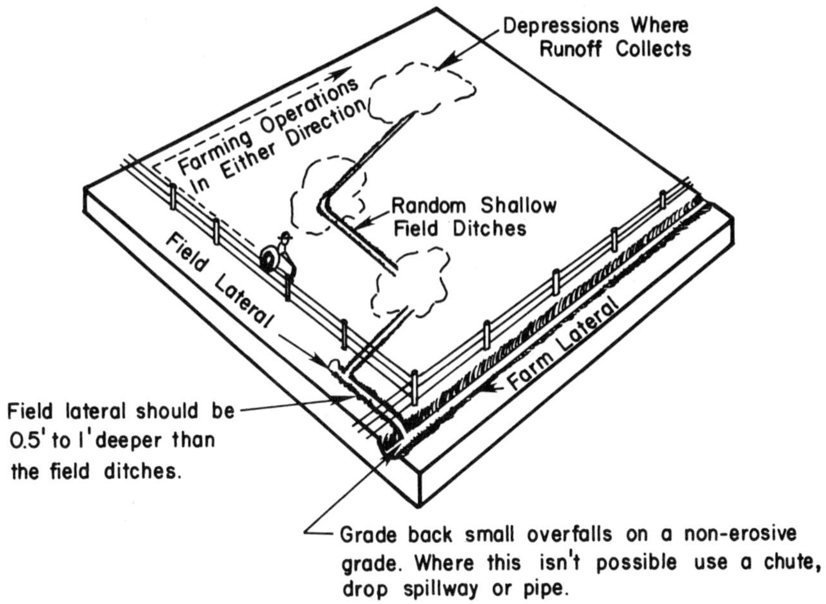

Figure 4. Random system of ditches draining surface depressions in a field (SCS, 1971).

area. These drains may be surface or subsurface installations. Figure 4 illustrates a surface random drain system, while Figure 5 shows a hypothetical layout utilizing both surface and subsurface drains. Whether or not the discharge from this type of drain is recycled through the treatment medium is dependent upon the water quality realized and the water quality standards of the receiving sink. Handling clean water through the waste application system constitutes an unnecessary cost which must be avoided where possible.

When subsurface installations are used to drain springs, seeps, or low spots the discharge may be recycled or released depending upon quality.

Parallel Drains for Water Table Control

Wastewater renovation is a function of the *amount of surface contact* with the treatment medium as well as a function of *contact time*. Some soils may not offer an acceptable depth for adequate renovation due to bedrock, hardpan layers, or high water table. Under high water table circumstances, drainage of an entire site may be advantageous. In lowering the water table, more of the soil profile may be used for wastewater renovation.

Lowering the water table may allow more of the soil profile to be used for treatment.

The depth of soil required for adequate renovation is dependent upon the influent waste characteristics, the renovation capacity of the soil profile, the required effluent quality, and the treatment alternative. The soil profile and its renovation capacity is highly variable. For this reason, many state laws and regulations require 5 to 10 feet depth to groundwater. However, most states reserve the option of a site specific investigation and evaluation.

The depth to groundwater for land treatment may be regulated by state laws or guidelines.

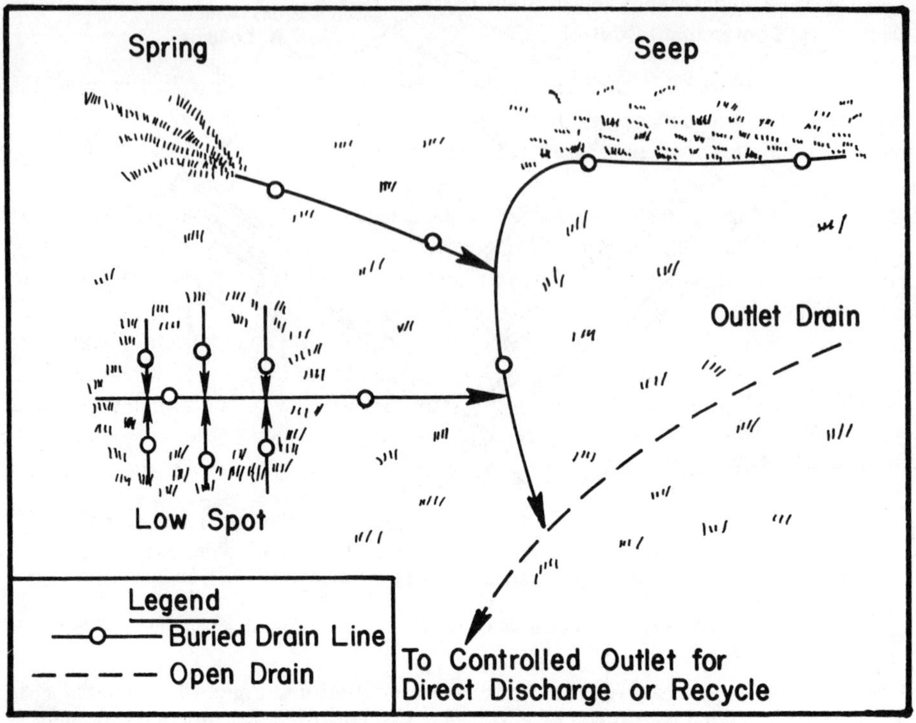

Figure 5. Random field drains servicing spring, seep, and low spot.

Subsurface drain lines are generally used to lower the water table over large areas. These systems utilize parallel drain laterals to collect the water. Transportation of this water from the site could be through open ditches as well as subsurface drains. Figure 6 outlines several configurations of drainage collection systems used for this purpose.

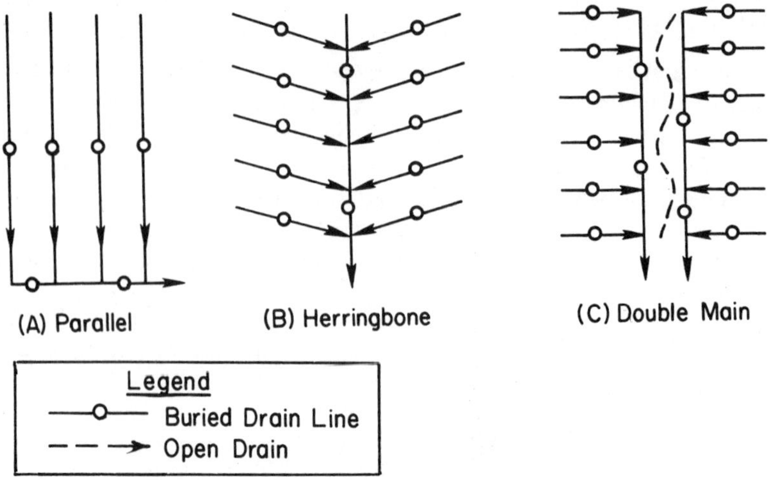

Figure 6. Several types of drainage collection systems utilizing parallel buried drains.

Drainage for Water Table Manipulation Resulting in Increased Renovation and Contaminant Control

Nitrification-Denitrification. Nitrification-denitrification may be controlled by manipulating the water table so as to enhance aerobic and anaerobic conditions in an alternating pattern in the soil profile. This can be accomplished by installing and operating a drainage system as discussed below.

The type of land application system selected will have a bearing on the feasibility of installing drainage for denitrification. Overland flow systems may require drainage for the control of wet areas due to seeps or springs. However, a large amount of water is not expected to penetrate the soil profile. Therefore, drainage for denitrification would not be used in most overland flow schemes. In systems such as rapid infiltration and crop irrigation, penetration of water into the soil profile is expected, and the soil water level may be manipulated to encourage nitrification-denitrification schemes. In rapid infiltration systems enough water can be added on a frequent application cycle to encourage anaerobic conditions within the soil profile. However, in crop irrigation systems relatively small amounts of water are added, usually on a once-weekly schedule.

Raising the water table by manipulating the drainage system encourages anaerobic conditions. These conditions favor denitrification.

The water table must be relatively close to the soil surface (5 to 10 feet) for its level to be readily raised or lowered to meet the nitrification-denitrification operational mode. This manipulation of the water table may be accomplished by varying the elevation of the drainage system outlet.

The amounts of nitrate nitrogen and readily oxidizable organic compounds within the wastewater and the soil profile are of interest in nitrification-denitrification systems. As discussed in Module II-1, "Nitrogen Considerations," the amount of denitrification obtainable is dependent upon the degree of initial nitrification and the amount of organic carbon left for denitrification.

Denitrification is dependent upon a supply of readily oxidizable organic carbon, high nitrate levels, proper pH and temperature, and low or zero oxygen concentration. The latter can result from restricted drainage. Under these conditions a facultative population of microbes utilizes the available carbon as a substrate and the oxygen in the nitrates for respiration, releasing the nitrogen as a gas.

Ion Exchange Important in Renovation Capacity

Ion exchange for the removal of metals and other substances is a major separation process that occurs in land treatment systems. Many soils consist of minute humus and clay particles which are negatively charged. Cations such as metals and ammonium are attracted to these cation exchange sites. The amount of actual exchange that takes place is dependent upon, among other things, the amount of contact time between applied wastewater and the soil matrix. Through manipulation of a drainage outlet, this contact time may be modified to increase the cation exchange potential of the soil profile.

Raising the water table can increase contact time between wastewater and the soil profile. This may facilitate cation exchange for retention of metals and ammonium.

Groundwater Quality

Groundwater Flow Control. After wastewater is applied to the soil surface, various phases of renovation are encountered. Pollutants are removed through filtering, ion exchange, microbial assimilation, and crop uptake. In some cases pollutants may not be removed adequately by land treatment and can enter the groundwater reservoir.

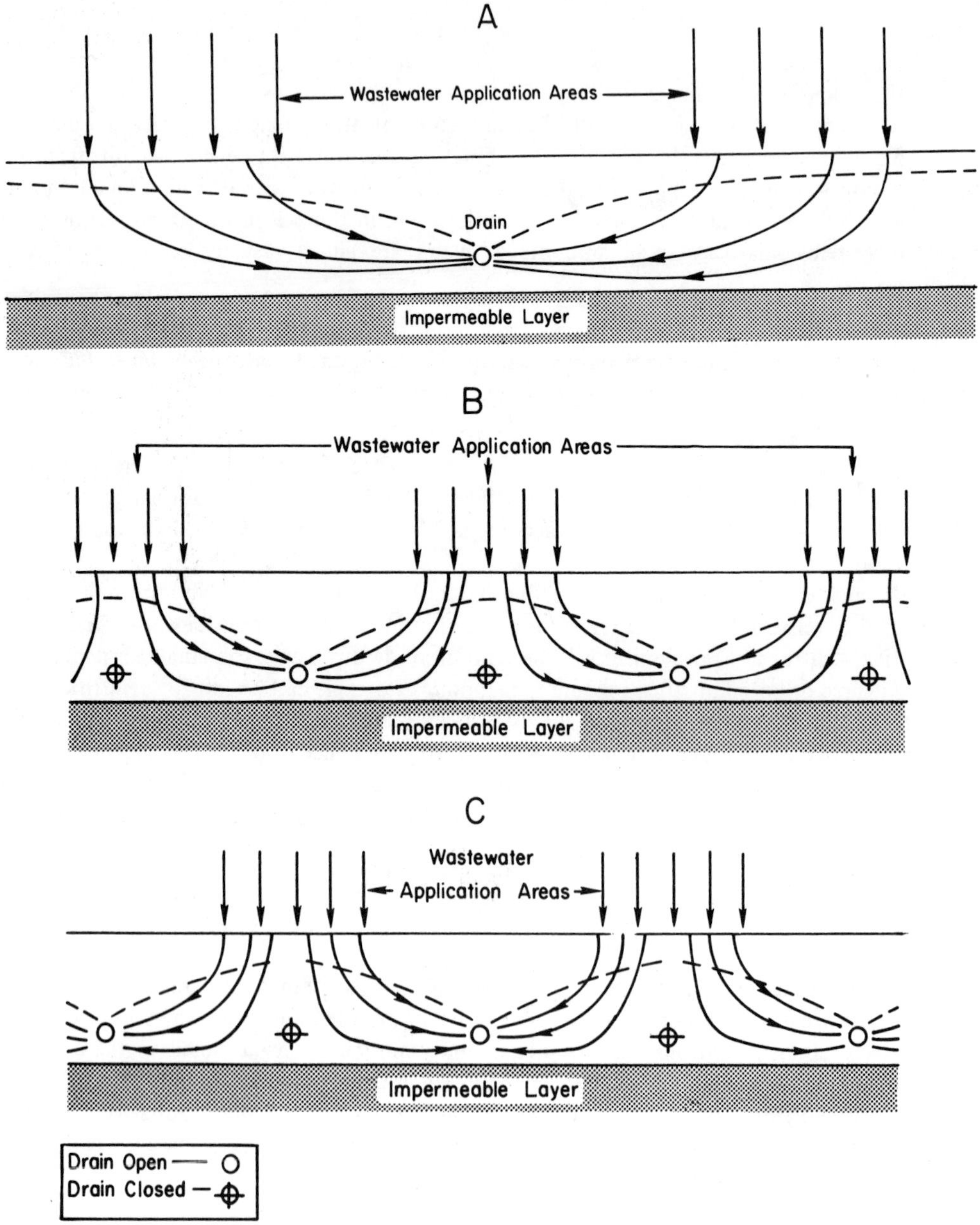

Figure 7. Rapid infiltration or crop irrigation scheme utilizing underdrains for renovated water recovery.

Bouwer (1974) has suggested a drainage system (Figure 7) to collect and recycle renovated wastewater from shallow aquifers. Figure 7A consists of two parallel strips from a rapid infiltration or crop irrigation system. Wastewater is applied in parallel strips along the soil surface. Upon infiltration, percolation occurs along the flow lines as illustrated in the figure. Once collected in the drain, the renovated wastewater may then be recycled to attain the desired effluent quality or it may be discharged directly into the receiving sink. Figures 7B and C illustrate an alternatively wet and dry cycling system utilizing the drainage collection system of Figure 7A. The flow of wastewater in the system is over a longer distance allowing for a greater travel time. In this manner, there is less danger of short circuiting. The exposure of the wastewater to the treatment medium is also increased.

Subsurface drainage can be used to collect wastewater for analysis and either reapplication or discharge. It can also be used to lengthen the soil contact distance, resulting in greater renovation.

Where deep aquifers are encountered, drainage through conventional drain systems is not feasible. Figure 8 illustrates an alternative utilizing parallel infiltration strips and recovery wells. Deep wells have been placed at strategic points in this system to control and collect the renovated wastewater. Although pumped drainage systems are high in initial and operating costs, the water table may be lowered to much greater depths than with other drainage methods, exposing deep strata which may be much more permeable than those nearer the surface (SCS, 1971).

Recovery wells can be used to collect renovated wastewater from great depths, therefore creating a water table lower than is feasible using drain lines.

For the most part, land application systems will be placed in rural areas. In these areas, shallow wells often are the main water supply for homes. A means of controlling the groundwater flow away from a land application site may be desirable where rural water supplies, streams and recreational water areas are nearby. Drain lines placed between the land application site and the water source to be protected, as shown in Figure 9, may be a feasible alternative for adequate protection.

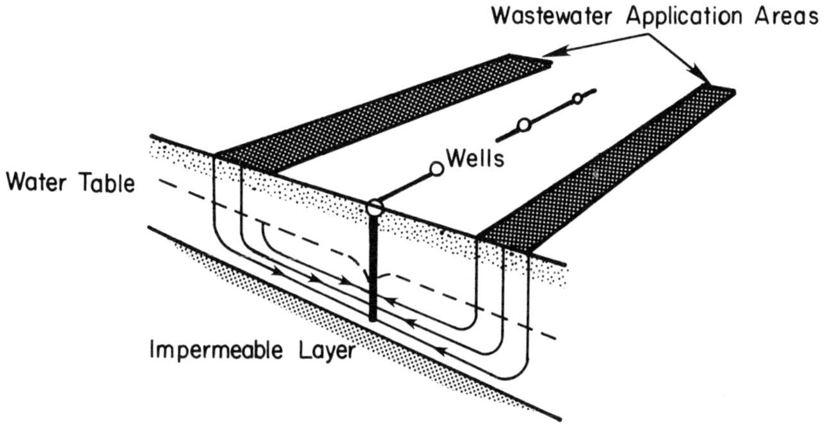

Figure 8. Rapid infiltration scheme utilizing wells for recovery of renovated wastewater.

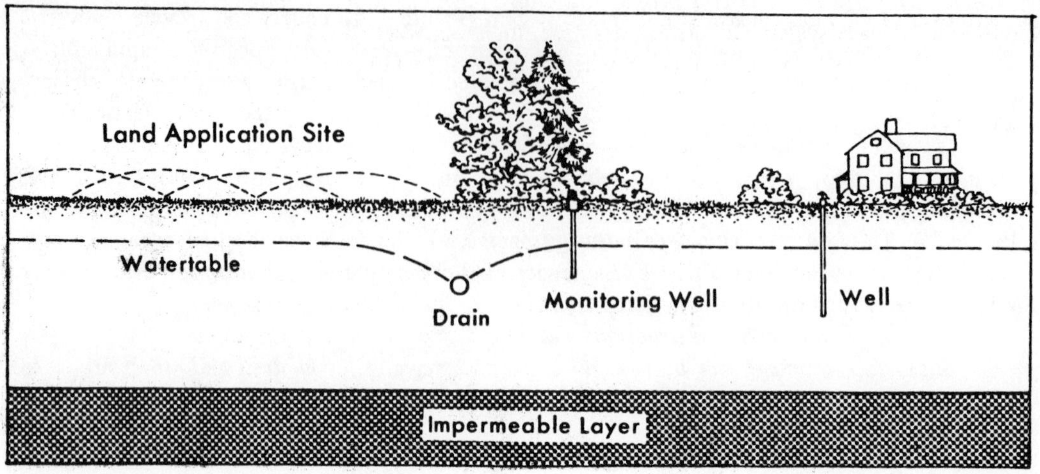

Figure 9. Drainage is provided around the perimeter of a land application site to protect a water supply.

Reconnaissance Survey—Factors Affecting Drainage

Besides volume of flow to be handled, drainage system design is most influenced by the topography, geology, man-made obstructions (e.g., roads), and soils of the site. These four areas of concern must be investigated during the various stages of drainage planning. Their investigation will reveal the intricacies of the site under consideration so that feasibility and cost analysis may be obtained.

The *topography* of a site must be known for several reasons. Some of these include the following:

1. An adequate outlet for the drainage system in mind must be available. Although a pumped outlet is always possible, gravitational flow would more than likely be preferable. Topographic maps would provide insight into this requirement for initial planning.
2. In surface drainage systems, enough elevation head must be available to move the water at the desired rate but not so fast as to cause erosive flows. Site topography must be known for estimates in the areas of overland flow, channel and waterway flows, and limiting nonerosive flows.
3. In subsurface drainage the grade of the drain line is generally set by the elevation head available and the size of drain pipe. However, the distance to the drain line from the soil surface is critical where heavy equipment may drive over and crush a shallow drain.

The *geology* of a site will reveal the type of material that must be moved to install drain lines as well as a profile of the site depicting probable causes of restricted flow of water. This information is necessary in the evaluation and design of a complete drainage system.

Man-made obstructions, such as roads, buried cables and pipes, diversions, dams, dikes, culverts, and bridges, must be pinpointed and assessed as to their role in the drainage scheme. Private and public roads, buried cables and pipes, dikes, and irrigation canals are not readily crossed without considerable expense. The capacities of existing water conveying systems such as road ditches, diversions, canals, culverts, and bridges may need to be increased as the result of discharging additional drainage water. The effect of an additional load on the existing drainage system must be evaluated with respect to the increased hydraulic load. Considerable costs could be incurred through the replacement of road culverts and bridges, and the enlarging of existing ditches.

Many properties of the soil profile must be known in order to intelligently evaluate and design a drainage system. In surface drainage, the erodability of the soil must be known to prevent erosive flows from accumulating, whereas in subsurface drainage, the hydraulic conductivity of the soil profile must be estimated. These kinds of judgments can best be made by persons with experience in the soils and geology of the area under consideration.

A survey of the site must indicate:

a. the availability of a drainage outlet
b. the availability of enough elevation head for surface drainage
c. the material that must be moved to install the drainage system
d. man-made obstructions that might affect the drainage system

Information Sources for Drainage Design Tie In With Those for General Site Design

Many of the sources of information mentioned in Module I-6, "Site Evaluation," also yield data necessary for drainage system design. Aerial photographs, USGS topographic maps, SCS county soil survey reports, precipitation data, and crop yield data are discussed in Module I-6. In addtion, other data useful for drainage investigations are streamflows, river stage, and tide runs to evaluate flood hazard; soil conservation district work plans and old drainage surveys for ground levels and other information; and geologic and ground water surveys for subsurface and pumped flow system information. Field surveys must also be done, in conjuction with the other on-site engineering work, to isolate problem drainage areas or plan the layout of the system.

Module I-6, "Site Evaluation" includes a full discussion of many of the information sources useful in drainage design.

Various texts deal with drainage. A recommended single source of information is the SCS National Engineering Handbook, Section 16, "Drainage of Agricultural Land." SCS drainage engineers may also be consulted. Drainage contractors have the necessary experience and knowledge in the field to fulfill the requirements for most land application site work.

BIBLIOGRAPHY

Bouwer, H. 1974. Design and operation of land treatment systems for minimum contamination of groundwater. *Groundwater*, **12**(3):140–147.

Demirjian, Y. A. 1975. Muskegon county wastewater management system. Prepared for U.S. EPA Tech. Transfer Program, Design Seminar for Land Treatment of Municipal Wastewater Effluents. U.S. Gov't. -EPA.

Linsley, R. K. and J. B. Franzini. 1964. Water-resources engineering. McGraw-Hill Book Co., New York. 654 p.

Schwab, G. O., R. K. Frevert, T. W. Edminster, and K. K. Barnes. 1966. Soil and water conservation engineering. John Wiley & Sons, New York. 683 p.

Soil Conservation Service (SCS). 1971. National engineering handbook, Section 16. Drainage of agricultural land. Engineering Division, SCS, USDA, Washington, D.C.

Module II-11

MONITORING AT LAND APPLICATION SITES

SUMMARY

This module summarizes four major reasons for employing monitoring during design and operation of a land application site: documentation of existing water quality and system performance, confirmation of design parameters, provision of data for future designs and for management decisions. Monitoring requirements are examined for different land treatment options—overland flow sites and others employing collection of surface runoff, underdrained sites, rapid infiltration sites with recovery wells, and downslope emergence of treated effluent from irrigation sites.

Methods of taking water samples and soil moisture measurements are discussed. Wells, lysimeters, tensiometers, and piezometers are described in conjunction with this sampling.

Monitoring of parameters other than water is discussed briefly. A very brief summary of governmental regulations pertaining to monitoring is included.

CONTENTS

GLOSSARY

bar–A unit of pressure equal to 10^5 pascals or 0.987 standard atmosphere.

methylene blue active substances–Surface active agents ("surfactants") contained in synthetic detergents, which produce a blue salt in the presence of methylene blue reagent.

PCB–Polychlorinated biphenyls, any of a number of organic compounds used in various industrial processes.

regolith–The blanket of unconsolidated rocky debris of any thickness that overlies bedrock and forms the surface of the land. Used in this sense to include the soil.

saturated soil–Soil in which all spaces, macropores, and micropores are filled with water. This condition is approached in the groundwater zone and sometimes above that after heavy rains or during irrigation.

soil moisture tension–The negative pressure which must be exerted on capillary water to extract it from soil sample.

suction lysimeter–An instrument for obtaining samples of water percolating through soil.

tensiometer–A device to measure soil moisture tension, very similar to a lysimeter which is fitted with a pressure gauge or manometer. Readings are made of the water tension developed in a porous cup at equilibrium with the surrounding soil moisture.

OBJECTIVES

On completion of this module, the reader should be able to:

1. Discuss the operation, costs, placement in a land application field, and usefulness of monitoring devices for groundwater and unsaturated soil water.
2. Discuss the three distinctions in groundwater made within the EPA guidelines and the implications each has for monitoring systems.
3. Discuss the relative response time of wells, lysimeters, and under-drainage and comment on advantages and disadvantages of each.
4. List and understand four reasons why monitoring is needed for waste treatment by land application.
5. List three parameters not connected with groundwater which might be monitored in at least some systems of land application.
6. Discuss the role monitoring plays in the management of land application systems.

INTRODUCTION

Monitoring Land Application Sites Serves Several Functions

Monitoring serves four distinct functions in land application systems:

1. Provides the documentation needed to prove that requirements of regulatory agencies are being met.
2. Confirms the adequacy of the original design.
3. Acts as source of data for future design.
4. Forms a basis for certain management decisions.

Documentation. Federal and state regulatory agencies specify minimum standards of water quality and environmental safety which must be maintained during the operation of the land application system. Most of these are reviewed in this module and analytical techniques are given in APHA, *et al.* (1975). The collection and analysis of representative samples, either by the site management personnel or regulatory agencies, is needed to document compliance with the law. In order to fully document the performance of a site, tests must be conducted prior to waste application in order to establish baseline characteristics.

**Monitoring provides data to prove that the land application system complies with standards of water quality and environmental safety.*

Confirmation. Land application sites must be carefully designed so that the anticipated waste treatment comes as close as possible to actual achievements. In such a scheme, monitoring serves to confirm the adequacy of design. Although monitoring cannot be used to take the "bugs" out of an inadequate design, results will point out any inadequacies which can then be dealt with through operational changes or future development of the system.

Monitoring will reveal any inadequacies in the original design of the land application system.

Future Design. The present knowledge of land application systems is not as sophisticated as that of conventional treatment systems. For example, using Figure 1 as a rough illustration, it is possible to predict the amount of a pollutant, say BOD, remaining in the wastewaters at various stages of a conventional treatment system, assuming perfect operation and no upsets in the system. Unfortunately, such precise information is not available for land application systems as yet (Figure 2). While it can be determined what is applied to land and usually an overall treatment efficiency, there is too much uncertainty to predict exactly what renovations are accomplished at various points as the wastewaters pass through land treatment. Therefore, good monitoring practice will provide more reliable data for the design of future systems.

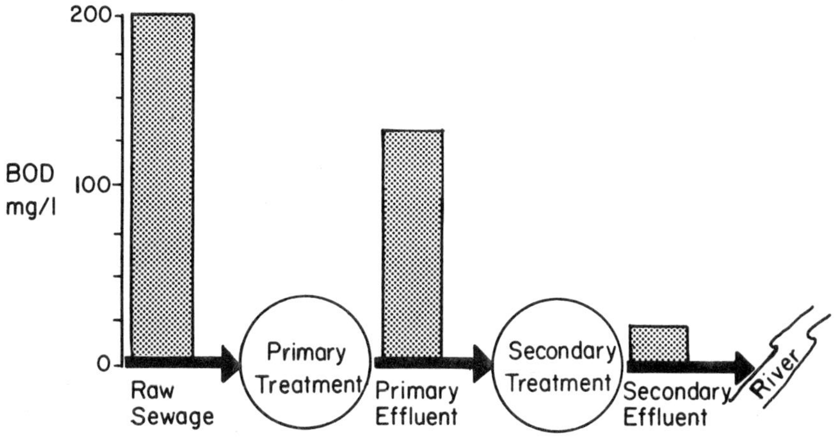

Figure 1. Conventional treatment systems are well researched and the amount of pollutant at each stage can be predicted with a certain degree of accuracy, assuming perfectly smooth operation.

*This and other italicized summaries are intended to highlight key ideas, provide a basis for later review or to aid in skimming sections that are relatively familiar. They can be ignored in a complete reading of the text.

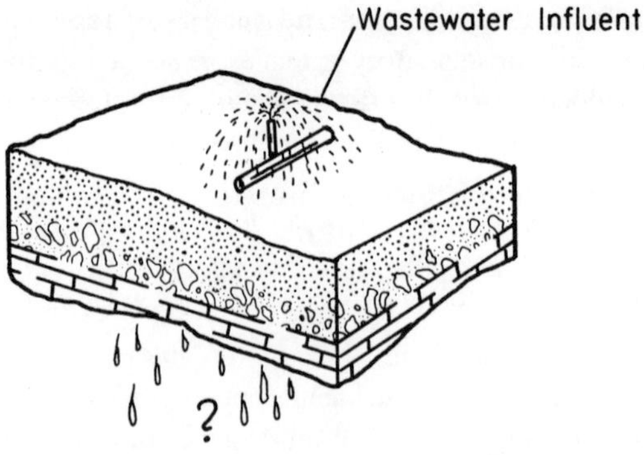

Figure 2. Land application systems have an inadequate backlog of empirical data needed for completely reliable predictions of performance. Dispersion of treated water in most systems leads to more interpretation of performance data.

Monitoring provides data which can be used in the design of future land application systems.

Management Decisions. Modern land application systems are major improvements over old fashioned "sewage farms." Those "farms" used land as a simple waste sink and from them arose the horror stories of disease and pestilence, contaminated wells, and foul odors (Egeland, 1973). Furthermore, deleterious effects from old "sewage farms" were observed at considerable distances from the site and displaced in time from the actual act of waste discharge.

However, even an adequately designed land application system can produce groundwater pollution, vegetation kills, water logged soils, runoff and soil erosion if subjected to poor management (Parizek, 1973 and Miller, *et al.*, 1974). While a well designed and adequately maintained monitoring program cannot improve a poor site design, conscientious monitoring can improve the performance of a good design by providing the information needed for day-to-day management.

Monitoring provides information needed for careful day-to-day management of the land application system.

Laws on Environmental Quality Are Limited

At present, statutory requirements on land application monitoring are generally limited to defining surface and groundwater sampling procedures and minimum water quality standards. In that light, it is fitting that the bulk of this module consider those issues. However, it is appropriate also to consider other parameters related to the general environment that may be monitored. These will be considered at the end of the module, recognizing that while research is needed on the impact of waste material on wildlife, vegetation, insects, and so on, such efforts probably are out of place if viewed as a continuing monitoring need at all facilities (Blakeslee, 1973).

Most laws on monitoring land application systems are limited to specifying the surface and groundwater sampling procedures and the minimum standards for water quality.

MONITORING WATER TREATED BY LAND APPLICATION

Water treated by land application falls into two convenient categories:

1. *Surface water.* Water that either remains on the surface or is returned to the surface.
2. *Sub-surface water.* Water in unsaturated and saturated soils that does not emerge at any identifiable point.

Each category offers distinct advantages and disadvantages regarding monitoring activities, so each will be discussed separately in this module.

Discharge of Surface Water Varies

When discussing surface waters, it is useful to identify four cases:

Case I: Overland flow. An overland flow system is illustrated in Figure 3. Wastewater is applied onto a terraced hillside and allowed to collect, ultimately, in a surface channel prior to discharge. Even systems not specifically designed for overland flow often have dikes or ditches around the periphery to contain runoff from exceptional rainfall or snowmelt. Monitoring requirements of these systems will be no less stringent than conventional discharge systems. A discharge permit also will be necessary for these systems.

In overland flow systems, treated surface water collects in a surface channel prior to discharge or recycling through the system.

Case II: Underdrain. Underdraining is common where water table levels rise close to the surface or where infiltration rates are to be improved, especially in less porous soils. A simple underdrain system is illustrated in Figure 4. Waters emerge from such systems as a point source discharge.

In an underdrained system, treated water emerges as a point source discharge.

Case III: Recovery wells. Certain systems, such as the rapid infiltration system used by Flushing Meadows in Phoenix, Arizona (Figure 5), seek to prevent contaminating natural

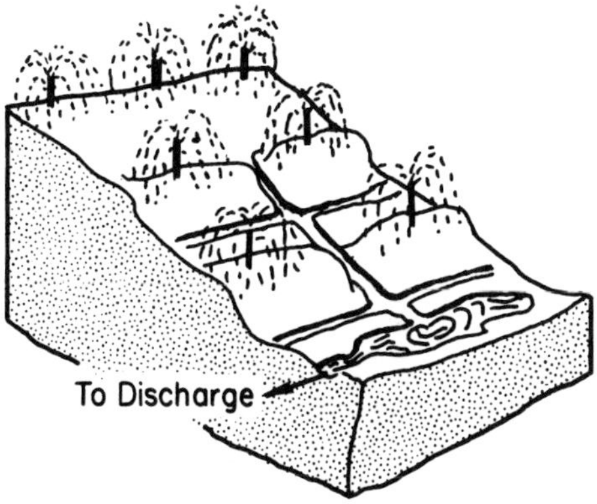

Figure 3. *Case 1:* Water sprayed onto overland flow system is collected in a receiving trench or lagoon before discharge.

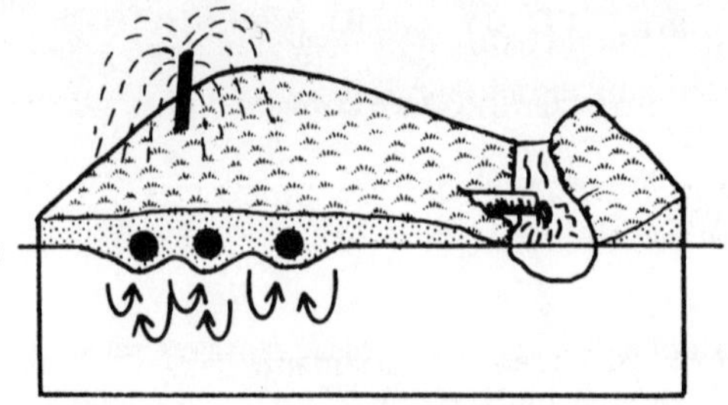

Figure 4. *Case 2:* Underdrains collect groundwater and return it to the surface.

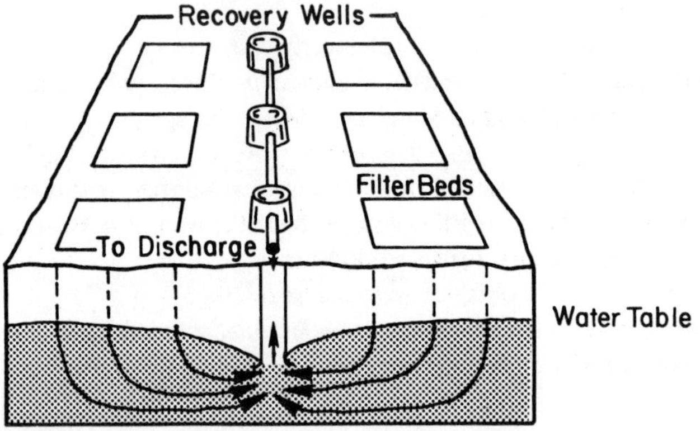

Figure 5. *Case 3:* Recovery wells may bring treated water back to the surface for storage or discharge (adapted from Stevens, 1974).

groundwater by sinking recovery wells and pumping treated waters back to the surface. Like Case II, the discharge from such systems serves as a point source.

> *In rapid infiltration systems with recovery wells, water emerges as a point source discharge.*

Case IV: Natural emergence. In some instances, treated wastewater may emerge downslope from the application site. That occurs when the water table intersects the surface contour (Figure 6). Unlike the three cases above, the emergent water in this circumstance may represent

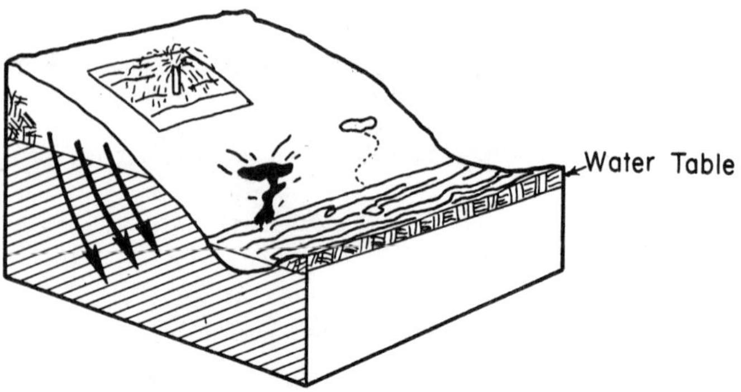

Figure 6. *Case 4:* Water emerges wherever the water table intersects the surface. It appears as springs, sloughs, bogs, swamps, or recharge for local rivers.

only a small portion of the total wastewater flow and may or may not be representative of the bulk of groundwater quality. In the situation diagrammed in Figure 6, renovated wastewater can emerge all along the river bank as a non-point source of river recharge water.

Treated water emerges downslope from the application site if the water table intersects the land surface contour; this is a nonpoint discharge.

Monitoring of Surface Waters Must Meet Requirements

In Cases I, II, and III the renovated wastewater is emerging from what is in effect a point source. Samples can be extracted with relative ease, analyzed, and recycled or otherwise treated if necessary to assure adequate renovation.

If treated wastewater emerges from a point source it can be monitored easily to determine renovation efficiency.

In Case IV, the emergent water quality may or may not represent the level of renovation that is occurring. Furthermore, there is little chance that these emergent waters can be recycled. Nevertheless, local surface waters which are likely to receive renovated wastewater need to be monitored regularly so that their water quality can be documented.

In all cases, a monitoring routine will have to be established in accordance with requirements set by local officials. Normally it can be expected that those requirements would specify sampling techniques and frequencies. Some of these should be similar to those used to sample outfalls from sewage treatment facilities or flows in natural surface waters. The water quality parameters which must be documented depend somewhat on the ultimate use of the water, but those discussions are reserved for the section on Water Quality Criteria (p. 417) of this module.

Treated wastewater emerging from point and non-point sources must be monitored according to the method and frequency specified by local officials.

Finally, it is extremely important to realize that even where water is discharged to surface waters, subsurface monitoring still may be very important. The Muskegon facility serves as a case in point. The renovated waters are discharged to Lake Michigan via one of two creeks, yet about 300 monitoring wells serve that site (Chaiken, *et al.*, 1973).

Monitoring Subsurface Water Presents a Challenge

Subsurface water must be monitored but the question of how to obtain representative samples of groundwater for analysis is challenging. It is convenient to approach this problem by addressing two arbitrary categories of subsurface water:

1. Water of saturated soils sampled through the use of wells driven below the permanent water table
2. Water of unsaturated soils sampled through the use of suction lysimeters.

Obtaining representative samples of groundwater for analysis to monitor the renovation efficiency of land treatment is often difficult.

Saturated Soil and the Use of Wells. A typical monitoring well, in place, is illustrated in Figure 7. Casings and other structures of such wells typically are made of plastics to avoid contaminat-

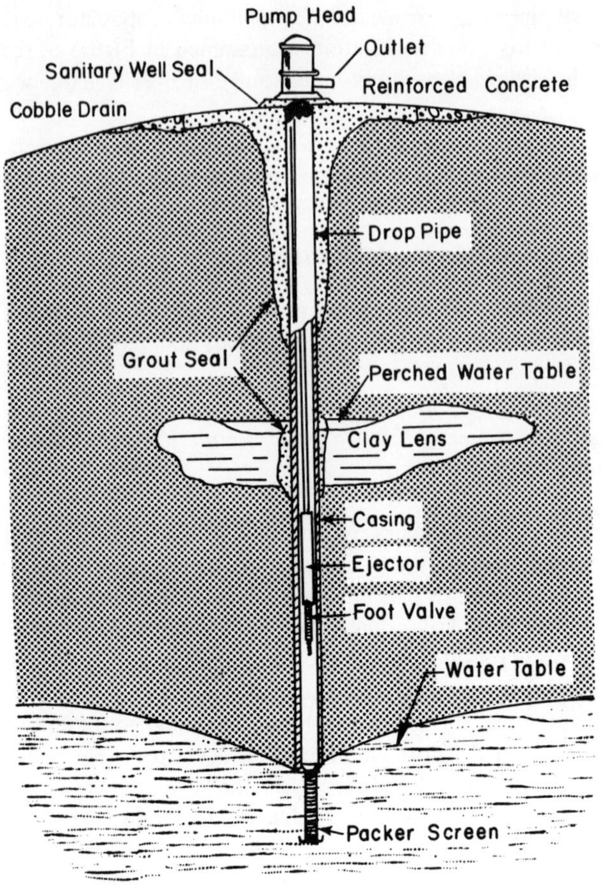

Figure 7. A drilled well extends below the water table and has several sanitation protection features to assure reliable sampling (adapted from U.S. EPA, 1975b).

ing samples with products from corrosion within the wells. Further protection is provided by proper sealing procedures at the well head and using cement grouting wherever necessary.

Water of saturated soils can be sampled through the use of wells driven below the permanent water table.

The cost of wells vary with depth. Approximate values, in 1975 dollars, are given in Figure 8. However, since monitoring wells often are not deep or required to penetrate thick layers of bedrock, their costs, relative to the development of the application site itself, are modest.

The placement of wells is, as yet (1979), not stipulated in any federal guidelines. Leadership in development of such guidelines has been taken by several states. The more conservative of these efforts recommend that wells be placed within and beyond the borders of the land application site and placed so that all subsurface waters flowing either into or out from the site can be sampled. Usually a single well in each direction of flow is adequate for small systems, but two and even three wells at the same distance and in a given direction gives information of greater statistical reliability. That precaution increases in importance in larger application systems as, for example, the Muskegon Project. At any rate, wells, or well clusters, are usually required to be

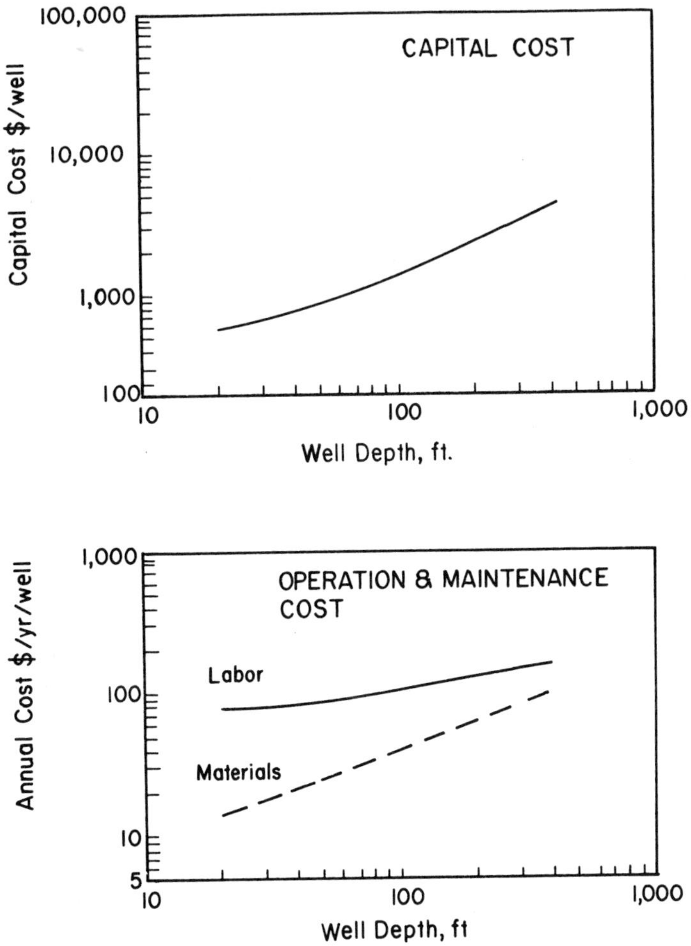

Figure 8. The cost of monitoring wells increases with depth (adapted from Pound, *et al.*, 1975).

no more than 250–500 feet apart at any given distance from the application site. The distance of points from each other and from the application area should depend on the permeability of the regolith. The higher the permeability, the greater the distance between wells. Also, it would be important to track the movement of groundwater with respect to time. If a pollution front were advancing, its position at some future date then could be predicted. That capability from a monitoring field would require more than one monitoring point up and down gradient from the site, and probably more than one depth of sampling. Other recommendations may stipulate that water be taken for analysis from neighboring private and public wells if contamination is a possible danger.

Precaution about the number and location of monitoring wells increases in larger land application systems; the Muskegon County system has 300 monitoring wells.

The distance between wells and from the application site to the wells should depend on the permeability of the regolith.

A hydrogeologic assessment of the site and its surroundings must be completed early in the design stages. That assessment should be done by a competent, experienced field consultant who can employ geologic maps, aerial photography, soil maps, seismic tests, borings, and physical geology. The coefficient of soil permeability needs to be determined (preferably using field methods). The usual test pit method, however, relying as it does on a steady state hydraulic head may be inadequate (Pound and Crites, 1973). More expensive methods can be necessary.

> *An experienced field consultant should do a hydrogeologic assessment of the site early in its design to determine the direction and rate of flow of groundwater both into and away from the site.*

The purpose of these studies is to determine the direction and rate of flow of groundwater both into and away from a land application site. The direction of groundwater flow depends on subsurface soil conditions and does not always approximate topographic relief. Channels and fractures may alter flow rates and affect the direction of flow of renovated water. However, even after the hydrogeology of the site is approximated, the actual flow of water through saturated soils must be confirmed by test borings and analytical studies well in advance of the actual wastewater application.

The layout of the monitoring field can be planned once the hydrogeologic assessment is completed. In general, the field will have four subdivisions:

1. "Up-gradient" to provide baseline data on groundwater moving into the land application area.
2. "Contact zone" to provide data for the evaluation of events occurring both above and at the point of mixing of the renovated water and the natural groundwater.
3. "Down gradient" to evaluate the quality of subsurface water moving off-site.
4. "Special locations" to cover, if and when needed, locations where inadequately treated wastewaters may present unusual hazards.

> *The layout of the monitoring field can be planned after the hydrogeologic assessment of the site.*

A simplified hypothetical monitoring field, with some built in errors, is illustrated in Figure 9.

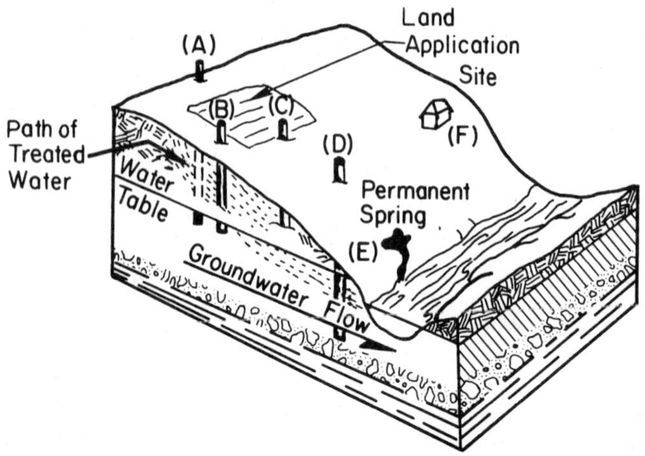

Figure 9. A hypothetical monitoring field which is critiqued in the text.

Notice that well (A) is situated to sample native groundwater flowing into the vicinity of the application field. Well (B) is poorly sited off to one side of the groundwater flow. Often it is very difficult to be sure that the well is sampling the effluent stream. Well (C) is properly sited, extending about 5 feet below the permanent water table and within the mixing zone for natural groundwater and renovated water. Well (D) may provide some interesting information on background quality of the groundwater, but would not be representative of effluent effects from the land application site. The problem here is that well (D) is too deep—almost to bedrock. Both the permanent spring (E) and the river receive renovated water. Clearly an effort must be made to monitor these points.

Other possible flows may exist. For example, two aerial views of the same application site are presented in Figure 10. It is assumed that most evidence suggests that subsurface flow will go as pictured in Figure 10-A and that wells could be placed as indicated. Yet if there are undetected features such as, say, the fracture zones mapped out in Figure 10-B, wastewater may be channeled undetected off-site with little renovation. Alternately, a fracture zone may bring large volumes of high quality native groundwater into the application area, diluting the wastewater and leading to overly optimistic monitoring data (Parizek, 1973). Of course, if the site is located where potential underground channels exist, the problems of sampling renovated wastewater are compounded. Figure 10 provides a good example of the value of a hydrogeologic assessment in the planning stages of a land treatment system.

Subsurface fracture zones can alter normal groundwater movement and complicate the monitoring of groundwater.

Sampling from wells can be a source of error. For example, surface-mounted suction pumps artificially decrease dissolved gases and volatile substances, and tend to leave solids behind. Where practical, submersible, corrosion-resistant pumps should be used to draw water samples. Even hand bailing could be used in certain circumstances, but that may lead to the collection of "old" water which does not represent actual conditions (Parizek, 1973).

Another point to emphasize is that wells often have a very long response time. For example, Parizek (1973) reports more than thirteen months were required for chloride ion to be detected in a 33 ft (10 m) well below the land application site at Penn State. Some 24 months passed

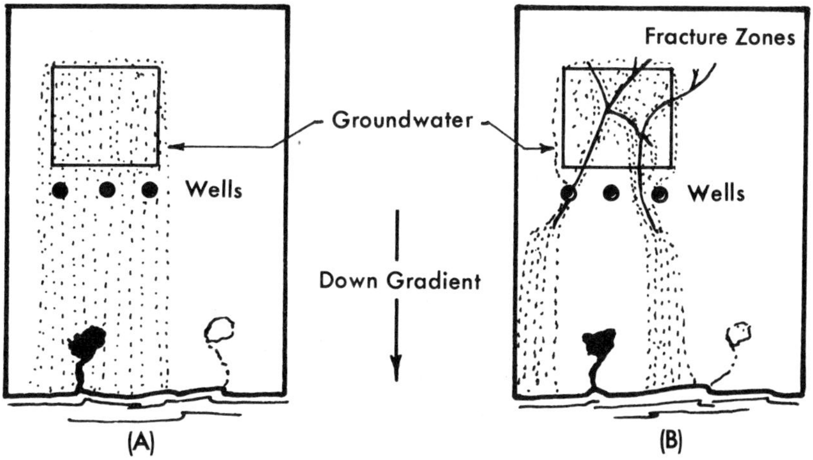

Figure 10. Groundwater flow may be expected to have a well defined, uniform flow (Fig. 10-A), but, in fact, be determined by channeling in subsurface fracture zones (Fig. 10-B) (adapted from Parizek, in site selection criteria for wastewater disposal—soil and hydrogeologic considerations, Kardos and Soffer, eds., the Pennsylvania State University Press, University Park, Pa., 1973.

before chloride ion began to be detected in concentrations above background in a well 275 ft (83 m) deep. Since chloride ion is not adsorbed into the soil and is considered a convenient "tracer" ion, it is apparent that in at least some cases well sampling may be several months, or even years, late in supplying information about renovation. A set of *specific guidelines* for well monitoring fields has been given by Blakeslee (1973) and is reproduced in the Appendix.

Unsaturated Soil and the Lysimeter. Unsaturated soils hold moisture, but that water does not flow rapidly into wells or down gradient unless displaced by further additions. On the other hand, vegetation can exert a tension on that moisture, pulling it away from soil particles and into the roots. To sample this water, it is necessary to use an instrument called the suction lysimeter which uses a vacuum to pull water away from the soil. One form of such a lysimeter is illustrated in Figure 11.

These lysimeters cost around $12.00 each (4 feet complete, 1975 prices) and are easily installed. The operating cost for such devices lies in the labor required to extract samples and in the laboratory analyses of the extracted water. Like wells, a cluster of lysimeters at a given sampling point provides greater statistical support for the data. In all likelihood, lysimeters would be more useful in spray irrigation fields than in overland flow or, particularly, rapid infiltration fields.

The advantages of lysimeters are that they provide means of obtaining rapid information on the quality of water before it reaches the water table. Since the major portion of wastewater renovation occurs in the first few feet of soil, water samples taken by lysimeter can give information on impending PO_4^{-3}, NO_3^{-1}, and toxic substance breakthrough long before wells

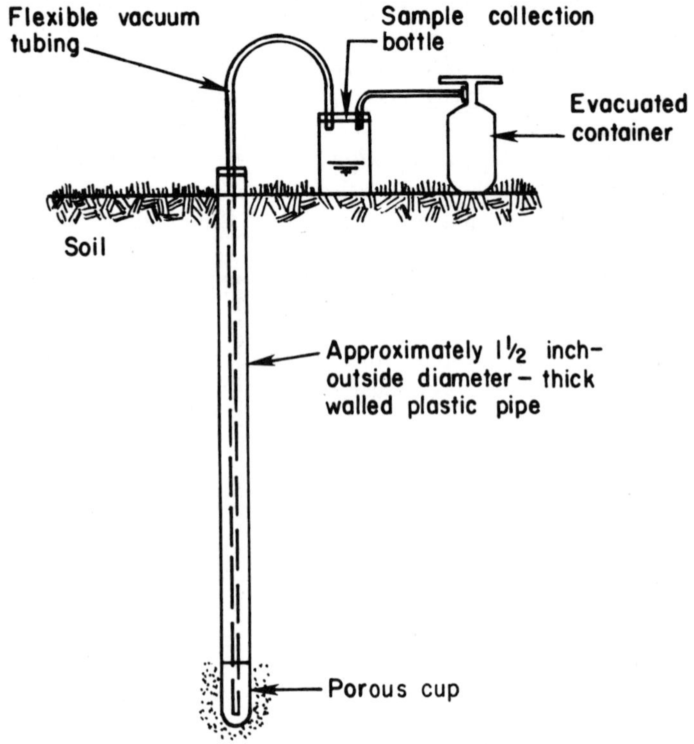

Figure 11. A suction lysimeter pulls available water from unsaturated soil through a porous cup and into a collection flask (Pound and Crites, 1973).

can. Lysimeters at various depths can provide the first indication of advancing fronts, give data that can help predict the rate of movement of contaminants, and allow early assessment of an impending exhaustion of soil CEC.

Lysimeters give rapid information on water quality before the wastewater reaches the groundwater, foretelling any undesirable breakthrough long before monitoring wells can.

The *number of lysimeter clusters* for a given site should be decided on the basis of what is needed to provide statistically acceptable information. Clusters must be protected from damage by machinery. One cluster for each spray field would not be unreasonable. Sampling schedules would have to be more frequent at first until the field behavior can be documented. Moving from weekly to bimonthly or longer schedules may be possible as monitoring data becomes available and is analyzed.

Land application sites developed for research purposes may use more elaborate lysimeters (e.g., pan lysimeters) and are more critical about lysimeter data (Hansen and Harris, 1975). Agronomic researchers should be consulted regarding the installation, costs, operation, and limitations of the various types of lysimeters (Parizek and Lane, 1970).

Measuring Soil Moisture

Application schedules for wastewaters should reflect a realistic evaluation of actual soil moisture conditions. Climatic variations may require that application periods be shortened or even suspended temporarily: alternately, longer application periods may be acceptable during hot, dry periods. In any event, soil moisture must be evaluated to prevent soil waterlogging and to ensure that crop roots experience optimal moisture conditions. The most widely used methods of measuring soil moisture are electrical resistance cells, neutron scattering devices, and tensiometers. Electrical resistance is the simplest method, but is not long-lasting or reliable. Accurate calibration is very difficult, and since the method relies on a block of gypsum which is dissolved by soil moisture, constant attention is required to maintain the units. Neutron scattering devices present exactly the opposite difficulty of being too sophisticated for most normal uses, although they could be considered for research applications or in irrigation projects when water application is to be minimized. These units bombard the surrounding soil with fast-moving neutrons which will be slowed only by hydrogen ions. The device incorporates a slow neutron detector to count reflected neutrons. This is an accurate method in mineral soils where water is the predominant source of combined hydrogen ions, but neutron scattering has some limitations in highly organic soils with other hydrogen ion sources. These devices are expensive.

A tensiometer is the most practical instrument to use to measure soil moisture at wastewater treatment sites.

Unsaturated Soil and Tensiometers. Tensiometers are by far the most practical method for treatment site management applications. The *tensiometer* uses the principle that water in the instrument will equilibrate with water in the surrounding soil, thus creating a suction in the tensiometer which will be equal to that in the soil. This pressure is read on a vacuum gauge. The general principles are illustrated in Figure 12. As the percent of soil moisture (grams of water per 100 grams of soil) decreases, greater vacuum pressure is needed to extract the water. The tensiometer, then, is much like a lysimeter but is fitted with a pressure gauge. However, in practice the tensiometer is filled with water and the negative pressure (soil moisture tension) is de-

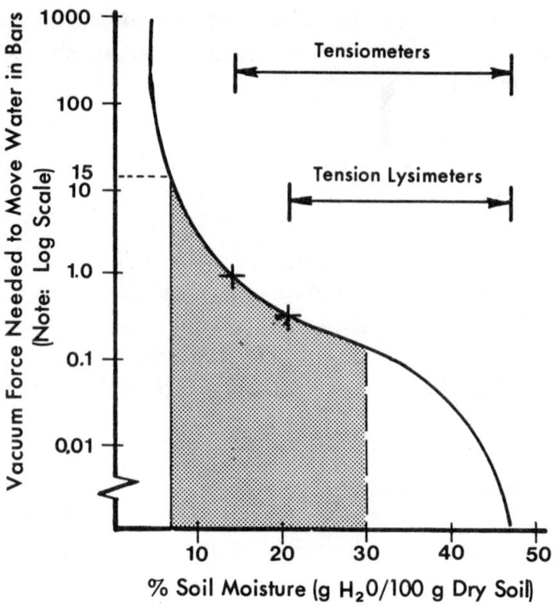

Figure 12. The force needed to move water from unsaturated soil is inversely related to the percent of soil moisture (adapted from Brady, 1974).

veloped as water flows from the porous cup to the soil. The sensitivity of the pressure gauge combined with the small total volume of water in the instrument allows a full scale deflection with the movement of only a few hundredths of a milliliter of water for some models. Figure 12 indicates that a typical tensiometer can read up to about 1 bar pressure, which would be a sufficient range for wastewater application sites. Their most useful range is between 0 and 0.8 bars, and they are applied successfully in monitoring irrigation sites where moisture levels are maintained near field capacity (Brady, 1974).

Costs of tensiometers vary with the length and complexity of the instrument, but in general are comparable to lysimeters. A 48 inch tensiometer similar to the one illustrated in Figure 13 was around $16.00 (1975 prices). A single probe with multiple tips, capable of measuring soil moisture at several depths simultaneously, can cost around $165.00 (1975 prices).

Tensiometer fields need to be protected from mechanical injury of course, but do not have to be clustered. The number of tensiometers needed will be very site specific, and should be placed in such a way that the range of conditions on the site can be monitored. It may be useful to place one tensiometer in each spot that may tend to become unusually dry and in one or two spots that can be expected to represent "average" soil conditions. Tensiometers are not difficult for one person to prepare and install.

The number of tensiometers is site specific; they should be placed so as to monitor the range of conditions on the site.

Unsaturated Soil and the Piezometer. A fourth instrument used to take measurements on soil moisture is the piezometer. The piezometer measures the relative pressure to which soil water is subjected at various depths. They may be used to examine subsurface flow to potential seep spots, to streams, swamps, and so forth. Piezometers may be employed in the initial stages of site assessment under the direction of competent soil scientists. Application sites designed for research

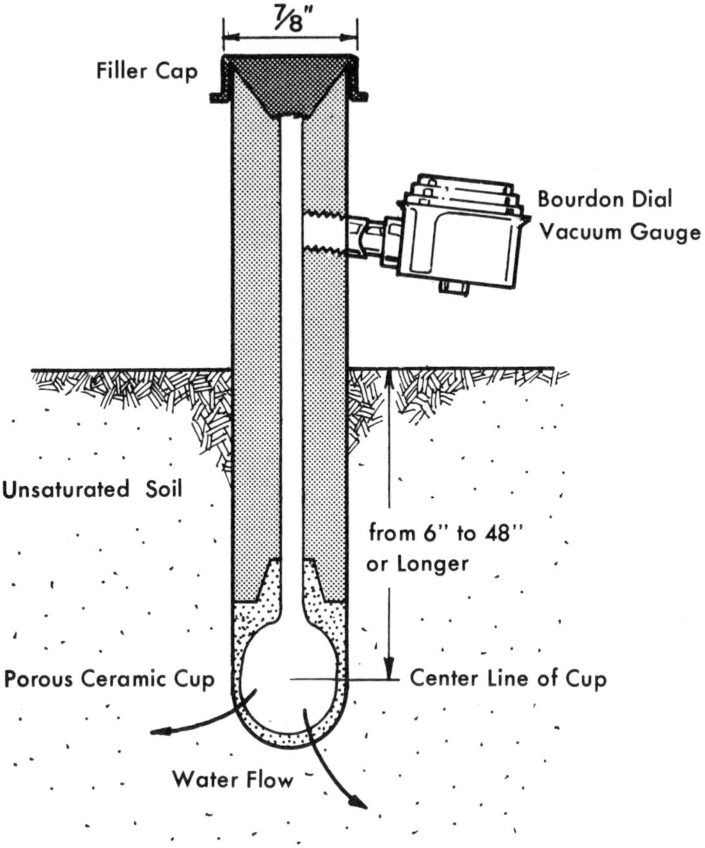

Figure 13. Tensiometers are filled with water and the flow from the soil is related to soil moisture tension by a vacuum gauge (adapted from Soil Moisture Equipment Co., 1975).

purposes may employ piezometers. However, piezometers are not suitable for wet soils and are of little interest to the usual working application site.

MONITORING PARAMETERS OTHER THAN WATER

While water quality remains a major concern of monitoring programs, other environmental factors may be of greater or lesser concern to individual sites. It is not unreasonable to expect that at least some of these factors may come under state (if not federal) regulation.

Air May be Monitored for Aerosols and Odors

Two distinct areas of possible air monitoring are aerosols and odors. Aerosols are tiny droplets of water, especially common to high pressure spray systems, and carried off-site by winds. Bacteria and viruses may be contained in aerosol droplets, and sampling and analysis of micro-organisms require competencies beyond routine managerial skills or training.

Conditions at land treatment sites that may have to be monitored in the future are: aerosols and odors, impact on soil organisms, appearance of the site, and quality of influent.

Experience from other systems will be useful in designing and predicting the performance of new land application projects. Aerosol generation and monitoring is discussed at some length in Module II-5, "Pathogens," including the results of monitoring studies which have been carried out.

Odors may be very hard to document or control. Land application systems usually are odorless or, at worst, odorous only occasionally and then only mildly so. That is particularly true when the applied wastes are from secondary pretreatment. It may be decidedly false when applied wastes are of high organic content (such as cannery waste) or are sludges which are surface applied rather than incorporated into the soil.

What constitutes an objectionable odor is highly subjective. Large animal feed lots may smell terrible to the neighbors, but may smell like money to the owner (so the saying goes). Nevertheless, when land application systems become waterlogged or suffer excessive loadings of putrescibles, malodorous conditions can be expected. Those conditions are usually correlated with an increase in incoming telephone calls as well as recognized by the site personnel, which obviates the need for sophisticated monitoring instrumentation.

Soil Should be Monitored Routinely

Soils should be subjected to routine, periodic analyses particularly if the community inventory and influent analyses indicate a significant loading of potentially toxic materials. The most important soil parameter in relation to uptake of metals and other waste constituents is the cation exchange capacity (CEC). The CEC of the site should be determined in the design stage as a factor in determining site area, then periodically after the site becomes operational to evaluate its remaining lifetime. Such analyses could occur on a semi-annual basis in most cases. Planting vegetation known to be particularly sensitive to suspected toxicants as a "poor man's" monitoring device can be very useful in eliminating the need for much more involved and costly laboratory analyses. The chemical analysis of soils is beyond the skills of most site managers and would be expected to be subcontracted to a competent analytical laboratory. Site managers need to know the rates of accumulation of toxic material, and the vertical distribution of toxic elements and phosphorus, as well as other soil changes.

Soil monitoring should include the rate of accumulation of toxic elements, and the vertical distribution of toxic elements and phosphorus.

Monitoring Plant and other Living Systems May Indicate Onset of Toxic Conditions

Land application systems have been called "living filters" (Kardos, *et al.*, 1973) because of their close connection with microbial populations, higher plants and higher animals. Nevertheless, the impact of waste application on living organisms is not routinely assessed except in research projects. There are many reasons for this.

The soil microflora is seldom monitored. Pathogens can exist in soils, but not proliferate, so monitoring them is not routine. Fecal coliform moving to the underlying water table is detected through well-water analyses, not soil analyses. Similarly, soil fungi and actinomycetes are not routinely monitored nor need they be.

Higher life forms are of more interest to the management of land application systems, although they also are not routinely monitored. Included in this category would be many of the worms, slugs, spiders, grubs, and insects that normally populate a healthy, well diversified soil community. The absence of such organisms from a site where they would normally be expected in profusion may be the first indication of the onset of toxic conditions.

The absence of life forms such as spiders, slugs, grubs, and insects from the land application site may be the first indication of the onset of toxic conditions.

Certainly it would be far more important to carefully monitor life systems where land application is being used to support specific life forms. For example, to develop a land application system for forests and then never establish a routine, however simple it may be, for assaying the health of that forest, would be quite irresponsible. Similarly, land application used to support crops should be surveyed in terms of the health of the crops and their suitability (namely, toxic element content) for consumption.

Here again it may be necessary to utilize the expertise of foresters, extension agents, or even knowledgeable farmers. Systems used for irrigating parks, golf courses, roadsides, spoils banks and so on should be routinely monitored (on a contractual basis if necessary) by experts in the respective areas.

Aesthetic Appeal May be Needed for Public Acceptance

Since land application systems can be quite controversial, it may be necessary to invest in surface landscaping and ground maintenance to gain public acceptance. This may be particularly true for public facilities as sensitivities to environmental quality grow in the future. Also, some noncrop, recreational facilities using wastewater [a golf course (Sullivan, 1970)] are very vulnerable to criticisms if appearances deteriorate.

Influent Quality Has Far-Reaching Effects

Throughout this module the assumption has been that the character of the influent used for land application has been ascertained through quantitative and qualitative analyses. Certainly that knowledge is needed in the very preliminary design stage. Since influent quality may change with time, it is necessary to maintain a regular schedule for influent analysis. Changes in influent quality may affect the functioning or even lifespan of a site. Influent to a land application site may be raw wastewater or effluent from conventional treatment processes. Characteristics of the various influent possibilities are discussed at length in Module I-4, "Treatment Systems, Effluent Qualities, and Costs," and I-5, "Waste Characteristics."

WATER QUALITY CRITERIA

Water quality criteria are established under federal law and, in some instances, further defined under state laws. Certain aspects of those requirements are reviewed in the following paragraphs.

Federal Requirements Apply to Groundwater

Federal requirements apply to "public water systems" as defined in the Public Health Service Act and amended by the Safe Drinking Water Act, PL 93-523 (Train, 1976). For administrative purposes, public water systems as defined by federal regulations are categorized as either "community water systems" or "non-community water systems". Furthermore, groundwater resulting from land application is arbitrarily divided into three cases:

Case 1: Groundwater which has the potential for being used as a drinking water supply.
Case 2: Groundwater which is used for drinking water supply.
Case 3: Groundwater which has either unpredictable uses or uses other than drinking water supply (Train, 1976).

> *Federal water quality requirements are the same whether groundwater is a potential or an actual source of drinking water. If the ground water is not used for drinking or if its future use is unknown, evaluation of such systems is on a site-by-site basis.*

For the most part, those delineations were created to allow flexibility in special situations. The basic water quality requirements promulgated by EPA for drinking water sources, however, apply across the board except for Case 3 which requires a site-by-site evaluation. Also, where natural groundwater quality does not meet federal groundwater requirements, the criteria that apply are that no further deterioration in water quality can result from land application. The specific federal water quality criteria are given in the "National Interim Primary Drinking Water Regulations" (EPA, 1976), which went into effect in June of 1977 and are summarized below.

Maximum Contaminant Levels for Inorganic Chemicals: Specific maximum contaminant levels for inorganic chemicals are given in Table 1.

Maximum Contamination Levels for Fluorides. A small margin separates beneficial levels of fluoride from levels which cause adverse effects. That fact prompted EPA to define maximum fluoride concentrations in terms of the climate served by the public water system as defined by the mean maximum daily air temperatures of the locality. The toxic effect of fluoride in drinking water is limited to mottling of dental enamel and minor changes in bone density, and there is evidence that these physiological effects are temperature dependent at various concentrations with small margins between beneficial and detrimental effects (Train, 1976). Maximum fluoride concentrations are shown in Table 2.

Table 1. Maximum Levels of Nine Inorganic Chemicals Are Stipulated for Public Water Systems (EPA, 1976).

Contaminant	Level (mg/1)
Arsenic	0.05
Barium	1.0
Cadmium	0.010
Chromium	0.05
Lead	0.05
Mercury	0.002
Nitrate (as N)	10.0
Selenium	0.01
Silver	0.05

Table 2. Maximum Fluoride Levels Are Established Based on the Climate in which the Public Water System is Located (EPA, 1976).

Temperature Degrees Fahrenheit[a]	Degrees Celsius	Level (mg/1)
53.7 and below	12 and below	2.4
53.8–58.3	12.1–14.6	2.2
58.4–63.8	14.7–17.6	2.0
63.9–70.6	17.7–21.4	1.8
70.7–79.2	21.5–26.2	1.6
79.3–90.5	26.3–32.5	1.4

[a]Annual average of the maximum daily air temperature.

Table 3. Maximum Levels of Specific Organic Chemicals Apply Only to Community Water Systems (EPA, 1976).

Organic Chemical	Maximum Level (mg/1)
Endrin	0.0002
Lindane	0.004
Methoxychlor	0.1
Toxaphene	0.005
2, 4-D	0.1
2, 4, 5-TP Silvex	0.01

Maximum Contaminant Levels for Organic Chemicals. The maximum contaminant levels of Table 3, applying only to public drinking water systems, are an adjunct to the "Special Monitoring Regulations for Organic Chemicals" (40 CFR Part 141, Subpart E) issued by the EPA. These regulations set out the tests to be performed and the procedures to be followed.

Maximum Contaminant Levels for Turbidity. Samples for contaminant levels for turbidity in drinking water are taken at a representative entry point(s) to the drinking water distribution system. In effect, this requirement specifies that the monthly average may not exceed one turbidity unit. Certain exceptions are recognized, however.

Maximum Microbial Contaminant Levels. Federal regulations specify microbial counts of coliform bacteria only. The actual number permitted per count depends on the techniques used and on a monthly averaging formula. To avoid irregularities in arriving at average values, the minimum number of samples to be taken for coliform analysis per month has been established. That samplifing routine is based on a sliding scale related to the population served as given in Figure 14.

The coliform bacterium is the only microbe that the federal government regulates in its safe drinking water standards.

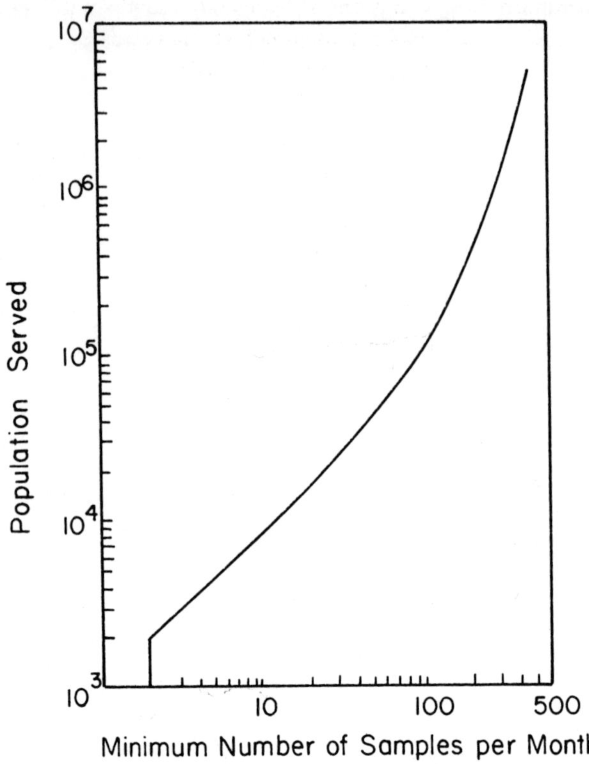

Figure 14. The recommended minimum monthly number of samples for coliform testing increases with the population served (adapted from U.S. EPA, 1975b).

Projected Changes. Further modifications of federal requirements are in the process of being enacted or may be expected in the near future. Some likely changes are reviewed below.

Sodium and sulfates are significant to diets of certain individuals and are discussed in the National Interim Primary Drinking Water Standards. Although it is recommended that the levels of these substances be made known so that certain people susceptible to ill effects can be alerted to high concentrations, not enough evidence is thought to be available to promulgate standards.

> *Not enough evidence is yet available to create standards for levels of sodium and sulfates in drinking water.*

PCB's and asbestos may be subjected to regulations when certain analytical methods, health effects, and environmental occurrence has been evaluated by EPA. Proposed rules were printed in the *Federal Register* of May 24, 1977.

> *Federal regulations on levels of PCB's and asbestos in drinking water have been proposed.*

No federal guidelines have been issued (1977) on groundwater monitoring protocol pertaining to the kinds, number, or implacement techniques of groundwater monitoring devices to be used at land application sites.

Table 4. The Maximum Concentrations of Inorganic Chemicals in Irrigation Waters Are Defined According to Two Standards (Adapted from EPA, 1975a).

Substance	Max. Conc. Continuous Use, All Soils (mg/1)	Max. Conc. for 20 yr Fine Textured Soil (pH 6.0–8.5) (mg/1)
Arsenic	0.10	2.0
Aluminum	5.0	20.0
Boron	0.75	2.0–10.0
Beryllium	0.10	0.50
Cadmium	0.01	0.05
Chromium	0.10	1.0
Cobalt	0.05	5.0
Copper	0.20	5.0
Lead	5.0	10.0
Manganese	0.20	10.0
Molybdenum	0.01	0.5
Nickel	0.20	2.0
Selenium	0.02	0.02
Zinc	2.0	10.0

The criteria for waters leaving land application sites has already been stated. But the following criteria for irrigation waters will usually apply to water being added to the land. The maximum permissible concentrations of inorganic chemicals in irrigation waters are given in Table 4.

Federal standards set the maximum permissible levels of inorganic chemicals in irrigation waters.

State Requirements May be More Restrictive than Federal

State water quality regulations for public drinking water or irrigation water may be more, but not less, restrictive than federal requirements. Specific parameters that may be controlled include some or all of the following: chlorine residuals, BOD, pH, NH_3, dissolved oxygen, total hardness, methylene blue active substances, phenols, conductance, and chloride ion.

State regulations on water may be more but not less restrictive than federal requirements.

Certain states specify minimum requirements for monitoring fields designed to sample groundwater related to land application fields while others issue non-obligatory "guidelines" (e.g., see Rhindress, 1973). The more conservative requirements specify that the groundwater at and around the proposed land treatment site be monitored for some time before start up to establish baseline data. Usually monitoring wells are required. Those wells must be placed where they can sample at several depths from the root zone to below the dry weather water table, sample in every direction of groundwater flow from a given site, and sample every direction from which natural groundwater may enter the site. Wells usually are to be located both within the site and off the site and no more than 250 to 500 feet apart. Sampling from wells of neighbors may be required. Sampling schedules vary from before and after each application to once a month or once an application season. Often some allowance must be made for the frequent recording of water table levels. Land application designs which return renovated water to surface water through underdrains, high volume recovery wells, or overland flow to temporary storage, usually must sample the recovered water on a routine basis and allow for recycling, if necessary. Further discussion of some specific state guidelines, including monitoring requirements, can be found in Module I-9, "Legal Aspects."

Some states require certain monitoring practices; others have non-obligatory guidelines for monitoring.

BIBLIOGRAPHY

American Public Health Association, American Water Works Association, and Water Pollution Control Federation. 1975. Standard methods for the examination of water and wastewater. 14th ed. American Public Health Association. Washington, D.C. 1193 p.

Blakeslee, P. 1973. Monitoring considerations for municipal wastewater effluent and sludge application to the land. *In* EPA/USDA/The National Association of State Universities and Land-Grant Colleges, Proceedings of the joint conference on recycling municipal sludges and effluents on land. Champaign, Ill. pp. 183–198.

Brady, N. C. 1974. The nature and properties of soils. 8th ed. McMillan Publishing Co., Inc., New York. 639 p.

Chaiken, E. I., *et al.* 1973. Muskegon sprays sewage effluents on land. *Civil Engineering*, **43**(5):49–53.

Egeland, D. R. 1973. Land disposal I: A giant step backward. *J. Water Pollut. Contr. Fed.*, **45**(7):1465–1475.

Environmental Protection Agency. 1975a. Evaluation of land application systems. Technical Bulletin 430/9–75–001. 182 p.

EPA. 1975b. Manual for evaluating public drinking water supplies. Office of Water and Hazardous Materials. EPA–430/9–75–001. pp. 15, 58.

EPA. 1976. National interim primary drinking water regulations. Office of Water Supply. EPA–570/9–76–003. 159 p.

Hansen, E. A. and A. R. Harris. 1975. Validity of soil water samples collected with porous ceramic cups. *Soil Sci. Soc. of Amer. Proc.*, **39**(3):528–536.

Miller, D. W., F. A. DeLuca, and T. L. Tessier. 1974. Groundwater contamination in the northeast states. EPA–660/2–74–056.

Parizek, R. R. 1973. Site selection criteria for wastewater disposal-soils and hydrogeologic considerations. *In* Kardos, W. E. and L. T. Sopper, Recycling treated municipal wastewater and sludge through forest and cropland. The Penn State University Press, University Park, Pa. pp. 95–147.

Parizek, R. R. and B. E. Lane. 1970. Soil-water sampling using pan and deep pressure-vacuum lysimeters. *Hydrology*, **11**:1–21.

Pound, C. E. and R. W. Crites. 1973. Wastewater treatment and reuse by land application, volume II. Office of Research and Development, EPA–660/2–73–006b. Washington, D.C. 249 p.

Pound, C. E., R. W. Crites, and D. A. Griffes. 1975. Costs of wastewater by land application. EPA technical report. EPA–430/9–75–003. pp. 116–117.

Rhindress, R. C. 1973. Spray irrigation—the regulatory agency view. *In* Kardos, W. E. and L. T. Sopper. Recycling treated municipal wastewater and sludges through forest and cropland. The Penn State University Press. University Park, Pa. pp. 440–462.

Soil Moisture Equipment Co. 1975. Catalogue 60. Santa Barbara, Calif. 38 p.

Stevens, L. A. 1974. Clean waters. E. P. Dutton & Co., Inc., New York. 289 p.

Sullivan, D. 1970. Wastewater for golf course irrigation. *Water and Sewage Works*, **117**(5):153–159.

Train, R. E. 1976. Alternate waste management techniques for the best practicable waste treatment. *Federal Register*, **41**(29):6190–6191.

APPENDIX
MONITORING CONSIDERATIONS FOR GROUNDWATER (Blakeslee, 1973)

GROUNDWATER MONITORING GUIDELINES ON LAND WASTE DISPOSAL FACILITIES

Monitoring Objectives

The function of a groundwater monitoring program for proposed land disposal facilities is to confirm judgements made during design. This is to be accomplished by a continuing long-term in-depth hydrogeologic study regarding the performance of the system and its influence on surrounding groundwater conditions. This applies to:

1. Wastewater treatment lagoons.
2. Wastewater storage lagoons.
3. Land irrigation systems.
4. Large subsurface disposal fields.
5. Wastewater sludge disposal sites.
6. Industrial waste concentrations disposal sites.

At all such existing sites groundwater monitoring programs are needed to determine the influence of disposal practices on the groundwater resource.

Design of Monitoring Wells

Monitoring wells must be designed and located to meet the specific geologic and hydrologic conditions at each site. Consideration must be given to the following:

1. Geological soil and rock formation existing at the specific site.
2. Depth to an impervious layer.
3. Direction of flow of groundwater and anticipated rate of movement.
4. Depth to seasonal high water table and an indication of seasonal variations in groundwater depth and direction of movement.
5. Nature, extent, and consequences of mounding of groundwater which can be anticipated to occur above the naturally occurring water table.
6. Location of nearby streams and swamps.
7. Potable and nonpotable water supply wells.
8. Other data as appropriate to the specific system design.

Groundwater quality should be monitored immediately below the water table surface near the application site as pollutional materials entering the groundwater system may have a tendency to remain in the upper few feet. Applied wastewater will generally be depressed within the groundwater system as the material travels away from the site. The need for sampling at more than one depth within a groundwater system will depend upon geologic conditions and distance from the pollutional source. Definition of the flow system with depth will be necessary to properly determine the depth to be monitored, especially when mounding is superimposed on the existing system.

Additional design and construction considerations are:

1. Monitoring wells in fine textured soils will require special construction such as gravel packing around the screen.
2. Wells constructed to a depth of 20 feet or more should be 4 inches in diameter to facilitate

use of submersible pump equipment for sample collection unless alternative sampling methods are approved by the reviewing agency.

3. Construction should be by a registered well driller or contractor using approved modern construction methods.
4. Casings shall be grouted and capped and a cap-locking device provided. Use of a vented cap is desirable but care must be taken to prevent introduction of contaminants through such vents.
5. The well casing should be protected against accidental damage and adequately marked to be clearly visible during winter and summer conditions.
6. Each well should be labeled and identified (owner, owner's address, well number, use of well and warning).
7. If a monitoring well is to be permanently abandoned, approved procedures are to be followed.

Location

Groundwater monitoring wells must be located so as to detect any influence of wastewater application on the groundwater resource. A minimum of one groundwater monitoring well must be provided in each direction of groundwater movement near the pollutional source with adequate consideration being given to possible changes in groundwater flow due to mounding effects. The orientation and spacing of multiple wells shall be determined by conditions at each site.

Water Level Measurement

The following considerations apply with regard to water level measurement:

1. Water levels should be determined by methods giving precision to $\frac{1}{8}$ inch or 0.01 feet. (Example—wetted tape method)
2. Measurements should be made from the top of the casing with the elevation of all casings in the monitoring wells system related to a permanent reference point, using USGS datum.
3. Water level measurements are to be made under static conditions prior to pumping for sample collection.
4. Monitoring wells should be installed early in the construction sequence and monthly water level readings obtained during the construction period and during the first two years of wastewater system operation to provide background information. Subsequent water level measurement frequency should be in accordance with a schedule established on a case by case basis.
5. All wells should be securely capped and locked when not in use.

Water Sampling

Background Water Quality. A minimum of three monthly samples should be collected from each monitoring well prior to placing the storage or disposal facility in operation. In cases where background water quality adjacent to the site may be influenced by prior waste applications, provision of monitoring wells or analysis of water quality from existing wells in the same aquifer beyond the area of influence will be necessary.

Operating Schedule. Samples should be collected monthly during the first two years of operation. After the accumulation of a minimum of two years of groundwater monitoring information, modification of the frequency of sampling may be considered upon written request.

Sample Collection

1. A measured amount of water equal to or greater than three times the amount of water in the well and/or gravel pack should be exhausted from the well before taking a sample for analysis. In the case of very low permeability soils the well may have to be exhausted and allowed to refill before a sample is collected.
2. Pumping equipment shall be thoroughly rinsed before use in each monitoring well.
3. A pressure tank shall not be used with a sampling system since the water in the pressure tank would be particularly difficult to exhaust.
4. Water pumped from each monitoring well should be discharged to the ground surface away from the wells to avoid recycling of flow in high permeability soil areas, or soil erosion.
5. Samples must be collected, stored, and transported to the laboratory in a manner so as to avoid contamination or interference with subsequent analyses.

Sample Analysis

Water samples collected for background water quality should be analyzed for the following: (Note: Parameters for groundwater monitoring at industrial waste disposal sites must be established on an individual basis depending on the composition of the wastes applied).

1. Chloride
2. Specific conductance
3. pH
4. Total hardness
5. Alkalinity
6. a. Ammonia nitrogen
   b. Nitrate nitrogen
   c. Nitrite nitrogen
7. Total phosphorus
8. Methylene blue active substances
9. Chemical oxygen demand*
10. Any heavy metals or toxic substances found in the applied wastes.

After adequate background water quality information has been obtained, a minimum of one sample per year, obtained at the end of the irrigation season in the case of seasonal operations, should be collected from each well and analyzed for the above constituents.

All other water samples collected in accordance with the operating schedule should be analyzed for chlorides and specific conductance as indicators of changes in groundwater quality resulting from the wastes applied. If significant changes are noted in chloride and/or specific conductance levels, samples should immediately be analyzed for the other parameters listed above to determine the extent of water quality deviation from background levels.

Groundwater Monitoring System Reports

Well Location Plan. The owner of the system is to provide a plan, drawn to scale, showing the location of each monitoring well and its relationship to the wastewater treatment lagoons, storage lagoon, irrigation area, sludge disposal site or subsurface disposal field and to other significant features such as municipal or private wells, surface streams, etc. It is suggested that individualized well location plan maps be prepared by the project consultant. The plan map

*Use of low concentration C.O.D. analysis methods per current edition of Standard Methods may be necessary.

shall include casing elevation information to facilitate conversion of water level measurements to datum elevations.

Reports. The owner of the system is to file standard reports of observations and sample analyses, obtained in accordance with the schedule listed above with the responsible state agency within 30 days of sample collection. Notification of significant deviations from background quality is to be given immediately.

INDEX

For definitions of technical terms, see also Volume I.